R1·90

*SCHAUM'S OUTLINE OF*

# THEORY AND PROBLEMS

OF

# VECTOR ANALYSIS

***and an introduction to***

# TENSOR ANALYSIS

# SI (METRIC) EDITION

BY

**MURRAY R. SPIEGEL, Ph.D.**

*Professor of Mathematics*
*Rensselaer Polytechnic Institute*



*McGraw-Hill International Book Company, New York*
*co-published with*
*McGraw-Hill Book Company (UK) Limited, London*
*McGraw-Hill Book Company GmbH, Düsseldorf*
*McGraw-Hill Book Company Australia Pty, Limited*
*McGraw-Hill Book Company (SA) (Pty) Limited, Johannesburg*

Printed and bound in Great Britain.


07 084378 3

# Preface

Vector analysis, which had its beginnings in the middle of the 19th century, has in recent years become an essential part of the mathematical background required of engineers, physicists, mathematicians and other scientists. This requirement is far from accidental, for not only does vector analysis provide a concise notation for presenting equations arising from mathematical formulations of physical and geometrical problems but it is also a natural aid in forming mental pictures of physical and geometrical ideas. In short, it might very well be considered a most rewarding language and mode of thought for the physical sciences.

This book is designed to be used either as a textbook for a formal course in vector analysis or as a very useful supplement to all current standard texts. It should also be of considerable value to those taking courses in physics, mechanics, electromagnetic theory, aerodynamics or any of the numerous other fields in which vector methods are employed.

Each chapter begins with a clear statement of pertinent definitions, principles and theorems together with illustrative and other descriptive material. This is followed by graded sets of solved and supplementary problems. The solved problems serve to illustrate and amplify the theory, bring into sharp focus those fine points without which the student continually feels himself on unsafe ground, and provide the repetition of basic principles so vital to effective teaching. Numerous proofs of theorems and derivations of formulae are included among the solved problems. The large number of supplementary problems with answers serve as a complete review of the material of each chapter.

Topics covered include the algebra and the differential and integral calculus of vectors, Stokes' theorem, the divergence theorem and other integral theorems together with many applications drawn from various fields. Added features are the chapters on curvilinear coordinates and tensor analysis which should prove extremely useful in the study of advanced engineering, physics and mathematics.

Considerably more material has been included here than can be covered in most first courses. This has been done to make the book more flexible, to provide a more useful book of reference, and to stimulate further interest in the topics.

The author gratefully acknowledges his indebtedness to Mr. Henry Hayden for typographical layout and art work for the figures. The realism of these figures adds greatly to the effectiveness of presentation in a subject where spatial visualizations play such an important role.

M. R. Spiegel

Rensselaer Polytechnic Institute

June, 1959

# Contents

# Chapter 1

# VECTORS and SCALARS

**A VECTOR** is a quantity having both magnitude and direction, such as displacement, velocity, force, and acceleration.

Graphically a vector is represented by an arrow $OP$ (Fig.1) defining the direction, the magnitude of the vector being indicated by the length of the arrow. The tail end $O$ of the arrow is called the *origin* or *initial point* of the vector, and the head $P$ is called the *terminal point* or *terminus*.

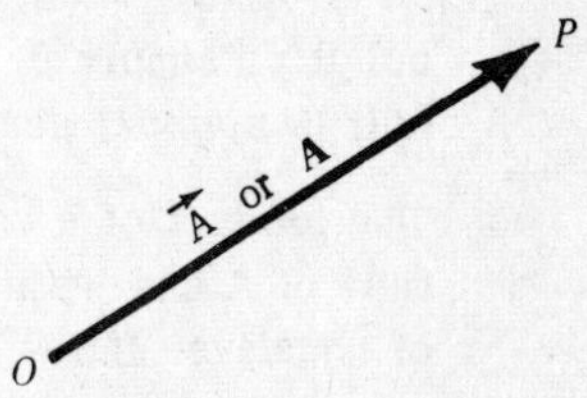


Fig.1

Analytically a vector is represented by a letter with an arrow over it, as $\vec{A}$ in Fig.1, and its magnitude is denoted by $|\vec{A}|$ or A. In printed works, bold faced type, such as **A**, is used to indicate the vector $\vec{A}$ while $|\mathbf{A}|$ or A indicates its magnitude. We shall use this bold faced notation in this book. The vector OP is also indicated as $\overrightarrow{OP}$ or **OP**; in such case we shall denote its magnitude by $\overline{OP}$, $|\overrightarrow{OP}|$, or $|\mathbf{OP}|$.

**A SCALAR** is a quantity having magnitude but no direction, e.g. mass, length, time, temperature, and any real number. Scalars are indicated by letters in ordinary type as in elementary algebra. Operations with scalars follow the same rules as in elementary algebra.

**VECTOR ALGEBRA**. The operations of addition, subtraction and multiplication familiar in the algebra of numbers or scalars are, with suitable definition, capable of extension to an algebra of vectors. The following definitions are fundamental.

1. Two vectors **A** and **B** are *equal* if they have the same magnitude and direction regardless of the position of their initial points. Thus $\mathbf{A} = \mathbf{B}$ in Fig.2.
2. A vector having direction opposite to that of vector **A** but having the same magnitude is denoted by $-\mathbf{A}$ (Fig.3).

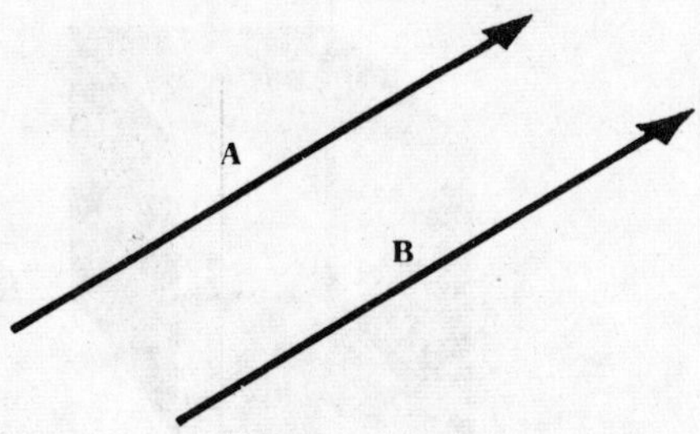


Fig. 2

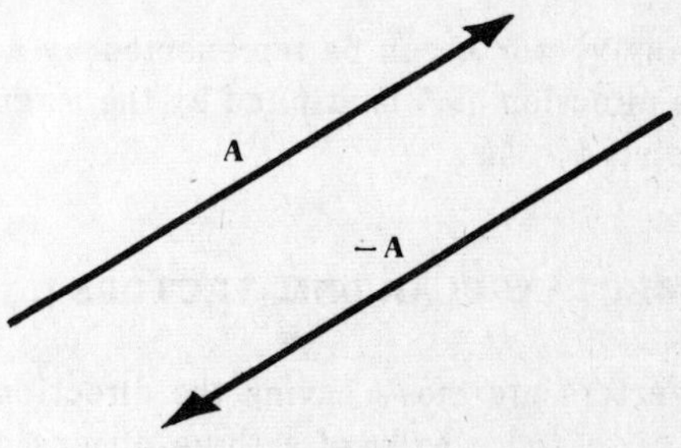


Fig. 3

3. The *sum* or *resultant* of vectors **A** and **B** is a vector **C** formed by placing the initial point of **B** on the terminal point of **A** and then joining the initial point of **A** to the terminal point of **B** (Fig.4). This sum is written **A**+**B**, i.e. **C** = **A**+**B**.

   The definition here is equivalent to the *parallelogram law* for vector addition (see Prob.3).

   Extensions to sums of more than two vectors are immediate (see Problem 4).

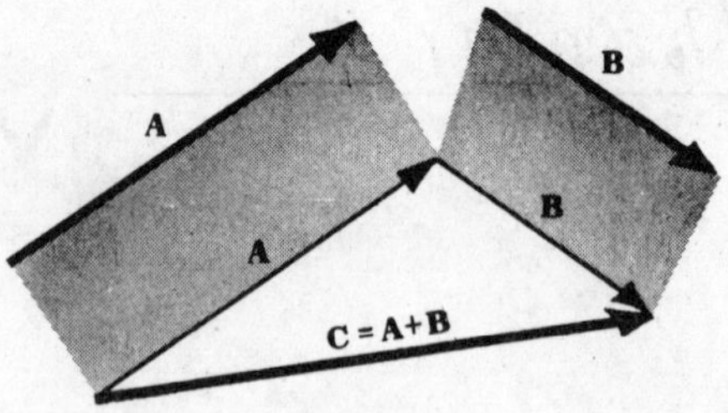

Fig. 4

4. The *difference* of vectors **A** and **B**, represented by **A** − **B**, is that vector **C** which added to **B** yields vector **A**. Equivalently, **A** − **B** can be defined as the sum **A** + (−**B**).

   If **A** = **B**, then **A** − **B** is defined as the *null* or *zero vector* and is represented by the symbol **0** or simply 0. It has zero magnitude and no specific direction. A vector which is not null is a *proper vector*. All vectors will be assumed proper unless otherwise stated.

5. The *product* of a vector **A** by a scalar $m$ is a vector $m$**A** with magnitude $|m|$ times the magnitude of **A** and with direction the same as or opposite to that of **A**, according as $m$ is positive or negative. If $m = 0$, $m$**A** is the null vector.

**LAWS OF VECTOR ALGEBRA.** If **A**, **B** and **C** are vectors and $m$ and $n$ are scalars, then

| | | |
|---|---|---|
| 1. | $\mathbf{A} + \mathbf{B} = \mathbf{B} + \mathbf{A}$ | Commutative Law for Addition |
| 2. | $\mathbf{A} + (\mathbf{B}+\mathbf{C}) = (\mathbf{A}+\mathbf{B}) + \mathbf{C}$ | Associative Law for Addition |
| 3. | $m\mathbf{A} = \mathbf{A}m$ | Commutative Law for Multiplication |
| 4. | $m(n\mathbf{A}) = (mn)\mathbf{A}$ | Associative Law for Multiplication |
| 5. | $(m+n)\mathbf{A} = m\mathbf{A} + n\mathbf{A}$ | Distributive Law |
| 6. | $m(\mathbf{A}+\mathbf{B}) = m\mathbf{A} + m\mathbf{B}$ | Distributive Law |

Note that in these laws only multiplication of a vector by one or more scalars is used. In Chapter 2, products of vectors are defined.

These laws enable us to treat vector equations in the same way as ordinary algebraic equations. For example, if **A**+**B** = **C** then by transposing **A** = **C** − **B**.

**A UNIT VECTOR** is a vector having unit magnitude, If **A** is a vector with magnitude $A \neq 0$, then **A**/$A$ is a unit vector having the same direction as **A**.

Any vector **A** can be represented by a unit vector **a** in the direction of **A** multiplied by the magnitude of **A**. In symbols, **A** = $A$**a**.

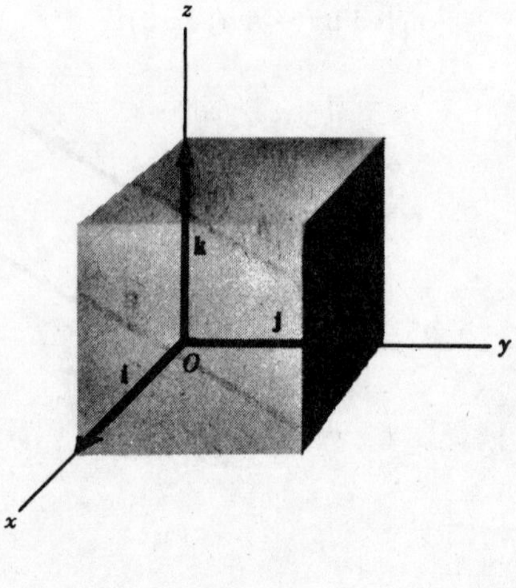

Fig. 5

**THE RECTANGULAR UNIT VECTORS i, j, k.** An important set of unit vectors are those having the directions of the positive $x$, $y$, and $z$ axes of a three dimensional rectangular coordinate system, and are denoted respectively by **i**, **j**, and **k** (Fig.5).

We shall use *right-handed rectangular coordinate systems* unless otherwise stated. Such a system derives its name from the fact that a right threaded screw rotated through 90° from $Ox$ to $Oy$ will advance in the positive $z$ direction, as in Fig.5 above.

In general, three vectors **A**, **B** and **C** which have coincident initial points and are not *coplanar*, i.e. do not lie in or are not parallel to the same plane, are said to form a *right-handed system* or *dextral system* if a right threaded screw rotated through an angle less than 180° from **A** to **B** will advance in the direction **C** as shown in Fig.6.

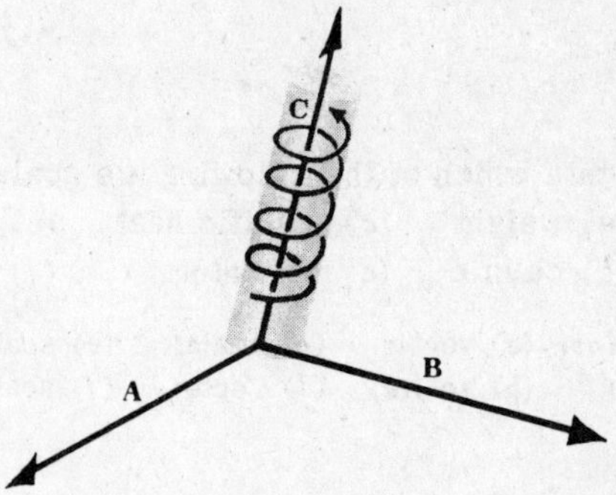

Fig. 6

**COMPONENTS OF A VECTOR.** Any vector **A** in 3 dimensions can be represented with initial point at the origin $O$ of a rectangular coordinate system (Fig.7). Let $(A_1, A_2, A_3)$ be the rectangular coordinates of the terminal point of vector **A** with initial point at $O$. The vectors $A_1\mathbf{i}$, $A_2\mathbf{j}$, and $A_3\mathbf{k}$ are called the *rectangular component vectors* or simply *component vectors* of **A** in the $x$, $y$ and $z$ directions respectively. $A_1$, $A_2$ and $A_3$ are called the *rectangular components* or simply *components* of **A** in the $x$, $y$ and $z$ directions respectively.

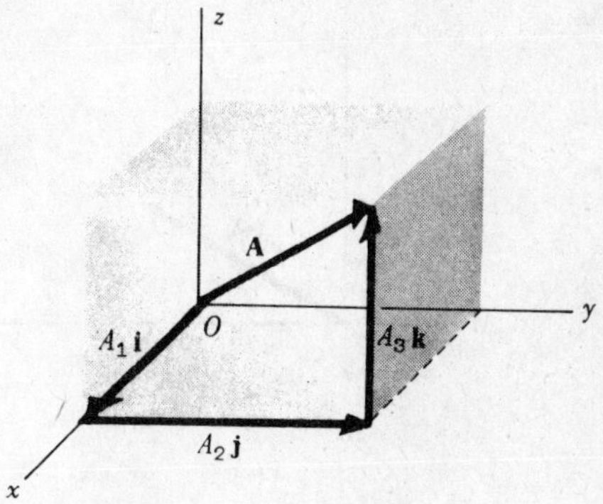

Fig. 7

The sum or resultant of $A_1\mathbf{i}$, $A_2\mathbf{j}$ and $A_3\mathbf{k}$ is the vector **A** so that we can write

$$\mathbf{A} = A_1\mathbf{i} + A_2\mathbf{j} + A_3\mathbf{k}$$

The magnitude of **A** is $$A = |\mathbf{A}| = \sqrt{A_1^2 + A_2^2 + A_3^2}$$

In particular, the *position vector* or *radius vector* **r** from $O$ to the point $(x,y,z)$ is written

$$\mathbf{r} = x\mathbf{i} + y\mathbf{j} + z\mathbf{k}$$

and has magnitude $r = |\mathbf{r}| = \sqrt{x^2 + y^2 + z^2}$.

**SCALAR FIELD.** If to each point $(x,y,z)$ of a region $R$ in space there corresponds a number or scalar $\phi(x,y,z)$, then $\phi$ is called a *scalar function of position* or *scalar point function* and we say that a *scalar field* $\phi$ has been defined in $R$.

**Examples.** *(1)* The temperature at any point within or on the earth's surface at a certain time defines a scalar field.

(2) $\phi(x,y,z) = x^3y - z^2$ defines a scalar field.

A scalar field which is independent of time is called a *stationary* or *steady-state scalar field.*

**VECTOR FIELD.** If to each point $(x,y,z)$ of a region $R$ in space there corresponds a vector $\mathbf{V}(x,y,z)$, then **V** is called a *vector function of position* or *vector point function* and we say that a *vector field* **V** has been defined in $R$.

**Examples.** *(1)* If the velocity at any point $(x,y,z)$ within a moving fluid is known at a certain time, then a vector field is defined.

(2) $\mathbf{V}(x,y,z) = xy^2\mathbf{i} - 2yz^3\mathbf{j} + x^2z\,\mathbf{k}$ defines a vector field.

A vector field which is independent of time is called a *stationary* or *steady-state vector field.*

## SOLVED PROBLEMS

**1.** State which of the following are scalars and which are vectors.

| | | | | |
|---|---|---|---|---|
| (*a*) weight | (*c*) specific heat | (*e*) density | (*g*) volume | (*i*) speed |
| (*b*) calorie | (*d*) momentum | (*f*) energy | (*h*) distance | (*j*) magnetic field intensity |

*Ans.*

| | | | | |
|---|---|---|---|---|
| (*a*) vector | (*c*) scalar | (*e*) scalar | (*g*) scalar | (*i*) scalar |
| (*b*) scalar | (*d*) vector | (*f*) scalar | (*h*) scalar | (*j*) vector |

**2.** Represent graphically (*a*) a force of 10 N in a direction 30° north of east
(*b*) a force of 15 N in a direction 30° east of north.

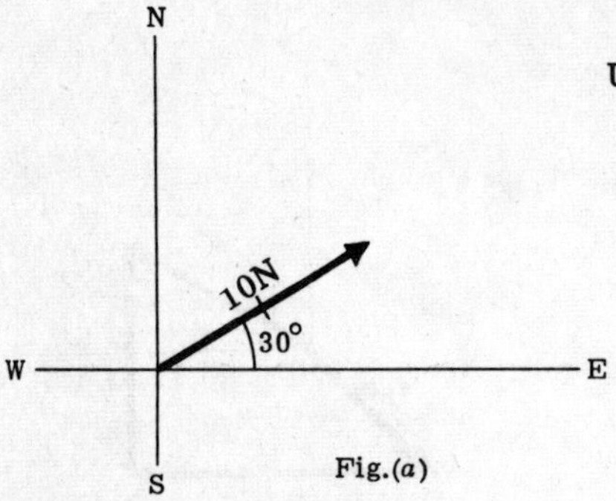


Fig.(*a*)

Unit = 5N

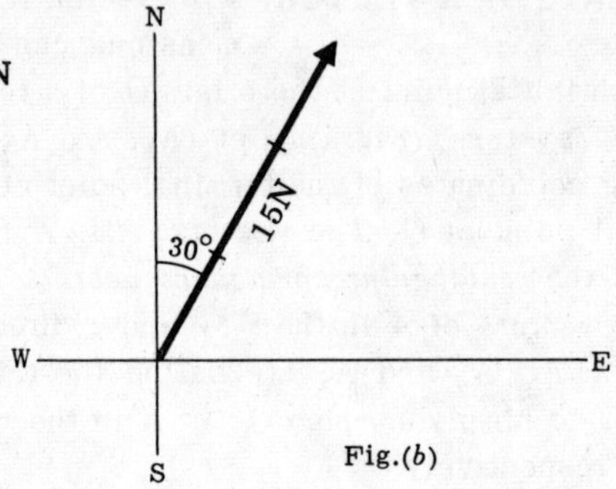


Fig.(*b*)

Choosing the unit of magnitude shown, the required vectors are as indicated above.

**3.** An automobile travels 3km due north, then 5km northeast. Represent these displacements graphically and determine the resultant displacement (*a*) graphically, (*b*) analytically.

Vector **OP** or **A** represents displacement of 3km due north.

Vector **PQ** or **B** represents displacement of 5km north east.

Vector **OQ** or **C** represents the resultant displacement or sum of vectors **A** and **B**, i.e. **C** = **A**+**B**. This is the *triangle law* of vector addition.

The resultant vector **OQ** can also be obtained by constructing the diagonal of the parallelogram *OPQR* having vectors **OP** = **A** and **OR** (equal to vector **PQ** or **B**) as sides. This is the *parallelogram law* of vector addition.

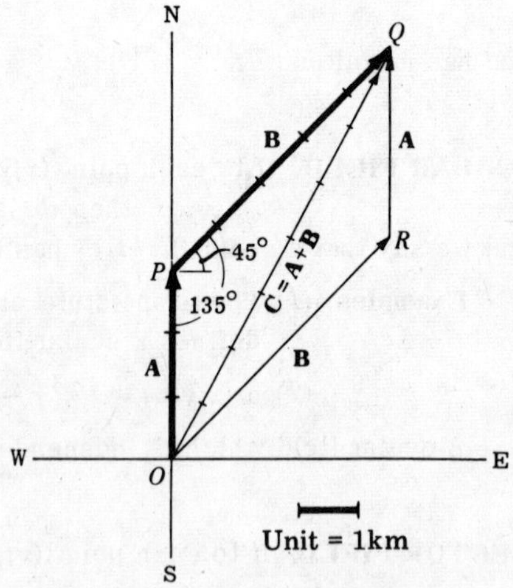


(*a*) *Graphical Determination of Resultant.* Lay off the 1 km unit on vector **OQ** to find the magnitude 7·4 km (approximately). Angle *EOQ*=61·5°, using a protractor. Then vector **OQ** has magnitude 7·4 km and direction 61·5° north of east.

(*b*) *Analytical Determination of Resultant.* From triangle *OPQ*, denoting the magnitudes of **A, B, C** by A, B, C, we have by the law of cosines

$$C^2 = A^2 + B^2 - 2AB\cos\angle OPQ = 3^2 + 5^2 - 2(3)(5)\cos 135^\circ = 34 + 15\sqrt{2} = 55{\cdot}21$$

and C = 7·43 (approximately).

By the law of sines, $\dfrac{A}{\sin\angle OQP} = \dfrac{C}{\sin\angle OPQ}$. Then

$$\sin \angle OQP = \frac{A \sin \angle OPQ}{C} = \frac{3(0{\cdot}707)}{7{\cdot}43} = 0{\cdot}2855 \quad \text{and} \quad \angle OQP = 16^\circ 35'.$$

Thus vector **OQ** has magnitude 7·43km and direction $(45^\circ + 16^\circ 35') = 61^\circ 35'$ north of east.

**4.** Find the sum or resultant of the following displacements:
**A**, 10m northwest; **B**, 20m 30° north of east; **C**, 35m due south. See Fig. (*a*) below.

At the terminal point of **A** place the initial point of **B**.

At the terminal point of **B** place the initial point of **C**.

The resultant **D** is formed by joining the initial point of **A** to the terminal point of **C**, i.e. **D** = **A**+**B**+**C**.

Graphically the resultant is measured to have magnitude of 4·1 units =20·5m and direction 60° south of E.

For an analytical method of addition of 3 or more vectors, either in a plane or in space see Problem 26.

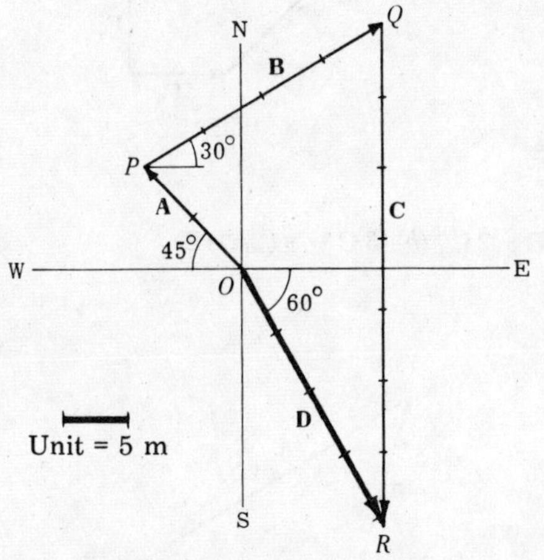


Fig.(*a*)

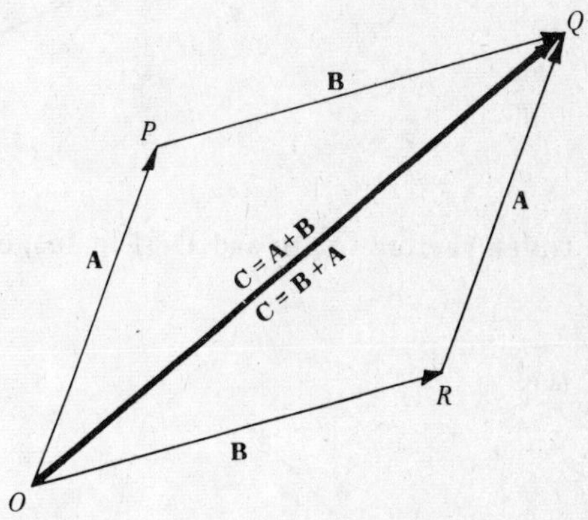


Fig.(*b*)

**5.** Show that addition of vectors is commutative, i.e. **A** + **B** = **B** + **A**. See Fig.(*b*) above.

**OP** + **PQ** = **OQ** or **A** + **B** = **C**,

and **OR** + **RQ** = **OQ** or **B** + **A** = **C**.

Then **A** + **B** = **B** + **A**.

**6.** Show that the addition of vectors is associative, i.e. **A** + (**B** + **C**) = (**A** + **B**) + **C** .

**OP** + **PQ** = **OQ** = (**A** + **B**),

and **PQ** + **QR** = **PR** = (**B** + **C**).

**OP** + **PR** = **OR** = **D**, i.e. **A** + (**B** + **C**) = **D**.

**OQ** + **QR** = **OR** = **D**, i.e. (**A** + **B**) + **C** = **D**.

Then **A** + (**B** + **C**) = (**A** + **B**) + **C** .

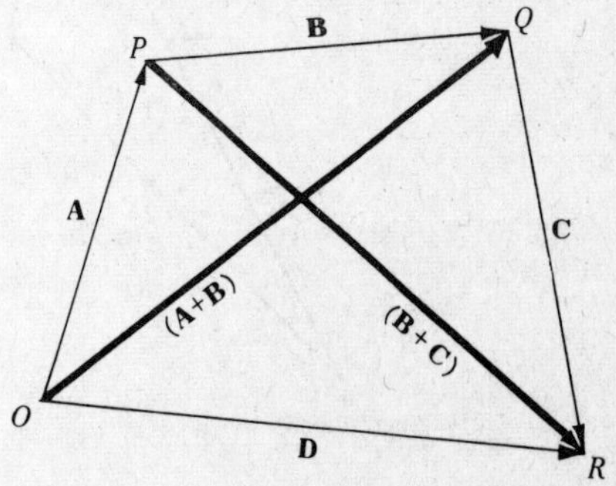


Extensions of the results of Problems 5 and 6 show that the order of addition of any number of vectors is immaterial.

**7.** Forces $\mathbf{F}_1, \mathbf{F}_2, \ldots, \mathbf{F}_6$ act as shown on object $P$. What force is needed to prevent $P$ from moving ?

Since the order of addition of vectors is immaterial, we may start with any vector, say $\mathbf{F}_1$. To $\mathbf{F}_1$ add $\mathbf{F}_2$, then $\mathbf{F}_3$, etc. The vector drawn from the initial point of $\mathbf{F}_1$ to the terminal point of $\mathbf{F}_6$ is the resultant $\mathbf{R}$, i.e. $\mathbf{R} = \mathbf{F}_1 + \mathbf{F}_2 + \mathbf{F}_3 + \mathbf{F}_4 + \mathbf{F}_5 + \mathbf{F}_6$.

The force needed to prevent $P$ from moving is $-\mathbf{R}$ which is a vector equal in magnitude to $\mathbf{R}$ but opposite in direction and sometimes called the *equilibrant*.

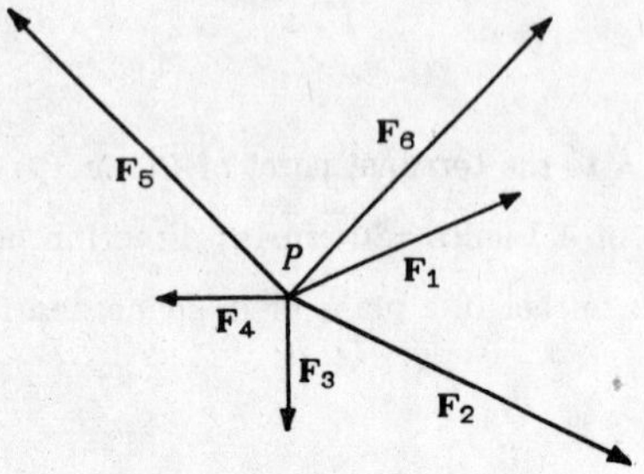


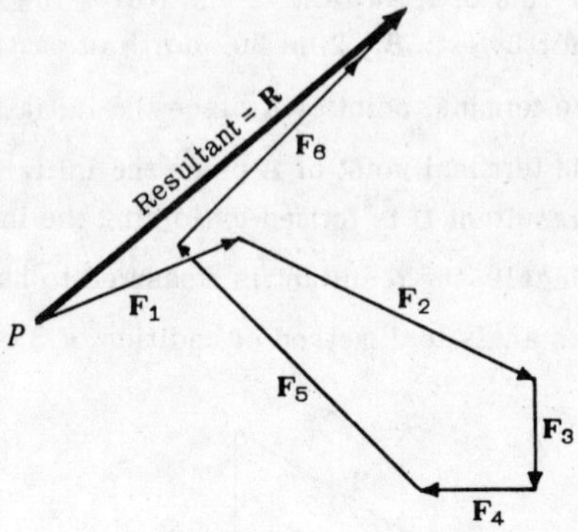


**8.** Given vectors **A**, **B** and **C** (Fig. 1*a*), construct (*a*) $\mathbf{A} - \mathbf{B} + 2\mathbf{C}$ (*b*) $3\mathbf{C} - \frac{1}{2}(2\mathbf{A} - \mathbf{B})$.

(*a*)

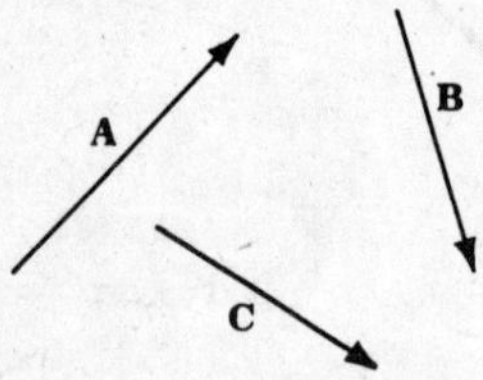


Fig. 1(*a*)

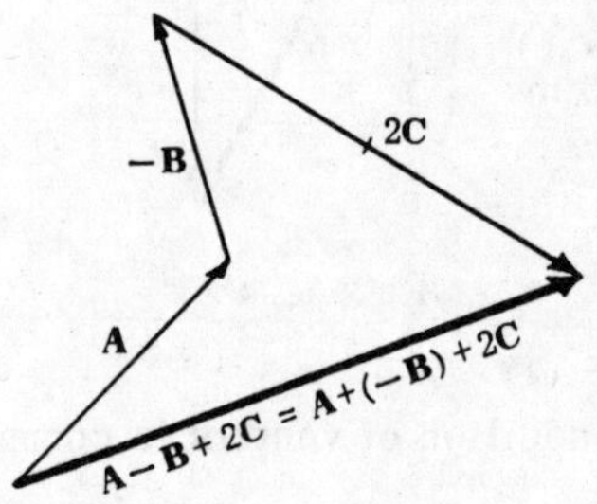


Fig. 2(*a*)

(*b*)

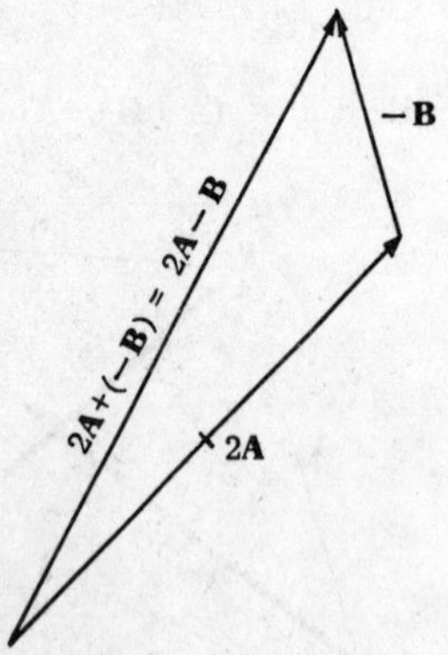


Fig. 1(*b*)

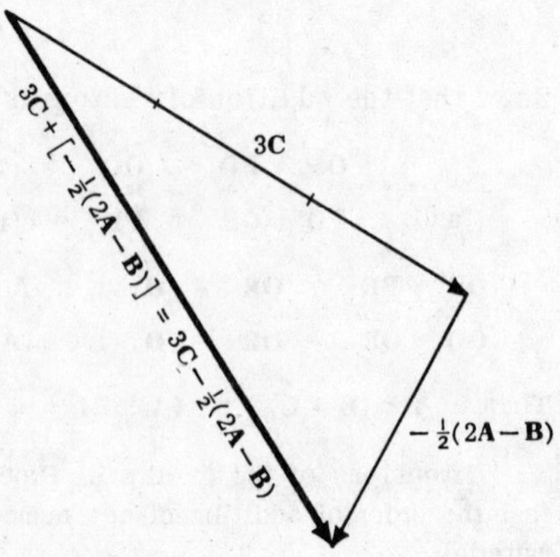


Fig. 2(*b*)

**9.** An aeroplane moves in a northwesterly direction at 125km/h relative to the ground, due to the fact there is a westerly wind of 50km/h relative to the ground. How fast and in what direction would the plane have travelled if there were no wind?

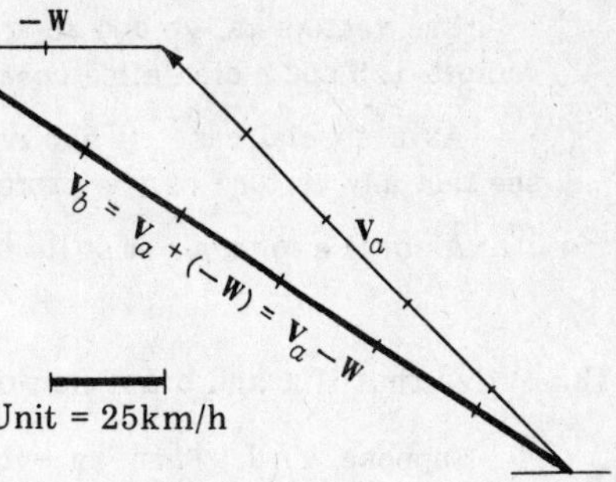


Let $\mathbf{W}$ = wind velocity

$\mathbf{V}_a$ = velocity of plane with wind

$\mathbf{V}_b$ = velocity of plane without wind

Then $\mathbf{V}_a = \mathbf{V}_b + \mathbf{W}$ or $\mathbf{V}_b = \mathbf{V}_a - \mathbf{W} = \mathbf{V}_a + (-\mathbf{W})$

$V_b$ has magnitude 6·5 units = 163km/h and direction 33° north of west.

**10.** Given two non-collinear vectors **a** and **b**, find an expression for any vector **r** lying in the plane determined by **a** and **b**.

Non-collinear vectors are vectors which are not parallel to the same line. Hence when their initial points coincide, they determine a plane. Let **r** be any vector lying in the plane of **a** and **b** and having its initial point coincident with the initial points of **a** and **b** at $O$. From the terminal point $R$ of **r** construct lines parallel to the vectors **a** and **b** and complete the parallelogram $ODRC$ by extension of the lines of action of **a** and **b** if necessary. From the adjoining figure

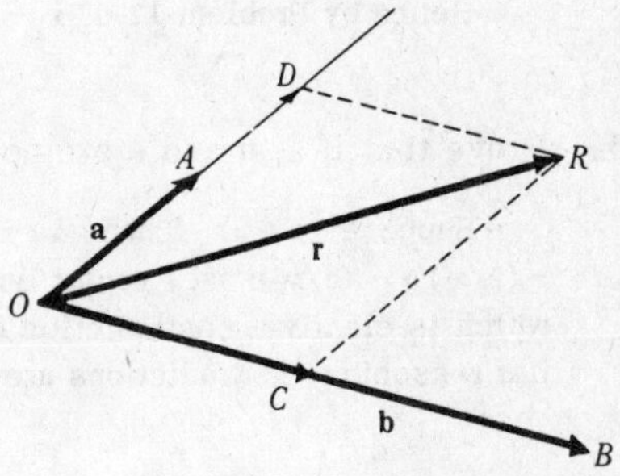


$\mathbf{OD} = x(\mathbf{OA}) = x\mathbf{a}$, where $x$ is a scalar

$\mathbf{OC} = y(\mathbf{OB}) = y\mathbf{b}$, where $y$ is a scalar.

But by the parallelogram law of vector addition

$\mathbf{OR} = \mathbf{OD} + \mathbf{OC}$ or $\mathbf{r} = x\mathbf{a} + y\mathbf{b}$

which is the required expression. The vectors $x\mathbf{a}$ and $y\mathbf{b}$ are called *component vectors* of **r** in the directions **a** and **b** respectively. The scalars $x$ and $y$ may be positive or negative depending on the relative orientations of the vectors. From the manner of construction it is clear that $x$ and $y$ are unique for a given **a**, **b**, and **r**. The vectors **a** and **b** are called *base vectors* in a plane.

**11.** Given three non-coplanar vectors **a**, **b**, and **c**, find an expression for any vector **r** in three dimensional space.

Non-coplanar vectors are vectors which are not parallel to the same plane. Hence when their initial points coincide they do not lie in the same plane.

Let **r** be any vector in space having its initial point coincident with the initial points of **a**, **b** and **c** at $O$. Through the terminal point of **r** pass planes parallel respectively to the planes determined by **a** and **b**, **b** and **c**, and **a** and **c**; and complete the parallelepiped $PQRSTUV$ by extension of the lines of action of **a**, **b** and **c** if necessary. From the adjoining figure,

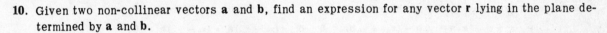

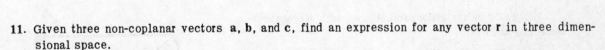

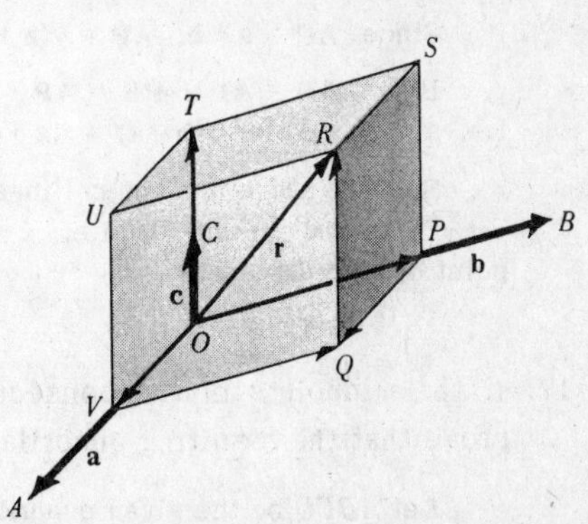


$\mathbf{OV} = x(\mathbf{OA}) = x\mathbf{a}$ where $x$ is a scalar

$\mathbf{OP} = y(\mathbf{OB}) = y\mathbf{b}$ where $y$ is a scalar

$\mathbf{OT} = z(\mathbf{OC}) = z\mathbf{c}$ where $z$ is a scalar.

But $\mathbf{OR} = \mathbf{OV} + \mathbf{VQ} + \mathbf{QR} = \mathbf{OV} + \mathbf{OP} + \mathbf{OT}$ or $\mathbf{r} = x\mathbf{a} + y\mathbf{b} + z\mathbf{c}$.

From the manner of construction it is clear that $x$, $y$ and $z$ are unique for a given **a**, **b**, **c** and **r**.

The vectors $x\mathbf{a}$, $y\mathbf{b}$ and $z\mathbf{c}$ are called *component vectors* of $\mathbf{r}$ in directions $\mathbf{a}$, $\mathbf{b}$ and $\mathbf{c}$ respectively. The vectors $\mathbf{a}$, $\mathbf{b}$ and $\mathbf{c}$ are called *base vectors* in three dimensions.

As a special case, if $\mathbf{a}$, $\mathbf{b}$ and $\mathbf{c}$ are the unit vectors $\mathbf{i}$, $\mathbf{j}$ and $\mathbf{k}$, which are mutually perpendicular, we see that any vector $\mathbf{r}$ can be expressed uniquely in terms of $\mathbf{i}$, $\mathbf{j}$, $\mathbf{k}$ by the expression $\mathbf{r} = x\mathbf{i} + y\mathbf{j} + z\mathbf{k}$.

Also, if $\mathbf{c} = \mathbf{0}$ then $\mathbf{r}$ must lie in the plane of $\mathbf{a}$ and $\mathbf{b}$ so the result of Problem 10 is obtained.

**12.** Prove that if $\mathbf{a}$ and $\mathbf{b}$ are non-collinear then $x\mathbf{a} + y\mathbf{b} = \mathbf{0}$ implies $x = y = 0$.

Suppose $x \neq 0$. Then $x\mathbf{a} + y\mathbf{b} = \mathbf{0}$ implies $x\mathbf{a} = -y\mathbf{b}$ or $\mathbf{a} = -(y/x)\mathbf{b}$, i.e. $\mathbf{a}$ and $\mathbf{b}$ must be parallel to to the same line (collinear) contrary to hypothesis. Thus $x = 0$; then $y\mathbf{b} = \mathbf{0}$, from which $y = 0$.

**13.** If $x_1\mathbf{a} + y_1\mathbf{b} = x_2\mathbf{a} + y_2\mathbf{b}$, where $\mathbf{a}$ and $\mathbf{b}$ are non-collinear, then $x_1 = x_2$ and $y_1 = y_2$.

$x_1\mathbf{a} + y_1\mathbf{b} = x_2\mathbf{a} + y_2\mathbf{b}$ can be written

$$x_1\mathbf{a} + y_1\mathbf{b} - (x_2\mathbf{a} + y_2\mathbf{b}) = \mathbf{0} \quad \text{or} \quad (x_1 - x_2)\mathbf{a} + (y_1 - y_2)\mathbf{b} = \mathbf{0}.$$

Hence by Problem 12, $x_1 - x_2 = 0$, $y_1 - y_2 = 0$ or $x_1 = x_2$, $y_1 = y_2$.

**14.** Prove that if $\mathbf{a}$, $\mathbf{b}$ and $\mathbf{c}$ are non-coplanar then $x\mathbf{a} + y\mathbf{b} + z\mathbf{c} = \mathbf{0}$ implies $x = y = z = 0$.

Suppose $x \neq 0$. Then $x\mathbf{a} + y\mathbf{b} + z\mathbf{c} = \mathbf{0}$ implies $x\mathbf{a} = -y\mathbf{b} - z\mathbf{c}$ or $\mathbf{a} = -(y/x)\mathbf{b} - (z/x)\mathbf{c}$. But $-(y/x)\mathbf{b} - (z/x)\mathbf{c}$ is a vector lying in the plane of $\mathbf{b}$ and $\mathbf{c}$ (Problem 10), i.e. $\mathbf{a}$ lies in the plane of $\mathbf{b}$ and $\mathbf{c}$ which is clearly a contradiction to the hypothesis that $\mathbf{a}$, $\mathbf{b}$ and $\mathbf{c}$ are non-coplanar. Hence $x = 0$. By similar reasoning, contradictions are obtained upon supposing $y \neq 0$ and $z \neq 0$.

**15.** If $x_1\mathbf{a} + y_1\mathbf{b} + z_1\mathbf{c} = x_2\mathbf{a} + y_2\mathbf{b} + z_2\mathbf{c}$, where $\mathbf{a}$, $\mathbf{b}$ and $\mathbf{c}$ are non-coplanar, then $x_1 = x_2$, $y_1 = y_2$, $z_1 = z_2$.

The equation can be written $(x_1 - x_2)\mathbf{a} + (y_1 - y_2)\mathbf{b} + (z_1 - z_2)\mathbf{c} = \mathbf{0}$. Then by Problem 14, $x_1 - x_2 = 0$, $y_1 - y_2 = 0$, $z_1 - z_2 = 0$ or $x_1 = x_2$, $y_1 = y_2$, $z_1 = z_2$.

**16.** Prove that the diagonals of a parallelogram bisect each other.

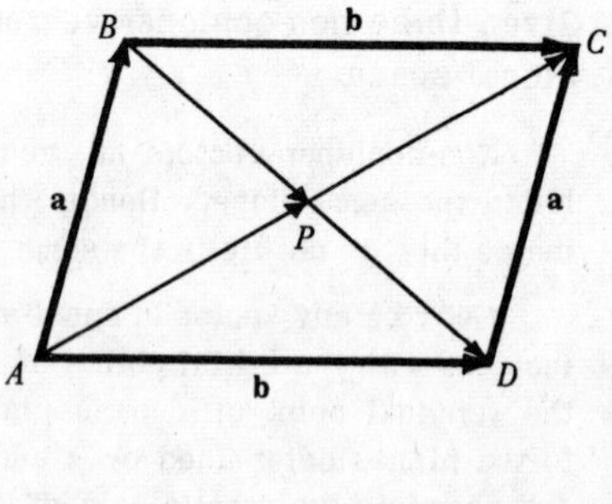


Let $ABCD$ be the given parallelogram with diagonals intersecting at $P$.

Since $\mathbf{BD} + \mathbf{a} = \mathbf{b}$, $\mathbf{BD} = \mathbf{b} - \mathbf{a}$. Then $\mathbf{BP} = x(\mathbf{b} - \mathbf{a})$.

Since $\mathbf{AC} = \mathbf{a} + \mathbf{b}$, $\mathbf{AP} = y(\mathbf{a} + \mathbf{b})$.

But $\mathbf{AB} = \mathbf{AP} + \mathbf{PB} = \mathbf{AP} - \mathbf{BP}$,
i.e. $\mathbf{a} = y(\mathbf{a} + \mathbf{b}) - x(\mathbf{b} - \mathbf{a}) = (x + y)\mathbf{a} + (y - x)\mathbf{b}$.

Since $\mathbf{a}$ and $\mathbf{b}$ are non-collinear we have by Problem 13, $x + y = 1$ and $y - x = 0$, i.e. $x = y = \frac{1}{2}$ and $P$ is the midpoint of both diagonals.

**17.** If the midpoints of the consecutive sides of any quadrilateral are connected by straight lines, prove that the resulting quadrilateral is a parallelogram.

Let $ABCD$ be the given quadrilateral and $P, Q, R, S$ the midpoints of its sides. Refer to Fig.(*a*) below.

Then $\mathbf{PQ} = \frac{1}{2}(\mathbf{a} + \mathbf{b})$, $\mathbf{QR} = \frac{1}{2}(\mathbf{b} + \mathbf{c})$, $\mathbf{RS} = \frac{1}{2}(\mathbf{c} + \mathbf{d})$, $\mathbf{SP} = \frac{1}{2}(\mathbf{d} + \mathbf{a})$.

But $\mathbf{a} + \mathbf{b} + \mathbf{c} + \mathbf{d} = \mathbf{0}$. Then

$$\mathbf{PQ} = \tfrac{1}{2}(\mathbf{a} + \mathbf{b}) = -\tfrac{1}{2}(\mathbf{c} + \mathbf{d}) = \mathbf{SR} \quad \text{and} \quad \mathbf{QR} = \tfrac{1}{2}(\mathbf{b} + \mathbf{c}) = -\tfrac{1}{2}(\mathbf{d} + \mathbf{a}) = \mathbf{PS}$$

Thus opposite sides are equal and parallel and $PQRS$ is a parallelogram.

**18.** Let $P_1, P_2, P_3$ be points fixed relative to an origin $O$ and let $\mathbf{r}_1, \mathbf{r}_2, \mathbf{r}_3$ be position vectors from $O$ to each point. Show that if the vector equation $a_1\mathbf{r}_1 + a_2\mathbf{r}_2 + a_3\mathbf{r}_3 = \mathbf{0}$ holds with respect to origin $O$ then it will hold with respect to any other origin $O'$ if and only if $a_1 + a_2 + a_3 = 0$.

Let $\mathbf{r}_1', \mathbf{r}_2'$ and $\mathbf{r}_3'$ be the position vectors of $P_1, P_2$ and $P_3$ with respect to $O'$ and let $\mathbf{v}$ be the position vector of $O'$ with respect to $O$. We seek conditions under which the equation $a_1\mathbf{r}_1' + a_2\mathbf{r}_2' + a_3\mathbf{r}_3' = \mathbf{0}$ will hold in the new reference system.

From Fig.(*b*) below, it is clear that $\mathbf{r}_1 = \mathbf{v} + \mathbf{r}_1'$, $\mathbf{r}_2 = \mathbf{v} + \mathbf{r}_2'$, $\mathbf{r}_3 = \mathbf{v} + \mathbf{r}_3'$ so that $a_1\mathbf{r}_1 + a_2\mathbf{r}_2 + a_3\mathbf{r}_3 = \mathbf{0}$ becomes

$$\begin{aligned} a_1\mathbf{r}_1 + a_2\mathbf{r}_2 + a_3\mathbf{r}_3 &= a_1(\mathbf{v} + \mathbf{r}_1') + a_2(\mathbf{v} + \mathbf{r}_2') + a_3(\mathbf{v} + \mathbf{r}_3') \\ &= (a_1 + a_2 + a_3)\mathbf{v} + a_1\mathbf{r}_1' + a_2\mathbf{r}_2' + a_3\mathbf{r}_3' = \mathbf{0} \end{aligned}$$

The result $a_1\mathbf{r}_1' + a_2\mathbf{r}_2' + a_3\mathbf{r}_3' = \mathbf{0}$ will hold if and only if

$$(a_1 + a_2 + a_3)\mathbf{v} = \mathbf{0}, \quad \text{i.e.} \quad a_1 + a_2 + a_3 = 0.$$

The result can be generalized.

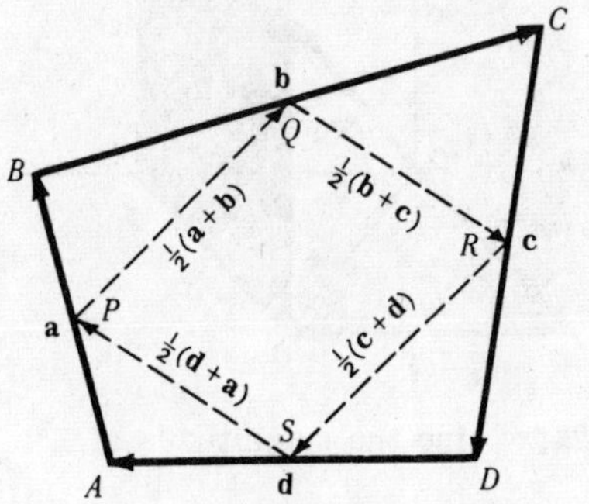


Fig.(*a*)

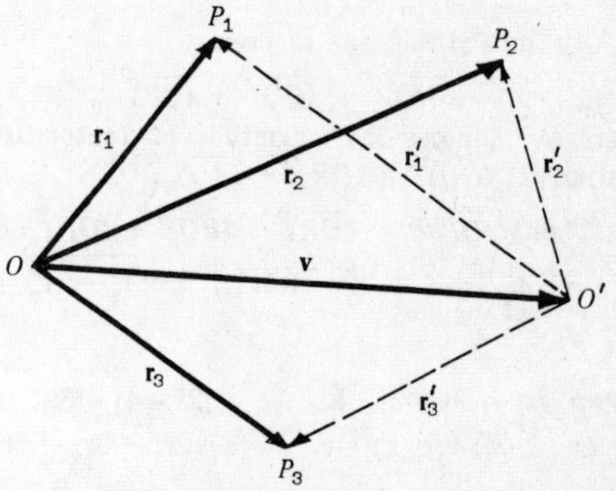


Fig.(*b*)

**19.** Find the equation of a straight line which passes through two given points $A$ and $B$ having position vectors $\mathbf{a}$ and $\mathbf{b}$ with respect to an origin $O$.

Let $\mathbf{r}$ be the position vector of any point $P$ on the line through $A$ and $B$.

From the adjoining figure,

$$\mathbf{OA} + \mathbf{AP} = \mathbf{OP} \quad \text{or} \quad \mathbf{a} + \mathbf{AP} = \mathbf{r}, \quad \text{i.e.} \quad \mathbf{AP} = \mathbf{r} - \mathbf{a}$$

and $$\mathbf{OA} + \mathbf{AB} = \mathbf{OB} \quad \text{or} \quad \mathbf{a} + \mathbf{AB} = \mathbf{b}, \quad \text{i.e.} \quad \mathbf{AB} = \mathbf{b} - \mathbf{a}$$

Since $\mathbf{AP}$ and $\mathbf{AB}$ are collinear, $\mathbf{AP} = t\mathbf{AB}$ or $\mathbf{r} - \mathbf{a} = t(\mathbf{b} - \mathbf{a})$. Then the required equation is

$$\mathbf{r} = \mathbf{a} + t(\mathbf{b} - \mathbf{a}) \quad \text{or} \quad \mathbf{r} = (1 - t)\mathbf{a} + t\mathbf{b}$$

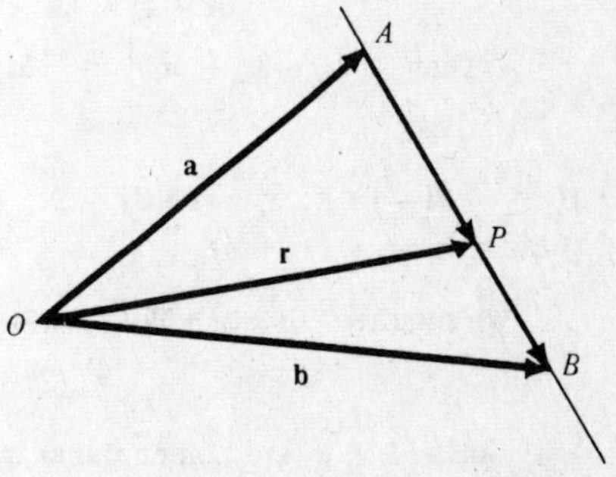


If the equation is written $(1 - t)\mathbf{a} + t\mathbf{b} - \mathbf{r} = \mathbf{0}$, the sum of the coefficients of $\mathbf{a}$, $\mathbf{b}$ and $\mathbf{r}$ is $1 - t + t - 1 = 0$. Hence by Problem 18 it is seen that the point $P$ is always on the line joining $A$ and $B$ and does not depend on the choice of origin $O$, which is of course as it should be.

*Another Method.* Since $\mathbf{AP}$ and $\mathbf{PB}$ are collinear, we have for scalars $m$ and $n$:

$$m\mathbf{AP} = n\mathbf{PB} \quad \text{or} \quad m(\mathbf{r} - \mathbf{a}) = n(\mathbf{b} - \mathbf{r})$$

Solving, $\mathbf{r} = \dfrac{m\mathbf{a} + n\mathbf{b}}{m + n}$ which is called the *symmetric form*.

**20.** (*a*) Find the position vectors $\mathbf{r}_1$ and $\mathbf{r}_2$ for the points $P(2,4,3)$ and $Q(1,-5,2)$ of a rectangular coordinate system in terms of the unit vectors **i, j, k**. (*b*) Determine graphically and analytically the resultant of these position vectors.

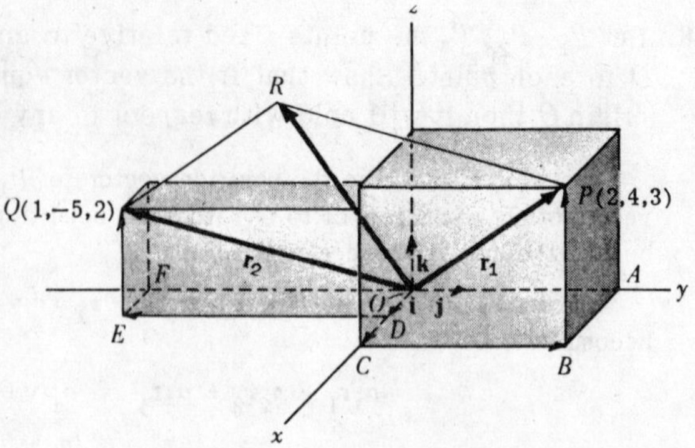


(*a*) $\mathbf{r}_1 = \mathbf{OP} = \mathbf{OC} + \mathbf{CB} + \mathbf{BP} = 2\mathbf{i} + 4\mathbf{j} + 3\mathbf{k}$

$\mathbf{r}_2 = \mathbf{OQ} = \mathbf{OD} + \mathbf{DE} + \mathbf{EQ} = \mathbf{i} - 5\mathbf{j} + 2\mathbf{k}$

(*b*) *Graphically*, the resultant of $\mathbf{r}_1$ and $\mathbf{r}_2$ is obtained as the diagonal **OR** of parallelogram $OPRQ$. *Analytically*, the resultant of $\mathbf{r}_1$ and $\mathbf{r}_2$ is given by

$$\mathbf{r}_1 + \mathbf{r}_2 = (2\mathbf{i} + 4\mathbf{j} + 3\mathbf{k}) + (\mathbf{i} - 5\mathbf{j} + 2\mathbf{k}) = 3\mathbf{i} - \mathbf{j} + 5\mathbf{k}$$

**21.** Prove that the magnitude $A$ of the vector $\mathbf{A} = A_1\mathbf{i} + A_2\mathbf{j} + A_3\mathbf{k}$ is $A = \sqrt{A_1^2 + A_2^2 + A_3^2}$.

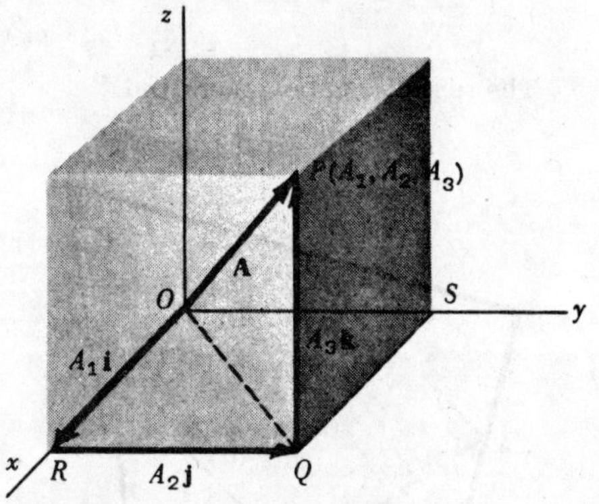


By the Pythagorean theorem,

$$(\overline{OP})^2 = (\overline{OQ})^2 + (\overline{QP})^2$$

where $\overline{OP}$ denotes the magnitude of vector **OP**, etc. Similarly, $(\overline{OQ})^2 = (\overline{OR})^2 + (\overline{RQ})^2$.

Then $(\overline{OP})^2 = (\overline{OR})^2 + (\overline{RQ})^2 + (\overline{QP})^2$ or

$$A^2 = A_1^2 + A_2^2 + A_3^2, \quad \text{i.e.} \quad A = \sqrt{A_1^2 + A_2^2 + A_3^2}.$$

**22.** Given $\mathbf{r}_1 = 3\mathbf{i} - 2\mathbf{j} + \mathbf{k}$, $\mathbf{r}_2 = 2\mathbf{i} - 4\mathbf{j} - 3\mathbf{k}$, $\mathbf{r}_3 = -\mathbf{i} + 2\mathbf{j} + 2\mathbf{k}$, find the magnitudes of (*a*) $\mathbf{r}_3$, (*b*) $\mathbf{r}_1 + \mathbf{r}_2 + \mathbf{r}_3$, (*c*) $2\mathbf{r}_1 - 3\mathbf{r}_2 - 5\mathbf{r}_3$.

(*a*) $|\mathbf{r}_3| = |-\mathbf{i} + 2\mathbf{j} + 2\mathbf{k}| = \sqrt{(-1)^2 + (2)^2 + (2)^2} = 3.$

(*b*) $\mathbf{r}_1 + \mathbf{r}_2 + \mathbf{r}_3 = (3\mathbf{i} - 2\mathbf{j} + \mathbf{k}) + (2\mathbf{i} - 4\mathbf{j} - 3\mathbf{k}) + (-\mathbf{i} + 2\mathbf{j} + 2\mathbf{k}) = 4\mathbf{i} - 4\mathbf{j} + 0\mathbf{k} = 4\mathbf{i} - 4\mathbf{j}$

Then $|\mathbf{r}_1 + \mathbf{r}_2 + \mathbf{r}_3| = |4\mathbf{i} - 4\mathbf{j} + 0\mathbf{k}| = \sqrt{(4)^2 + (-4)^2 + (0)^2} = \sqrt{32} = 4\sqrt{2}$.

(*c*) $2\mathbf{r}_1 - 3\mathbf{r}_2 - 5\mathbf{r}_3 = 2(3\mathbf{i} - 2\mathbf{j} + \mathbf{k}) - 3(2\mathbf{i} - 4\mathbf{j} - 3\mathbf{k}) - 5(-\mathbf{i} + 2\mathbf{j} + 2\mathbf{k})$

$= 6\mathbf{i} - 4\mathbf{j} + 2\mathbf{k} - 6\mathbf{i} + 12\mathbf{j} + 9\mathbf{k} + 5\mathbf{i} - 10\mathbf{j} - 10\mathbf{k} = 5\mathbf{i} - 2\mathbf{j} + \mathbf{k}$.

Then $|2\mathbf{r}_1 - 3\mathbf{r}_2 - 5\mathbf{r}_3| = |5\mathbf{i} - 2\mathbf{j} + \mathbf{k}| = \sqrt{(5)^2 + (-2)^2 + (1)^2} = \sqrt{30}$.

**23.** If $\mathbf{r}_1 = 2\mathbf{i} - \mathbf{j} + \mathbf{k}$, $\mathbf{r}_2 = \mathbf{i} + 3\mathbf{j} - 2\mathbf{k}$, $\mathbf{r}_3 = -2\mathbf{i} + \mathbf{j} - 3\mathbf{k}$ and $\mathbf{r}_4 = 3\mathbf{i} + 2\mathbf{j} + 5\mathbf{k}$, find scalars $a, b, c$ such that $\mathbf{r}_4 = a\mathbf{r}_1 + b\mathbf{r}_2 + c\mathbf{r}_3$.

We require $3\mathbf{i} + 2\mathbf{j} + 5\mathbf{k} = a(2\mathbf{i} - \mathbf{j} + \mathbf{k}) + b(\mathbf{i} + 3\mathbf{j} - 2\mathbf{k}) + c(-2\mathbf{i} + \mathbf{j} - 3\mathbf{k})$

$= (2a + b - 2c)\mathbf{i} + (-a + 3b + c)\mathbf{j} + (a - 2b - 3c)\mathbf{k}$.

Since **i, j, k** are non-coplanar we have by Problem 15,

$$2a + b - 2c = 3, \quad -a + 3b + c = 2, \quad a - 2b - 3c = 5.$$

Solving, $a = -2$, $b = 1$, $c = -3$ and $\mathbf{r}_4 = -2\mathbf{r}_1 + \mathbf{r}_2 - 3\mathbf{r}_3$.

The vector $\mathbf{r}_4$ is said to be *linearly dependent* on $\mathbf{r}_1$, $\mathbf{r}_2$, and $\mathbf{r}_3$; in other words $\mathbf{r}_1$, $\mathbf{r}_2$, $\mathbf{r}_3$ and $\mathbf{r}_4$ constitute a *linearly dependent* set of vectors. On the other hand any three (or fewer) of these vectors are *linearly independent*.

In general the vectors **A, B, C, ...** are called linearly dependent if we can find a set of scalars, $a, b, c, \ldots$, not all zero, so that $a\mathbf{A} + b\mathbf{B} + c\mathbf{C} + \ldots = \mathbf{0}$, otherwise they are linearly independent.

**24.** Find a unit vector parallel to the resultant of vectors $\mathbf{r}_1 = 2\mathbf{i} + 4\mathbf{j} - 5\mathbf{k}$, $\mathbf{r}_2 = \mathbf{i} + 2\mathbf{j} + 3\mathbf{k}$.

Resultant $\mathbf{R} = \mathbf{r}_1 + \mathbf{r}_2 = (2\mathbf{i}+4\mathbf{j}-5\mathbf{k}) + (\mathbf{i}+2\mathbf{j}+3\mathbf{k}) = 3\mathbf{i} + 6\mathbf{j} - 2\mathbf{k}$.

$$R = |\mathbf{R}| = |3\mathbf{i}+6\mathbf{j}-2\mathbf{k}| = \sqrt{(3)^2+(6)^2+(-2)^2} = 7.$$

Then a unit vector parallel to $\mathbf{R}$ is $\dfrac{\mathbf{R}}{R} = \dfrac{3\mathbf{i}+6\mathbf{j}-2\mathbf{k}}{7} = \dfrac{3}{7}\mathbf{i} + \dfrac{6}{7}\mathbf{j} - \dfrac{2}{7}\mathbf{k}$.

Check: $\left|\dfrac{3}{7}\mathbf{i} + \dfrac{6}{7}\mathbf{j} - \dfrac{2}{7}\mathbf{k}\right| = \sqrt{(\frac{3}{7})^2 + (\frac{6}{7})^2 + (-\frac{2}{7})^2} = 1.$

**25.** Determine the vector having initial point $P(x_1, y_1, z_1)$ and terminal point $Q(x_2, y_2, z_2)$ and find its magnitude.

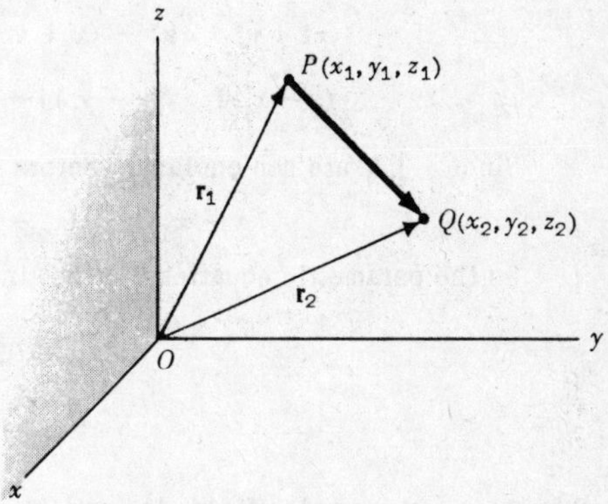


The position vector of $P$ is $\mathbf{r}_1 = x_1\mathbf{i} + y_1\mathbf{j} + z_1\mathbf{k}$.

The position vector of $Q$ is $\mathbf{r}_2 = x_2\mathbf{i} + y_2\mathbf{j} + z_2\mathbf{k}$.

$\mathbf{r}_1 + \mathbf{PQ} = \mathbf{r}_2$ or

$$\mathbf{PQ} = \mathbf{r}_2 - \mathbf{r}_1 = (x_2\mathbf{i}+y_2\mathbf{j}+z_2\mathbf{k}) - (x_1\mathbf{i}+y_1\mathbf{j}+z_1\mathbf{k})$$
$$= (x_2 - x_1)\mathbf{i} + (y_2 - y_1)\mathbf{j} + (z_2 - z_1)\mathbf{k}.$$

Magnitude of $\mathbf{PQ} = \overline{PQ} = \sqrt{(x_2-x_1)^2 + (y_2-y_1)^2 + (z_2-z_1)^2}$.

Note that this is the distance between points $P$ and $Q$.

**26.** Forces $\mathbf{A}$, $\mathbf{B}$ and $\mathbf{C}$ acting on an object are given in terms of their components by the vector equations $\mathbf{A} = A_1\mathbf{i} + A_2\mathbf{j} + A_3\mathbf{k}$, $\mathbf{B} = B_1\mathbf{i} + B_2\mathbf{j} + B_3\mathbf{k}$, $\mathbf{C} = C_1\mathbf{i} + C_2\mathbf{j} + C_3\mathbf{k}$. Find the magnitude of the resultant of these forces.

Resultant force $\mathbf{R} = \mathbf{A}+\mathbf{B}+\mathbf{C} = (A_1+B_1+C_1)\mathbf{i} + (A_2+B_2+C_2)\mathbf{j} + (A_3+B_3+C_3)\mathbf{k}$.

Magnitude of resultant $= \sqrt{(A_1+B_1+C_1)^2 + (A_2+B_2+C_2)^2 + (A_3+B_3+C_3)^2}$.

The result is easily extended to more than three forces.

**27.** Determine the angles $\alpha$, $\beta$ and $\gamma$ which the vector $\mathbf{r} = x\mathbf{i} + y\mathbf{j} + z\mathbf{k}$ makes with the positive directions of the coordinate axes and show that

$$\cos^2\alpha + \cos^2\beta + \cos^2\gamma = 1.$$

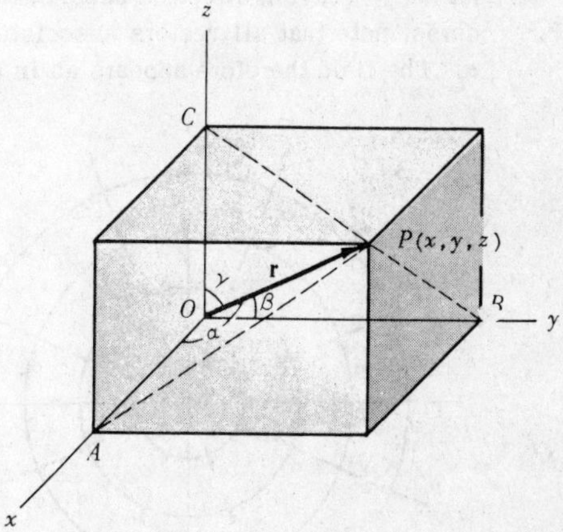


Referring to the figure, triangle $OAP$ is a right-angled triangle with right angle at $A$; then $\cos\alpha = \dfrac{x}{|\mathbf{r}|}$. Similarly from right-angled triangles $OBP$ and $OCP$, $\cos\beta = \dfrac{y}{|\mathbf{r}|}$ and $\cos\gamma = \dfrac{z}{|\mathbf{r}|}$. Also, $|\mathbf{r}| = r = \sqrt{x^2+y^2+z^2}$.

Then $\cos\alpha = \dfrac{x}{r}$, $\cos\beta = \dfrac{y}{r}$, $\cos\gamma = \dfrac{z}{r}$ from which $\alpha, \beta, \gamma$ can be obtained. From these it follows that

$$\cos^2\alpha + \cos^2\beta + \cos^2\gamma = \frac{x^2+y^2+z^2}{r^2} = 1.$$

The numbers $\cos\alpha$, $\cos\beta$, $\cos\gamma$ are called the *direction cosines* of the vector $\mathbf{OP}$.

**28.** Determine a set of equations for the straight line passing through the points $P(x_1, y_1, z_1)$ and $Q(x_2, y_2, z_2)$.

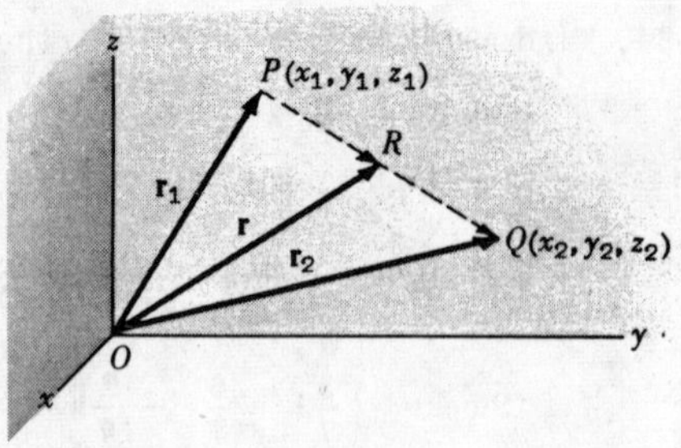


Let $\mathbf{r}_1$ and $\mathbf{r}_2$ be the position vectors of $P$ and $Q$ respectively, and $\mathbf{r}$ the position vector of any point $R$ on the line joining $P$ and $Q$.

$$\mathbf{r}_1 + \mathbf{PR} = \mathbf{r} \quad \text{or} \quad \mathbf{PR} = \mathbf{r} - \mathbf{r}_1$$
$$\mathbf{r}_1 + \mathbf{PQ} = \mathbf{r}_2 \quad \text{or} \quad \mathbf{PQ} = \mathbf{r}_2 - \mathbf{r}_1$$

But $\mathbf{PR} = t\,\mathbf{PQ}$ where $t$ is a scalar. Then $\mathbf{r} - \mathbf{r}_1 = t(\mathbf{r}_2 - \mathbf{r}_1)$ is the required vector equation of the straight line (compare with Problem 19).

In rectangular coordinates we have, since $\mathbf{r} = x\mathbf{i} + y\mathbf{j} + z\mathbf{k}$,

$$(x\mathbf{i} + y\mathbf{j} + z\mathbf{k}) - (x_1\mathbf{i} + y_1\mathbf{j} + z_1\mathbf{k}) = t[(x_2\mathbf{i} + y_2\mathbf{j} + z_2\mathbf{k}) - (x_1\mathbf{i} + y_1\mathbf{j} + z_1\mathbf{k})]$$

or

$$(x - x_1)\mathbf{i} + (y - y_1)\mathbf{j} + (z - z_1)\mathbf{k} = t[(x_2 - x_1)\mathbf{i} + (y_2 - y_1)\mathbf{j} + (z_2 - z_1)\mathbf{k}]$$

Since $\mathbf{i}, \mathbf{j}, \mathbf{k}$ are non-coplanar vectors we have by Problem 15,

$$x - x_1 = t(x_2 - x_1), \quad y - y_1 = t(y_2 - y_1), \quad z - z_1 = t(z_2 - z_1)$$

as the parametric equations of the line, $t$ being the parameter. Eliminating $t$, the equations become

$$\frac{x - x_1}{x_2 - x_1} = \frac{y - y_1}{y_2 - y_1} = \frac{z - z_1}{z_2 - z_1}.$$

**29.** Given the scalar field defined by $\phi(x,y,z) = 3x^2z - xy^3 + 5$, find $\phi$ at the points (*a*) (0, 0, 0), (*b*) (1, −2, 2) (*c*) (−1, −2, −3).

(*a*) $\phi(0,0,0) = 3(0)^2(0) - (0)(0)^3 + 5 = 0 - 0 + 5 = 5$

(*b*) $\phi(1,-2,2) = 3(1)^2(2) - (1)(-2)^3 + 5 = 6 + 8 + 5 = 19$

(*c*) $\phi(-1,-2,-3) = 3(-1)^2(-3) - (-1)(-2)^3 + 5 = -9 - 8 + 5 = -12$

**30.** Graph the vector fields defined by:
(*a*) $\mathbf{V}(x,y) = x\mathbf{i} + y\mathbf{j}$, (*b*) $\mathbf{V}(x,y) = -x\mathbf{i} - y\mathbf{j}$, (*c*) $\mathbf{V}(x,y,z) = x\mathbf{i} + y\mathbf{j} + z\mathbf{k}$.

(*a*) At each point $(x, y)$, except (0, 0), of the $xy$ plane there is defined a unique vector $x\mathbf{i} + y\mathbf{j}$ of magnitude $\sqrt{x^2 + y^2}$ having direction passing through the origin and outward from it. To simplify graphing procedures, note that all vectors associated with points on the circles $x^2 + y^2 = a^2$, $a > 0$ have magnitude $a$. The field therefore appears as in Figure (*a*) where an appropriate scale is used.

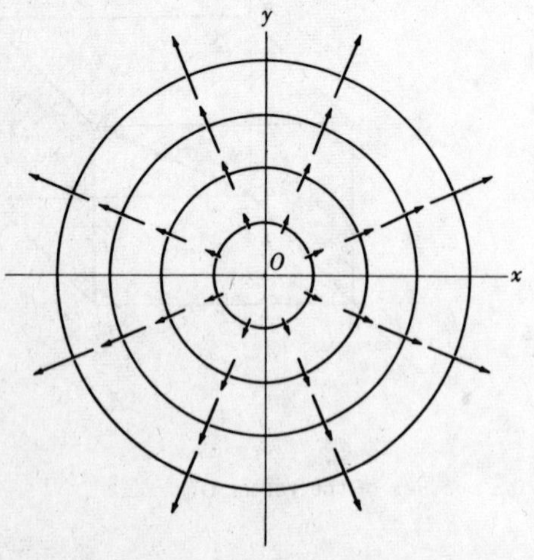


Fig.(*a*)

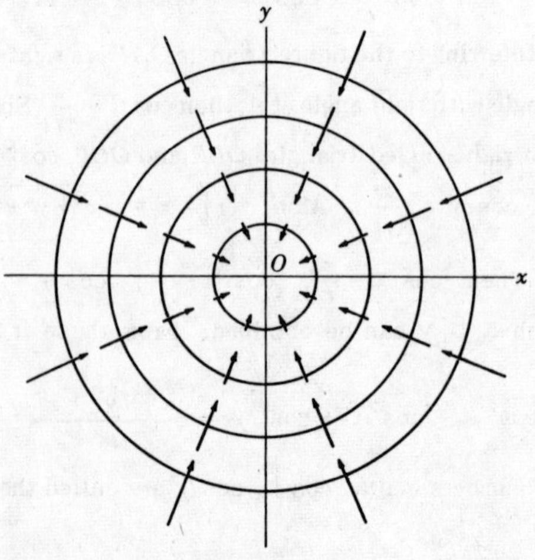


Fig.(*b*)

(*b*) Here each vector is equal to but opposite in direction to the corresponding one in (*a*). The field therefore appears as in Fig.(*b*).

In Fig.(*a*) the field has the appearance of a fluid emerging from a point source $O$ and flowing in the directions indicated. For this reason the field is called a *source field* and $O$ is a *source*.

In Fig.(*b*) the field seems to be flowing toward $O$, and the field is therefore called a *sink field* and $O$ is a *sink*.

In three dimensions the corresponding interpretation is that a fluid is emerging radially from (or proceeding radially toward) a line source (or line sink).

The vector field is called two dimensional since it is independent of $z$.

(*c*) Since the magnitude of each vector is $\sqrt{x^2 + y^2 + z^2}$, all points on the sphere $x^2 + y^2 + z^2 = a^2$, $a > 0$ have vectors of magnitude $a$ associated with them. The field therefore takes on the appearance of that of a fluid emerging from source $O$ and proceeding in all directions in space. This is a *three dimensional source field.*

# SUPPLEMENTARY PROBLEMS

**31.** Which of the following are scalars and which are vectors? (*a*) Kinetic energy, (*b*) electric field intensity, (*c*) entropy, (*d*) work, (*e*) centrifugal force, (*f*) temperature, (*g*) gravitational potential, (*h*) charge, (*i*) shearing stress, (*j*) frequency.
*Ans.* (*a*) scalar, (*b*) vector, (*c*) scalar, (*d*) scalar, (*e*) vector, (*f*) scalar, (*g*) scalar, (*h*) scalar, (*i*) vector (*j*) scalar

**32.** An aeroplane travels 200km due west and then 150km 60° north of west. Determine the resultant displacement (*a*) graphically, (*b*) analytically.
*Ans.* magnitude 304·1km ($50\sqrt{37}$), direction 25°17′ north of east (arc sin $3\sqrt{111}/74$)

**33.** Find the resultant of the following displacements: A, 20km 30° south of east; B, 50km due west; C, 40km northeast; D, 30km 60° south of west.
*Ans.* magnitude 20·9km, direction 21°39′ south of west

**34.** Show graphically that $-(\mathbf{A}-\mathbf{B}) = -\mathbf{A} + \mathbf{B}$.

**35.** An object $P$ is acted upon by three coplanar forces as shown in Fig.(*a*) below. Determine the force needed to prevent $P$ from moving. *Ans.* 323N directly opposite force of 150N.

**36.** Given vectors **A, B, C** and **D** (Fig.(*b*) below). Construct (*a*) $3\mathbf{A} - 2\mathbf{B} - (\mathbf{C} - \mathbf{D})$ (*b*) $\frac{1}{2}\mathbf{C} + \frac{2}{3}(\mathbf{A} - \mathbf{B} + 2\mathbf{D})$.

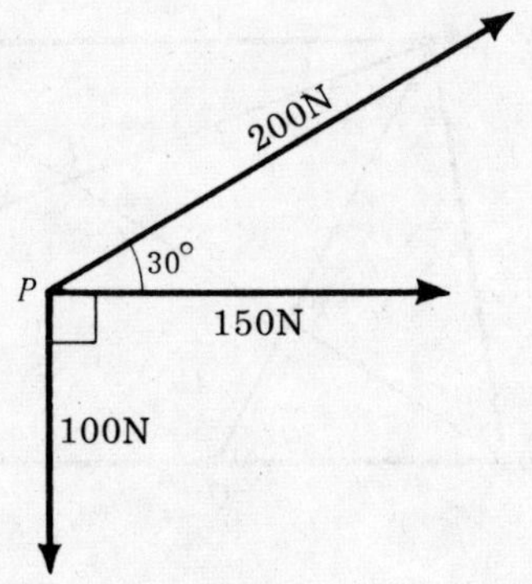


Fig.(*a*)

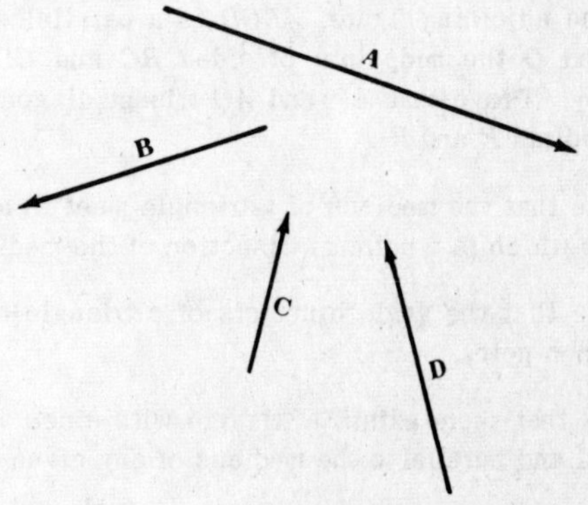


Fig.(*b*)

**37.** If $ABCDEF$ are the vertices of a regular hexagon, find the resultant of the forces represented by the vectors **AB, AC, AD, AE** and **AF**. *Ans.* 3 **AD**

**38.** If **A** and **B** are given vectors show that (*a*) $|\mathbf{A}+\mathbf{B}| \leqq |\mathbf{A}| + |\mathbf{B}|$, (*b*) $|\mathbf{A}-\mathbf{B}| \geqq |\mathbf{A}| - |\mathbf{B}|$.

**39.** Show that $|\mathbf{A}+\mathbf{B}+\mathbf{C}| \leqq |\mathbf{A}| + |\mathbf{B}| + |\mathbf{C}|$.

**40.** Two towns $A$ and $B$ are situated directly opposite each other on the banks of a river whose width is 8km and which flows at a speed of 4km/h. A man located at $A$ wishes to reach town $C$ which is 6km upstream from and on the same side of the river as town $B$. If his boat can travel at a maximum speed of 10 km/h and if he wishes to reach $C$ in the shortest possible time what course must he follow and how long will the trip take?
*Ans.* A straight line course upstream making an angle of $34°28'$ with the shore line. 1h 25min.

**41.** A man travelling southward at 15 km/h observes that the wind appears to be coming from the west. On increasing his speed to 25 km/h it appears to be coming from the southwest. Find the direction and speed of the wind. *Ans.* The wind is coming from a direction $56°18'$ north of west at 18 km/h

**42.** A 100kg weight is suspended from the centre of a rope as shown in the adjoining figure. Determine the tension $T$ in the rope. *Ans.* 100kg

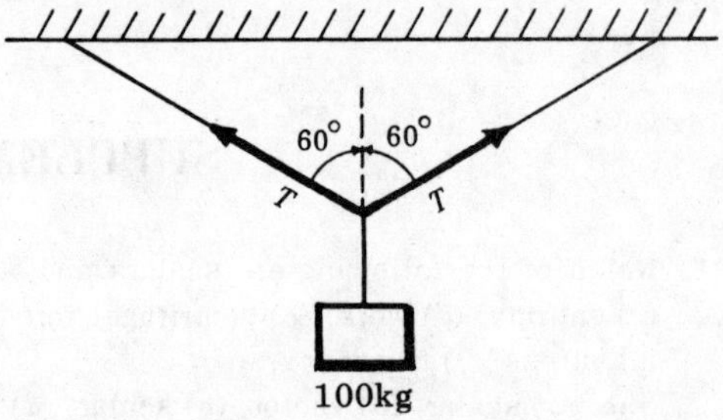


**43.** Simplify $2\mathbf{A}+\mathbf{B}+3\mathbf{C} - \{\mathbf{A}-2\mathbf{B}-2(2\mathbf{A}-3\mathbf{B}-\mathbf{C})\}$.
*Ans.* $5\mathbf{A}-3\mathbf{B}+\mathbf{C}$

**44.** If **a** and **b** are non-collinear vectors and $\mathbf{A} = (x+4y)\mathbf{a} + (2x+y+1)\mathbf{b}$ and $\mathbf{B} = (y-2x+2)\mathbf{a} + (2x-3y-1)\mathbf{b}$, find $x$ and $y$ such that $3\mathbf{A} = 2\mathbf{B}$.
*Ans.* $x = 2$, $y = -1$

**45.** The base vectors $\mathbf{a}_1, \mathbf{a}_2, \mathbf{a}_3$ are given in terms of the base vectors $\mathbf{b}_1, \mathbf{b}_2, \mathbf{b}_3$ by the relations

$$\mathbf{a}_1 = 2\mathbf{b}_1 + 3\mathbf{b}_2 - \mathbf{b}_3, \quad \mathbf{a}_2 = \mathbf{b}_1 - 2\mathbf{b}_2 + 2\mathbf{b}_3, \quad \mathbf{a}_3 = -2\mathbf{b}_1 + \mathbf{b}_2 - 2\mathbf{b}_3$$

If $\mathbf{F} = 3\mathbf{b}_1 - \mathbf{b}_2 + 2\mathbf{b}_3$, express **F** in terms of $\mathbf{a}_1, \mathbf{a}_2$ and $\mathbf{a}_3$. *Ans.* $2\mathbf{a}_1 + 5\mathbf{a}_2 + 3\mathbf{a}_3$

**46.** If **a, b, c** are non-coplanar vectors determine whether the vectors $\mathbf{r}_1 = 2\mathbf{a}-3\mathbf{b}+\mathbf{c}$, $\mathbf{r}_2 = 3\mathbf{a}-5\mathbf{b}+2\mathbf{c}$, and $\mathbf{r}_3 = 4\mathbf{a}-5\mathbf{b}+\mathbf{c}$ are linearly independent or dependent. *Ans.* Linearly dependent since $\mathbf{r}_3 = 5\mathbf{r}_1 - 2\mathbf{r}_2$.

**47.** If **A** and **B** are given vectors representing the diagonals of a parallelogram, construct the parallelogram.

**48.** Prove that the line joining the midpoints of two sides of a triangle is parallel to the third side and has one half of its magnitude.

**49.** (*a*) If $O$ is any point within triangle $ABC$ and $P, Q, R$ are midpoints of the sides $AB, BC, CA$ respectively, prove that $\mathbf{OA}+\mathbf{OB}+\mathbf{OC} = \mathbf{OP}+\mathbf{OQ}+\mathbf{OR}$.
(*b*) Does the result hold if $O$ is any point outside the triangle? Prove your result. *Ans.* Yes

**50.** In the adjoining figure, $ABCD$ is a parallelogram with $P$ and $Q$ the midpoints of sides $BC$ and $CD$ respectively. Prove that $AP$ and $AQ$ trisect diagonal $BD$ at the points $E$ and $F$.

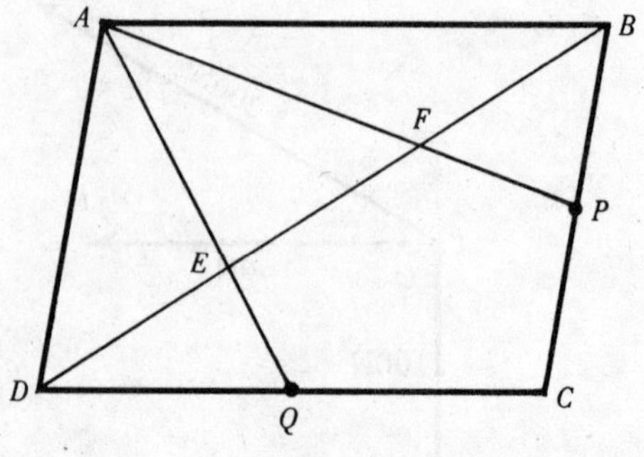


**51.** Prove that the medians of a triangle meet in a common point which is a point of trisection of the medians.

**52.** Prove that the angle bisectors of a triangle meet in a common point.

**53.** Show that there exists a triangle with sides which are equal and parallel to the medians of any given triangle.

**54.** Let the position vectors of points $P$ and $Q$ relative to an origin $O$ be given by **p** and **q** respectively. If $R$ is a point which divides line $PQ$ into segments which are in the ratio $m:n$ show that the position vector of $R$

is given by $\mathbf{r} = \dfrac{m\mathbf{p} + n\mathbf{q}}{m+n}$ and that this is independent of the origin.

**55.** If $\mathbf{r}_1, \mathbf{r}_2, \ldots, \mathbf{r}_n$ are the position vectors of masses $m_1, m_2, \ldots, m_n$ respectively relative to an origin $O$, show that the position vector of the centroid is given by

$$\mathbf{r} = \frac{m_1\mathbf{r}_1 + m_2\mathbf{r}_2 + \ldots + m_n\mathbf{r}_n}{m_1 + m_2 + \ldots + m_n}$$

and that this is independent of the origin.

**56.** A quadrilateral $ABCD$ has masses of 1, 2, 3 and 4 units located respectively at its vertices $A(-1,-2,2)$, $B(3,2,-1)$, $C(1,-2,4)$, and $D(3,1,2)$. Find the coordinates of the centroid. *Ans.* $(2,0,2)$

**57.** Show that the equation of a plane which passes through three given points $A, B, C$ not in the same straight line and having position vectors $\mathbf{a}, \mathbf{b}, \mathbf{c}$ relative to an origin $O$, can be written

$$\mathbf{r} = \frac{m\mathbf{a} + n\mathbf{b} + p\mathbf{c}}{m + n + p}$$

where $m, n, p$ are scalars. Verify that the equation is independent of the origin.

**58.** The position vectors of points $P$ and $Q$ are given by $\mathbf{r}_1 = 2\mathbf{i} + 3\mathbf{j} - \mathbf{k}$, $\mathbf{r}_2 = 4\mathbf{i} - 3\mathbf{j} + 2\mathbf{k}$. Determine $\mathbf{PQ}$ in terms of $\mathbf{i}, \mathbf{j}, \mathbf{k}$ and find its magnitude. *Ans.* $2\mathbf{i} - 6\mathbf{j} + 3\mathbf{k}$, 7

**59.** If $\mathbf{A} = 3\mathbf{i} - \mathbf{j} - 4\mathbf{k}$, $\mathbf{B} = -2\mathbf{i} + 4\mathbf{j} - 3\mathbf{k}$, $\mathbf{C} = \mathbf{i} + 2\mathbf{j} - \mathbf{k}$, find
(*a*) $2\mathbf{A} - \mathbf{B} + 3\mathbf{C}$, (*b*) $|\mathbf{A} + \mathbf{B} + \mathbf{C}|$, (*c*) $|3\mathbf{A} - 2\mathbf{B} + 4\mathbf{C}|$, (*d*) a unit vector parallel to $3\mathbf{A} - 2\mathbf{B} + 4\mathbf{C}$.
*Ans.* (*a*) $11\mathbf{i} - 8\mathbf{k}$ (*b*) $\sqrt{93}$ (*c*) $\sqrt{398}$ (*d*) $\dfrac{3\mathbf{A} - 2\mathbf{B} + 4\mathbf{C}}{\sqrt{398}}$

**60.** The following forces act on a particle $P$: $\mathbf{F}_1 = 2\mathbf{i} + 3\mathbf{j} - 5\mathbf{k}$, $\mathbf{F}_2 = -5\mathbf{i} + \mathbf{j} + 3\mathbf{k}$, $\mathbf{F}_3 = \mathbf{i} - 2\mathbf{j} + 4\mathbf{k}$, $\mathbf{F}_4 = 4\mathbf{i} - 3\mathbf{j} - 2\mathbf{k}$, measured in newtons. Find (*a*) the resultant of the forces, (*b*) the magnitude of the resultant.
*Ans.* (*a*) $2\mathbf{i} - \mathbf{j}$ (*b*) $\sqrt{5}$

**61.** In each case determine whether the vectors are linearly independent or linearly dependent:
(*a*) $\mathbf{A} = 2\mathbf{i} + \mathbf{j} - 3\mathbf{k}$, $\mathbf{B} = \mathbf{i} - 4\mathbf{k}$, $\mathbf{C} = 4\mathbf{i} + 3\mathbf{j} - \mathbf{k}$, (*b*) $\mathbf{A} = \mathbf{i} - 3\mathbf{j} + 2\mathbf{k}$, $\mathbf{B} = 2\mathbf{i} - 4\mathbf{j} - \mathbf{k}$, $\mathbf{C} = 3\mathbf{i} + 2\mathbf{j} - \mathbf{k}$.
*Ans.* (*a*) linearly dependent, (*b*) linearly independent

**62.** Prove that any four vectors in three dimensions must be linearly dependent.

**63.** Show that a necessary and sufficient condition that the vectors $\mathbf{A} = A_1\mathbf{i} + A_2\mathbf{j} + A_3\mathbf{k}$, $\mathbf{B} = B_1\mathbf{i} + B_2\mathbf{j} + B_3\mathbf{k}$, $\mathbf{C} = C_1\mathbf{i} + C_2\mathbf{j} + C_3\mathbf{k}$ be linearly independent is that the determinant $\begin{vmatrix} A_1 & A_2 & A_3 \\ B_1 & B_2 & B_3 \\ C_1 & C_2 & C_3 \end{vmatrix}$ be different from zero.

**64.** (*a*) Prove that the vectors $\mathbf{A} = 3\mathbf{i} + \mathbf{j} - 2\mathbf{k}$, $\mathbf{B} = -\mathbf{i} + 3\mathbf{j} + 4\mathbf{k}$, $\mathbf{C} = 4\mathbf{i} - 2\mathbf{j} - 6\mathbf{k}$ can form the sides of a triangle.
(*b*) Find the lengths of the medians of the triangle.
*Ans.* (*b*) $\sqrt{6}$, $\frac{1}{2}\sqrt{114}$, $\frac{1}{2}\sqrt{150}$

**65.** Given the scalar field defined by $\phi(x,y,z) = 4yz^3 + 3xyz - z^2 + 2$. Find (*a*) $\phi(1,-1,-2)$, (*b*) $\phi(0,-3,1)$.
*Ans.* (*a*) 36 (*b*) $-11$

**66.** Graph the vector fields defined by
(*a*) $\mathbf{V}(x,y) = x\mathbf{i} - y\mathbf{j}$, (*b*) $\mathbf{V}(x,y) = y\mathbf{i} - x\mathbf{j}$, (*c*) $\mathbf{V}(x,y,z) = \dfrac{x\mathbf{i} + y\mathbf{i} + z\mathbf{k}}{\sqrt{x^2 + y^2 + z^2}}$.

# Chapter 2

# The DOT and CROSS PRODUCT

**THE DOT OR SCALAR PRODUCT** of two vectors **A** and **B**, denoted by $\mathbf{A}\cdot\mathbf{B}$ (read **A** dot **B**), is defined as the product of the magnitudes of **A** and **B** and the cosine of the angle $\theta$ between them. In symbols,

$$\mathbf{A}\cdot\mathbf{B} = AB\cos\theta, \qquad 0 \leqq \theta \leqq \pi$$

Note that $\mathbf{A}\cdot\mathbf{B}$ is a scalar and not a vector.

The following laws are valid:

1. $\mathbf{A}\cdot\mathbf{B} = \mathbf{B}\cdot\mathbf{A}$ — Commutative Law for Dot Products
2. $\mathbf{A}\cdot(\mathbf{B}+\mathbf{C}) = \mathbf{A}\cdot\mathbf{B} + \mathbf{A}\cdot\mathbf{C}$ — Distributive Law
3. $m(\mathbf{A}\cdot\mathbf{B}) = (m\mathbf{A})\cdot\mathbf{B} = \mathbf{A}\cdot(m\mathbf{B}) = (\mathbf{A}\cdot\mathbf{B})m$, where $m$ is a scalar.
4. $\mathbf{i}\cdot\mathbf{i} = \mathbf{j}\cdot\mathbf{j} = \mathbf{k}\cdot\mathbf{k} = 1, \quad \mathbf{i}\cdot\mathbf{j} = \mathbf{j}\cdot\mathbf{k} = \mathbf{k}\cdot\mathbf{i} = 0$
5. If $\mathbf{A} = A_1\mathbf{i} + A_2\mathbf{j} + A_3\mathbf{k}$ and $\mathbf{B} = B_1\mathbf{i} + B_2\mathbf{j} + B_3\mathbf{k}$, then

$$\mathbf{A}\cdot\mathbf{B} = A_1B_1 + A_2B_2 + A_3B_3$$

$$\mathbf{A}\cdot\mathbf{A} = A^2 = A_1^2 + A_2^2 + A_3^2$$

$$\mathbf{B}\cdot\mathbf{B} = B^2 = B_1^2 + B_2^2 + B_3^2$$

6. If $\mathbf{A}\cdot\mathbf{B} = 0$ and **A** and **B** are not null vectors, then **A** and **B** are perpendicular.

**THE CROSS OR VECTOR PRODUCT** of **A** and **B** is a vector $\mathbf{C} = \mathbf{A}\times\mathbf{B}$ (read **A** cross **B**). The magnitude of $\mathbf{A}\times\mathbf{B}$ is defined as the product of the magnitudes of **A** and **B** and the sine of the angle $\theta$ between them. The direction of the vector $\mathbf{C} = \mathbf{A}\times\mathbf{B}$ is perpendicular to the plane of **A** and **B** and such that **A**, **B** and **C** form a right-handed system. In symbols,

$$\mathbf{A}\times\mathbf{B} = AB\sin\theta\,\mathbf{u}, \qquad 0 \leqq \theta \leqq \pi$$

where **u** is a unit vector indicating the direction of $\mathbf{A}\times\mathbf{B}$. If $\mathbf{A} = \mathbf{B}$, or if **A** is parallel to **B**, then $\sin\theta = 0$ and we define $\mathbf{A}\times\mathbf{B} = \mathbf{0}$.

The following laws are valid:

1. $\mathbf{A}\times\mathbf{B} = -\mathbf{B}\times\mathbf{A}$ — (Commutative Law for Cross Products Fails.)
2. $\mathbf{A}\times(\mathbf{B}+\mathbf{C}) = \mathbf{A}\times\mathbf{B} + \mathbf{A}\times\mathbf{C}$ — Distributive Law
3. $m(\mathbf{A}\times\mathbf{B}) = (m\mathbf{A})\times\mathbf{B} = \mathbf{A}\times(m\mathbf{B}) = (\mathbf{A}\times\mathbf{B})m$, where $m$ is a scalar.
4. $\mathbf{i}\times\mathbf{i} = \mathbf{j}\times\mathbf{j} = \mathbf{k}\times\mathbf{k} = \mathbf{0}, \quad \mathbf{i}\times\mathbf{j} = \mathbf{k},\ \mathbf{j}\times\mathbf{k} = \mathbf{i},\ \mathbf{k}\times\mathbf{i} = \mathbf{j}$
5. If $\mathbf{A} = A_1\mathbf{i} + A_2\mathbf{j} + A_3\mathbf{k}$ and $\mathbf{B} = B_1\mathbf{i} + B_2\mathbf{j} + B_3\mathbf{k}$, then

$$\mathbf{A}\times\mathbf{B} = \begin{vmatrix} \mathbf{i} & \mathbf{j} & \mathbf{k} \\ A_1 & A_2 & A_3 \\ B_1 & B_2 & B_3 \end{vmatrix}$$

6. The magnitude of $\mathbf{A}\times\mathbf{B}$ is the same as the area of a parallelogram with sides $\mathbf{A}$ and $\mathbf{B}$.

7. If $\mathbf{A}\times\mathbf{B} = \mathbf{0}$, and $\mathbf{A}$ and $\mathbf{B}$ are not null vectors, then $\mathbf{A}$ and $\mathbf{B}$ are parallel.

**TRIPLE PRODUCTS.** Dot and cross multiplication of three vectors $\mathbf{A}, \mathbf{B}$ and $\mathbf{C}$ may produce meaningful products of the form $(\mathbf{A}\cdot\mathbf{B})\mathbf{C}$, $\mathbf{A}\cdot(\mathbf{B}\times\mathbf{C})$ and $\mathbf{A}\times(\mathbf{B}\times\mathbf{C})$. The following laws are valid:

1. $(\mathbf{A}\cdot\mathbf{B})\mathbf{C} \neq \mathbf{A}(\mathbf{B}\cdot\mathbf{C})$

2. $\mathbf{A}\cdot(\mathbf{B}\times\mathbf{C}) = \mathbf{B}\cdot(\mathbf{C}\times\mathbf{A}) = \mathbf{C}\cdot(\mathbf{A}\times\mathbf{B})$ = volume of a parallelepiped having $\mathbf{A}, \mathbf{B}$ and $\mathbf{C}$ as edges, or the negative of this volume, according as $\mathbf{A}, \mathbf{B}$ and $\mathbf{C}$ do or do not form a right-handed system. If $\mathbf{A} = A_1\mathbf{i} + A_2\mathbf{j} + A_3\mathbf{k}$, $\mathbf{B} = B_1\mathbf{i} + B_2\mathbf{j} + B_3\mathbf{k}$ and $\mathbf{C} = C_1\mathbf{i} + C_2\mathbf{j} + C_3\mathbf{k}$, then

$$\mathbf{A}\cdot(\mathbf{B}\times\mathbf{C}) = \begin{vmatrix} A_1 & A_2 & A_3 \\ B_1 & B_2 & B_3 \\ C_1 & C_2 & C_3 \end{vmatrix}$$

3. $\mathbf{A}\times(\mathbf{B}\times\mathbf{C}) \neq (\mathbf{A}\times\mathbf{B})\times\mathbf{C}$ (Associative Law for Cross Products Fails.)

4. $\mathbf{A}\times(\mathbf{B}\times\mathbf{C}) = (\mathbf{A}\cdot\mathbf{C})\mathbf{B} - (\mathbf{A}\cdot\mathbf{B})\mathbf{C}$
   $(\mathbf{A}\times\mathbf{B})\times\mathbf{C} = (\mathbf{A}\cdot\mathbf{C})\mathbf{B} - (\mathbf{B}\cdot\mathbf{C})\mathbf{A}$

The product $\mathbf{A}\cdot(\mathbf{B}\times\mathbf{C})$ is sometimes called the *scalar triple product* or *box product* and may be denoted by $[\mathbf{ABC}]$. The product $\mathbf{A}\times(\mathbf{B}\times\mathbf{C})$ is called the *vector triple product*.

In $\mathbf{A}\cdot(\mathbf{B}\times\mathbf{C})$ parentheses are sometimes omitted and we write $\mathbf{A}\cdot\mathbf{B}\times\mathbf{C}$ (see Problem 41). However, parentheses must be used in $\mathbf{A}\times(\mathbf{B}\times\mathbf{C})$ (see Problems 29 and 47).

**RECIPROCAL SETS OF VECTORS.** The sets of vectors $\mathbf{a}, \mathbf{b}, \mathbf{c}$ and $\mathbf{a}', \mathbf{b}', \mathbf{c}'$ are called *reciprocal sets or systems of vectors* if

$$\mathbf{a}\cdot\mathbf{a}' = \mathbf{b}\cdot\mathbf{b}' = \mathbf{c}\cdot\mathbf{c}' = 1$$

$$\mathbf{a}'\cdot\mathbf{b} = \mathbf{a}'\cdot\mathbf{c} = \mathbf{b}'\cdot\mathbf{a} = \mathbf{b}'\cdot\mathbf{c} = \mathbf{c}'\cdot\mathbf{a} = \mathbf{c}'\cdot\mathbf{b} = 0$$

The sets $\mathbf{a}, \mathbf{b}, \mathbf{c}$ and $\mathbf{a}', \mathbf{b}', \mathbf{c}'$ are reciprocal sets of vectors if and only if

$$\mathbf{a}' = \frac{\mathbf{b}\times\mathbf{c}}{\mathbf{a}\cdot\mathbf{b}\times\mathbf{c}}, \qquad \mathbf{b}' = \frac{\mathbf{c}\times\mathbf{a}}{\mathbf{a}\cdot\mathbf{b}\times\mathbf{c}}, \qquad \mathbf{c}' = \frac{\mathbf{a}\times\mathbf{b}}{\mathbf{a}\cdot\mathbf{b}\times\mathbf{c}}$$

where $\mathbf{a}\cdot\mathbf{b}\times\mathbf{c} \neq 0$. See Problems 53 and 54.

## SOLVED PROBLEMS

### THE DOT OR SCALAR PRODUCT.

**1.** Prove $\mathbf{A}\cdot\mathbf{B} = \mathbf{B}\cdot\mathbf{A}$.

$$\mathbf{A}\cdot\mathbf{B} = AB\cos\theta = BA\cos\theta = \mathbf{B}\cdot\mathbf{A}$$

Then the commutative law for dot products is valid.

**2.** Prove that the projection of **A** on **B** is equal to $\mathbf{A}\cdot\mathbf{b}$, where **b** is a unit vector in the direction of **B**.

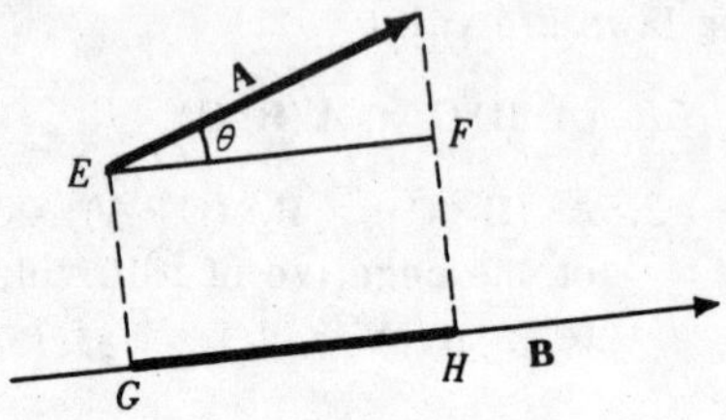


Through the initial and terminal points of **A** pass planes perpendicular to **B** at $G$ and $H$ respectively as in the adjacent figure; then

$$\text{Projection of } \mathbf{A} \text{ on } \mathbf{B} = \overline{GH} = \overline{EF} = A\cos\theta = \mathbf{A}\cdot\mathbf{b}$$

**3.** Prove $\mathbf{A}\cdot(\mathbf{B}+\mathbf{C}) = \mathbf{A}\cdot\mathbf{B} + \mathbf{A}\cdot\mathbf{C}$.

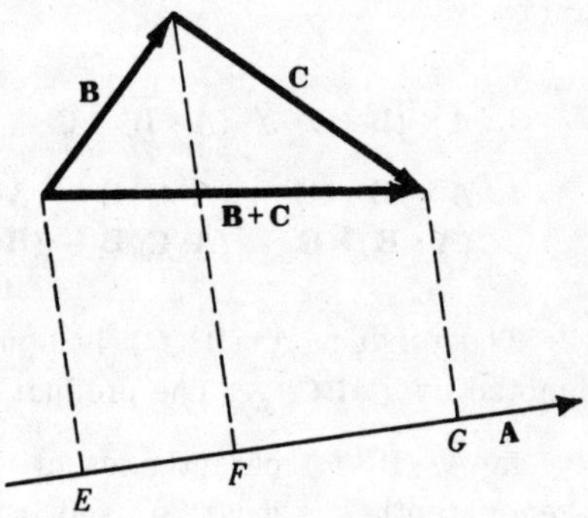


Let **a** be a unit vector in the direction of **A**; then

Projection of (**B** + **C**) on **A** = proj. of **B** on **A** + proj. of **C** on **A**

$$(\mathbf{B}+\mathbf{C})\cdot\mathbf{a} = \mathbf{B}\cdot\mathbf{a} + \mathbf{C}\cdot\mathbf{a}$$

Multiplying by $A$,

$$(\mathbf{B}+\mathbf{C})\cdot A\mathbf{a} = \mathbf{B}\cdot A\mathbf{a} + \mathbf{C}\cdot A\mathbf{a}$$

and
$$(\mathbf{B}+\mathbf{C})\cdot\mathbf{A} = \mathbf{B}\cdot\mathbf{A} + \mathbf{C}\cdot\mathbf{A}$$

Then by the commutative law for dot products,

$$\mathbf{A}\cdot(\mathbf{B}+\mathbf{C}) = \mathbf{A}\cdot\mathbf{B} + \mathbf{A}\cdot\mathbf{C}$$

and the distributive law is valid.

**4.** Prove that $(\mathbf{A}+\mathbf{B})\cdot(\mathbf{C}+\mathbf{D}) = \mathbf{A}\cdot\mathbf{C} + \mathbf{A}\cdot\mathbf{D} + \mathbf{B}\cdot\mathbf{C} + \mathbf{B}\cdot\mathbf{D}$.

By Problem 3, $(\mathbf{A}+\mathbf{B})\cdot(\mathbf{C}+\mathbf{D}) = \mathbf{A}\cdot(\mathbf{C}+\mathbf{D}) + \mathbf{B}\cdot(\mathbf{C}+\mathbf{D}) = \mathbf{A}\cdot\mathbf{C} + \mathbf{A}\cdot\mathbf{D} + \mathbf{B}\cdot\mathbf{C} + \mathbf{B}\cdot\mathbf{D}$

The ordinary laws of algebra are valid for dot products.

**5.** Evaluate each of the following.

(a) $\mathbf{i}\cdot\mathbf{i} = |\mathbf{i}|\,|\mathbf{i}|\cos 0^\circ = (1)(1)(1) = 1$

(b) $\mathbf{i}\cdot\mathbf{k} = |\mathbf{i}|\,|\mathbf{k}|\cos 90^\circ = (1)(1)(0) = 0$

(c) $\mathbf{k}\cdot\mathbf{j} = |\mathbf{k}|\,|\mathbf{j}|\cos 90^\circ = (1)(1)(0) = 0$

(d) $\mathbf{j}\cdot(2\mathbf{i}-3\mathbf{j}+\mathbf{k}) = 2\mathbf{j}\cdot\mathbf{i} - 3\mathbf{j}\cdot\mathbf{j} + \mathbf{j}\cdot\mathbf{k} = 0 - 3 + 0 = -3$

(e) $(2\mathbf{i}-\mathbf{j})\cdot(3\mathbf{i}+\mathbf{k}) = 2\mathbf{i}\cdot(3\mathbf{i}+\mathbf{k}) - \mathbf{j}\cdot(3\mathbf{i}+\mathbf{k}) = 6\mathbf{i}\cdot\mathbf{i} + 2\mathbf{i}\cdot\mathbf{k} - 3\mathbf{j}\cdot\mathbf{i} - \mathbf{j}\cdot\mathbf{k} = 6 + 0 - 0 - 0 = 6$

**6.** If $\mathbf{A} = A_1\mathbf{i} + A_2\mathbf{j} + A_3\mathbf{k}$ and $\mathbf{B} = B_1\mathbf{i} + B_2\mathbf{j} + B_3\mathbf{k}$, prove that $\mathbf{A}\cdot\mathbf{B} = A_1B_1 + A_2B_2 + A_3B_3$.

$$\begin{aligned}\mathbf{A}\cdot\mathbf{B} &= (A_1\mathbf{i}+A_2\mathbf{j}+A_3\mathbf{k})\cdot(B_1\mathbf{i}+B_2\mathbf{j}+B_3\mathbf{k}) \\ &= A_1\mathbf{i}\cdot(B_1\mathbf{i}+B_2\mathbf{j}+B_3\mathbf{k}) + A_2\mathbf{j}\cdot(B_1\mathbf{i}+B_2\mathbf{j}+B_3\mathbf{k}) + A_3\mathbf{k}\cdot(B_1\mathbf{i}+B_2\mathbf{j}+B_3\mathbf{k}) \\ &= A_1B_1\mathbf{i}\cdot\mathbf{i} + A_1B_2\mathbf{i}\cdot\mathbf{j} + A_1B_3\mathbf{i}\cdot\mathbf{k} + A_2B_1\mathbf{j}\cdot\mathbf{i} + A_2B_2\mathbf{j}\cdot\mathbf{j} + A_2B_3\mathbf{j}\cdot\mathbf{k} + A_3B_1\mathbf{k}\cdot\mathbf{i} + A_3B_2\mathbf{k}\cdot\mathbf{j} + A_3B_3\mathbf{k}\cdot\mathbf{k}\end{aligned}$$

$= A_1B_1 + A_2B_2 + A_3B_3$

since $\mathbf{i\cdot i} = \mathbf{j\cdot j} = \mathbf{k\cdot k} = 1$ and all other dot products are zero.

**7.** If $\mathbf{A} = A_1\mathbf{i} + A_2\mathbf{j} + A_3\mathbf{k}$, show that $A = \sqrt{\mathbf{A\cdot A}} = \sqrt{A_1^2 + A_2^2 + A_3^2}$.

$\mathbf{A\cdot A} = (A)(A)\cos 0° = A^2$. Then $A = \sqrt{\mathbf{A\cdot A}}$.

Also, $\mathbf{A\cdot A} = (A_1\mathbf{i} + A_2\mathbf{j} + A_3\mathbf{k})\cdot(A_1\mathbf{i} + A_2\mathbf{j} + A_3\mathbf{k})$

$= (A_1)(A_1) + (A_2)(A_2) + (A_3)(A_3) = A_1^2 + A_2^2 + A_3^2$

by Problem 6, taking $\mathbf{B} = \mathbf{A}$.

Then $A = \sqrt{\mathbf{A\cdot A}} = \sqrt{A_1^2 + A_2^2 + A_3^2}$ is the magnitude of $\mathbf{A}$. Sometimes $\mathbf{A\cdot A}$ is written $\mathbf{A}^2$.

**8.** Find the angle between $\mathbf{A} = 2\mathbf{i} + 2\mathbf{j} - \mathbf{k}$ and $\mathbf{B} = 6\mathbf{i} - 3\mathbf{j} + 2\mathbf{k}$.

$\mathbf{A\cdot B} = AB\cos\theta$, $\quad A = \sqrt{(2)^2 + (2)^2 + (-1)^2} = 3$, $\quad B = \sqrt{(6)^2 + (-3)^2 + (2)^2} = 7$

$\mathbf{A\cdot B} = (2)(6) + (2)(-3) + (-1)(2) = 12 - 6 - 2 = 4$

Then $\cos\theta = \dfrac{\mathbf{A\cdot B}}{AB} = \dfrac{4}{(3)(7)} = \dfrac{4}{21} = 0{\cdot}1905$ and $\theta = 79°$ approximately.

**9.** If $\mathbf{A\cdot B} = 0$ and if $A$ and $B$ are not zero, show that $\mathbf{A}$ is perpendicular to $\mathbf{B}$.

If $\mathbf{A\cdot B} = AB\cos\theta = 0$, then $\cos\theta = 0$ or $\theta = 90°$. Conversely, if $\theta = 90°$, $\mathbf{A\cdot B} = 0$.

**10.** Determine the value of $a$ so that $\mathbf{A} = 2\mathbf{i} + a\mathbf{j} + \mathbf{k}$ and $\mathbf{B} = 4\mathbf{i} - 2\mathbf{j} - 2\mathbf{k}$ are perpendicular.

From Problem 9, $\mathbf{A}$ and $\mathbf{B}$ are perpendicular if $\mathbf{A\cdot B} = 0$.

Then $\mathbf{A\cdot B} = (2)(4) + (a)(-2) + (1)(-2) = 8 - 2a - 2 = 0$ for $a = 3$.

**11.** Show that the vectors $\mathbf{A} = 3\mathbf{i} - 2\mathbf{j} + \mathbf{k}$, $\mathbf{B} = \mathbf{i} - 3\mathbf{j} + 5\mathbf{k}$, $\mathbf{C} = 2\mathbf{i} + \mathbf{j} - 4\mathbf{k}$ form a right-angled triangle.

We first have to show that the vectors form a triangle.

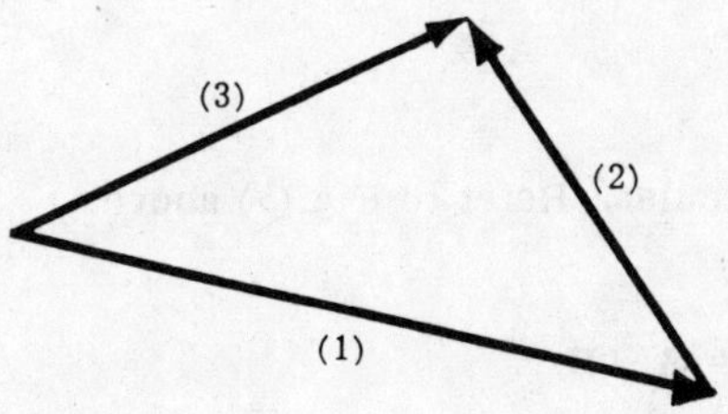


(a)

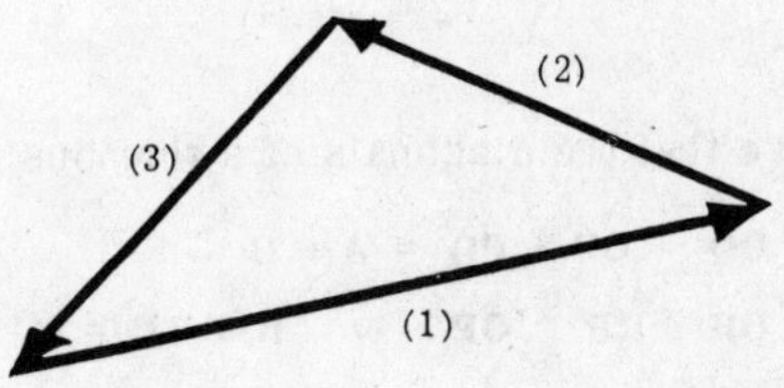


(b)

From the figures it is seen that the vectors will form a triangle if

(a) one of the vectors, say (3), is the resultant or sum of (1) and (2),
(b) the sum or resultant of the vectors (1) + (2) + (3) is zero,

according as (a) two vectors have a common terminal point or (b) none of the vectors have a common terminal point. By trial we find $\mathbf{A} = \mathbf{B} + \mathbf{C}$ so that the vectors do form a triangle.

Since $\mathbf{A\cdot B} = (3)(1) + (-2)(-3) + (1)(5) = 14$, $\mathbf{A\cdot C} = (3)(2) + (-2)(1) + (1)(-4) = 0$, and $\mathbf{B\cdot C} = (1)(2) + (-3)(1) + (5)(-4) = -21$, it follows that $\mathbf{A}$ and $\mathbf{C}$ are perpendicular and the triangle is a right-angled triangle.

12. Find the angles which the vector $\mathbf{A} = 3\mathbf{i} - 6\mathbf{j} + 2\mathbf{k}$ makes with the coordinate axes.

Let $\alpha, \beta, \gamma$ be the angles which $\mathbf{A}$ makes with the positive $x, y, z$ axes respectively.

$$\mathbf{A}\cdot\mathbf{i} = (A)(1)\cos\alpha = \sqrt{(3)^2+(-6)^2+(2)^2}\cos\alpha = 7\cos\alpha$$

$$\mathbf{A}\cdot\mathbf{i} = (3\mathbf{i}-6\mathbf{j}+2\mathbf{k})\cdot\mathbf{i} = 3\mathbf{i}\cdot\mathbf{i} - 6\mathbf{j}\cdot\mathbf{i} + 2\mathbf{k}\cdot\mathbf{i} = 3$$

Then $\cos\alpha = 3/7 = 0.4286$, and $\alpha = 64.6^\circ$ approximately.

Similarly, $\cos\beta = -6/7$, $\beta = 149^\circ$ and $\cos\gamma = 2/7$, $\gamma = 73.4^\circ$.

The cosines of $\alpha$, $\beta$, and $\gamma$ are called the *direction cosines* of $\mathbf{A}$. (See Prob. 27, Chap. 1).

13. Find the projection of the vector $\mathbf{A} = \mathbf{i} - 2\mathbf{j} + \mathbf{k}$ on the vector $\mathbf{B} = 4\mathbf{i} - 4\mathbf{j} + 7\mathbf{k}$.

A unit vector in the direction $\mathbf{B}$ is $\mathbf{b} = \dfrac{\mathbf{B}}{B} = \dfrac{4\mathbf{i}-4\mathbf{j}+7\mathbf{k}}{\sqrt{(4)^2+(-4)^2+(7)^2}} = \dfrac{4}{9}\mathbf{i} - \dfrac{4}{9}\mathbf{j} + \dfrac{7}{9}\mathbf{k}$.

Projection of $\mathbf{A}$ on the vector $\mathbf{B}$ $= \mathbf{A}\cdot\mathbf{b} = (\mathbf{i}-2\mathbf{j}+\mathbf{k})\cdot(\frac{4}{9}\mathbf{i}-\frac{4}{9}\mathbf{j}+\frac{7}{9}\mathbf{k})$

$$= (1)(\tfrac{4}{9}) + (-2)(-\tfrac{4}{9}) + (1)(\tfrac{7}{9}) = \tfrac{19}{9}.$$

14. Prove the law of cosines for plane triangles.

From Fig.(*a*) below, $\mathbf{B} + \mathbf{C} = \mathbf{A}$ or $\mathbf{C} = \mathbf{A} - \mathbf{B}$.

Then $\mathbf{C}\cdot\mathbf{C} = (\mathbf{A}-\mathbf{B})\cdot(\mathbf{A}-\mathbf{B}) = \mathbf{A}\cdot\mathbf{A} + \mathbf{B}\cdot\mathbf{B} - 2\mathbf{A}\cdot\mathbf{B}$

and $C^2 = A^2 + B^2 - 2AB\cos\theta$.

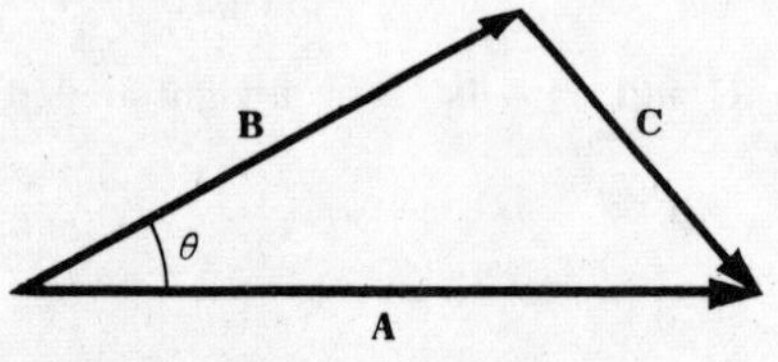

Fig.(*a*)

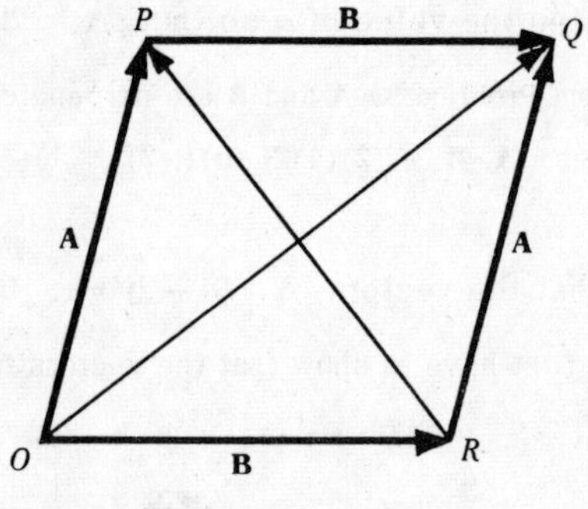

Fig.(*b*)

15. Prove that the diagonals of a rhombus are perpendicular. Refer to Fig.(*b*) above.

$\mathbf{OQ} = \mathbf{OP} + \mathbf{PQ} = \mathbf{A} + \mathbf{B}$

$\mathbf{OR} + \mathbf{RP} = \mathbf{OP}$ or $\mathbf{B} + \mathbf{RP} = \mathbf{A}$ and $\mathbf{RP} = \mathbf{A} - \mathbf{B}$

Then $\mathbf{OQ}\cdot\mathbf{RP} = (\mathbf{A}+\mathbf{B})\cdot(\mathbf{A}-\mathbf{B}) = A^2 - B^2 = 0$, since $A = B$.

Hence $\mathbf{OQ}$ is perpendicular to $\mathbf{RP}$.

16. Determine a unit vector perpendicular to the plane of $\mathbf{A} = 2\mathbf{i} - 6\mathbf{j} - 3\mathbf{k}$ and $\mathbf{B} = 4\mathbf{i} + 3\mathbf{j} - \mathbf{k}$.

Let vector $\mathbf{C} = c_1\mathbf{i} + c_2\mathbf{j} + c_3\mathbf{k}$ be perpendicular to the plane of $\mathbf{A}$ and $\mathbf{B}$. Then $\mathbf{C}$ is perpendicular to $\mathbf{A}$ and also to $\mathbf{B}$. Hence,

$$\mathbf{C}\cdot\mathbf{A} = 2c_1 - 6c_2 - 3c_3 = 0 \quad\text{or}\quad (1)\ 2c_1 - 6c_2 = 3c_3$$

$$\mathbf{C}\cdot\mathbf{B} = 4c_1 + 3c_2 - c_3 = 0 \quad\text{or}\quad (2)\ 4c_1 + 3c_2 = c_3$$

Solving (*1*) and (*2*) simultaneously: $c_1 = \frac{1}{2}c_3$, $c_2 = -\frac{1}{3}c_3$, $\mathbf{C} = c_3(\frac{1}{2}\mathbf{i} - \frac{1}{3}\mathbf{j} + \mathbf{k})$.

Then a unit vector in the direction of **C** is $\frac{\mathbf{C}}{C} = \frac{c_3(\frac{1}{2}\mathbf{i} - \frac{1}{3}\mathbf{j} + \mathbf{k})}{\sqrt{c_3^2[(\frac{1}{2})^2 + (-\frac{1}{3})^2 + (1)^2]}} = \pm(\frac{3}{7}\mathbf{i} - \frac{2}{7}\mathbf{j} + \frac{6}{7}\mathbf{k})$.

**17.** Find the work done in moving an object along a vector $\mathbf{r} = 3\mathbf{i} + 2\mathbf{j} - 5\mathbf{k}$ if the applied force is $\mathbf{F} = 2\mathbf{i} - \mathbf{j} - \mathbf{k}$. Refer to Fig.(*a*) below.

$$\begin{aligned} \text{Work done} &= (\text{magnitude of force in direction of motion})(\text{distance moved}) \\ &= (F\cos\theta)(r) = \mathbf{F}\cdot\mathbf{r} \\ &= (2\mathbf{i} - \mathbf{j} - \mathbf{k})\cdot(3\mathbf{i} + 2\mathbf{j} - 5\mathbf{k}) = 6 - 2 + 5 = 9. \end{aligned}$$

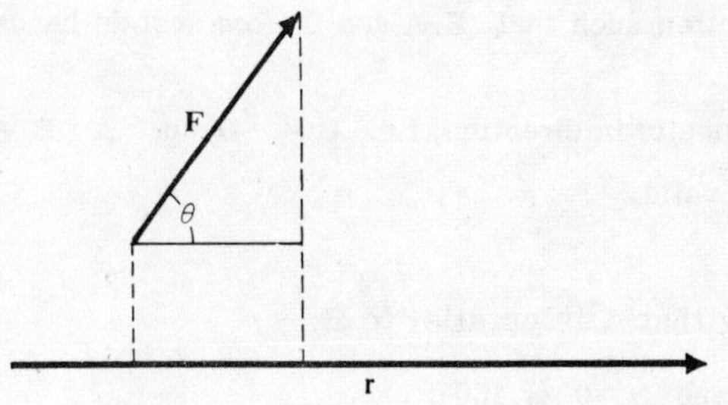

Fig.(*a*)

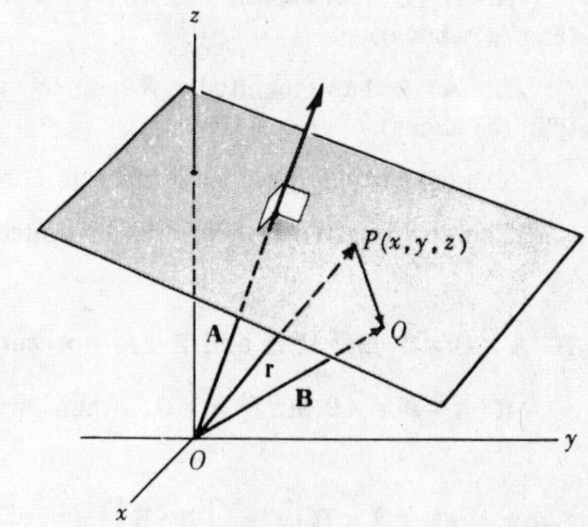

Fig.(*b*)

**18.** Find an equation for the plane perpendicular to the vector $\mathbf{A} = 2\mathbf{i} + 3\mathbf{j} + 6\mathbf{k}$ and passing through the terminal point of the vector $\mathbf{B} = \mathbf{i} + 5\mathbf{j} + 3\mathbf{k}$ (see Fig.(*b*) above).

Let **r** be the position vector of point $P$, and $Q$ the terminal point of **B**.

Since $\mathbf{PQ} = \mathbf{B} - \mathbf{r}$ is perpendicular to **A**, $(\mathbf{B} - \mathbf{r})\cdot\mathbf{A} = 0$ or $\mathbf{r}\cdot\mathbf{A} = \mathbf{B}\cdot\mathbf{A}$ is the required equation of the plane in vector form. In rectangular form this becomes

$$(x\mathbf{i} + y\mathbf{j} + z\mathbf{k})\cdot(2\mathbf{i} + 3\mathbf{j} + 6\mathbf{k}) = (\mathbf{i} + 5\mathbf{j} + 3\mathbf{k})\cdot(2\mathbf{i} + 3\mathbf{j} + 6\mathbf{k})$$

or

$$2x + 3y + 6z = (1)(2) + (5)(3) + (3)(6) = 35$$

**19.** In Problem 18 find the distance from the origin to the plane.

The distance from the origin to the plane is the projection of **B** on **A**.

A unit vector in direction **A** is $\mathbf{a} = \frac{\mathbf{A}}{A} = \frac{2\mathbf{i} + 3\mathbf{j} + 6\mathbf{k}}{\sqrt{(2)^2 + (3)^2 + (6)^2}} = \frac{2}{7}\mathbf{i} + \frac{3}{7}\mathbf{j} + \frac{6}{7}\mathbf{k}$.

Then, projection of **B** on **A** $= \mathbf{B}\cdot\mathbf{a} = (\mathbf{i} + 5\mathbf{j} + 3\mathbf{k})\cdot(\frac{2}{7}\mathbf{i} + \frac{3}{7}\mathbf{j} + \frac{6}{7}\mathbf{k}) = 1(\frac{2}{7}) + 5(\frac{3}{7}) + 3(\frac{6}{7}) = 5$.

**20.** If **A** is any vector, prove that $\mathbf{A} = (\mathbf{A}\cdot\mathbf{i})\mathbf{i} + (\mathbf{A}\cdot\mathbf{j})\mathbf{j} + (\mathbf{A}\cdot\mathbf{k})\mathbf{k}$.

Since $\mathbf{A} = A_1\mathbf{i} + A_2\mathbf{j} + A_3\mathbf{k}$, $\mathbf{A}\cdot\mathbf{i} = A_1\mathbf{i}\cdot\mathbf{i} + A_2\mathbf{j}\cdot\mathbf{i} + A_3\mathbf{k}\cdot\mathbf{i} = A_1$

Similarly, $\mathbf{A}\cdot\mathbf{j} = A_2$ and $\mathbf{A}\cdot\mathbf{k} = A_3$.

Then $\mathbf{A} = A_1\mathbf{i} + A_2\mathbf{j} + A_3\mathbf{k} = (\mathbf{A}\cdot\mathbf{i})\mathbf{i} + (\mathbf{A}\cdot\mathbf{j})\mathbf{j} + (\mathbf{A}\cdot\mathbf{k})\mathbf{k}$.

## THE CROSS OR VECTOR PRODUCT.

**21.** Prove $\mathbf{A}\times\mathbf{B} = -\mathbf{B}\times\mathbf{A}$.

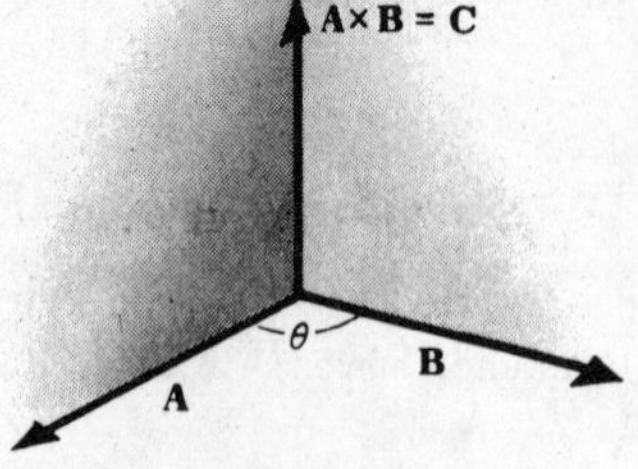


Fig.(*a*)

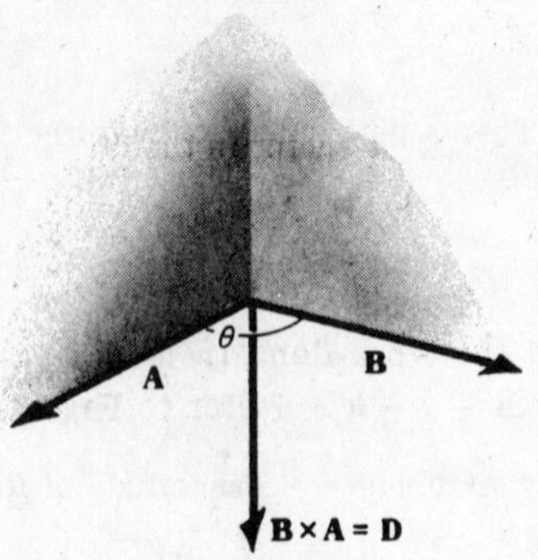


Fig.(*b*)

$\mathbf{A}\times\mathbf{B} = \mathbf{C}$ has magnitude $AB \sin\theta$ and direction such that **A, B** and **C** form a right-handed system (Fig.(*a*) above).

$\mathbf{B}\times\mathbf{A} = \mathbf{D}$ has magnitude $BA \sin\theta$ and direction such that **B, A** and **D** form a right-handed system (Fig.(*b*) above).

Then **D** has the same magnitude as **C** but is opposite in direction, i.e. $\mathbf{C} = -\mathbf{D}$ or $\mathbf{A}\times\mathbf{B} = -\mathbf{B}\times\mathbf{A}$.

The commutative law for cross products is not valid.

**22.** If $\mathbf{A}\times\mathbf{B} = \mathbf{0}$ and if **A** and **B** are not zero, show that **A** is parallel to **B**.

If $\mathbf{A}\times\mathbf{B} = AB\sin\theta\,\mathbf{u} = \mathbf{0}$, then $\sin\theta = 0$ and $\theta = 0^\circ$ or $180^\circ$.

**23.** Show that $|\mathbf{A}\times\mathbf{B}|^2 + |\mathbf{A}\cdot\mathbf{B}|^2 = |\mathbf{A}|^2\,|\mathbf{B}|^2$.

$$|\mathbf{A}\times\mathbf{B}|^2 + |\mathbf{A}\cdot\mathbf{B}|^2 = |AB\sin\theta\,\mathbf{u}|^2 + |AB\cos\theta|^2 = A^2B^2\sin^2\theta + A^2B^2\cos^2\theta$$
$$= A^2B^2 = |\mathbf{A}|^2\,|\mathbf{B}|^2$$

**24.** Evaluate each of the following.

(*a*) $\mathbf{i}\times\mathbf{j} = \mathbf{k}$

(*b*) $\mathbf{j}\times\mathbf{k} = \mathbf{i}$

(*c*) $\mathbf{k}\times\mathbf{i} = \mathbf{j}$

(*d*) $\mathbf{k}\times\mathbf{j} = -\mathbf{j}\times\mathbf{k} = -\mathbf{i}$

(*e*) $\mathbf{i}\times\mathbf{i} = \mathbf{0}$

(*f*) $\mathbf{j}\times\mathbf{j} = \mathbf{0}$

(*g*) $\mathbf{i}\times\mathbf{k} = -\mathbf{k}\times\mathbf{i} = -\mathbf{j}$

(*h*) $(2\mathbf{j})\times(3\mathbf{k}) = 6\,\mathbf{j}\times\mathbf{k} = 6\mathbf{i}$

(*i*) $(3\mathbf{i})\times(-2\mathbf{k}) = -6\,\mathbf{i}\times\mathbf{k} = 6\mathbf{j}$

(*j*) $2\mathbf{j}\times\mathbf{i} - 3\mathbf{k} = -2\mathbf{k} - 3\mathbf{k} = -5\mathbf{k}$

**25.** Prove that $\mathbf{A}\times(\mathbf{B}+\mathbf{C}) = \mathbf{A}\times\mathbf{B} + \mathbf{A}\times\mathbf{C}$ for the case where **A** is perpendicular to **B** and also to **C**.

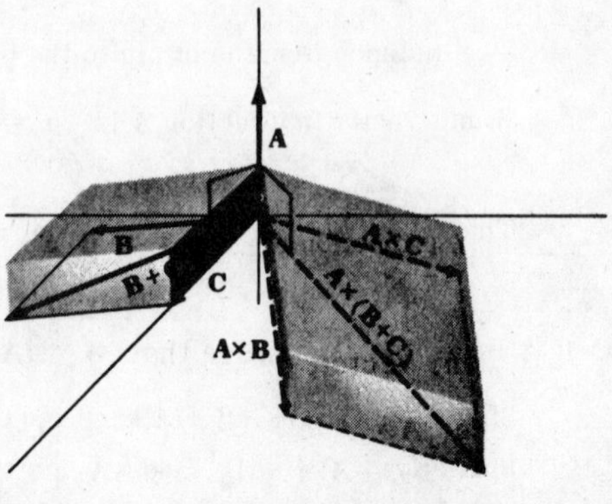


Since **A** is perpendicular to **B**, $\mathbf{A}\times\mathbf{B}$ is a vector perpendicular to the plane of **A** and **B** and having magnitude $AB\sin 90^\circ = AB$ or magnitude of $A\mathbf{B}$. This is equivalent to multiplying vector **B** by $A$ and rotating the resultant vector through $90^\circ$ to the position shown in the adjoining diagram.

Similarly, $\mathbf{A}\times\mathbf{C}$ is the vector obtained by multiplying **C** by $A$ and rotating the resultant vector through $90^\circ$ to the position shown.

In like manner, $\mathbf{A}\times(\mathbf{B}+\mathbf{C})$ is the vector obtained by multiplying $\mathbf{B}+\mathbf{C}$ by $A$ and rotating the resultant vector through $90^\circ$ to the position shown.

Since $\mathbf{A}\times(\mathbf{B}+\mathbf{C})$ is the diagonal of the parallelogram with $\mathbf{A}\times\mathbf{B}$ and $\mathbf{A}\times\mathbf{C}$ as sides, we have $\mathbf{A}\times(\mathbf{B}+\mathbf{C}) = \mathbf{A}\times\mathbf{B} + \mathbf{A}\times\mathbf{C}$.

**26.** Prove that $\mathbf{A}\times(\mathbf{B}+\mathbf{C}) = \mathbf{A}\times\mathbf{B} + \mathbf{A}\times\mathbf{C}$ in the general case where $\mathbf{A}, \mathbf{B}$ and $\mathbf{C}$ are non-coplanar.

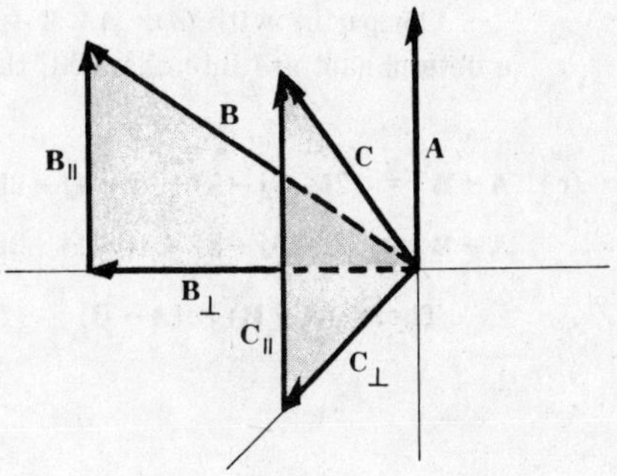


Resolve $\mathbf{B}$ into two component vectors, one perpendicular to $\mathbf{A}$ and the other parallel to $\mathbf{A}$, and denote them by $\mathbf{B}_\perp$ and $\mathbf{B}_\parallel$ respectively. Then $\mathbf{B} = \mathbf{B}_\perp + \mathbf{B}_\parallel$.

If $\theta$ is the angle between $\mathbf{A}$ and $\mathbf{B}$, then $B_\perp = B\sin\theta$. Thus the magnitude of $\mathbf{A}\times\mathbf{B}_\perp$ is $AB\sin\theta$, the same as the magnitude of $\mathbf{A}\times\mathbf{B}$. Also, the direction of $\mathbf{A}\times\mathbf{B}_\perp$ is the same as the direction of $\mathbf{A}\times\mathbf{B}$. Hence $\mathbf{A}\times\mathbf{B}_\perp = \mathbf{A}\times\mathbf{B}$.

Similarly if $\mathbf{C}$ is resolved into two component vectors $\mathbf{C}_\parallel$ and $\mathbf{C}_\perp$, parallel and perpendicular respectively to $\mathbf{A}$, then $\mathbf{A}\times\mathbf{C}_\perp = \mathbf{A}\times\mathbf{C}$.

Also, since $\mathbf{B}+\mathbf{C} = \mathbf{B}_\perp+\mathbf{B}_\parallel+\mathbf{C}_\perp+\mathbf{C}_\parallel = (\mathbf{B}_\perp+\mathbf{C}_\perp)+(\mathbf{B}_\parallel+\mathbf{C}_\parallel)$ it follows that

$$\mathbf{A}\times(\mathbf{B}_\perp+\mathbf{C}_\perp) = \mathbf{A}\times(\mathbf{B}+\mathbf{C}).$$

Now $\mathbf{B}_\perp$ and $\mathbf{C}_\perp$ are vectors perpendicular to $\mathbf{A}$ and so by Problem 25,

$$\mathbf{A}\times(\mathbf{B}_\perp+\mathbf{C}_\perp) = \mathbf{A}\times\mathbf{B}_\perp + \mathbf{A}\times\mathbf{C}_\perp$$

Then

$$\mathbf{A}\times(\mathbf{B}+\mathbf{C}) = \mathbf{A}\times\mathbf{B} + \mathbf{A}\times\mathbf{C}$$

and the distributive law holds. Multiplying by $-1$, using Prob. 21, this becomes $(\mathbf{B}+\mathbf{C})\times\mathbf{A} = \mathbf{B}\times\mathbf{A} + \mathbf{C}\times\mathbf{A}$. Note that the order of factors in cross products is important. The usual laws of algebra apply only if proper order is maintained.

**27.** If $\mathbf{A} = A_1\mathbf{i} + A_2\mathbf{j} + A_3\mathbf{k}$ and $\mathbf{B} = B_1\mathbf{i} + B_2\mathbf{j} + B_3\mathbf{k}$, prove that $\mathbf{A}\times\mathbf{B} = \begin{vmatrix} \mathbf{i} & \mathbf{j} & \mathbf{k} \\ A_1 & A_2 & A_3 \\ B_1 & B_2 & B_3 \end{vmatrix}$.

$$\begin{aligned}
\mathbf{A}\times\mathbf{B} &= (A_1\mathbf{i}+A_2\mathbf{j}+A_3\mathbf{k})\times(B_1\mathbf{i}+B_2\mathbf{j}+B_3\mathbf{k}) \\
&= A_1\mathbf{i}\times(B_1\mathbf{i}+B_2\mathbf{j}+B_3\mathbf{k}) + A_2\mathbf{j}\times(B_1\mathbf{i}+B_2\mathbf{j}+B_3\mathbf{k}) + A_3\mathbf{k}\times(B_1\mathbf{i}+B_2\mathbf{j}+B_3\mathbf{k}) \\
&= A_1B_1\mathbf{i}\times\mathbf{i} + A_1B_2\mathbf{i}\times\mathbf{j} + A_1B_3\mathbf{i}\times\mathbf{k} + A_2B_1\mathbf{j}\times\mathbf{i} + A_2B_2\mathbf{j}\times\mathbf{j} + A_2B_3\mathbf{j}\times\mathbf{k} + A_3B_1\mathbf{k}\times\mathbf{i} + A_3B_2\mathbf{k}\times\mathbf{j} + A_3B_3\mathbf{k}\times\mathbf{k} \\
&= (A_2B_3 - A_3B_2)\mathbf{i} + (A_3B_1 - A_1B_3)\mathbf{j} + (A_1B_2 - A_2B_1)\mathbf{k} = \begin{vmatrix} \mathbf{i} & \mathbf{j} & \mathbf{k} \\ A_1 & A_2 & A_3 \\ B_1 & B_2 & B_3 \end{vmatrix}.
\end{aligned}$$

**28.** If $\mathbf{A} = 2\mathbf{i} - 3\mathbf{j} - \mathbf{k}$ and $\mathbf{B} = \mathbf{i} + 4\mathbf{j} - 2\mathbf{k}$, find (*a*) $\mathbf{A}\times\mathbf{B}$, (*b*) $\mathbf{B}\times\mathbf{A}$, (*c*) $(\mathbf{A}+\mathbf{B})\times(\mathbf{A}-\mathbf{B})$.

(*a*) $$\mathbf{A}\times\mathbf{B} = (2\mathbf{i}-3\mathbf{j}-\mathbf{k})\times(\mathbf{i}+4\mathbf{j}-2\mathbf{k}) = \begin{vmatrix} \mathbf{i} & \mathbf{j} & \mathbf{k} \\ 2 & -3 & -1 \\ 1 & 4 & -2 \end{vmatrix}$$

$$= \mathbf{i}\begin{vmatrix} -3 & -1 \\ 4 & -2 \end{vmatrix} - \mathbf{j}\begin{vmatrix} 2 & -1 \\ 1 & -2 \end{vmatrix} + \mathbf{k}\begin{vmatrix} 2 & -3 \\ 1 & 4 \end{vmatrix} = 10\mathbf{i} + 3\mathbf{j} + 11\mathbf{k}$$

*Another Method.*

$$\begin{aligned}
(2\mathbf{i}-3\mathbf{j}-\mathbf{k})\times(\mathbf{i}+4\mathbf{j}-2\mathbf{k}) &= 2\mathbf{i}\times(\mathbf{i}+4\mathbf{j}-2\mathbf{k}) - 3\mathbf{j}\times(\mathbf{i}+4\mathbf{j}-2\mathbf{k}) - \mathbf{k}\times(\mathbf{i}+4\mathbf{j}-2\mathbf{k}) \\
&= 2\mathbf{i}\times\mathbf{i} + 8\mathbf{i}\times\mathbf{j} - 4\mathbf{i}\times\mathbf{k} - 3\mathbf{j}\times\mathbf{i} - 12\mathbf{j}\times\mathbf{j} + 6\mathbf{j}\times\mathbf{k} - \mathbf{k}\times\mathbf{i} - 4\mathbf{k}\times\mathbf{j} + 2\mathbf{k}\times\mathbf{k} \\
&= \mathbf{0} + 8\mathbf{k} + 4\mathbf{j} + 3\mathbf{k} - \mathbf{0} + 6\mathbf{i} - \mathbf{j} + 4\mathbf{i} + \mathbf{0} = 10\mathbf{i} + 3\mathbf{j} + 11\mathbf{k}
\end{aligned}$$

(*b*) $\mathbf{B}\times\mathbf{A} = (\mathbf{i}+4\mathbf{j}-2\mathbf{k})\times(2\mathbf{i}-3\mathbf{j}-\mathbf{k}) = \begin{vmatrix}\mathbf{i} & \mathbf{j} & \mathbf{k}\\ 1 & 4 & -2\\ 2 & -3 & -1\end{vmatrix}$

$$= \mathbf{i}\begin{vmatrix}4 & -2\\ -3 & -1\end{vmatrix} - \mathbf{j}\begin{vmatrix}1 & -2\\ 2 & -1\end{vmatrix} + \mathbf{k}\begin{vmatrix}1 & 4\\ 2 & -3\end{vmatrix} = -10\mathbf{i} - 3\mathbf{j} - 11\mathbf{k}.$$

Comparing with (*a*), $\mathbf{A}\times\mathbf{B} = -\mathbf{B}\times\mathbf{A}$. Note that this is equivalent to the theorem: If two rows of a determinant are interchanged, the determinant changes sign.

(*c*) $\mathbf{A}+\mathbf{B} = (2\mathbf{i}-3\mathbf{j}-\mathbf{k}) + (\mathbf{i}+4\mathbf{j}-2\mathbf{k}) = 3\mathbf{i}+\mathbf{j}-3\mathbf{k}$

$\mathbf{A}-\mathbf{B} = (2\mathbf{i}-3\mathbf{j}-\mathbf{k}) - (\mathbf{i}+4\mathbf{j}-2\mathbf{k}) = \mathbf{i}-7\mathbf{j}+\mathbf{k}$

Then $(\mathbf{A}+\mathbf{B})\times(\mathbf{A}-\mathbf{B}) = (3\mathbf{i}+\mathbf{j}-3\mathbf{k})\times(\mathbf{i}-7\mathbf{j}+\mathbf{k}) = \begin{vmatrix}\mathbf{i} & \mathbf{j} & \mathbf{k}\\ 3 & 1 & -3\\ 1 & -7 & 1\end{vmatrix}$

$$= \mathbf{i}\begin{vmatrix}1 & -3\\ -7 & 1\end{vmatrix} - \mathbf{j}\begin{vmatrix}3 & -3\\ 1 & 1\end{vmatrix} + \mathbf{k}\begin{vmatrix}3 & 1\\ 1 & -7\end{vmatrix} = -20\mathbf{i} - 6\mathbf{j} - 22\mathbf{k}.$$

*Another Method.*

$$(\mathbf{A}+\mathbf{B})\times(\mathbf{A}-\mathbf{B}) = \mathbf{A}\times(\mathbf{A}-\mathbf{B}) + \mathbf{B}\times(\mathbf{A}-\mathbf{B})$$

$$= \mathbf{A}\times\mathbf{A} - \mathbf{A}\times\mathbf{B} + \mathbf{B}\times\mathbf{A} - \mathbf{B}\times\mathbf{B} = \mathbf{0} - \mathbf{A}\times\mathbf{B} - \mathbf{A}\times\mathbf{B} - \mathbf{0} = -2\mathbf{A}\times\mathbf{B}$$

$$= -2(10\mathbf{i}+3\mathbf{j}+11\mathbf{k}) = -20\mathbf{i} - 6\mathbf{j} - 22\mathbf{k}, \text{ using } (a).$$

**29.** If $\mathbf{A} = 3\mathbf{i}-\mathbf{j}+2\mathbf{k}$, $\mathbf{B} = 2\mathbf{i}+\mathbf{j}-\mathbf{k}$, and $\mathbf{C} = \mathbf{i}-2\mathbf{j}+2\mathbf{k}$, find (*a*) $(\mathbf{A}\times\mathbf{B})\times\mathbf{C}$, (*b*) $\mathbf{A}\times(\mathbf{B}\times\mathbf{C})$.

(*a*) $\mathbf{A}\times\mathbf{B} = \begin{vmatrix}\mathbf{i} & \mathbf{j} & \mathbf{k}\\ 3 & -1 & 2\\ 2 & 1 & -1\end{vmatrix} = -\mathbf{i} + 7\mathbf{j} + 5\mathbf{k}.$

Then $(\mathbf{A}\times\mathbf{B})\times\mathbf{C} = (-\mathbf{i}+7\mathbf{j}+5\mathbf{k})\times(\mathbf{i}-2\mathbf{j}+2\mathbf{k}) = \begin{vmatrix}\mathbf{i} & \mathbf{j} & \mathbf{k}\\ -1 & 7 & 5\\ 1 & -2 & 2\end{vmatrix} = 24\mathbf{i} + 7\mathbf{j} - 5\mathbf{k}.$

(*b*) $\mathbf{B}\times\mathbf{C} = \begin{vmatrix}\mathbf{i} & \mathbf{j} & \mathbf{k}\\ 2 & 1 & -1\\ 1 & -2 & 2\end{vmatrix} = 0\mathbf{i} - 5\mathbf{j} - 5\mathbf{k} = -5\mathbf{j} - 5\mathbf{k}.$

Then $\mathbf{A}\times(\mathbf{B}\times\mathbf{C}) = (3\mathbf{i}-\mathbf{j}+2\mathbf{k})\times(-5\mathbf{j}-5\mathbf{k}) = \begin{vmatrix}\mathbf{i} & \mathbf{j} & \mathbf{k}\\ 3 & -1 & 2\\ 0 & -5 & -5\end{vmatrix} = 15\mathbf{i} + 15\mathbf{j} - 15\mathbf{k}.$

Thus $(\mathbf{A}\times\mathbf{B})\times\mathbf{C} \neq \mathbf{A}\times(\mathbf{B}\times\mathbf{C})$, showing the need for parentheses in $\mathbf{A}\times\mathbf{B}\times\mathbf{C}$ to avoid ambiguity.

**30.** Prove that the area of a parallelogram with sides **A** and **B** is $|\mathbf{A}\times\mathbf{B}|$.

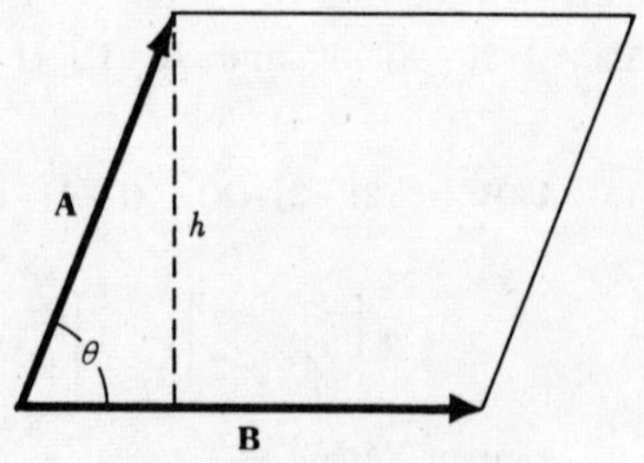


$$\text{Area of parallelogram} = h\,|\mathbf{B}| = |\mathbf{A}|\sin\theta\,|\mathbf{B}| = |\mathbf{A}\times\mathbf{B}|.$$

Note that the area of the triangle with sides **A** and $\mathbf{B} = \frac{1}{2}|\mathbf{A}\times\mathbf{B}|$.

**31.** Find the area of the triangle having vertices at $P(1, 3, 2)$, $Q(2, -1, 1)$, $R(-1, 2, 3)$.

$$\mathbf{PQ} = (2-1)\mathbf{i} + (-1-3)\mathbf{j} + (1-2)\mathbf{k} = \mathbf{i} - 4\mathbf{j} - \mathbf{k}$$

$$\mathbf{PR} = (-1-1)\mathbf{i} + (2-3)\mathbf{j} + (3-2)\mathbf{k} = -2\mathbf{i} - \mathbf{j} + \mathbf{k}$$

From Problem 30,

$$\text{area of triangle} = \tfrac{1}{2}|\mathbf{PQ}\times\mathbf{PR}| = \tfrac{1}{2}|(\mathbf{i}-4\mathbf{j}-\mathbf{k})\times(-2\mathbf{i}-\mathbf{j}+\mathbf{k})|$$

$$= \tfrac{1}{2}\left|\begin{vmatrix}\mathbf{i} & \mathbf{j} & \mathbf{k}\\ 1 & -4 & -1\\ -2 & -1 & 1\end{vmatrix}\right| = \tfrac{1}{2}|-5\mathbf{i}+\mathbf{j}-9\mathbf{k}| = \tfrac{1}{2}\sqrt{(-5)^2+(1)^2+(-9)^2} = \tfrac{1}{2}\sqrt{107}.$$

**32.** Determine a unit vector perpendicular to the plane of $\mathbf{A} = 2\mathbf{i}-6\mathbf{j}-3\mathbf{k}$ and $\mathbf{B} = 4\mathbf{i}+3\mathbf{j}-\mathbf{k}$.

$\mathbf{A}\times\mathbf{B}$ is a vector perpendicular to the plane of $\mathbf{A}$ and $\mathbf{B}$.

$$\mathbf{A}\times\mathbf{B} = \begin{vmatrix}\mathbf{i} & \mathbf{j} & \mathbf{k}\\ 2 & -6 & -3\\ 4 & 3 & -1\end{vmatrix} = 15\mathbf{i}-10\mathbf{j}+30\mathbf{k}$$

A unit vector parallel to $\mathbf{A}\times\mathbf{B}$ is $\dfrac{\mathbf{A}\times\mathbf{B}}{|\mathbf{A}\times\mathbf{B}|} = \dfrac{15\mathbf{i}-10\mathbf{j}+30\mathbf{k}}{\sqrt{(15)^2+(-10)^2+(30)^2}} = \dfrac{3}{7}\mathbf{i}-\dfrac{2}{7}\mathbf{j}+\dfrac{6}{7}\mathbf{k}$.

Another unit vector, opposite in direction, is $(-3\mathbf{i}+2\mathbf{j}-6\mathbf{k})/7$.

Compare with Problem 16.

**33.** Prove the law of sines for plane triangles.

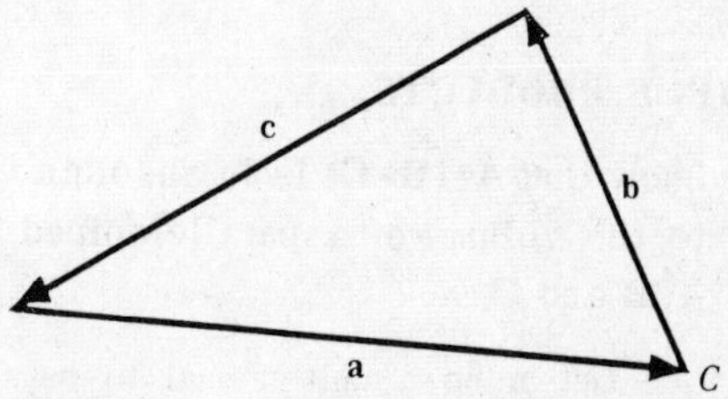


Let $\mathbf{a}, \mathbf{b}$ and $\mathbf{c}$ represent the sides of triangle $ABC$ as shown in the adjoining figure; then $\mathbf{a}+\mathbf{b}+\mathbf{c} = \mathbf{0}$. Multiplying by $\mathbf{a}\times$, $\mathbf{b}\times$ and $\mathbf{c}\times$ in succession, we find

$$\mathbf{a}\times\mathbf{b} = \mathbf{b}\times\mathbf{c} = \mathbf{c}\times\mathbf{a}$$

i.e. $\quad ab\sin C = bc\sin A = ca\sin B$

or $\quad \dfrac{\sin A}{a} = \dfrac{\sin B}{b} = \dfrac{\sin C}{c}$.

**34.** Consider a tetrahedron with faces $F_1, F_2, F_3, F_4$. Let $\mathbf{V}_1, \mathbf{V}_2, \mathbf{V}_3, \mathbf{V}_4$ be vectors whose magnitudes are respectively equal to the areas of $F_1, F_2, F_3, F_4$ and whose directions are perpendicular to these faces in the outward direction. Show that $\mathbf{V}_1+\mathbf{V}_2+\mathbf{V}_3+\mathbf{V}_4 = \mathbf{0}$.

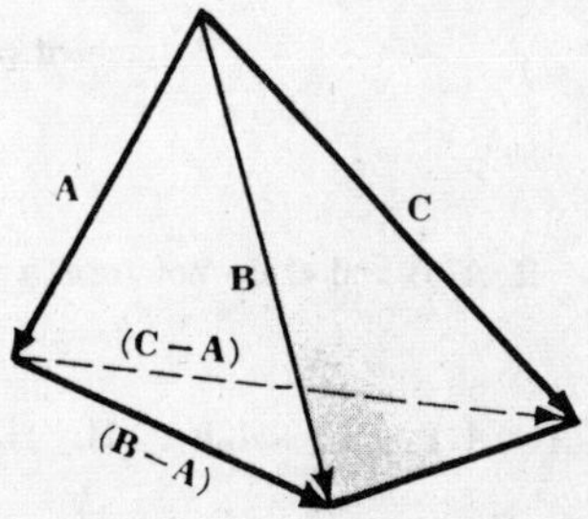


By Problem 30, the area of a triangular face determined by $\mathbf{R}$ and $\mathbf{S}$ is $\tfrac{1}{2}|\mathbf{R}\times\mathbf{S}|$.

The vectors associated with each of the faces of the tetrahedron are

$$\mathbf{V}_1 = \tfrac{1}{2}\mathbf{A}\times\mathbf{B}, \quad \mathbf{V}_2 = \tfrac{1}{2}\mathbf{B}\times\mathbf{C}, \quad \mathbf{V}_3 = \tfrac{1}{2}\mathbf{C}\times\mathbf{A}, \quad \mathbf{V}_4 = \tfrac{1}{2}(\mathbf{C}-\mathbf{A})\times(\mathbf{B}-\mathbf{A})$$

Then

$$\mathbf{V}_1+\mathbf{V}_2+\mathbf{V}_3+\mathbf{V}_4 = \tfrac{1}{2}[\mathbf{A}\times\mathbf{B}+\mathbf{B}\times\mathbf{C}+\mathbf{C}\times\mathbf{A}+(\mathbf{C}-\mathbf{A})\times(\mathbf{B}-\mathbf{A})]$$

$$= \tfrac{1}{2}[\mathbf{A}\times\mathbf{B}+\mathbf{B}\times\mathbf{C}+\mathbf{C}\times\mathbf{A}+\mathbf{C}\times\mathbf{B}-\mathbf{C}\times\mathbf{A}-\mathbf{A}\times\mathbf{B}+\mathbf{A}\times\mathbf{A}] = \mathbf{0}.$$

This result can be generalized to closed polyhedra and in the limiting case to any closed surface.

Because of the application presented here it is sometimes convenient to assign a direction to area and we speak of the *vector area*.

**35.** Find an expression for the moment of a force $\mathbf{F}$ about a point $P$.

The moment $\mathbf{M}$ of $\mathbf{F}$ about $P$ is in magnitude equal to $F$ times the perpendicular distance from $P$ to the line of action of **F**. Then if **r** is the vector from $P$ to the initial point $Q$ of **F**,

$$M = F(r \sin\theta) = rF\sin\theta = |\mathbf{r}\times\mathbf{F}|$$

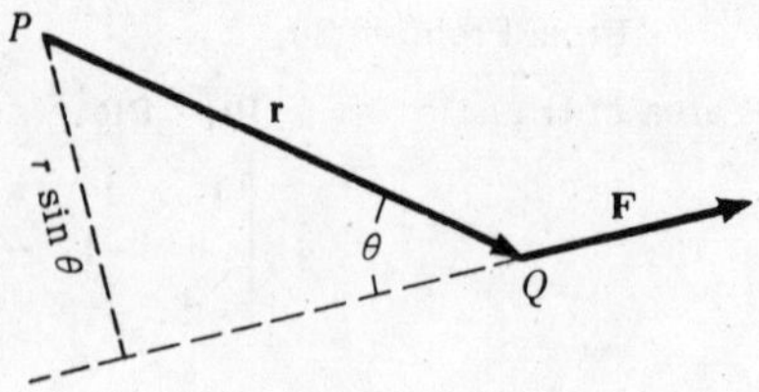


If we think of a right-threaded screw at $P$ perpendicular to the plane of **r** and **F**, then when the force **F** acts the screw will move in the direction of $\mathbf{r}\times\mathbf{F}$. Because of this it is convenient to define the moment as the vector $\mathbf{M} = \mathbf{r}\times\mathbf{F}$.

**36.** A rigid body rotates about an axis through point $O$ with angular speed $\omega$. Prove that the linear velocity **v** of a point $P$ of the body with position vector **r** is given by $\mathbf{v} = \boldsymbol{\omega}\times\mathbf{r}$, where $\boldsymbol{\omega}$ is the vector with magnitude $\omega$ whose direction is that in which a right-handed screw would advance under the given rotation.

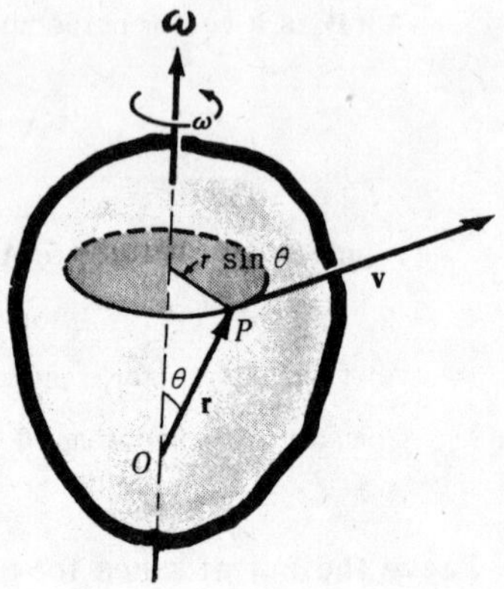


Since $P$ travels in a circle of radius $r\sin\theta$, the magnitude of the linear velocity **v** is $\omega(r\sin\theta) = |\boldsymbol{\omega}\times\mathbf{r}|$. Also, **v** must be perpendicular to both $\boldsymbol{\omega}$ and **r** and is such that $\mathbf{r}, \boldsymbol{\omega}$ and **v** form a right-handed system.

Then **v** agrees both in magnitude and direction with $\boldsymbol{\omega}\times\mathbf{r}$; hence $\mathbf{v} = \boldsymbol{\omega}\times\mathbf{r}$. The vector $\boldsymbol{\omega}$ is called the *angular velocity*.

**TRIPLE PRODUCTS.**

**37.** Show that $\mathbf{A}\cdot(\mathbf{B}\times\mathbf{C})$ is in absolute value equal to the volume of a parallelepiped with sides **A**, **B** and **C**.

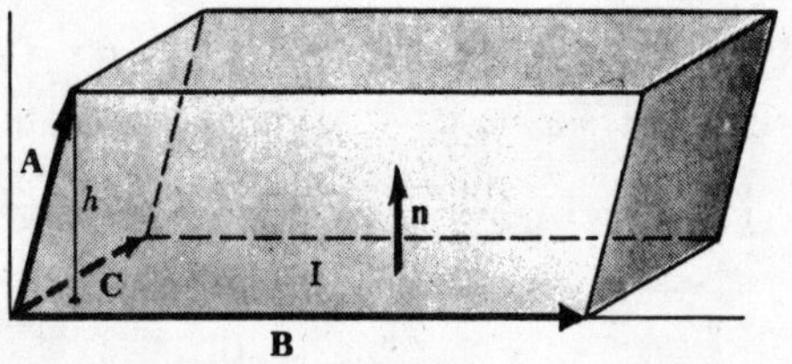


Let **n** be a unit normal to parallelogram $I$, having the direction of $\mathbf{B}\times\mathbf{C}$, and let $h$ be the height of the terminal point of **A** above the parallelogram $I$.

$$\begin{aligned}\text{Volume of parallelepiped} &= (\text{height } h)(\text{area of parallelogram } I)\\ &= (\mathbf{A}\cdot\mathbf{n})(|\mathbf{B}\times\mathbf{C}|)\\ &= \mathbf{A}\cdot\{|\mathbf{B}\times\mathbf{C}|\,\mathbf{n}\} = \mathbf{A}\cdot(\mathbf{B}\times\mathbf{C})\end{aligned}$$

If **A**, **B** and **C** do not form a right-handed system, $\mathbf{A}\cdot\mathbf{n} < 0$ and the volume $= |\mathbf{A}\cdot(\mathbf{B}\times\mathbf{C})|$.

**38.** If $\mathbf{A} = A_1\mathbf{i} + A_2\mathbf{j} + A_3\mathbf{k}$, $\mathbf{B} = B_1\mathbf{i} + B_2\mathbf{j} + B_3\mathbf{k}$, $\mathbf{C} = C_1\mathbf{i} + C_2\mathbf{j} + C_3\mathbf{k}$ show that

$$\mathbf{A}\cdot(\mathbf{B}\times\mathbf{C}) = \begin{vmatrix} A_1 & A_2 & A_3 \\ B_1 & B_2 & B_3 \\ C_1 & C_2 & C_3 \end{vmatrix}$$

$$\mathbf{A}\cdot(\mathbf{B}\times\mathbf{C}) = \mathbf{A}\cdot\begin{vmatrix} \mathbf{i} & \mathbf{j} & \mathbf{k} \\ B_1 & B_2 & B_3 \\ C_1 & C_2 & C_3 \end{vmatrix}$$

$$= (A_1\mathbf{i} + A_2\mathbf{j} + A_3\mathbf{k})\cdot[(B_2C_3 - B_3C_2)\mathbf{i} + (B_3C_1 - B_1C_3)\mathbf{j} + (B_1C_2 - B_2C_1)\mathbf{k}]$$

$$= A_1(B_2C_3 - B_3C_2) + A_2(B_3C_1 - B_1C_3) + A_3(B_1C_2 - B_2C_1) = \begin{vmatrix} A_1 & A_2 & A_3 \\ B_1 & B_2 & B_3 \\ C_1 & C_2 & C_3 \end{vmatrix}.$$

**39.** Evaluate $(2\mathbf{i}-3\mathbf{j})\cdot[(\mathbf{i}+\mathbf{j}-\mathbf{k})\times(3\mathbf{i}-\mathbf{k})]$.

By Problem 38, the result is $\begin{vmatrix} 2 & -3 & 0 \\ 1 & 1 & -1 \\ 3 & 0 & -1 \end{vmatrix} = 4$.

*Another Method.* The result is equal to

$$\begin{aligned}
&(2\mathbf{i}-3\mathbf{j})\cdot[\mathbf{i}\times(3\mathbf{i}-\mathbf{k}) + \mathbf{j}\times(3\mathbf{i}-\mathbf{k}) - \mathbf{k}\times(3\mathbf{i}-\mathbf{k})] \\
&\quad = (2\mathbf{i}-3\mathbf{j})\cdot[3\mathbf{i}\times\mathbf{i} - \mathbf{i}\times\mathbf{k} + 3\mathbf{j}\times\mathbf{i} - \mathbf{j}\times\mathbf{k} - 3\mathbf{k}\times\mathbf{i} + \mathbf{k}\times\mathbf{k}] \\
&\quad = (2\mathbf{i}-3\mathbf{j})\cdot(\mathbf{0} + \mathbf{j} - 3\mathbf{k} - \mathbf{i} - 3\mathbf{j} + \mathbf{0}) \\
&\quad = (2\mathbf{i}-3\mathbf{j})\cdot(-\mathbf{i}-2\mathbf{j}-3\mathbf{k}) = (2)(-1)+(-3)(-2)+(0)(-3) = 4.
\end{aligned}$$

**40.** Prove that $\mathbf{A}\cdot(\mathbf{B}\times\mathbf{C}) = \mathbf{B}\cdot(\mathbf{C}\times\mathbf{A}) = \mathbf{C}\cdot(\mathbf{A}\times\mathbf{B})$.

By Problem 38, $\quad \mathbf{A}\cdot(\mathbf{B}\times\mathbf{C}) = \begin{vmatrix} A_1 & A_2 & A_3 \\ B_1 & B_2 & B_3 \\ C_1 & C_2 & C_3 \end{vmatrix}$.

By a theorem of determinants which states that interchange of two rows of a determinant changes its sign, we have

$$\begin{vmatrix} A_1 & A_2 & A_3 \\ B_1 & B_2 & B_3 \\ C_1 & C_2 & C_3 \end{vmatrix} = -\begin{vmatrix} B_1 & B_2 & B_3 \\ A_1 & A_2 & A_3 \\ C_1 & C_2 & C_3 \end{vmatrix} = \begin{vmatrix} B_1 & B_2 & B_3 \\ C_1 & C_2 & C_3 \\ A_1 & A_2 & A_3 \end{vmatrix} = \mathbf{B}\cdot(\mathbf{C}\times\mathbf{A})$$

$$\begin{vmatrix} A_1 & A_2 & A_3 \\ B_1 & B_2 & B_3 \\ C_1 & C_2 & C_3 \end{vmatrix} = -\begin{vmatrix} C_1 & C_2 & C_3 \\ B_1 & B_2 & B_3 \\ A_1 & A_2 & A_3 \end{vmatrix} = \begin{vmatrix} C_1 & C_2 & C_3 \\ A_1 & A_2 & A_3 \\ B_1 & B_2 & B_3 \end{vmatrix} = \mathbf{C}\cdot(\mathbf{A}\times\mathbf{B})$$

**41.** Show that $\mathbf{A}\cdot(\mathbf{B}\times\mathbf{C}) = (\mathbf{A}\times\mathbf{B})\cdot\mathbf{C}$.

From Problem 40, $\quad \mathbf{A}\cdot(\mathbf{B}\times\mathbf{C}) = \mathbf{C}\cdot(\mathbf{A}\times\mathbf{B}) = (\mathbf{A}\times\mathbf{B})\cdot\mathbf{C}$

Occasionally $\mathbf{A}\cdot(\mathbf{B}\times\mathbf{C})$ is written without parentheses as $\mathbf{A}\cdot\mathbf{B}\times\mathbf{C}$. In such case there cannot be any ambiguity since the only possible interpretations are $\mathbf{A}\cdot(\mathbf{B}\times\mathbf{C})$ and $(\mathbf{A}\cdot\mathbf{B})\times\mathbf{C}$. The latter however has no meaning since the cross product of a scalar with a vector is undefined.

The result $\mathbf{A}\cdot\mathbf{B}\times\mathbf{C} = \mathbf{A}\times\mathbf{B}\cdot\mathbf{C}$ is sometimes summarized in the statement that the dot and cross can be interchanged without affecting the result.

**42.** Prove that $\mathbf{A}\cdot(\mathbf{A}\times\mathbf{C}) = 0$.

From Problem 41, $\quad \mathbf{A}\cdot(\mathbf{A}\times\mathbf{C}) = (\mathbf{A}\times\mathbf{A})\cdot\mathbf{C} = 0$.

**43.** Prove that a necessary and sufficient condition for the vectors **A, B** and **C** to be coplanar is that $\mathbf{A}\cdot\mathbf{B}\times\mathbf{C} = 0$.

Note that $\mathbf{A}\cdot\mathbf{B}\times\mathbf{C}$ can have no meaning other than $\mathbf{A}\cdot(\mathbf{B}\times\mathbf{C})$.

If **A, B** and **C** are coplanar the volume of the parallelepiped formed by them is zero. Then by Problem 37, $\mathbf{A}\cdot\mathbf{B}\times\mathbf{C} = 0$.

Conversely, if $\mathbf{A}\cdot\mathbf{B}\times\mathbf{C} = 0$ the volume of the parallelepiped formed by vectors **A, B** and **C** is zero, and so the vectors must lie in a plane.

**44.** Let $\mathbf{r}_1 = x_1\mathbf{i} + y_1\mathbf{j} + z_1\mathbf{k}$, $\mathbf{r}_2 = x_2\mathbf{i} + y_2\mathbf{j} + z_2\mathbf{k}$ and $\mathbf{r}_3 = x_3\mathbf{i} + y_3\mathbf{j} + z_3\mathbf{k}$ be the position vectors of points $P_1(x_1, y_1, z_1)$, $P_2(x_2, y_2, z_2)$ and $P_3(x_3, y_3, z_3)$. Find an equation for the plane passing through $P_1$, $P_2$ and $P_3$.

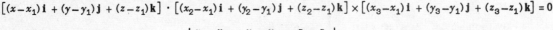

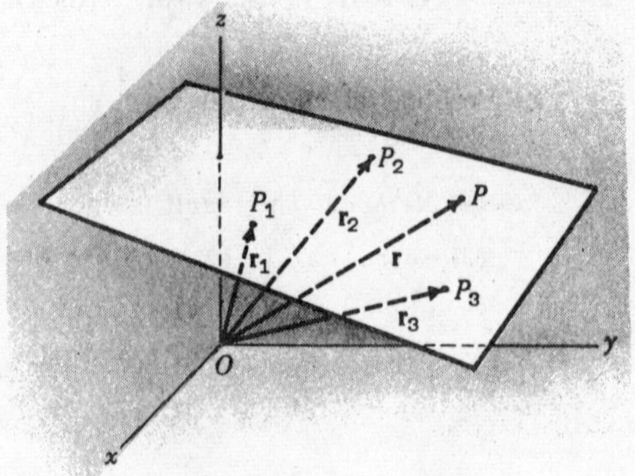


We assume that $P_1, P_2$ and $P_3$ do not lie in the same straight line; hence they determine a plane.

Let $\mathbf{r} = x\mathbf{i} + y\mathbf{j} + z\mathbf{k}$ denote the position vector of any point $P(x, y, z)$ in the plane. Consider vectors $\mathbf{P_1P_2} = \mathbf{r}_2 - \mathbf{r}_1$, $\mathbf{P_1P_3} = \mathbf{r}_3 - \mathbf{r}_1$ and $\mathbf{P_1P} = \mathbf{r} - \mathbf{r}_1$ which all lie in the plane.

By Problem 43, $\mathbf{P_1P} \cdot \mathbf{P_1P_2} \times \mathbf{P_1P_3} = 0$ or

$$(\mathbf{r} - \mathbf{r}_1) \cdot (\mathbf{r}_2 - \mathbf{r}_1) \times (\mathbf{r}_3 - \mathbf{r}_1) = 0$$

In terms of rectangular coordinates this becomes

$$[(x-x_1)\mathbf{i} + (y-y_1)\mathbf{j} + (z-z_1)\mathbf{k}] \cdot [(x_2-x_1)\mathbf{i} + (y_2-y_1)\mathbf{j} + (z_2-z_1)\mathbf{k}] \times [(x_3-x_1)\mathbf{i} + (y_3-y_1)\mathbf{j} + (z_3-z_1)\mathbf{k}] = 0$$

or, using Problem 38,
$$\begin{vmatrix} x-x_1 & y-y_1 & z-z_1 \\ x_2-x_1 & y_2-y_1 & z_2-z_1 \\ x_3-x_1 & y_3-y_1 & z_3-z_1 \end{vmatrix} = 0.$$

**45.** Find an equation for the plane determined by the points $P_1(2, -1, 1)$, $P_2(3, 2, -1)$ and $P_3(-1, 3, 2)$.

The position vectors of $P_1, P_2, P_3$ and any point $P(x, y, z)$ are respectively $\mathbf{r}_1 = 2\mathbf{i} - \mathbf{j} + \mathbf{k}$, $\mathbf{r}_2 = 3\mathbf{i} + 2\mathbf{j} - \mathbf{k}$, $\mathbf{r}_3 = -\mathbf{i} + 3\mathbf{j} + 2\mathbf{k}$ and $\mathbf{r} = x\mathbf{i} + y\mathbf{j} + z\mathbf{k}$.

Then $\mathbf{PP_1} = \mathbf{r} - \mathbf{r}_1$, $\mathbf{P_2P_1} = \mathbf{r}_2 - \mathbf{r}_1$, $\mathbf{P_3P_1} = \mathbf{r}_3 - \mathbf{r}_1$ all lie in the required plane, so that

$$(\mathbf{r} - \mathbf{r}_1) \cdot (\mathbf{r}_2 - \mathbf{r}_1) \times (\mathbf{r}_3 - \mathbf{r}_1) = 0$$

i.e.
$$[(x-2)\mathbf{i} + (y+1)\mathbf{j} + (z-1)\mathbf{k}] \cdot [\mathbf{i} + 3\mathbf{j} - 2\mathbf{k}] \times [-3\mathbf{i} + 4\mathbf{j} + \mathbf{k}] = 0$$
$$[(x-2)\mathbf{i} + (y+1)\mathbf{j} + (z-1)\mathbf{k}] \cdot [11\mathbf{i} + 5\mathbf{j} + 13\mathbf{k}] = 0$$
$$11(x-2) + 5(y+1) + 13(z-1) = 0 \quad \text{or} \quad 11x + 5y + 13z = 30.$$

**46.** If the points $P, Q$ and $R$, not all lying on the same straight line, have position vectors **a**, **b** and **c** relative to a given origin, show that $\mathbf{a}\times\mathbf{b} + \mathbf{b}\times\mathbf{c} + \mathbf{c}\times\mathbf{a}$ is a vector perpendicular to the plane of $P, Q$ and $R$.

Let **r** be the position vector of any point in the plane of $P, Q$ and $R$. Then the vectors $\mathbf{r} - \mathbf{a}$, $\mathbf{b} - \mathbf{a}$ and $\mathbf{c} - \mathbf{a}$ are coplanar, so that by Problem 43

$$(\mathbf{r}-\mathbf{a}) \cdot (\mathbf{b}-\mathbf{a}) \times (\mathbf{c}-\mathbf{a}) = 0 \quad \text{or} \quad (\mathbf{r}-\mathbf{a}) \cdot (\mathbf{a}\times\mathbf{b} + \mathbf{b}\times\mathbf{c} + \mathbf{c}\times\mathbf{a}) = 0.$$

Thus $\mathbf{a}\times\mathbf{b} + \mathbf{b}\times\mathbf{c} + \mathbf{c}\times\mathbf{a}$ is perpendicular to $\mathbf{r} - \mathbf{a}$ and is therefore perpendicular to the plane of $P, Q$ and $R$.

**47.** Prove: (*a*) $\mathbf{A}\times(\mathbf{B}\times\mathbf{C}) = \mathbf{B}(\mathbf{A}\cdot\mathbf{C}) - \mathbf{C}(\mathbf{A}\cdot\mathbf{B})$, (*b*) $(\mathbf{A}\times\mathbf{B})\times\mathbf{C} = \mathbf{B}(\mathbf{A}\cdot\mathbf{C}) - \mathbf{A}(\mathbf{B}\cdot\mathbf{C})$.

(*a*) Let $\mathbf{A} = A_1\mathbf{i} + A_2\mathbf{j} + A_3\mathbf{k}$, $\mathbf{B} = B_1\mathbf{i} + B_2\mathbf{j} + B_3\mathbf{k}$, $\mathbf{C} = C_1\mathbf{i} + C_2\mathbf{j} + C_3\mathbf{k}$.

Then
$$\mathbf{A}\times(\mathbf{B}\times\mathbf{C}) = (A_1\mathbf{i} + A_2\mathbf{j} + A_3\mathbf{k}) \times \begin{vmatrix} \mathbf{i} & \mathbf{j} & \mathbf{k} \\ B_1 & B_2 & B_3 \\ C_1 & C_2 & C_3 \end{vmatrix}$$
$$= (A_1\mathbf{i} + A_2\mathbf{j} + A_3\mathbf{k}) \times ([B_2C_3 - B_3C_2]\mathbf{i} + [B_3C_1 - B_1C_3]\mathbf{j} + [B_1C_2 - B_2C_1]\mathbf{k})$$

$$= \begin{vmatrix} \mathbf{i} & \mathbf{j} & \mathbf{k} \\ A_1 & A_2 & A_3 \\ B_2C_3 - B_3C_2 & B_3C_1 - B_1C_3 & B_1C_2 - B_2C_1 \end{vmatrix}$$

$$= (A_2B_1C_2 - A_2B_2C_1 - A_3B_3C_1 + A_3B_1C_3)\,\mathbf{i} + (A_3B_2C_3 - A_3B_3C_2 - A_1B_1C_2 + A_1B_2C_1)\,\mathbf{j} + (A_1B_3C_1 - A_1B_1C_3 - A_2B_2C_3 + A_2B_3C_2)\,\mathbf{k}$$

Also $\mathbf{B}(\mathbf{A}\cdot\mathbf{C}) - \mathbf{C}(\mathbf{A}\cdot\mathbf{B})$

$$= (B_1\mathbf{i} + B_2\mathbf{j} + B_3\mathbf{k})(A_1C_1 + A_2C_2 + A_3C_3) - (C_1\mathbf{i} + C_2\mathbf{j} + C_3\mathbf{k})(A_1B_1 + A_2B_2 + A_3B_3)$$

$$= (A_2B_1C_2 + A_3B_1C_3 - A_2C_1B_2 - A_3C_1B_3)\,\mathbf{i} + (B_2A_1C_1 + B_2A_3C_3 - C_2A_1B_1 - C_2A_3B_3)\,\mathbf{j} + (B_3A_1C_1 + B_3A_2C_2 - C_3A_1B_1 - C_3A_2B_2)\,\mathbf{k}$$

and the result follows.

(*b*) $(\mathbf{A}\times\mathbf{B})\times\mathbf{C} = -\mathbf{C}\times(\mathbf{A}\times\mathbf{B}) = -\{\mathbf{A}(\mathbf{C}\cdot\mathbf{B}) - \mathbf{B}(\mathbf{C}\cdot\mathbf{A})\} = \mathbf{B}(\mathbf{A}\cdot\mathbf{C}) - \mathbf{A}(\mathbf{B}\cdot\mathbf{C})$ upon replacing **A**, **B** and **C** in (*a*) by **C**, **A** and **B** respectively.

Note that $\mathbf{A}\times(\mathbf{B}\times\mathbf{C}) \neq (\mathbf{A}\times\mathbf{B})\times\mathbf{C}$, i.e. the associative law for vector cross products is not valid for all vectors **A**, **B**, **C**.

**48.** Prove: $(\mathbf{A}\times\mathbf{B})\cdot(\mathbf{C}\times\mathbf{D}) = (\mathbf{A}\cdot\mathbf{C})(\mathbf{B}\cdot\mathbf{D}) - (\mathbf{A}\cdot\mathbf{D})(\mathbf{B}\cdot\mathbf{C})$.

From Problem 41, $\mathbf{X}\cdot(\mathbf{C}\times\mathbf{D}) = (\mathbf{X}\times\mathbf{C})\cdot\mathbf{D}$. Let $\mathbf{X} = \mathbf{A}\times\mathbf{B}$; then

$$(\mathbf{A}\times\mathbf{B})\cdot(\mathbf{C}\times\mathbf{D}) = \{(\mathbf{A}\times\mathbf{B})\times\mathbf{C}\}\cdot\mathbf{D} = \{\mathbf{B}(\mathbf{A}\cdot\mathbf{C}) - \mathbf{A}(\mathbf{B}\cdot\mathbf{C})\}\cdot\mathbf{D}$$

$$= (\mathbf{A}\cdot\mathbf{C})(\mathbf{B}\cdot\mathbf{D}) - (\mathbf{A}\cdot\mathbf{D})(\mathbf{B}\cdot\mathbf{C}), \quad \text{using Problem 47}(b).$$

**49.** Prove: $\mathbf{A}\times(\mathbf{B}\times\mathbf{C}) + \mathbf{B}\times(\mathbf{C}\times\mathbf{A}) + \mathbf{C}\times(\mathbf{A}\times\mathbf{B}) = \mathbf{0}$.

By Problem 47(*a*),

$$\mathbf{A}\times(\mathbf{B}\times\mathbf{C}) = \mathbf{B}(\mathbf{A}\cdot\mathbf{C}) - \mathbf{C}(\mathbf{A}\cdot\mathbf{B})$$
$$\mathbf{B}\times(\mathbf{C}\times\mathbf{A}) = \mathbf{C}(\mathbf{B}\cdot\mathbf{A}) - \mathbf{A}(\mathbf{B}\cdot\mathbf{C})$$
$$\mathbf{C}\times(\mathbf{A}\times\mathbf{B}) = \mathbf{A}(\mathbf{C}\cdot\mathbf{B}) - \mathbf{B}(\mathbf{C}\cdot\mathbf{A})$$

Adding, the result follows.

**50.** Prove: $(\mathbf{A}\times\mathbf{B})\times(\mathbf{C}\times\mathbf{D}) = \mathbf{B}(\mathbf{A}\cdot\mathbf{C}\times\mathbf{D}) - \mathbf{A}(\mathbf{B}\cdot\mathbf{C}\times\mathbf{D}) = \mathbf{C}(\mathbf{A}\cdot\mathbf{B}\times\mathbf{D}) - \mathbf{D}(\mathbf{A}\cdot\mathbf{B}\times\mathbf{C})$.

By Problem 47(*a*), $\mathbf{X}\times(\mathbf{C}\times\mathbf{D}) = \mathbf{C}(\mathbf{X}\cdot\mathbf{D}) - \mathbf{D}(\mathbf{X}\cdot\mathbf{C})$. Let $\mathbf{X} = \mathbf{A}\times\mathbf{B}$; then

$$(\mathbf{A}\times\mathbf{B})\times(\mathbf{C}\times\mathbf{D}) = \mathbf{C}(\mathbf{A}\times\mathbf{B}\cdot\mathbf{D}) - \mathbf{D}(\mathbf{A}\times\mathbf{B}\cdot\mathbf{C})$$
$$= \mathbf{C}(\mathbf{A}\cdot\mathbf{B}\times\mathbf{D}) - \mathbf{D}(\mathbf{A}\cdot\mathbf{B}\times\mathbf{C})$$

By Problem 47(*b*), $(\mathbf{A}\times\mathbf{B})\times\mathbf{Y} = \mathbf{B}(\mathbf{A}\cdot\mathbf{Y}) - \mathbf{A}(\mathbf{B}\cdot\mathbf{Y})$. Let $\mathbf{Y} = \mathbf{C}\times\mathbf{D}$; then

$$(\mathbf{A}\times\mathbf{B})\times(\mathbf{C}\times\mathbf{D}) = \mathbf{B}(\mathbf{A}\cdot\mathbf{C}\times\mathbf{D}) - \mathbf{A}(\mathbf{B}\cdot\mathbf{C}\times\mathbf{D})$$

**51.** Let $PQR$ be a spherical triangle whose sides $p, q, r$ are arcs of great circles. Prove that

$$\frac{\sin P}{\sin p} = \frac{\sin Q}{\sin q} = \frac{\sin R}{\sin r}$$

Suppose that the sphere (see figure below) has unit radius, and let unit vectors **A**, **B** and **C** be drawn from the centre $O$ of the sphere to $P$, $Q$ and $R$ respectively. From Problem 50,

(*1*) $$(\mathbf{A}\times\mathbf{B})\times(\mathbf{A}\times\mathbf{C}) = (\mathbf{A}\cdot\mathbf{B}\times\mathbf{C})\mathbf{A}$$

A unit vector perpendicular to $\mathbf{A}\times\mathbf{B}$ and $\mathbf{A}\times\mathbf{C}$ is $\mathbf{A}$, so that (1) becomes

(2) $\sin r \ \sin q \ \sin P \ \mathbf{A} = (\mathbf{A}\cdot\mathbf{B}\times\mathbf{C})\mathbf{A}$ or

(3) $\sin r \ \sin q \ \sin P = \mathbf{A}\cdot\mathbf{B}\times\mathbf{C}$

By cyclic permutation of $p,q,r$, $P,Q,R$ and $\mathbf{A},\mathbf{B},\mathbf{C}$ we obtain

(4) $\sin p \ \sin r \ \sin Q = \mathbf{B}\cdot\mathbf{C}\times\mathbf{A}$

(5) $\sin q \ \sin p \ \sin R = \mathbf{C}\cdot\mathbf{A}\times\mathbf{B}$

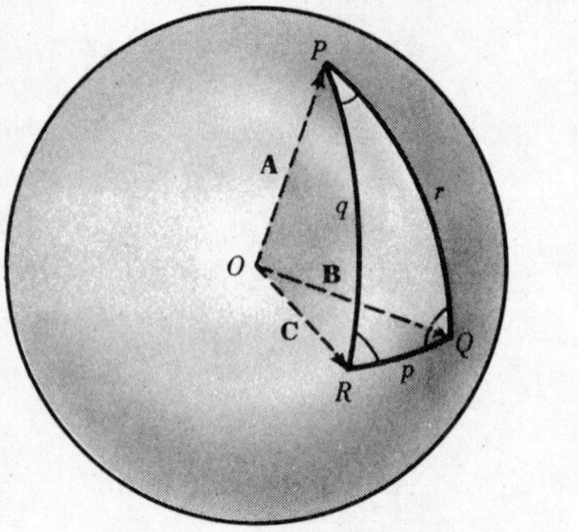


Then since the right hand sides of (3), (4) and (5) are equal (Problem 40)

$$\sin r \ \sin q \ \sin P = \sin p \ \sin r \ \sin Q = \sin q \ \sin p \ \sin R$$

from which we find $$\frac{\sin P}{\sin p} = \frac{\sin Q}{\sin q} = \frac{\sin R}{\sin r}$$

This is called the *law of sines* for spherical triangles.

**52.** Prove: $(\mathbf{A}\times\mathbf{B})\cdot(\mathbf{B}\times\mathbf{C})\times(\mathbf{C}\times\mathbf{A}) = (\mathbf{A}\cdot\mathbf{B}\times\mathbf{C})^2$.

By Problem 47(*a*), $\mathbf{X}\times(\mathbf{C}\times\mathbf{A}) = \mathbf{C}(\mathbf{X}\cdot\mathbf{A}) - \mathbf{A}(\mathbf{X}\cdot\mathbf{C})$. Let $\mathbf{X} = \mathbf{B}\times\mathbf{C}$; then

$$(\mathbf{B}\times\mathbf{C})\times(\mathbf{C}\times\mathbf{A}) = \mathbf{C}(\mathbf{B}\times\mathbf{C}\cdot\mathbf{A}) - \mathbf{A}(\mathbf{B}\times\mathbf{C}\cdot\mathbf{C})$$
$$= \mathbf{C}(\mathbf{A}\cdot\mathbf{B}\times\mathbf{C}) - \mathbf{A}(\mathbf{B}\cdot\mathbf{C}\times\mathbf{C}) = \mathbf{C}(\mathbf{A}\cdot\mathbf{B}\times\mathbf{C})$$

Thus $$(\mathbf{A}\times\mathbf{B})\cdot(\mathbf{B}\times\mathbf{C})\times(\mathbf{C}\times\mathbf{A}) = (\mathbf{A}\times\mathbf{B})\cdot\mathbf{C}(\mathbf{A}\cdot\mathbf{B}\times\mathbf{C})$$
$$= (\mathbf{A}\times\mathbf{B}\cdot\mathbf{C})(\mathbf{A}\cdot\mathbf{B}\times\mathbf{C}) = (\mathbf{A}\cdot\mathbf{B}\times\mathbf{C})^2$$

**53.** Given the vectors $\mathbf{a}' = \dfrac{\mathbf{b}\times\mathbf{c}}{\mathbf{a}\cdot\mathbf{b}\times\mathbf{c}}$, $\mathbf{b}' = \dfrac{\mathbf{c}\times\mathbf{a}}{\mathbf{a}\cdot\mathbf{b}\times\mathbf{c}}$ and $\mathbf{c}' = \dfrac{\mathbf{a}\times\mathbf{b}}{\mathbf{a}\cdot\mathbf{b}\times\mathbf{c}}$, show that if $\mathbf{a}\cdot\mathbf{b}\times\mathbf{c} \neq 0$,

(*a*) $\mathbf{a}'\cdot\mathbf{a} = \mathbf{b}'\cdot\mathbf{b} = \mathbf{c}'\cdot\mathbf{c} = 1$,

(*b*) $\mathbf{a}'\cdot\mathbf{b} = \mathbf{a}'\cdot\mathbf{c} = 0$, $\mathbf{b}'\cdot\mathbf{a} = \mathbf{b}'\cdot\mathbf{c} = 0$, $\mathbf{c}'\cdot\mathbf{a} = \mathbf{c}'\cdot\mathbf{b} = 0$,

(*c*) if $\mathbf{a}\cdot\mathbf{b}\times\mathbf{c} = V$ then $\mathbf{a}'\cdot\mathbf{b}'\times\mathbf{c}' = 1/V$,

(*d*) $\mathbf{a}', \mathbf{b}'$, and $\mathbf{c}'$ are non-coplanar if $\mathbf{a}, \mathbf{b}$ and $\mathbf{c}$ are non-coplanar.

(*a*) $$\mathbf{a}'\cdot\mathbf{a} = \mathbf{a}\cdot\mathbf{a}' = \mathbf{a}\cdot\frac{\mathbf{b}\times\mathbf{c}}{\mathbf{a}\cdot\mathbf{b}\times\mathbf{c}} = \frac{\mathbf{a}\cdot\mathbf{b}\times\mathbf{c}}{\mathbf{a}\cdot\mathbf{b}\times\mathbf{c}} = 1$$

$$\mathbf{b}'\cdot\mathbf{b} = \mathbf{b}\cdot\mathbf{b}' = \mathbf{b}\cdot\frac{\mathbf{c}\times\mathbf{a}}{\mathbf{a}\cdot\mathbf{b}\times\mathbf{c}} = \frac{\mathbf{b}\cdot\mathbf{c}\times\mathbf{a}}{\mathbf{a}\cdot\mathbf{b}\times\mathbf{c}} = \frac{\mathbf{a}\cdot\mathbf{b}\times\mathbf{c}}{\mathbf{a}\cdot\mathbf{b}\times\mathbf{c}} = 1$$

$$\mathbf{c}'\cdot\mathbf{c} = \mathbf{c}\cdot\mathbf{c}' = \mathbf{c}\cdot\frac{\mathbf{a}\times\mathbf{b}}{\mathbf{a}\cdot\mathbf{b}\times\mathbf{c}} = \frac{\mathbf{c}\cdot\mathbf{a}\times\mathbf{b}}{\mathbf{a}\cdot\mathbf{b}\times\mathbf{c}} = \frac{\mathbf{a}\cdot\mathbf{b}\times\mathbf{c}}{\mathbf{a}\cdot\mathbf{b}\times\mathbf{c}} = 1$$

(*b*) $$\mathbf{a}'\cdot\mathbf{b} = \mathbf{b}\cdot\mathbf{a}' = \mathbf{b}\cdot\frac{\mathbf{b}\times\mathbf{c}}{\mathbf{a}\cdot\mathbf{b}\times\mathbf{c}} = \frac{\mathbf{b}\cdot\mathbf{b}\times\mathbf{c}}{\mathbf{a}\cdot\mathbf{b}\times\mathbf{c}} = \frac{\mathbf{b}\times\mathbf{b}\cdot\mathbf{c}}{\mathbf{a}\cdot\mathbf{b}\times\mathbf{c}} = 0$$

Similarly the other results follow. The results can also be seen by noting, for example, that $\mathbf{a}'$ has the direction of $\mathbf{b}\times\mathbf{c}$ and so must be perpendicular to both $\mathbf{b}$ and $\mathbf{c}$, from which $\mathbf{a}'\cdot\mathbf{b} = 0$ and $\mathbf{a}'\cdot\mathbf{c} = 0$.

From (*a*) and (*b*) we see that the sets of vectors $\mathbf{a},\mathbf{b},\mathbf{c}$ and $\mathbf{a}',\mathbf{b}',\mathbf{c}'$ are reciprocal vectors. See also Supplementary Problems 104 and 106.

(c) $$\mathbf{a}' = \frac{\mathbf{b}\times\mathbf{c}}{V}, \quad \mathbf{b}' = \frac{\mathbf{c}\times\mathbf{a}}{V}, \quad \mathbf{c}' = \frac{\mathbf{a}\times\mathbf{b}}{V}$$

Then $$\mathbf{a}'\cdot\mathbf{b}'\times\mathbf{c}' = \frac{(\mathbf{b}\times\mathbf{c})\cdot(\mathbf{c}\times\mathbf{a})\times(\mathbf{a}\times\mathbf{b})}{V^3} = \frac{(\mathbf{a}\times\mathbf{b})\cdot(\mathbf{b}\times\mathbf{c})\times(\mathbf{c}\times\mathbf{a})}{V^3}$$

$$= \frac{(\mathbf{a}\cdot\mathbf{b}\times\mathbf{c})^2}{V^3} = \frac{V^2}{V^3} = \frac{1}{V} \quad \text{using Problem 52.}$$

(d) By Problem 43, if $\mathbf{a}, \mathbf{b}$ and $\mathbf{c}$ are non-coplanar $\mathbf{a}\cdot\mathbf{b}\times\mathbf{c} \neq 0$. Then from part (c) it follows that $\mathbf{a}'\cdot\mathbf{b}'\times\mathbf{c}' \neq 0$, so that $\mathbf{a}', \mathbf{b}'$ and $\mathbf{c}'$ are also non-coplanar.

**54.** Show that any vector $\mathbf{r}$ can be expressed in terms of the reciprocal vectors of Problem 53 as

$$\mathbf{r} = (\mathbf{r}\cdot\mathbf{a}')\mathbf{a} + (\mathbf{r}\cdot\mathbf{b}')\mathbf{b} + (\mathbf{r}\cdot\mathbf{c}')\mathbf{c}.$$

From Problem 50, $\mathbf{B}(\mathbf{A}\cdot\mathbf{C}\times\mathbf{D}) - \mathbf{A}(\mathbf{B}\cdot\mathbf{C}\times\mathbf{D}) = \mathbf{C}(\mathbf{A}\cdot\mathbf{B}\times\mathbf{D}) - \mathbf{D}(\mathbf{A}\cdot\mathbf{B}\times\mathbf{C})$

Then $$\mathbf{D} = \frac{\mathbf{A}(\mathbf{B}\cdot\mathbf{C}\times\mathbf{D})}{\mathbf{A}\cdot\mathbf{B}\times\mathbf{C}} - \frac{\mathbf{B}(\mathbf{A}\cdot\mathbf{C}\times\mathbf{D})}{\mathbf{A}\cdot\mathbf{B}\times\mathbf{C}} + \frac{\mathbf{C}(\mathbf{A}\cdot\mathbf{B}\times\mathbf{D})}{\mathbf{A}\cdot\mathbf{B}\times\mathbf{C}}$$

Let $\mathbf{A} = \mathbf{a}$, $\mathbf{B} = \mathbf{b}$, $\mathbf{C} = \mathbf{c}$ and $\mathbf{D} = \mathbf{r}$. Then

$$\mathbf{r} = \frac{\mathbf{r}\cdot\mathbf{b}\times\mathbf{c}}{\mathbf{a}\cdot\mathbf{b}\times\mathbf{c}}\mathbf{a} + \frac{\mathbf{r}\cdot\mathbf{c}\times\mathbf{a}}{\mathbf{a}\cdot\mathbf{b}\times\mathbf{c}}\mathbf{b} + \frac{\mathbf{r}\cdot\mathbf{a}\times\mathbf{b}}{\mathbf{a}\cdot\mathbf{b}\times\mathbf{c}}\mathbf{c}$$

$$= \mathbf{r}\cdot\left(\frac{\mathbf{b}\times\mathbf{c}}{\mathbf{a}\cdot\mathbf{b}\times\mathbf{c}}\right)\mathbf{a} + \mathbf{r}\cdot\left(\frac{\mathbf{c}\times\mathbf{a}}{\mathbf{a}\cdot\mathbf{b}\times\mathbf{c}}\right)\mathbf{b} + \mathbf{r}\cdot\left(\frac{\mathbf{a}\times\mathbf{b}}{\mathbf{a}\cdot\mathbf{b}\times\mathbf{c}}\right)\mathbf{c}$$

$$= (\mathbf{r}\cdot\mathbf{a}')\mathbf{a} + (\mathbf{r}\cdot\mathbf{b}')\mathbf{b} + (\mathbf{r}\cdot\mathbf{c}')\mathbf{c}$$

## SUPPLEMENTARY PROBLEMS

**55.** Evaluate: (a) $\mathbf{k}\cdot(\mathbf{i}+\mathbf{j})$, (b) $(\mathbf{i}-2\mathbf{k})\cdot(\mathbf{j}+3\mathbf{k})$, (c) $(2\mathbf{i}-\mathbf{j}+3\mathbf{k})\cdot(3\mathbf{i}+2\mathbf{j}-\mathbf{k})$.
*Ans.* (a) 0 (b) $-6$ (c) 1

**56.** If $\mathbf{A} = \mathbf{i}+3\mathbf{j}-2\mathbf{k}$ and $\mathbf{B} = 4\mathbf{i}-2\mathbf{j}+4\mathbf{k}$, find:
(a) $\mathbf{A}\cdot\mathbf{B}$, (b) $A$, (c) $B$, (d) $|3\mathbf{A}+2\mathbf{B}|$, (e) $(2\mathbf{A}+\mathbf{B})\cdot(\mathbf{A}-2\mathbf{B})$.
*Ans.* (a) $-10$ (b) $\sqrt{14}$ (c) 6 (d) $\sqrt{150}$ (e) $-14$

**57.** Find the angle between: (a) $\mathbf{A} = 3\mathbf{i}+2\mathbf{j}-6\mathbf{k}$ and $\mathbf{B} = 4\mathbf{i}-3\mathbf{j}+\mathbf{k}$, (b) $\mathbf{C} = 4\mathbf{i}-2\mathbf{j}+4\mathbf{k}$ and $\mathbf{D} = 3\mathbf{i}-6\mathbf{j}-2\mathbf{k}$.
*Ans.* (a) $90^\circ$ (b) arc cos 8/21 = $67^\circ 36'$

**58.** For what values of $a$ are $\mathbf{A} = a\mathbf{i}-2\mathbf{j}+\mathbf{k}$ and $\mathbf{B} = 2a\mathbf{i}+a\mathbf{j}-4\mathbf{k}$ perpendicular? *Ans.* $a = 2, -1$

**59.** Find the acute angles which the line joining the points (1,–3,2) and (3,–5,1) makes with the coordinate axes. *Ans.* arc cos 2/3, arc cos 2/3, arc cos 1/3 or $48^\circ 12'$, $48^\circ 12'$, $70^\circ 32'$

**60.** Find the direction cosines of the line joining the points (3,2,–4) and (1,–1,2).
*Ans.* 2/7, 3/7, –6/7 or –2/7, –3/7, 6/7

**61.** Two sides of a triangle are formed by the vectors $\mathbf{A} = 3\mathbf{i}+6\mathbf{j}-2\mathbf{k}$ and $\mathbf{B} = 4\mathbf{i}-\mathbf{j}+3\mathbf{k}$. Determine the angles of the triangle. *Ans.* arc cos $7/\sqrt{75}$, arc cos $\sqrt{26}/\sqrt{75}$, $90^\circ$ or $36^\circ 4'$, $53^\circ 56'$, $90^\circ$

**62.** The diagonals of a parallelogram are given by $\mathbf{A} = 3\mathbf{i}-4\mathbf{j}-\mathbf{k}$ and $\mathbf{B} = 2\mathbf{i}+3\mathbf{j}-6\mathbf{k}$. Show that the parallelogram is a rhombus and determine the length of its sides and its angles.
*Ans.* $5\sqrt{3}/2$, arc cos 23/75, $180^\circ$ – arc cos 23/75 or 4.33, $72^\circ 8'$, $107^\circ 52'$

63. Find the projection of the vector $2\mathbf{i}-3\mathbf{j}+6\mathbf{k}$ on the vector $\mathbf{i}+2\mathbf{j}+2\mathbf{k}$. *Ans.* 8/3

64. Find the projection of the vector $4\mathbf{i}-3\mathbf{j}+\mathbf{k}$ on the line passing through the points (2,3,−1) and (−2,−4,3).
*Ans.* 1

65. If $\mathbf{A}=4\mathbf{i}-\mathbf{j}+3\mathbf{k}$ and $\mathbf{B}=-2\mathbf{i}+\mathbf{j}-2\mathbf{k}$, find a unit vector perpendicular to both **A** and **B**.
*Ans.* $\pm(\mathbf{i}-2\mathbf{j}-2\mathbf{k})/3$

66. Find the acute angle formed by two diagonals of a cube. *Ans.* arc cos 1/3 or $70°32'$

67. Find a unit vector parallel to the $xy$ plane and perpendicular to the vector $4\mathbf{i}-3\mathbf{j}+\mathbf{k}$. *Ans.* $\pm(3\mathbf{i}+4\mathbf{j})/5$

68. Show that $\mathbf{A}=(2\mathbf{i}-2\mathbf{j}+\mathbf{k})/3$, $\mathbf{B}=(\mathbf{i}+2\mathbf{j}+2\mathbf{k})/3$ and $\mathbf{C}=(2\mathbf{i}+\mathbf{j}-2\mathbf{k})/3$ are mutually orthogonal unit vectors.

69. Find the work done in moving an object along a straight line from (3,2,−1) to (2,−1,4) in a force field given by $\mathbf{F}=4\mathbf{i}-3\mathbf{j}+2\mathbf{k}$. *Ans.* 15

70. Let **F** be a constant vector force field. Show that the work done in moving an object around any closed polygon in this force field is zero.

71. Prove that an angle inscribed in a semi-circle is a right angle.

72. Let $ABCD$ be a parallelogram. Prove that $\overline{AB}^2+\overline{BC}^2+\overline{CD}^2+\overline{DA}^2 = \overline{AC}^2+\overline{BD}^2$.

73. If $ABCD$ is any quadrilateral and $P$ and $Q$ are the midpoints of its diagonals, prove that

$$\overline{AB}^2+\overline{BC}^2+\overline{CD}^2+\overline{DA}^2 = \overline{AC}^2+\overline{BD}^2+4\overline{PQ}^2$$

This is a generalization of the preceding problem.

74. (*a*) Find an equation of a plane perpendicular to a given vector **A** and distant $p$ from the origin.
(*b*) Express the equation of (*a*) in rectangular coordinates.
*Ans.* (*a*) $\mathbf{r}\cdot\mathbf{n}=p$, where $\mathbf{n}=\mathbf{A}/A$; (*b*) $A_1x+A_2y+A_3z = Ap$

75. Let $\mathbf{r}_1$ and $\mathbf{r}_2$ be unit vectors in the $xy$ plane making angles $\alpha$ and $\beta$ with the positive $x$-axis.
(*a*) Prove that $\mathbf{r}_1=\cos\alpha\,\mathbf{i}+\sin\alpha\,\mathbf{j}$, $\mathbf{r}_2=\cos\beta\,\mathbf{i}+\sin\beta\,\mathbf{j}$.
(*b*) By considering $\mathbf{r}_1\cdot\mathbf{r}_2$ prove the trigonometric formulas

$$\cos(\alpha-\beta) = \cos\alpha\cos\beta+\sin\alpha\sin\beta, \qquad \cos(\alpha+\beta) = \cos\alpha\cos\beta-\sin\alpha\sin\beta$$

76. Let **a** be the position vector of a given point $(x_1,y_1,z_1)$, and **r** the position vector of any point $(x,y,z)$. Describe the locus of **r** if (*a*) $|\mathbf{r}-\mathbf{a}|=3$, (*b*) $(\mathbf{r}-\mathbf{a})\cdot\mathbf{a}=0$, (*c*) $(\mathbf{r}-\mathbf{a})\cdot\mathbf{r}=0$.
*Ans.* (*a*) Sphere, centre at $(x_1,y_1,z_1)$ and radius 3.
(*b*) Plane perpendicular to **a** and passing through its terminal point.
(*c*) Sphere with centre at $(x_1/2,\ y_1/2,\ z_1/2)$ and radius $\frac{1}{2}\sqrt{x_1^2+y_1^2+z_1^2}$, or a sphere with **a** as diameter.

77. Given that $\mathbf{A}=3\mathbf{i}+\mathbf{j}+2\mathbf{k}$ and $\mathbf{B}=\mathbf{i}-2\mathbf{j}-4\mathbf{k}$ are the position vectors of points $P$ and $Q$ respectively.
(*a*) Find an equation for the plane passing through $Q$ and perpendicular to line $PQ$.
(*b*) What is the distance from the point (−1,1,1) to the plane?
*Ans.* (*a*) $(\mathbf{r}-\mathbf{B})\cdot(\mathbf{A}-\mathbf{B})=0$ or $2x+3y+6z=-28$; (*b*) 5

78. Evaluate each of the following:
(*a*) $2\mathbf{j}\times(3\mathbf{i}-4\mathbf{k})$, (*b*) $(\mathbf{i}+2\mathbf{j})\times\mathbf{k}$, (*c*) $(2\mathbf{i}-4\mathbf{k})\times(\mathbf{i}+2\mathbf{j})$, (*d*) $(4\mathbf{i}+\mathbf{j}-2\mathbf{k})\times(3\mathbf{i}+\mathbf{k})$, (*e*) $(2\mathbf{i}+\mathbf{j}-\mathbf{k})\times(3\mathbf{i}-2\mathbf{j}+4\mathbf{k})$.
*Ans.* (*a*) $-8\mathbf{i}-6\mathbf{k}$, (*b*) $2\mathbf{i}-\mathbf{j}$, (*c*) $8\mathbf{i}-4\mathbf{j}+4\mathbf{k}$, (*d*) $\mathbf{i}-10\mathbf{j}-3\mathbf{k}$, (*e*) $2\mathbf{i}-11\mathbf{j}-7\mathbf{k}$

79. If $\mathbf{A}=3\mathbf{i}-\mathbf{j}-2\mathbf{k}$ and $\mathbf{B}=2\mathbf{i}+3\mathbf{j}+\mathbf{k}$, find: (*a*) $|\mathbf{A}\times\mathbf{B}|$, (*b*) $(\mathbf{A}+2\mathbf{B})\times(2\mathbf{A}-\mathbf{B})$, (*c*) $|(\mathbf{A}+\mathbf{B})\times(\mathbf{A}-\mathbf{B})|$.
*Ans.* (*a*) $\sqrt{195}$, (*b*) $-25\mathbf{i}+35\mathbf{j}-55\mathbf{k}$, (*c*) $2\sqrt{195}$

80. If $\mathbf{A}=\mathbf{i}-2\mathbf{j}-3\mathbf{k}$, $\mathbf{B}=2\mathbf{i}+\mathbf{j}-\mathbf{k}$ and $\mathbf{C}=\mathbf{i}+3\mathbf{j}-2\mathbf{k}$, find:
(*a*) $|(\mathbf{A}\times\mathbf{B})\times\mathbf{C}|$, (*c*) $\mathbf{A}\cdot(\mathbf{B}\times\mathbf{C})$, (*e*) $(\mathbf{A}\times\mathbf{B})\times(\mathbf{B}\times\mathbf{C})$
(*b*) $|\mathbf{A}\times(\mathbf{B}\times\mathbf{C})|$, (*d*) $(\mathbf{A}\times\mathbf{B})\cdot\mathbf{C}$, (*f*) $(\mathbf{A}\times\mathbf{B})(\mathbf{B}\cdot\mathbf{C})$
*Ans.* (*a*) $5\sqrt{26}$, (*b*) $3\sqrt{10}$, (*c*) −20, (*d*) −20, (*e*) $-40\mathbf{i}-20\mathbf{j}+20\mathbf{k}$, ((*f*) $35\mathbf{i}-35\mathbf{j}+35\mathbf{k}$

81. Show that if $\mathbf{A}\neq\mathbf{0}$ and both of the conditions (*a*) $\mathbf{A}\cdot\mathbf{B}=\mathbf{A}\cdot\mathbf{C}$ and (*b*) $\mathbf{A}\times\mathbf{B}=\mathbf{A}\times\mathbf{C}$ hold simultaneously then $\mathbf{B}=\mathbf{C}$, but if only one of these conditions holds then $\mathbf{B}\neq\mathbf{C}$ necessarily.

82. Find the area of a parallelogram having diagonals $\mathbf{A}=3\mathbf{i}+\mathbf{j}-2\mathbf{k}$ and $\mathbf{B}=\mathbf{i}-3\mathbf{j}+4\mathbf{k}$. *Ans.* $5\sqrt{3}$

**83.** Find the area of a triangle with vertices at (3,–1,2), (1,–1,–3) and (4,–3,1). *Ans.* $\frac{1}{2}\sqrt{165}$

**84.** If $\mathbf{A} = 2\mathbf{i}+\mathbf{j}-3\mathbf{k}$ and $\mathbf{B} = \mathbf{i}-2\mathbf{j}+\mathbf{k}$, find a vector of magnitude 5 perpendicular to both **A** and **B**.
*Ans.* $\pm\dfrac{5\sqrt{3}}{3}(\mathbf{i}+\mathbf{j}+\mathbf{k})$

**85.** Use Problem 75 to derive the formulas

$$\sin(\alpha-\beta) = \sin\alpha\cos\beta - \cos\alpha\sin\beta, \qquad \sin(\alpha+\beta) = \sin\alpha\cos\beta + \cos\alpha\sin\beta$$

**86.** A force given by $\mathbf{F} = 3\mathbf{i}+2\mathbf{j}-4\mathbf{k}$ is applied at the point (1, –1, 2). Find the moment of **F** about the point (2, –1, 3). *Ans.* $2\mathbf{i}-7\mathbf{j}-2\mathbf{k}$

**87.** The angular velocity of a rotating rigid body about an axis of rotation is given by $\boldsymbol{\omega} = 4\mathbf{i}+\mathbf{j}-2\mathbf{k}$. Find the linear velocity of a point $P$ on the body whose position vector relative to a point on the axis of rotation is $2\mathbf{i}-3\mathbf{j}+\mathbf{k}$. *Ans.* $-5\mathbf{i}-8\mathbf{j}-14\mathbf{k}$

**88.** Simplify $(\mathbf{A}+\mathbf{B})\cdot(\mathbf{B}+\mathbf{C})\times(\mathbf{C}+\mathbf{A})$. *Ans.* $2\mathbf{A}\cdot\mathbf{B}\times\mathbf{C}$

**89.** Prove that $(\mathbf{A}\cdot\mathbf{B}\times\mathbf{C})(\mathbf{a}\cdot\mathbf{b}\times\mathbf{c}) = \begin{vmatrix} \mathbf{A}\cdot\mathbf{a} & \mathbf{A}\cdot\mathbf{b} & \mathbf{A}\cdot\mathbf{c} \\ \mathbf{B}\cdot\mathbf{a} & \mathbf{B}\cdot\mathbf{b} & \mathbf{B}\cdot\mathbf{c} \\ \mathbf{C}\cdot\mathbf{a} & \mathbf{C}\cdot\mathbf{b} & \mathbf{C}\cdot\mathbf{c} \end{vmatrix}$

**90.** Find the volume of the parallelepiped whose edges are represented by $\mathbf{A} = 2\mathbf{i}-3\mathbf{j}+4\mathbf{k}$, $\mathbf{B} = \mathbf{i}+2\mathbf{j}-\mathbf{k}$, $\mathbf{C} = 3\mathbf{i}-\mathbf{j}+2\mathbf{k}$. *Ans.* 7

**91.** If $\mathbf{A}\cdot\mathbf{B}\times\mathbf{C} = 0$, show that either (*a*) **A**, **B** and **C** are coplanar but no two of them are collinear, or (*b*) two of the vectors **A**, **B** and **C** are collinear, or (*c*) all of the vectors **A**, **B** and **C** are collinear.

**92.** Find the constant $a$ such that the vectors $2\mathbf{i}-\mathbf{j}+\mathbf{k}$, $\mathbf{i}+2\mathbf{j}-3\mathbf{k}$ and $3\mathbf{i}+a\mathbf{j}+5\mathbf{k}$ are coplanar. *Ans.* $a = -4$

**93.** If $\mathbf{A} = x_1\mathbf{a}+y_1\mathbf{b}+z_1\mathbf{c}$, $\mathbf{B} = x_2\mathbf{a}+y_2\mathbf{b}+z_2\mathbf{c}$ and $\mathbf{C} = x_3\mathbf{a}+y_3\mathbf{b}+z_3\mathbf{c}$, prove that

$$\mathbf{A}\cdot\mathbf{B}\times\mathbf{C} = \begin{vmatrix} x_1 & y_1 & z_1 \\ x_2 & y_2 & z_2 \\ x_3 & y_3 & z_3 \end{vmatrix} (\mathbf{a}\cdot\mathbf{b}\times\mathbf{c})$$

**94.** Prove that a necessary and sufficient condition that $\mathbf{A}\times(\mathbf{B}\times\mathbf{C}) = (\mathbf{A}\times\mathbf{B})\times\mathbf{C}$ is $(\mathbf{A}\times\mathbf{C})\times\mathbf{B} = \mathbf{0}$. Discuss the cases where $\mathbf{A}\cdot\mathbf{B} = 0$ or $\mathbf{B}\cdot\mathbf{C} = 0$.

**95.** Let points $P, Q$ and $R$ have position vectors $\mathbf{r}_1 = 3\mathbf{i}-2\mathbf{j}-\mathbf{k}$, $\mathbf{r}_2 = \mathbf{i}+3\mathbf{j}+4\mathbf{k}$ and $\mathbf{r}_3 = 2\mathbf{i}+\mathbf{j}-2\mathbf{k}$ relative to an origin $O$. Find the distance from $P$ to the plane $OQR$. *Ans.* 3

**96.** Find the shortest distance from (6,–4,4) to the line joining (2,1,2) and (3,–1,4). *Ans.* 3

**97.** Given points $P(2,1,3)$, $Q(1,2,1)$, $R(-1,-2,-2)$ and $S(1,-4,0)$, find the shortest distance between lines $PQ$ and $RS$. *Ans.* $3\sqrt{2}$

**98.** Prove that the perpendiculars from the vertices of a triangle to the opposite sides (extended if necessary) meet in a point (the *orthocentre* of the triangle).

**99.** Prove that the perpendicular bisectors of the sides of a triangle meet in a point (the *circumcentre* of the triangle).

**100.** Prove that $(\mathbf{A}\times\mathbf{B})\cdot(\mathbf{C}\times\mathbf{D}) + (\mathbf{B}\times\mathbf{C})\cdot(\mathbf{A}\times\mathbf{D}) + (\mathbf{C}\times\mathbf{A})\cdot(\mathbf{B}\times\mathbf{D}) = 0$.

**101.** Let $PQR$ be a spherical triangle whose sides $p, q, r$ are arcs of great circles. Prove the *law of cosines for spherical triangles*,

$$\cos p = \cos q \cos r + \sin q \sin r \cos P$$

with analogous formulas for $\cos q$ and $\cos r$ obtained by cyclic permutation of the letters.
[Hint: Interpret both sides of the identity $(\mathbf{A}\times\mathbf{B})\cdot(\mathbf{A}\times\mathbf{C}) = (\mathbf{B}\cdot\mathbf{C})(\mathbf{A}\cdot\mathbf{A}) - (\mathbf{A}\cdot\mathbf{C})(\mathbf{B}\cdot\mathbf{A})$.]

**102.** Find a set of vectors reciprocal to the set $2\mathbf{i}+3\mathbf{j}-\mathbf{k}$, $\mathbf{i}-\mathbf{j}-2\mathbf{k}$, $-\mathbf{i}+2\mathbf{j}+2\mathbf{k}$.

*Ans.* $\frac{2}{3}\mathbf{i}+\frac{1}{3}\mathbf{k}$, $-\frac{8}{3}\mathbf{i}+\mathbf{j}-\frac{7}{3}\mathbf{k}$, $-\frac{7}{3}\mathbf{i}+\mathbf{j}-\frac{5}{3}\mathbf{k}$

**103.** If $\mathbf{a}'=\dfrac{\mathbf{b}\times\mathbf{c}}{\mathbf{a}\cdot\mathbf{b}\times\mathbf{c}}$, $\mathbf{b}'=\dfrac{\mathbf{c}\times\mathbf{a}}{\mathbf{a}\cdot\mathbf{b}\times\mathbf{c}}$ and $\mathbf{c}'=\dfrac{\mathbf{a}\times\mathbf{b}}{\mathbf{a}\cdot\mathbf{b}\times\mathbf{c}}$, prove that

$$\mathbf{a}=\frac{\mathbf{b}'\times\mathbf{c}'}{\mathbf{a}'\cdot\mathbf{b}'\times\mathbf{c}'},\qquad \mathbf{b}=\frac{\mathbf{c}'\times\mathbf{a}'}{\mathbf{a}'\cdot\mathbf{b}'\times\mathbf{c}'},\qquad \mathbf{c}=\frac{\mathbf{a}'\times\mathbf{b}'}{\mathbf{a}'\cdot\mathbf{b}'\times\mathbf{c}'}$$

**104.** If $\mathbf{a},\mathbf{b},\mathbf{c}$ and $\mathbf{a}',\mathbf{b}',\mathbf{c}'$ are such that

$$\mathbf{a}'\cdot\mathbf{a}=\mathbf{b}'\cdot\mathbf{b}=\mathbf{c}'\cdot\mathbf{c}=1$$
$$\mathbf{a}'\cdot\mathbf{b}=\mathbf{a}'\cdot\mathbf{c}=\mathbf{b}'\cdot\mathbf{a}=\mathbf{b}'\cdot\mathbf{c}=\mathbf{c}'\cdot\mathbf{a}=\mathbf{c}'\cdot\mathbf{b}=0$$

prove that it necessarily follows that

$$\mathbf{a}'=\frac{\mathbf{b}\times\mathbf{c}}{\mathbf{a}\cdot\mathbf{b}\times\mathbf{c}},\qquad \mathbf{b}'=\frac{\mathbf{c}\times\mathbf{a}}{\mathbf{a}\cdot\mathbf{b}\times\mathbf{c}},\qquad \mathbf{c}'=\frac{\mathbf{a}\times\mathbf{b}}{\mathbf{a}\cdot\mathbf{b}\times\mathbf{c}}$$

**105.** Prove that the only right-handed self-reciprocal sets of vectors are the unit vectors $\mathbf{i},\mathbf{j},\mathbf{k}$.

**106.** Prove that there is one and only one set of vectors reciprocal to a given set of non-coplanar vectors $\mathbf{a},\mathbf{b},\mathbf{c}$.

# Chapter 3 VECTOR DIFFERENTIATION

**ORDINARY DERIVATIVES OF VECTORS**. Let $\mathbf{R}(u)$ be a vector depending on a single scalar variable $u$. Then

$$\frac{\Delta\mathbf{R}}{\Delta u} = \frac{\mathbf{R}(u+\Delta u) - \mathbf{R}(u)}{\Delta u}$$

where $\Delta u$ denotes an increment in $u$ (see adjoining figure).

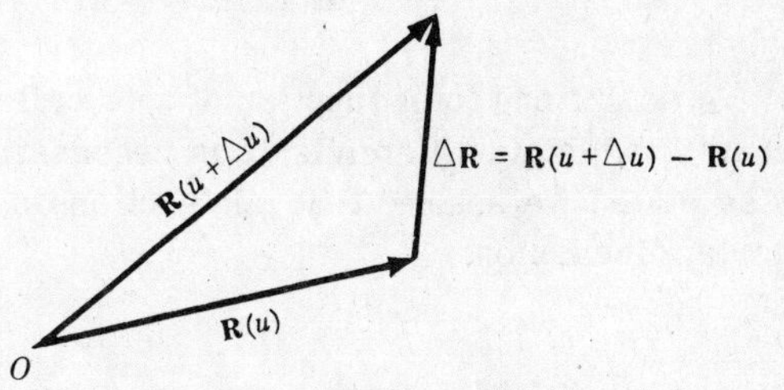


The ordinary derivative of the vector $\mathbf{R}(u)$ with respect to the scalar $u$ is given by

$$\frac{d\mathbf{R}}{du} = \lim_{\Delta u\to 0}\frac{\Delta\mathbf{R}}{\Delta u} = \lim_{\Delta u\to 0}\frac{\mathbf{R}(u+\Delta u) - \mathbf{R}(u)}{\Delta u}$$

if the limit exists.

Since $\dfrac{d\mathbf{R}}{du}$ is itself a vector depending on $u$, we can consider its derivative with respect to $u$. If this derivative exists it is denoted by $\dfrac{d^2\mathbf{R}}{du^2}$. In like manner higher order derivatives are described.

**SPACE CURVES.** If in particular $\mathbf{R}(u)$ is the position vector $\mathbf{r}(u)$ joining the origin $O$ of a coordinate system and any point $(x,y,z)$, then

$$\mathbf{r}(u) = x(u)\mathbf{i} + y(u)\mathbf{j} + z(u)\mathbf{k}$$

and specification of the vector function $\mathbf{r}(u)$ defines $x, y$ and $z$ as functions of $u$.

As $u$ changes, the terminal point of $\mathbf{r}$ describes a *space curve* having parametric equations

$$x = x(u), \quad y = y(u), \quad z = z(u)$$

Then $\dfrac{\Delta\mathbf{r}}{\Delta u} = \dfrac{\mathbf{r}(u+\Delta u) - \mathbf{r}(u)}{\Delta u}$ is a vector in the direction of $\Delta\mathbf{r}$ (see adjacent figure). If $\lim\limits_{\Delta u\to 0}\dfrac{\Delta\mathbf{r}}{\Delta u} = \dfrac{d\mathbf{r}}{du}$ exists, the limit will be a vector in the direction of the tangent to the space curve at $(x,y,z)$ and is given by

$$\frac{d\mathbf{r}}{du} = \frac{dx}{du}\mathbf{i} + \frac{dy}{du}\mathbf{j} + \frac{dz}{du}\mathbf{k}$$

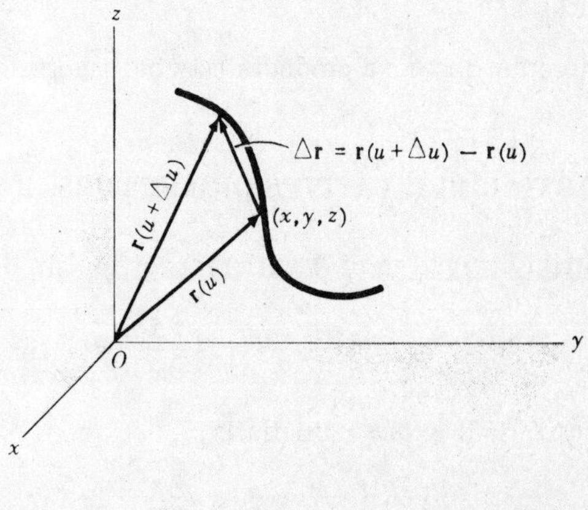


If $u$ is the time $t$, $\dfrac{d\mathbf{r}}{dt}$ represents the *velocity* $\mathbf{v}$ with which the terminal point of $\mathbf{r}$ describes the curve. Similarly, $\dfrac{d\mathbf{v}}{dt} = \dfrac{d^2\mathbf{r}}{dt^2}$ represents its *acceleration* $\mathbf{a}$ along the curve.

**CONTINUITY AND DIFFERENTIABILITY.** A scalar function $\phi(u)$ is called *continuous* at $u$ if $\lim_{\Delta u \to 0} \phi(u+\Delta u) = \phi(u)$. Equivalently, $\phi(u)$ is continuous at $u$ if for each positive number $\epsilon$ we can find some positive number $\delta$ such that

$$|\phi(u+\Delta u) - \phi(u)| < \epsilon \quad \text{whenever} \quad |\Delta u| < \delta.$$

A vector function $\mathbf{R}(u) = R_1(u)\mathbf{i} + R_2(u)\mathbf{j} + R_3(u)\mathbf{k}$ is called *continuous* at $u$ if the three scalar functions $R_1(u)$, $R_2(u)$ and $R_3(u)$ are continuous at $u$ or if $\lim_{\Delta u \to 0} \mathbf{R}(u+\Delta u) = \mathbf{R}(u)$. Equivalently, $\mathbf{R}(u)$ is continuous at $u$ if for each positive number $\epsilon$ we can find some positive number $\delta$ such that

$$|\mathbf{R}(u+\Delta u) - \mathbf{R}(u)| < \epsilon \quad \text{whenever} \quad |\Delta u| < \delta.$$

A scalar or vector function of $u$ is called *differentiable of order n* if its $n$th derivative exists. A function which is differentiable is necessarily continuous but the converse is not true. Unless otherwise stated we assume that all functions considered are differentiable to any order needed in a particular discussion.

**DIFFERENTIATION FORMULAE.** If $\mathbf{A}$, $\mathbf{B}$ and $\mathbf{C}$ are differentiable vector functions of a scalar $u$, and $\phi$ is a differentiable scalar function of $u$, then

1. $$\frac{d}{du}(\mathbf{A}+\mathbf{B}) = \frac{d\mathbf{A}}{du} + \frac{d\mathbf{B}}{du}$$

2. $$\frac{d}{du}(\mathbf{A}\cdot\mathbf{B}) = \mathbf{A}\cdot\frac{d\mathbf{B}}{du} + \frac{d\mathbf{A}}{du}\cdot\mathbf{B}$$

3. $$\frac{d}{du}(\mathbf{A}\times\mathbf{B}) = \mathbf{A}\times\frac{d\mathbf{B}}{du} + \frac{d\mathbf{A}}{du}\times\mathbf{B}$$

4. $$\frac{d}{du}(\phi\mathbf{A}) = \phi\frac{d\mathbf{A}}{du} + \frac{d\phi}{du}\mathbf{A}$$

5. $$\frac{d}{du}(\mathbf{A}\cdot\mathbf{B}\times\mathbf{C}) = \mathbf{A}\cdot\mathbf{B}\times\frac{d\mathbf{C}}{du} + \mathbf{A}\cdot\frac{d\mathbf{B}}{du}\times\mathbf{C} + \frac{d\mathbf{A}}{du}\cdot\mathbf{B}\times\mathbf{C}$$

6. $$\frac{d}{du}\{\mathbf{A}\times(\mathbf{B}\times\mathbf{C})\} = \mathbf{A}\times(\mathbf{B}\times\frac{d\mathbf{C}}{du}) + \mathbf{A}\times(\frac{d\mathbf{B}}{du}\times\mathbf{C}) + \frac{d\mathbf{A}}{du}\times(\mathbf{B}\times\mathbf{C})$$

The order in these products may be important.

**PARTIAL DERIVATIVES OF VECTORS.** If $\mathbf{A}$ is a vector depending on more than one scalar variable, say $x, y, z$ for example, then we write $\mathbf{A} = \mathbf{A}(x,y,z)$. The partial derivative of $\mathbf{A}$ with respect to $x$ is defined as

$$\frac{\partial\mathbf{A}}{\partial x} = \lim_{\Delta x \to 0} \frac{\mathbf{A}(x+\Delta x, y, z) - \mathbf{A}(x,y,z)}{\Delta x}$$

if this limit exists. Similarly,

$$\frac{\partial\mathbf{A}}{\partial y} = \lim_{\Delta y \to 0} \frac{\mathbf{A}(x, y+\Delta y, z) - \mathbf{A}(x,y,z)}{\Delta y}$$

$$\frac{\partial\mathbf{A}}{\partial z} = \lim_{\Delta z \to 0} \frac{\mathbf{A}(x,y, z+\Delta z) - \mathbf{A}(x,y,z)}{\Delta z}$$

are the partial derivatives of **A** with respect to $y$ and $z$ respectively if these limits exist.

The remarks on continuity and differentiability for functions of one variable can be extended to functions of two or more variables. For example, $\phi(x,y)$ is called continuous at $(x,y)$ if $\lim_{\substack{\Delta x \to 0 \\ \Delta y \to 0}} \phi(x+\Delta x, y+\Delta y) = \phi(x,y)$, or if for each positive number $\epsilon$ we can find some positive number $\delta$ such that $|\phi(x+\Delta x, y+\Delta y) - \phi(x,y)| < \epsilon$ whenever $|\Delta x| < \delta$ and $|\Delta y| < \delta$. Similar definitions hold for vector functions.

For functions of two or more variables we use the term *differentiable* to mean that the function has continuous first partial derivatives. (The term is used by others in a slightly weaker sense.)

Higher derivatives can be defined as in the calculus. Thus, for example,

$$\frac{\partial^2 \mathbf{A}}{\partial x^2} = \frac{\partial}{\partial x}\left(\frac{\partial \mathbf{A}}{\partial x}\right), \quad \frac{\partial^2 \mathbf{A}}{\partial y^2} = \frac{\partial}{\partial y}\left(\frac{\partial \mathbf{A}}{\partial y}\right), \quad \frac{\partial^2 \mathbf{A}}{\partial z^2} = \frac{\partial}{\partial z}\left(\frac{\partial \mathbf{A}}{\partial z}\right)$$

$$\frac{\partial^2 \mathbf{A}}{\partial x\,\partial y} = \frac{\partial}{\partial x}\left(\frac{\partial \mathbf{A}}{\partial y}\right), \quad \frac{\partial^2 \mathbf{A}}{\partial y\,\partial x} = \frac{\partial}{\partial y}\left(\frac{\partial \mathbf{A}}{\partial x}\right), \quad \frac{\partial^3 \mathbf{A}}{\partial x\,\partial z^2} = \frac{\partial}{\partial x}\left(\frac{\partial^2 \mathbf{A}}{\partial z^2}\right)$$

If **A** has continuous partial derivatives of the second order at least, then $\dfrac{\partial^2 \mathbf{A}}{\partial x\,\partial y} = \dfrac{\partial^2 \mathbf{A}}{\partial y\,\partial x}$, i.e. the order of differentiation does not matter.

Rules for partial differentiation of vectors are similar to those used in elementary calculus for scalar functions. Thus if **A** and **B** are functions of $x,y,z$ then, for example,

1. $\dfrac{\partial}{\partial x}(\mathbf{A}\cdot\mathbf{B}) = \mathbf{A}\cdot\dfrac{\partial \mathbf{B}}{\partial x} + \dfrac{\partial \mathbf{A}}{\partial x}\cdot\mathbf{B}$

2. $\dfrac{\partial}{\partial x}(\mathbf{A}\times\mathbf{B}) = \mathbf{A}\times\dfrac{\partial \mathbf{B}}{\partial x} + \dfrac{\partial \mathbf{A}}{\partial x}\times\mathbf{B}$

3. $\dfrac{\partial^2}{\partial y\,\partial x}(\mathbf{A}\cdot\mathbf{B}) = \dfrac{\partial}{\partial y}\left\{\dfrac{\partial}{\partial x}(\mathbf{A}\cdot\mathbf{B})\right\} = \dfrac{\partial}{\partial y}\left\{\mathbf{A}\cdot\dfrac{\partial \mathbf{B}}{\partial x} + \dfrac{\partial \mathbf{A}}{\partial x}\cdot\mathbf{B}\right\}$

$$= \mathbf{A}\cdot\frac{\partial^2 \mathbf{B}}{\partial y\,\partial x} + \frac{\partial \mathbf{A}}{\partial y}\cdot\frac{\partial \mathbf{B}}{\partial x} + \frac{\partial \mathbf{A}}{\partial x}\cdot\frac{\partial \mathbf{B}}{\partial y} + \frac{\partial^2 \mathbf{A}}{\partial y\,\partial x}\cdot\mathbf{B}, \quad \text{etc.}$$

**DIFFERENTIALS OF VECTORS** follow rules similar to those of elementary calculus. For example,

1. If $\mathbf{A} = A_1\mathbf{i} + A_2\mathbf{j} + A_3\mathbf{k}$, then $d\mathbf{A} = dA_1\mathbf{i} + dA_2\mathbf{j} + dA_3\mathbf{k}$

2. $d(\mathbf{A}\cdot\mathbf{B}) = \mathbf{A}\cdot d\mathbf{B} + d\mathbf{A}\cdot\mathbf{B}$

3. $d(\mathbf{A}\times\mathbf{B}) = \mathbf{A}\times d\mathbf{B} + d\mathbf{A}\times\mathbf{B}$

4. If $\mathbf{A} = \mathbf{A}(x,y,z)$, then $d\mathbf{A} = \dfrac{\partial \mathbf{A}}{\partial x}dx + \dfrac{\partial \mathbf{A}}{\partial y}dy + \dfrac{\partial \mathbf{A}}{\partial z}dz$, etc.

**DIFFERENTIAL GEOMETRY** involves a study of space curves and surfaces. If $C$ is a space curve defined by the function $\mathbf{r}(u)$, then we have seen that $\dfrac{d\mathbf{r}}{du}$ is a vector in the direction of the tangent to $C$. If the scalar $u$ is taken as the arc length $s$ measured from some fixed point on $C$, then $\dfrac{d\mathbf{r}}{ds}$ is a unit tangent vector to $C$ and is denoted by **T** (see diagram below). The rate at which **T** changes with respect to $s$ is a measure of the curvature of $C$ and is given by $\frac{d\mathbf{T}}{ds}$. The direction of $\frac{d\mathbf{T}}{ds}$ at any given point on $C$ is normal to the curve at that point (see Problem 9). If **N** is a unit vector in this normal direction, it is called the *principal normal* to the curve. Then $\frac{d\mathbf{T}}{ds} = \kappa\mathbf{N}$, where $\kappa$ is called the *curvature* of $C$ at the specified point. The quantity $\rho = 1/\kappa$ is called the *radius of curvature*.

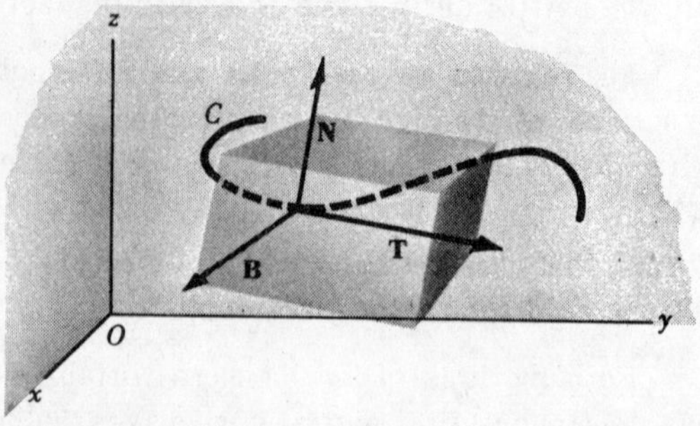


A unit vector **B** perpendicular to the plane of **T** and **N** and such that $\mathbf{B} = \mathbf{T}\times\mathbf{N}$, is called the *binormal* to the curve. It follows that directions **T**, **N**, **B** form a localized right-handed rectangular coordinate system at any specified point of $C$. This coordinate system is called the *trihedral* or *triad* at the point. As $s$ changes, the coordinate system moves and is known as the *moving trihedral.*

A set of relations involving derivatives of the fundamental vectors **T**, **N** and **B** is known collectively as the *Frenet-Serret formulae* given by

$$\frac{d\mathbf{T}}{ds} = \kappa\mathbf{N}, \qquad \frac{d\mathbf{N}}{ds} = \tau\mathbf{B} - \kappa\mathbf{T}, \qquad \frac{d\mathbf{B}}{ds} = -\tau\mathbf{N}$$

where $\tau$ is a scalar called the *torsion*. The quantity $\sigma = 1/\tau$ is called the *radius of torsion.*

The *osculating plane* to a curve at a point $P$ is the plane containing the tangent and principal normal at $P$. The *normal plane* is the plane through $P$ perpendicular to the tangent. The *rectifying plane* is the plane through $P$ which is perpendicular to the principal normal.

**MECHANICS** often includes a study of the motion of particles along curves, this study being known as *kinematics*. In this connection some of the results of differential geometry can be of value.

A study of forces on moving objects is considered in *dynamics*. Fundamental to this study is Newton's famous law which states that if **F** is the net force acting on an object of mass $m$ moving with velocity **v**, then

$$\mathbf{F} = \frac{d}{dt}(m\mathbf{v})$$

where $m\mathbf{v}$ is the momentum of the object. If $m$ is constant this becomes $\mathbf{F} = m\frac{d\mathbf{v}}{dt} = m\mathbf{a}$, where **a** is the acceleration of the object.

## SOLVED PROBLEMS

**1.** If $\mathbf{R}(u) = x(u)\mathbf{i} + y(u)\mathbf{j} + z(u)\mathbf{k}$, where $x, y$ and $z$ are differentiable functions of a scalar $u$, prove that $\dfrac{d\mathbf{R}}{du} = \dfrac{dx}{du}\mathbf{i} + \dfrac{dy}{du}\mathbf{j} + \dfrac{dz}{du}\mathbf{k}$.

$$\frac{d\mathbf{R}}{du} = \lim_{\Delta u \to 0} \frac{\mathbf{R}(u+\Delta u) - \mathbf{R}(u)}{\Delta u}$$

$$= \lim_{\Delta u \to 0} \frac{[x(u+\Delta u)\mathbf{i} + y(u+\Delta u)\mathbf{j} + z(u+\Delta u)\mathbf{k}] - [x(u)\mathbf{i} + y(u)\mathbf{j} + z(u)\mathbf{k}]}{\Delta u}$$

$$= \lim_{\Delta u \to 0} \frac{x(u+\Delta u) - x(u)}{\Delta u}\mathbf{i} + \frac{y(u+\Delta u) - y(u)}{\Delta u}\mathbf{j} + \frac{z(u+\Delta u) - z(u)}{\Delta u}\mathbf{k}$$

$$= \frac{dx}{du}\mathbf{i} + \frac{dy}{du}\mathbf{j} + \frac{dz}{du}\mathbf{k}$$

**2.** Given $\mathbf{R} = \sin t\,\mathbf{i} + \cos t\,\mathbf{j} + t\mathbf{k}$, find (a) $\dfrac{d\mathbf{R}}{dt}$, (b) $\dfrac{d^2\mathbf{R}}{dt^2}$, (c) $\left|\dfrac{d\mathbf{R}}{dt}\right|$, (d) $\left|\dfrac{d^2\mathbf{R}}{dt^2}\right|$.

(a) $\dfrac{d\mathbf{R}}{dt} = \dfrac{d}{dt}(\sin t)\mathbf{i} + \dfrac{d}{dt}(\cos t)\mathbf{j} + \dfrac{d}{dt}(t)\mathbf{k} = \cos t\,\mathbf{i} - \sin t\,\mathbf{j} + \mathbf{k}$

(b) $\dfrac{d^2\mathbf{R}}{dt^2} = \dfrac{d}{dt}\left(\dfrac{d\mathbf{R}}{dt}\right) = \dfrac{d}{dt}(\cos t)\mathbf{i} - \dfrac{d}{dt}(\sin t)\mathbf{j} + \dfrac{d}{dt}(1)\mathbf{k} = -\sin t\,\mathbf{i} - \cos t\,\mathbf{j}$

(c) $\left|\dfrac{d\mathbf{R}}{dt}\right| = \sqrt{(\cos t)^2 + (-\sin t)^2 + (1)^2} = \sqrt{2}$

(d) $\left|\dfrac{d^2\mathbf{R}}{dt^2}\right| = \sqrt{(-\sin t)^2 + (-\cos t)^2} = 1$

**3.** A particle moves along a curve whose parametric equations are $x = e^{-t}$, $y = 2\cos 3t$, $z = 2\sin 3t$, where $t$ is the time.
(a) Determine its velocity and acceleration at any time.
(b) Find the magnitudes of the velocity and acceleration at $t = 0$.

(a) The position vector $\mathbf{r}$ of the particle is $\mathbf{r} = x\mathbf{i} + y\mathbf{j} + z\mathbf{k} = e^{-t}\mathbf{i} + 2\cos 3t\,\mathbf{j} + 2\sin 3t\,\mathbf{k}$.

Then the velocity is $\mathbf{v} = \dfrac{d\mathbf{r}}{dt} = -e^{-t}\mathbf{i} - 6\sin 3t\,\mathbf{j} + 6\cos 3t\,\mathbf{k}$

and the acceleration is $\mathbf{a} = \dfrac{d^2\mathbf{r}}{dt^2} = e^{-t}\mathbf{i} - 18\cos 3t\,\mathbf{j} - 18\sin 3t\,\mathbf{k}$

(b) At $t = 0$, $\dfrac{d\mathbf{r}}{dt} = -\mathbf{i} + 6\mathbf{k}$ and $\dfrac{d^2\mathbf{r}}{dt^2} = \mathbf{i} - 18\mathbf{j}$. Then

magnitude of velocity at $t = 0$ is $\sqrt{(-1)^2 + (6)^2} = \sqrt{37}$

magnitude of acceleration at $t = 0$ is $\sqrt{(1)^2 + (-18)^2} = \sqrt{325}$.

**4.** A particle moves along the curve $x = 2t^2$, $y = t^2 - 4t$, $z = 3t - 5$, where $t$ is the time. Find the components of its velocity and acceleration at time $t = 1$ in the direction $\mathbf{i} - 3\mathbf{j} + 2\mathbf{k}$.

$$\text{Velocity} = \frac{d\mathbf{r}}{dt} = \frac{d}{dt}[2t^2\mathbf{i} + (t^2-4t)\mathbf{j} + (3t-5)\mathbf{k}]$$
$$= 4t\,\mathbf{i} + (2t-4)\mathbf{j} + 3\mathbf{k} = 4\mathbf{i} - 2\mathbf{j} + 3\mathbf{k} \quad \text{at } t = 1.$$

Unit vector in direction $\mathbf{i} - 3\mathbf{j} + 2\mathbf{k}$ is $\dfrac{\mathbf{i} - 3\mathbf{j} + 2\mathbf{k}}{\sqrt{(1)^2+(-3)^2+(2)^2}} = \dfrac{\mathbf{i} - 3\mathbf{j} + 2\mathbf{k}}{\sqrt{14}}$.

Then the component of the velocity in the given direction is

$$\frac{(4\mathbf{i} - 2\mathbf{j} + 3\mathbf{k})\cdot(\mathbf{i} - 3\mathbf{j} + 2\mathbf{k})}{\sqrt{14}} = \frac{(4)(1) + (-2)(-3) + (3)(2)}{\sqrt{14}} = \frac{16}{\sqrt{14}} = \frac{8\sqrt{14}}{7}$$

$$\text{Acceleration} = \frac{d^2\mathbf{r}}{dt^2} = \frac{d}{dt}\left(\frac{d\mathbf{r}}{dt}\right) = \frac{d}{dt}[4t\,\mathbf{i} + (2t-4)\mathbf{j} + 3\mathbf{k}] = 4\mathbf{i} + 2\mathbf{j} + 0\mathbf{k}.$$

Then the component of the acceleration in the given direction is

$$\frac{(4\mathbf{i} + 2\mathbf{j} + 0\mathbf{k})\cdot(\mathbf{i} - 3\mathbf{j} + 2\mathbf{k})}{\sqrt{14}} = \frac{(4)(1) + (2)(-3) + (0)(2)}{\sqrt{14}} = \frac{-2}{\sqrt{14}} = \frac{-\sqrt{14}}{7}$$

**5.** A curve $C$ is defined by parametric equations $x = x(s)$, $y = y(s)$, $z = z(s)$, where $s$ is the arc length of $C$ measured from a fixed point on $C$. If $\mathbf{r}$ is the position vector of any point on $C$, show that $d\mathbf{r}/ds$ is a unit vector tangent to $C$.

The vector $\dfrac{d\mathbf{r}}{ds} = \dfrac{d}{ds}(x\mathbf{i} + y\mathbf{j} + z\mathbf{k}) = \dfrac{dx}{ds}\mathbf{i} + \dfrac{dy}{ds}\mathbf{j} + \dfrac{dz}{ds}\mathbf{k}$ is tangent to the curve $x = x(s)$, $y = y(s)$, $z = z(s)$. To show that it has unit magnitude we note that

$$\left|\frac{d\mathbf{r}}{ds}\right| = \sqrt{\left(\frac{dx}{ds}\right)^2 + \left(\frac{dy}{ds}\right)^2 + \left(\frac{dz}{ds}\right)^2} = \sqrt{\frac{(dx)^2 + (dy)^2 + (dz)^2}{(ds)^2}} = 1$$

since $(ds)^2 = (dx)^2 + (dy)^2 + (dz)^2$ from the calculus.

**6.** (*a*) Find the unit tangent vector to any point on the curve $x = t^2+1$, $y = 4t-3$, $z = 2t^2-6t$.
(*b*) Determine the unit tangent at the point where $t = 2$.

(*a*) A tangent vector to the curve at any point is

$$\frac{d\mathbf{r}}{dt} = \frac{d}{dt}[(t^2+1)\mathbf{i} + (4t-3)\mathbf{j} + (2t^2-6t)\mathbf{k}] = 2t\,\mathbf{i} + 4\mathbf{j} + (4t-6)\mathbf{k}$$

The magnitude of the vector is $\left|\dfrac{d\mathbf{r}}{dt}\right| = \sqrt{(2t)^2 + (4)^2 + (4t-6)^2}$.

Then the required unit tangent vector is $\mathbf{T} = \dfrac{2t\,\mathbf{i} + 4\mathbf{j} + (4t-6)\mathbf{k}}{\sqrt{(2t)^2 + (4)^2 + (4t-6)^2}}$

Note that since $\left|\dfrac{d\mathbf{r}}{dt}\right| = \dfrac{ds}{dt}$, $\mathbf{T} = \dfrac{d\mathbf{r}/dt}{ds/dt} = \dfrac{d\mathbf{r}}{ds}$.

(*b*) At $t = 2$, the unit tangent vector is $\mathbf{T} = \dfrac{4\mathbf{i} + 4\mathbf{j} + 2\mathbf{k}}{\sqrt{(4)^2 + (4)^2 + (2)^2}} = \dfrac{2}{3}\mathbf{i} + \dfrac{2}{3}\mathbf{j} + \dfrac{1}{3}\mathbf{k}$.

**7.** If $\mathbf{A}$ and $\mathbf{B}$ are differentiable functions of a scalar $u$, prove:

(*a*) $\dfrac{d}{du}(\mathbf{A}\cdot\mathbf{B}) = \mathbf{A}\cdot\dfrac{d\mathbf{B}}{du} + \dfrac{d\mathbf{A}}{du}\cdot\mathbf{B}$, (*b*) $\dfrac{d}{du}(\mathbf{A}\times\mathbf{B}) = \mathbf{A}\times\dfrac{d\mathbf{B}}{du} + \dfrac{d\mathbf{A}}{du}\times\mathbf{B}$

(a) $$\frac{d}{du}(\mathbf{A}\cdot\mathbf{B}) = \lim_{\Delta u\to 0}\frac{(\mathbf{A}+\Delta\mathbf{A})\cdot(\mathbf{B}+\Delta\mathbf{B}) - \mathbf{A}\cdot\mathbf{B}}{\Delta u}$$

$$= \lim_{\Delta u\to 0}\frac{\mathbf{A}\cdot\Delta\mathbf{B} + \Delta\mathbf{A}\cdot\mathbf{B} + \Delta\mathbf{A}\cdot\Delta\mathbf{B}}{\Delta u}$$

$$= \lim_{\Delta u\to 0}\ \mathbf{A}\cdot\frac{\Delta\mathbf{B}}{\Delta u} + \frac{\Delta\mathbf{A}}{\Delta u}\cdot\mathbf{B} + \frac{\Delta\mathbf{A}}{\Delta u}\cdot\Delta\mathbf{B} = \mathbf{A}\cdot\frac{d\mathbf{B}}{du} + \frac{d\mathbf{A}}{du}\cdot\mathbf{B}$$

*Another Method.* Let $\mathbf{A} = A_1\mathbf{i} + A_2\mathbf{j} + A_3\mathbf{k}$, $\mathbf{B} = B_1\mathbf{i} + B_2\mathbf{j} + B_3\mathbf{k}$. Then

$$\frac{d}{du}(\mathbf{A}\cdot\mathbf{B}) = \frac{d}{du}(A_1B_1 + A_2B_2 + A_3B_3)$$

$$= (A_1\frac{dB_1}{du} + A_2\frac{dB_2}{du} + A_3\frac{dB_3}{du}) + (\frac{dA_1}{du}B_1 + \frac{dA_2}{du}B_2 + \frac{dA_3}{du}B_3) = \mathbf{A}\cdot\frac{d\mathbf{B}}{du} + \frac{d\mathbf{A}}{du}\cdot\mathbf{B}$$

(b) $$\frac{d}{du}(\mathbf{A}\times\mathbf{B}) = \lim_{\Delta u\to 0}\frac{(\mathbf{A}+\Delta\mathbf{A})\times(\mathbf{B}+\Delta\mathbf{B}) - \mathbf{A}\times\mathbf{B}}{\Delta u}$$

$$= \lim_{\Delta u\to 0}\frac{\mathbf{A}\times\Delta\mathbf{B} + \Delta\mathbf{A}\times\mathbf{B} + \Delta\mathbf{A}\times\Delta\mathbf{B}}{\Delta u}$$

$$= \lim_{\Delta u\to 0}\ \mathbf{A}\times\frac{\Delta\mathbf{B}}{\Delta u} + \frac{\Delta\mathbf{A}}{\Delta u}\times\mathbf{B} + \frac{\Delta\mathbf{A}}{\Delta u}\times\Delta\mathbf{B} = \mathbf{A}\times\frac{d\mathbf{B}}{du} + \frac{d\mathbf{A}}{du}\times\mathbf{B}$$

*Another Method.*

$$\frac{d}{du}(\mathbf{A}\times\mathbf{B}) = \frac{d}{du}\begin{vmatrix}\mathbf{i} & \mathbf{j} & \mathbf{k}\\ A_1 & A_2 & A_3\\ B_1 & B_2 & B_3\end{vmatrix}$$

Using a theorem on differentiation of a determinant, this becomes

$$\begin{vmatrix}\mathbf{i} & \mathbf{j} & \mathbf{k}\\ A_1 & A_2 & A_3\\ \frac{dB_1}{du} & \frac{dB_2}{du} & \frac{dB_3}{du}\end{vmatrix} + \begin{vmatrix}\mathbf{i} & \mathbf{j} & \mathbf{k}\\ \frac{dA_1}{du} & \frac{dA_2}{du} & \frac{dA_3}{du}\\ B_1 & B_2 & B_3\end{vmatrix} = \mathbf{A}\times\frac{d\mathbf{B}}{du} + \frac{d\mathbf{A}}{du}\times\mathbf{B}$$

8. If $\mathbf{A} = 5t^2\mathbf{i} + t\mathbf{j} - t^3\mathbf{k}$ and $\mathbf{B} = \sin t\ \mathbf{i} - \cos t\ \mathbf{j}$, find (a) $\frac{d}{dt}(\mathbf{A}\cdot\mathbf{B})$, (b) $\frac{d}{dt}(\mathbf{A}\times\mathbf{B})$, (c) $\frac{d}{dt}(\mathbf{A}\cdot\mathbf{A})$.

(a) $$\frac{d}{dt}(\mathbf{A}\cdot\mathbf{B}) = \mathbf{A}\cdot\frac{d\mathbf{B}}{dt} + \frac{d\mathbf{A}}{dt}\cdot\mathbf{B}$$

$$= (5t^2\mathbf{i} + t\mathbf{j} - t^3\mathbf{k})\cdot(\cos t\ \mathbf{i} + \sin t\ \mathbf{j}) + (10t\,\mathbf{i} + \mathbf{j} - 3t^2\mathbf{k})\cdot(\sin t\ \mathbf{i} - \cos t\ \mathbf{j})$$

$$= 5t^2\cos t + t\sin t + 10t\sin t - \cos t = (5t^2-1)\cos t + 11t\sin t$$

*Another Method.* $\mathbf{A}\cdot\mathbf{B} = 5t^2\sin t - t\cos t$. Then

$$\frac{d}{dt}(\mathbf{A}\cdot\mathbf{B}) = \frac{d}{dt}(5t^2\sin t - t\cos t) = 5t^2\cos t + 10t\sin t + t\sin t - \cos t$$

$$= (5t^2-1)\cos t + 11t\sin t$$

(b) $$\frac{d}{dt}(\mathbf{A}\times\mathbf{B}) = \mathbf{A}\times\frac{d\mathbf{B}}{dt} + \frac{d\mathbf{A}}{dt}\times\mathbf{B} = \begin{vmatrix}\mathbf{i} & \mathbf{j} & \mathbf{k}\\ 5t^2 & t & -t^3\\ \cos t & \sin t & 0\end{vmatrix} + \begin{vmatrix}\mathbf{i} & \mathbf{j} & \mathbf{k}\\ 10t & 1 & -3t^2\\ \sin t & -\cos t & 0\end{vmatrix}$$

$$= [t^3 \sin t\,\mathbf{i} - t^3 \cos t\,\mathbf{j} + (5t^2 \sin t - t\cos t)\mathbf{k}]$$
$$+ [-3t^2 \cos t\,\mathbf{i} - 3t^2 \sin t\,\mathbf{j} + (-10t \cos t - \sin t)\mathbf{k}]$$
$$= (t^3 \sin t - 3t^2 \cos t)\mathbf{i} - (t^3 \cos t + 3t^2 \sin t)\mathbf{j} + (5t^2 \sin t - \sin t - 11t \cos t)\mathbf{k}$$

*Another Method.*

$$\mathbf{A}\times\mathbf{B} = \begin{vmatrix} \mathbf{i} & \mathbf{j} & \mathbf{k} \\ 5t^2 & t & -t^3 \\ \sin t & -\cos t & 0 \end{vmatrix} = -t^3 \cos t\,\mathbf{i} - t^3 \sin t\,\mathbf{j} + (-5t^2 \cos t - t \sin t)\mathbf{k}$$

Then $\dfrac{d}{dt}(\mathbf{A}\times\mathbf{B}) = (t^3 \sin t - 3t^2 \cos t)\mathbf{i} - (t^3 \cos t + 3t^2 \sin t)\mathbf{j} + (5t^2 \sin t - 11t \cos t - \sin t)\mathbf{k}$

(c) $\dfrac{d}{dt}(\mathbf{A}\cdot\mathbf{A}) = \mathbf{A}\cdot\dfrac{d\mathbf{A}}{dt} + \dfrac{d\mathbf{A}}{dt}\cdot\mathbf{A} = 2\mathbf{A}\cdot\dfrac{d\mathbf{A}}{dt}$

$$= 2(5t^2\mathbf{i} + t\mathbf{j} - t^3\mathbf{k})\cdot(10t\,\mathbf{i} + \mathbf{j} - 3t^2\mathbf{k}) = 100t^3 + 2t + 6t^5$$

*Another Method.* $\mathbf{A}\cdot\mathbf{A} = (5t^2)^2 + (t)^2 + (-t^3)^2 = 25t^4 + t^2 + t^6$

Then $\dfrac{d}{dt}(25t^4 + t^2 + t^6) = 100t^3 + 2t + 6t^5$.

**9.** If $\mathbf{A}$ has constant magnitude show that $\mathbf{A}$ and $d\mathbf{A}/dt$ are perpendicular provided $|d\mathbf{A}/dt| \neq 0$.

Since $\mathbf{A}$ has constant magnitude, $\mathbf{A}\cdot\mathbf{A}$ = constant.

Then $\dfrac{d}{dt}(\mathbf{A}\cdot\mathbf{A}) = \mathbf{A}\cdot\dfrac{d\mathbf{A}}{dt} + \dfrac{d\mathbf{A}}{dt}\cdot\mathbf{A} = 2\mathbf{A}\cdot\dfrac{d\mathbf{A}}{dt} = 0$.

Thus $\mathbf{A}\cdot\dfrac{d\mathbf{A}}{dt} = 0$ and $\mathbf{A}$ is perpendicular to $\dfrac{d\mathbf{A}}{dt}$ provided $\left|\dfrac{d\mathbf{A}}{dt}\right| \neq 0$.

**10.** Prove $\dfrac{d}{du}(\mathbf{A}\cdot\mathbf{B}\times\mathbf{C}) = \mathbf{A}\cdot\mathbf{B}\times\dfrac{d\mathbf{C}}{du} + \mathbf{A}\cdot\dfrac{d\mathbf{B}}{du}\times\mathbf{C} + \dfrac{d\mathbf{A}}{du}\cdot\mathbf{B}\times\mathbf{C}$, where $\mathbf{A}, \mathbf{B}, \mathbf{C}$ are differentiable functions of a scalar $u$.

By Problems 7(*a*) and 7(*b*), $\dfrac{d}{du}\mathbf{A}\cdot(\mathbf{B}\times\mathbf{C}) = \mathbf{A}\cdot\dfrac{d}{du}(\mathbf{B}\times\mathbf{C}) + \dfrac{d\mathbf{A}}{du}\cdot\mathbf{B}\times\mathbf{C}$

$$= \mathbf{A}\cdot\left[\mathbf{B}\times\frac{d\mathbf{C}}{du} + \frac{d\mathbf{B}}{du}\times\mathbf{C}\right] + \frac{d\mathbf{A}}{du}\cdot\mathbf{B}\times\mathbf{C}$$
$$= \mathbf{A}\cdot\mathbf{B}\times\frac{d\mathbf{C}}{du} + \mathbf{A}\cdot\frac{d\mathbf{B}}{du}\times\mathbf{C} + \frac{d\mathbf{A}}{du}\cdot\mathbf{B}\times\mathbf{C}$$

**11.** Evaluate $\dfrac{d}{dt}\left(\mathbf{V}\cdot\dfrac{d\mathbf{V}}{dt}\times\dfrac{d^2\mathbf{V}}{dt^2}\right)$.

By Problem 10, $\dfrac{d}{dt}\left(\mathbf{V}\cdot\dfrac{d\mathbf{V}}{dt}\times\dfrac{d^2\mathbf{V}}{dt^2}\right) = \mathbf{V}\cdot\dfrac{d\mathbf{V}}{dt}\times\dfrac{d^3\mathbf{V}}{dt^3} + \mathbf{V}\cdot\dfrac{d^2\mathbf{V}}{dt^2}\times\dfrac{d^2\mathbf{V}}{dt^2} + \dfrac{d\mathbf{V}}{dt}\cdot\dfrac{d\mathbf{V}}{dt}\times\dfrac{d^2\mathbf{V}}{dt^2}$

$$= \mathbf{V}\cdot\frac{d\mathbf{V}}{dt}\times\frac{d^3\mathbf{V}}{dt^3} + 0 + 0 = \mathbf{V}\cdot\frac{d\mathbf{V}}{dt}\times\frac{d^3\mathbf{V}}{dt^3}$$

**12.** A particle moves so that its position vector is given by $\mathbf{r} = \cos\omega t\,\mathbf{i} + \sin\omega t\,\mathbf{j}$ where $\omega$ is a constant. Show that (*a*) the velocity $\mathbf{v}$ of the particle is perpendicular to $\mathbf{r}$, (*b*) the acceleration $\mathbf{a}$ is directed toward the origin and has magnitude proportional to the distance from the origin, (*c*) $\mathbf{r}\times\mathbf{v}$ = a constant vector.

(a) $\mathbf{v} = \frac{d\mathbf{r}}{dt} = -\omega \sin \omega t\, \mathbf{i} + \omega \cos \omega t\, \mathbf{j}$

Then $\mathbf{r}\cdot\mathbf{v} = [\cos \omega t\, \mathbf{i} + \sin \omega t\, \mathbf{j}] \cdot [-\omega \sin \omega t\, \mathbf{i} + \omega \cos \omega t\, \mathbf{j}]$

$= (\cos \omega t)(-\omega \sin \omega t) + (\sin \omega t)(\omega \cos \omega t) = 0$

and **r** and **v** are perpendicular.

(b) $\frac{d^2\mathbf{r}}{dt^2} = \frac{d\mathbf{v}}{dt} = -\omega^2 \cos \omega t\, \mathbf{i} - \omega^2 \sin \omega t\, \mathbf{j}$

$= -\omega^2 [\cos \omega t\, \mathbf{i} + \sin \omega t\, \mathbf{j}] = -\omega^2 \mathbf{r}$

Then the acceleration is opposite to the direction of **r**, i.e. it is directed toward the origin. Its magnitude is proportional to $|\mathbf{r}|$ which is the distance from the origin.

(c) $\mathbf{r}\times\mathbf{v} = [\cos \omega t\, \mathbf{i} + \sin \omega t\, \mathbf{j}] \times [-\omega \sin \omega t\, \mathbf{i} + \omega \cos \omega t\, \mathbf{j}]$

$$= \begin{vmatrix} \mathbf{i} & \mathbf{j} & \mathbf{k} \\ \cos \omega t & \sin \omega t & 0 \\ -\omega \sin \omega t & \omega \cos \omega t & 0 \end{vmatrix} = \omega(\cos^2 \omega t + \sin^2 \omega t)\mathbf{k} = \omega \mathbf{k}, \text{ a constant vector.}$$

Physically, the motion is that of a particle moving on the circumference of a circle with constant angular speed $\omega$. The acceleration, directed toward the centre of the circle, is the *centripetal acceleration.*

13. Prove: $\mathbf{A}\times\frac{d^2\mathbf{B}}{dt^2} - \frac{d^2\mathbf{A}}{dt^2}\times\mathbf{B} = \frac{d}{dt}(\mathbf{A}\times\frac{d\mathbf{B}}{dt} - \frac{d\mathbf{A}}{dt}\times\mathbf{B})$.

$$\frac{d}{dt}(\mathbf{A}\times\frac{d\mathbf{B}}{dt} - \frac{d\mathbf{A}}{dt}\times\mathbf{B}) = \frac{d}{dt}(\mathbf{A}\times\frac{d\mathbf{B}}{dt}) - \frac{d}{dt}(\frac{d\mathbf{A}}{dt}\times\mathbf{B})$$

$$= \mathbf{A}\times\frac{d^2\mathbf{B}}{dt^2} + \frac{d\mathbf{A}}{dt}\times\frac{d\mathbf{B}}{dt} - [\frac{d\mathbf{A}}{dt}\times\frac{d\mathbf{B}}{dt} + \frac{d^2\mathbf{A}}{dt^2}\times\mathbf{B}] = \mathbf{A}\times\frac{d^2\mathbf{B}}{dt^2} - \frac{d^2\mathbf{A}}{dt^2}\times\mathbf{B}$$

14. Show that $\mathbf{A}\cdot\frac{d\mathbf{A}}{dt} = A\frac{dA}{dt}$.

Let $\mathbf{A} = A_1\mathbf{i} + A_2\mathbf{j} + A_3\mathbf{k}$. Then $A = \sqrt{A_1^2 + A_2^2 + A_3^2}$.

$$\frac{dA}{dt} = \frac{1}{2}(A_1^2 + A_2^2 + A_3^2)^{-1/2}(2A_1\frac{dA_1}{dt} + 2A_2\frac{dA_2}{dt} + 2A_3\frac{dA_3}{dt})$$

$$= \frac{A_1\frac{dA_1}{dt} + A_2\frac{dA_2}{dt} + A_3\frac{dA_3}{dt}}{(A_1^2 + A_2^2 + A_3^2)^{1/2}} = \frac{\mathbf{A}\cdot\frac{d\mathbf{A}}{dt}}{A}, \quad \text{i.e. } A\frac{dA}{dt} = \mathbf{A}\cdot\frac{d\mathbf{A}}{dt}.$$

*Another Method.*

Since $\mathbf{A}\cdot\mathbf{A} = A^2$, $\frac{d}{dt}(\mathbf{A}\cdot\mathbf{A}) = \frac{d}{dt}(A^2)$.

$$\frac{d}{dt}(\mathbf{A}\cdot\mathbf{A}) = \mathbf{A}\cdot\frac{d\mathbf{A}}{dt} + \frac{d\mathbf{A}}{dt}\cdot\mathbf{A} = 2\mathbf{A}\cdot\frac{d\mathbf{A}}{dt} \quad \text{and} \quad \frac{d}{dt}(A^2) = 2A\frac{dA}{dt}$$

Then $2\mathbf{A}\cdot\frac{d\mathbf{A}}{dt} = 2A\frac{dA}{dt}$ or $\mathbf{A}\cdot\frac{d\mathbf{A}}{dt} = A\frac{dA}{dt}$.

Note that if **A** is a constant vector $\mathbf{A}\cdot\frac{d\mathbf{A}}{dt} = 0$ as in Problem 9.

**15.** If $\mathbf{A} = (2x^2y - x^4)\mathbf{i} + (e^{xy} - y\sin x)\mathbf{j} + (x^2\cos y)\mathbf{k}$, find: $\dfrac{\partial \mathbf{A}}{\partial x}, \dfrac{\partial \mathbf{A}}{\partial y}, \dfrac{\partial^2 \mathbf{A}}{\partial x^2}, \dfrac{\partial^2 \mathbf{A}}{\partial y^2}, \dfrac{\partial^2 \mathbf{A}}{\partial x\,\partial y}, \dfrac{\partial^2 \mathbf{A}}{\partial y\,\partial x}$.

$$\frac{\partial \mathbf{A}}{\partial x} = \frac{\partial}{\partial x}(2x^2y - x^4)\mathbf{i} + \frac{\partial}{\partial x}(e^{xy} - y\sin x)\mathbf{j} + \frac{\partial}{\partial x}(x^2\cos y)\mathbf{k}$$

$$= (4xy - 4x^3)\mathbf{i} + (ye^{xy} - y\cos x)\mathbf{j} + 2x\cos y\,\mathbf{k}$$

$$\frac{\partial \mathbf{A}}{\partial y} = \frac{\partial}{\partial y}(2x^2y - x^4)\mathbf{i} + \frac{\partial}{\partial y}(e^{xy} - y\sin x)\mathbf{j} + \frac{\partial}{\partial y}(x^2\cos y)\mathbf{k}$$

$$= 2x^2\,\mathbf{i} + (xe^{xy} - \sin x)\mathbf{j} - x^2\sin y\,\mathbf{k}$$

$$\frac{\partial^2 \mathbf{A}}{\partial x^2} = \frac{\partial}{\partial x}(4xy - 4x^3)\mathbf{i} + \frac{\partial}{\partial x}(ye^{xy} - y\cos x)\mathbf{j} + \frac{\partial}{\partial x}(2x\cos y)\mathbf{k}$$

$$= (4y - 12x^2)\mathbf{i} + (y^2e^{xy} + y\sin x)\mathbf{j} + 2\cos y\,\mathbf{k}$$

$$\frac{\partial^2 \mathbf{A}}{\partial y^2} = \frac{\partial}{\partial y}(2x^2)\mathbf{i} + \frac{\partial}{\partial y}(xe^{xy} - \sin x)\mathbf{j} - \frac{\partial}{\partial y}(x^2\sin y)\mathbf{k}$$

$$= \mathbf{0} + x^2e^{xy}\,\mathbf{j} - x^2\cos y\,\mathbf{k} = x^2e^{xy}\,\mathbf{j} - x^2\cos y\,\mathbf{k}$$

$$\frac{\partial^2 \mathbf{A}}{\partial x\,\partial y} = \frac{\partial}{\partial x}\left(\frac{\partial \mathbf{A}}{\partial y}\right) = \frac{\partial}{\partial x}(2x^2)\mathbf{i} + \frac{\partial}{\partial x}(xe^{xy} - \sin x)\mathbf{j} - \frac{\partial}{\partial x}(x^2\sin y)\mathbf{k}$$

$$= 4x\,\mathbf{i} + (xye^{xy} + e^{xy} - \cos x)\mathbf{j} - 2x\sin y\,\mathbf{k}$$

$$\frac{\partial^2 \mathbf{A}}{\partial y\,\partial x} = \frac{\partial}{\partial y}\left(\frac{\partial \mathbf{A}}{\partial x}\right) = \frac{\partial}{\partial y}(4xy - 4x^3)\mathbf{i} + \frac{\partial}{\partial y}(ye^{xy} - y\cos x)\mathbf{j} + \frac{\partial}{\partial y}(2x\cos y)\mathbf{k}$$

$$= 4x\,\mathbf{i} + (xye^{xy} + e^{xy} - \cos x)\mathbf{j} - 2x\sin y\,\mathbf{k}$$

Note that $\dfrac{\partial^2 \mathbf{A}}{\partial y\,\partial x} = \dfrac{\partial^2 \mathbf{A}}{\partial x\,\partial y}$, i.e. the order of differentiation is immaterial. This is true in general if $\mathbf{A}$ has continuous partial derivatives of the second order at least.

**16.** If $\phi(x,y,z) = xy^2z$ and $\mathbf{A} = xz\,\mathbf{i} - xy^2\,\mathbf{j} + yz^2\mathbf{k}$, find $\dfrac{\partial^3}{\partial x^2\,\partial z}(\phi\mathbf{A})$ at the point $(2,-1,1)$.

$$\phi\mathbf{A} = (xy^2z)(xz\,\mathbf{i} - xy^2\,\mathbf{j} + yz^2\,\mathbf{k}) = x^2y^2z^2\,\mathbf{i} - x^2y^4z\,\mathbf{j} + xy^3z^3\,\mathbf{k}$$

$$\frac{\partial}{\partial z}(\phi\mathbf{A}) = \frac{\partial}{\partial z}(x^2y^2z^2\,\mathbf{i} - x^2y^4z\,\mathbf{j} + xy^3z^3\,\mathbf{k}) = 2x^2y^2z\,\mathbf{i} - x^2y^4\,\mathbf{j} + 3xy^3z^2\,\mathbf{k}$$

$$\frac{\partial^2}{\partial x\,\partial z}(\phi\mathbf{A}) = \frac{\partial}{\partial x}(2x^2y^2z\,\mathbf{i} - x^2y^4\,\mathbf{j} + 3xy^3z^2\,\mathbf{k}) = 4xy^2z\,\mathbf{i} - 2xy^4\,\mathbf{j} + 3y^3z^2\,\mathbf{k}$$

$$\frac{\partial^3}{\partial x^2\,\partial z}(\phi\mathbf{A}) = \frac{\partial}{\partial x}(4xy^2z\,\mathbf{i} - 2xy^4\,\mathbf{j} + 3y^3z^2\,\mathbf{k}) = 4y^2z\,\mathbf{i} - 2y^4\,\mathbf{j}$$

If $x = 2$, $y = -1$, $z = 1$ this becomes $4(-1)^2(1)\mathbf{i} - 2(-1)^4\,\mathbf{j} = 4\mathbf{i} - 2\mathbf{j}$.

**17.** Let $\mathbf{F}$ depend on $x, y, z, t$ where $x, y$ and $z$ depend on $t$. Prove that

$$\frac{d\mathbf{F}}{dt} = \frac{\partial \mathbf{F}}{\partial t} + \frac{\partial \mathbf{F}}{\partial x}\frac{dx}{dt} + \frac{\partial \mathbf{F}}{\partial y}\frac{dy}{dt} + \frac{\partial \mathbf{F}}{\partial z}\frac{dz}{dt}$$

under suitable assumptions of differentiability.

Suppose that $\mathbf{F} = F_1(x,y,z,t)\,\mathbf{i} + F_2(x,y,z,t)\,\mathbf{j} + F_3(x,y,z,t)\,\mathbf{k}$. Then

$$\begin{aligned} d\mathbf{F} &= dF_1\,\mathbf{i} + dF_2\,\mathbf{j} + dF_3\,\mathbf{k} \\ &= [\frac{\partial F_1}{\partial t}dt + \frac{\partial F_1}{\partial x}dx + \frac{\partial F_1}{\partial y}dy + \frac{\partial F_1}{\partial z}dz]\,\mathbf{i} + [\frac{\partial F_2}{\partial t}dt + \frac{\partial F_2}{\partial x}dx + \frac{\partial F_2}{\partial y}dy + \frac{\partial F_2}{\partial z}dz]\,\mathbf{j} \\ &\quad + [\frac{\partial F_3}{\partial t}dt + \frac{\partial F_3}{\partial x}dx + \frac{\partial F_3}{\partial y}dy + \frac{\partial F_3}{\partial z}dz]\,\mathbf{k} \\ &= (\frac{\partial F_1}{\partial t}\mathbf{i} + \frac{\partial F_2}{\partial t}\mathbf{j} + \frac{\partial F_3}{\partial t}\mathbf{k})dt + (\frac{\partial F_1}{\partial x}\mathbf{i} + \frac{\partial F_2}{\partial x}\mathbf{j} + \frac{\partial F_3}{\partial x}\mathbf{k})dx \\ &\quad + (\frac{\partial F_1}{\partial y}\mathbf{i} + \frac{\partial F_2}{\partial y}\mathbf{j} + \frac{\partial F_3}{\partial y}\mathbf{k})dy + (\frac{\partial F_1}{\partial z}\mathbf{i} + \frac{\partial F_2}{\partial z}\mathbf{j} + \frac{\partial F_3}{\partial z}\mathbf{k})dz \\ &= \frac{\partial \mathbf{F}}{\partial t}dt + \frac{\partial \mathbf{F}}{\partial x}dx + \frac{\partial \mathbf{F}}{\partial y}dy + \frac{\partial \mathbf{F}}{\partial z}dz \end{aligned}$$

and so $$\frac{d\mathbf{F}}{dt} = \frac{\partial \mathbf{F}}{\partial t} + \frac{\partial \mathbf{F}}{\partial x}\frac{dx}{dt} + \frac{\partial \mathbf{F}}{\partial y}\frac{dy}{dt} + \frac{\partial \mathbf{F}}{\partial z}\frac{dz}{dt}.$$

## DIFFERENTIAL GEOMETRY.

**18.** Prove the Frenet-Serret formulae (*a*) $\frac{d\mathbf{T}}{ds} = \kappa\mathbf{N}$, (*b*) $\frac{d\mathbf{B}}{ds} = -\tau\mathbf{N}$, (*c*) $\frac{d\mathbf{N}}{ds} = \tau\mathbf{B} - \kappa\mathbf{T}$.

(*a*) Since $\mathbf{T}\cdot\mathbf{T} = 1$, it follows from Problem 9 that $\mathbf{T}\cdot\frac{d\mathbf{T}}{ds} = 0$, i.e. $\frac{d\mathbf{T}}{ds}$ is perpendicular to $\mathbf{T}$.

If $\mathbf{N}$ is a unit vector in the direction $\frac{d\mathbf{T}}{ds}$, then $\frac{d\mathbf{T}}{ds} = \kappa\mathbf{N}$. We call $\mathbf{N}$ the *principal normal*, $\kappa$ the *curvature* and $\rho = 1/\kappa$ the *radius of curvature*.

(*b*) Let $\mathbf{B} = \mathbf{T}\times\mathbf{N}$, so that $\frac{d\mathbf{B}}{ds} = \mathbf{T}\times\frac{d\mathbf{N}}{ds} + \frac{d\mathbf{T}}{ds}\times\mathbf{N} = \mathbf{T}\times\frac{d\mathbf{N}}{ds} + \kappa\mathbf{N}\times\mathbf{N} = \mathbf{T}\times\frac{d\mathbf{N}}{ds}$.

Then $\mathbf{T}\cdot\frac{d\mathbf{B}}{ds} = \mathbf{T}\cdot\mathbf{T}\times\frac{d\mathbf{N}}{ds} = 0$, so that $\mathbf{T}$ is perpendicular to $\frac{d\mathbf{B}}{ds}$.

But from $\mathbf{B}\cdot\mathbf{B} = 1$ it follows that $\mathbf{B}\cdot\frac{d\mathbf{B}}{ds} = 0$ (Problem 9), so that $\frac{d\mathbf{B}}{ds}$ is perpendicular to $\mathbf{B}$ and is thus in the plane of $\mathbf{T}$ and $\mathbf{N}$.

Since $\frac{d\mathbf{B}}{ds}$ is in the plane of $\mathbf{T}$ and $\mathbf{N}$ and is perpendicular to $\mathbf{T}$, it must be parallel to $\mathbf{N}$; then $\frac{d\mathbf{B}}{ds} = -\tau\mathbf{N}$. We call $\mathbf{B}$ the *binormal*, $\tau$ the *torsion*, and $\sigma = 1/\tau$ the *radius of torsion*.

(*c*) Since $\mathbf{T}, \mathbf{N}, \mathbf{B}$ form a right-handed system, so do $\mathbf{N}, \mathbf{B}$ and $\mathbf{T}$, i.e. $\mathbf{N} = \mathbf{B}\times\mathbf{T}$.

Then $\frac{d\mathbf{N}}{ds} = \mathbf{B}\times\frac{d\mathbf{T}}{ds} + \frac{d\mathbf{B}}{ds}\times\mathbf{T} = \mathbf{B}\times\kappa\mathbf{N} - \tau\mathbf{N}\times\mathbf{T} = -\kappa\mathbf{T} + \tau\mathbf{B} = \tau\mathbf{B} - \kappa\mathbf{T}$.

**19.** Sketch the space curve $x = 3\cos t$, $y = 3\sin t$, $z = 4t$ and find (*a*) the unit tangent $\mathbf{T}$, (*b*) the principal normal $\mathbf{N}$, curvature $\kappa$ and radius of curvature $\rho$, (*c*) the binormal $\mathbf{B}$, torsion $\tau$ and radius of torsion $\sigma$.

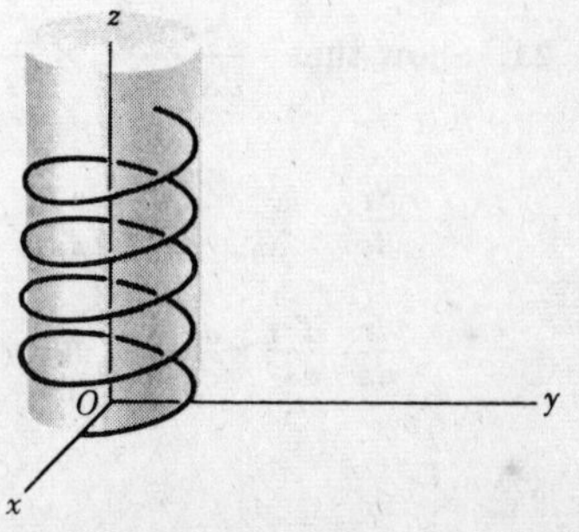


The space curve is a *circular helix* (see adjacent figure). Since $t = z/4$, the curve has equations $x = 3\cos(z/4)$, $y = 3\sin(z/4)$ and therefore lies on the cylinder $x^2 + y^2 = 9$.

(*a*) The position vector for any point on the curve is

$$\mathbf{r} = 3\cos t\,\mathbf{i} + 3\sin t\,\mathbf{j} + 4t\,\mathbf{k}$$

Then $\dfrac{d\mathbf{r}}{dt} = -3\sin t\,\mathbf{i} + 3\cos t\,\mathbf{j} + 4\mathbf{k}$

$$\frac{ds}{dt} = \left|\frac{d\mathbf{r}}{dt}\right| = \sqrt{\frac{d\mathbf{r}}{dt}\cdot\frac{d\mathbf{r}}{dt}} = \sqrt{(-3\sin t)^2 + (3\cos t)^2 + 4^2} = 5$$

Thus $\mathbf{T} = \dfrac{d\mathbf{r}}{ds} = \dfrac{d\mathbf{r}/dt}{ds/dt} = -\dfrac{3}{5}\sin t\,\mathbf{i} + \dfrac{3}{5}\cos t\,\mathbf{j} + \dfrac{4}{5}\mathbf{k}.$

(*b*) $\dfrac{d\mathbf{T}}{dt} = \dfrac{d}{dt}(-\dfrac{3}{5}\sin t\,\mathbf{i} + \dfrac{3}{5}\cos t\,\mathbf{j} + \dfrac{4}{5}\mathbf{k}) = -\dfrac{3}{5}\cos t\,\mathbf{i} - \dfrac{3}{5}\sin t\,\mathbf{j}$

$$\frac{d\mathbf{T}}{ds} = \frac{d\mathbf{T}/dt}{ds/dt} = -\frac{3}{25}\cos t\,\mathbf{i} - \frac{3}{25}\sin t\,\mathbf{j}$$

Since $\dfrac{d\mathbf{T}}{ds} = \kappa\mathbf{N}$, $\left|\dfrac{d\mathbf{T}}{ds}\right| = |\kappa|\,|\mathbf{N}| = \kappa$ as $\kappa \geqq 0$.

Then $\kappa = \left|\dfrac{d\mathbf{T}}{ds}\right| = \sqrt{(-\dfrac{3}{25}\cos t)^2 + (-\dfrac{3}{25}\sin t)^2} = \dfrac{3}{25}$ and $\rho = \dfrac{1}{\kappa} = \dfrac{25}{3}$.

From $\dfrac{d\mathbf{T}}{ds} = \kappa\mathbf{N}$, we obtain $\mathbf{N} = \dfrac{1}{\kappa}\dfrac{d\mathbf{T}}{ds} = -\cos t\,\mathbf{i} - \sin t\,\mathbf{j}$.

(*c*) $\mathbf{B} = \mathbf{T}\times\mathbf{N} = \begin{vmatrix} \mathbf{i} & \mathbf{j} & \mathbf{k} \\ -\frac{3}{5}\sin t & \frac{3}{5}\cos t & \frac{4}{5} \\ -\cos t & -\sin t & 0 \end{vmatrix} = \dfrac{4}{5}\sin t\,\mathbf{i} - \dfrac{4}{5}\cos t\,\mathbf{j} + \dfrac{3}{5}\mathbf{k}$

$$\frac{d\mathbf{B}}{dt} = \frac{4}{5}\cos t\,\mathbf{i} + \frac{4}{5}\sin t\,\mathbf{j}, \qquad \frac{d\mathbf{B}}{ds} = \frac{d\mathbf{B}/dt}{ds/dt} = \frac{4}{25}\cos t\,\mathbf{i} + \frac{4}{25}\sin t\,\mathbf{j}$$

$$-\tau\mathbf{N} = -\tau(-\cos t\,\mathbf{i} - \sin t\,\mathbf{j}) = \frac{4}{25}\cos t\,\mathbf{i} + \frac{4}{25}\sin t\,\mathbf{j} \quad\text{or}\quad \tau = \frac{4}{25} \quad\text{and}\quad \sigma = \frac{1}{\tau} = \frac{25}{4}$$

**20.** Prove that the radius of curvature of the curve with parametric equations $x = x(s)$, $y = y(s)$, $z = z(s)$ is given by $\rho = \left[\left(\dfrac{d^2x}{ds^2}\right)^2 + \left(\dfrac{d^2y}{ds^2}\right)^2 + \left(\dfrac{d^2z}{ds^2}\right)^2\right]^{-1/2}$.

The position vector of any point on the curve is $\mathbf{r} = x(s)\mathbf{i} + y(s)\mathbf{j} + z(s)\mathbf{k}$.

Then $\mathbf{T} = \dfrac{d\mathbf{r}}{ds} = \dfrac{dx}{ds}\mathbf{i} + \dfrac{dy}{ds}\mathbf{j} + \dfrac{dz}{ds}\mathbf{k}$ and $\dfrac{d\mathbf{T}}{ds} = \dfrac{d^2x}{ds^2}\mathbf{i} + \dfrac{d^2y}{ds^2}\mathbf{j} + \dfrac{d^2z}{ds^2}\mathbf{k}$.

But $\dfrac{d\mathbf{T}}{ds} = \kappa\mathbf{N}$ so that $\kappa = \left|\dfrac{d\mathbf{T}}{ds}\right| = \sqrt{\left(\dfrac{d^2x}{ds^2}\right)^2 + \left(\dfrac{d^2y}{ds^2}\right)^2 + \left(\dfrac{d^2z}{ds^2}\right)^2}$ and the result follows since $\rho = \dfrac{1}{\kappa}$.

**21.** Show that $\dfrac{d\mathbf{r}}{ds}\cdot\dfrac{d^2\mathbf{r}}{ds^2}\times\dfrac{d^3\mathbf{r}}{ds^3} = \dfrac{\tau}{\rho^2}$.

$$\frac{d\mathbf{r}}{ds} = \mathbf{T}, \quad \frac{d^2\mathbf{r}}{ds^2} = \frac{d\mathbf{T}}{ds} = \kappa\mathbf{N}, \quad \frac{d^3\mathbf{r}}{ds^3} = \kappa\frac{d\mathbf{N}}{ds} + \frac{d\kappa}{ds}\mathbf{N} = \kappa(\tau\mathbf{B} - \kappa\mathbf{T}) + \frac{d\kappa}{ds}\mathbf{N} = \kappa\tau\mathbf{B} - \kappa^2\mathbf{T} + \frac{d\kappa}{ds}\mathbf{N}$$

$$\frac{d\mathbf{r}}{ds}\cdot\frac{d^2\mathbf{r}}{ds^2}\times\frac{d^3\mathbf{r}}{ds^3} = \mathbf{T}\cdot\kappa\mathbf{N}\times(\kappa\tau\mathbf{B} - \kappa^2\mathbf{T} + \frac{d\kappa}{ds}\mathbf{N})$$

$$= \mathbf{T}\cdot(\kappa^2\tau\,\mathbf{N}\times\mathbf{B} - \kappa^3\mathbf{N}\times\mathbf{T} + \kappa\frac{d\kappa}{ds}\mathbf{N}\times\mathbf{N}) = \mathbf{T}\cdot(\kappa^2\tau\mathbf{T} + \kappa^3\mathbf{B}) = \kappa^2\tau = \frac{\tau}{\rho^2}$$

The result can be written

$$\tau = [(x'')^2 + (y'')^2 + (z'')^2]^{-1} \begin{vmatrix} x' & y' & z' \\ x'' & y'' & z'' \\ x''' & y''' & z''' \end{vmatrix}$$

where primes denote derivatives with respect to $s$, by using the result of Problem 20.

**22.** Given the space curve $x = t,\ y = t^2,\ z = \frac{2}{3}t^3$, find (*a*) the curvature $\kappa$, (*b*) the torsion $\tau$.

(*a*) The position vector is $\mathbf{r} = t\,\mathbf{i} + t^2\,\mathbf{j} + \frac{2}{3}t^3\,\mathbf{k}$.

Then $\dfrac{d\mathbf{r}}{dt} = \mathbf{i} + 2t\,\mathbf{j} + 2t^2\,\mathbf{k}$

$$\frac{ds}{dt} = \left|\frac{d\mathbf{r}}{dt}\right| = \sqrt{\frac{d\mathbf{r}}{dt}\cdot\frac{d\mathbf{r}}{dt}} = \sqrt{(1)^2 + (2t)^2 + (2t^2)^2} = 1 + 2t^2$$

and $$\mathbf{T} = \frac{d\mathbf{r}}{ds} = \frac{d\mathbf{r}/dt}{ds/dt} = \frac{\mathbf{i} + 2t\,\mathbf{j} + 2t^2\,\mathbf{k}}{1 + 2t^2}.$$

$$\frac{d\mathbf{T}}{dt} = \frac{(1 + 2t^2)(2\mathbf{j} + 4t\,\mathbf{k}) - (\mathbf{i} + 2t\,\mathbf{j} + 2t^2\,\mathbf{k})(4t)}{(1 + 2t^2)^2} = \frac{-4t\,\mathbf{i} + (2 - 4t^2)\mathbf{j} + 4t\,\mathbf{k}}{(1 + 2t^2)^2}$$

Then $$\frac{d\mathbf{T}}{ds} = \frac{d\mathbf{T}/dt}{ds/dt} = \frac{-4t\,\mathbf{i} + (2 - 4t^2)\mathbf{j} + 4t\,\mathbf{k}}{(1 + 2t^2)^3}.$$

Since $\dfrac{d\mathbf{T}}{ds} = \kappa\mathbf{N}$, $$\kappa = \left|\frac{d\mathbf{T}}{ds}\right| = \frac{\sqrt{(-4t)^2 + (2 - 4t^2)^2 + (4t)^2}}{(1 + 2t^2)^3} = \frac{2}{(1 + 2t^2)^2}$$

(*b*) From (*a*), $$\mathbf{N} = \frac{1}{\kappa}\frac{d\mathbf{T}}{ds} = \frac{-2t\,\mathbf{i} + (1 - 2t^2)\mathbf{j} + 2t\,\mathbf{k}}{1 + 2t^2}$$

Then $$\mathbf{B} = \mathbf{T}\times\mathbf{N} = \begin{vmatrix} \mathbf{i} & \mathbf{j} & \mathbf{k} \\ \dfrac{1}{1 + 2t^2} & \dfrac{2t}{1 + 2t^2} & \dfrac{2t^2}{1 + 2t^2} \\ \dfrac{-2t}{1 + 2t^2} & \dfrac{1 - 2t^2}{1 + 2t^2} & \dfrac{2t}{1 + 2t^2} \end{vmatrix} = \frac{2t^2\,\mathbf{i} - 2t\,\mathbf{j} + \mathbf{k}}{1 + 2t^2}$$

Now $\dfrac{d\mathbf{B}}{dt} = \dfrac{4t\,\mathbf{i} + (4t^2 - 2)\mathbf{j} - 4t\,\mathbf{k}}{(1 + 2t^2)^2}$ and $\dfrac{d\mathbf{B}}{ds} = \dfrac{d\mathbf{B}/dt}{ds/dt} = \dfrac{4t\,\mathbf{i} + (4t^2 - 2)\mathbf{j} - 4t\,\mathbf{k}}{(1 + 2t^2)^3}$

Also, $-\tau\mathbf{N} = -\tau\left[\dfrac{-2t\,\mathbf{i} + (1 - 2t^2)\mathbf{j} + 2t\,\mathbf{k}}{1 + 2t^2}\right]$. Since $\dfrac{d\mathbf{B}}{ds} = -\tau\mathbf{N}$, we find $\tau = \dfrac{2}{(1 + 2t^2)^2}$.

Note that $\kappa = \tau$ for this curve.

**23.** Find equations in vector and rectangular form for the (*a*) tangent, (*b*) principal normal, and (*c*) binormal to the curve of Problem 22 at the point where $t = 1$.

Let $\mathbf{T}_0, \mathbf{N}_0$ and $\mathbf{B}_0$ denote the tangent, principal normal and binormal vectors at the required point. Then from Problem 22,

$$\mathbf{T}_0 = \frac{\mathbf{i} + 2\mathbf{j} + 2\mathbf{k}}{3}, \quad \mathbf{N}_0 = \frac{-2\mathbf{i} - \mathbf{j} + 2\mathbf{k}}{3}, \quad \mathbf{B}_0 = \frac{2\mathbf{i} - 2\mathbf{j} + \mathbf{k}}{3}$$

If **A** denotes a given vector while $\mathbf{r}_0$ and $\mathbf{r}$ denote respectively the position vectors of the initial point and an arbitrary point of **A**, then $\mathbf{r}-\mathbf{r}_0$ is parallel to **A** and so the equation of **A** is $(\mathbf{r}-\mathbf{r}_0)\times\mathbf{A} = \mathbf{0}$.

| | | |
|---|---|---|
| | Equation of tangent is | $(\mathbf{r}-\mathbf{r}_0)\times\mathbf{T}_0 = \mathbf{0}$ |
| Then: | Equation of principal normal is | $(\mathbf{r}-\mathbf{r}_0)\times\mathbf{N}_0 = \mathbf{0}$ |
| | Equation of binormal is | $(\mathbf{r}-\mathbf{r}_0)\times\mathbf{B}_0 = \mathbf{0}$ |

In rectangular form, with $\mathbf{r} = x\mathbf{i} + y\mathbf{j} + z\mathbf{k}$, $\mathbf{r}_0 = \mathbf{i} + \mathbf{j} + \frac{2}{3}\mathbf{k}$ these become respectively

$$\frac{x-1}{1} = \frac{y-1}{2} = \frac{z-2/3}{2}, \quad \frac{x-1}{-2} = \frac{y-1}{-1} = \frac{z-2/3}{2}, \quad \frac{x-1}{2} = \frac{y-1}{-2} = \frac{z-2/3}{1}.$$

These equations can also be written in parametric form (see Problem 28, Chapter 1).

**24.** Find equations in vector and rectangular form for the (*a*) osculating plane, (*b*) normal plane, and (*c*) rectifying plane to the curve of Problems 22 and 23 at the point where $t = 1$.

(*a*) The osculating plane is the plane which contains the tangent and principal normal. If **r** is the position vector of any point in this plane and $\mathbf{r}_0$ is the position vector of the point $t=1$, then $\mathbf{r}-\mathbf{r}_0$ is perpendicular to $\mathbf{B}_0$, the binormal at the point $t=1$, i.e. $(\mathbf{r}-\mathbf{r}_0)\cdot\mathbf{B}_0 = 0$.

(*b*) The normal plane is the plane which is perpendicular to the tangent vector at the given point. Then the required equation is $(\mathbf{r}-\mathbf{r}_0)\cdot\mathbf{T}_0 = 0$.

(*c*) The rectifying plane is the plane which is perpendicular to the principal normal at the given point. The required equation is $(\mathbf{r}-\mathbf{r}_0)\cdot\mathbf{N}_0 = 0$.

In rectangular form the equations of (*a*), (*b*) and (*c*) become respectively,

$$2(x-1) - 2(y-1) + 1(z-2/3) = 0,$$
$$1(x-1) + 2(y-1) + 2(z-2/3) = 0,$$
$$-2(x-1) - 1(y-1) + 2(z-2/3) = 0.$$

The adjoining figure shows the osculating, normal and rectifying planes to a curve $C$ at the point $P$.

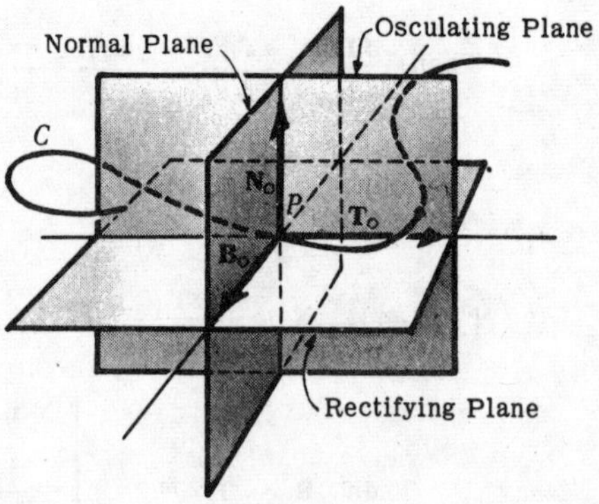


**25.** (*a*) Show that the equation $\mathbf{r} = \mathbf{r}(u,v)$ represents a surface.
(*b*) Show that $\dfrac{\partial\mathbf{r}}{\partial u}\times\dfrac{\partial\mathbf{r}}{\partial v}$ represents a vector normal to the surface.
(*c*) Determine a unit normal to the following surface, where $a > 0$:

$$\mathbf{r} = a\cos u\,\sin v\,\mathbf{i} + a\sin u\,\sin v\,\mathbf{j} + a\cos v\,\mathbf{k}$$

(*a*) If we consider $u$ to have a fixed value, say $u_0$, then $\mathbf{r} = \mathbf{r}(u_0,v)$ represents a curve which can be denoted by $u = u_0$. Similarly $u = u_1$ defines another curve $\mathbf{r} = \mathbf{r}(u_1,v)$. As $u$ varies, therefore, $\mathbf{r} = \mathbf{r}(u,v)$ represents a curve which moves in space and generates a surface $S$. Then $\mathbf{r} = \mathbf{r}(u,v)$ represents the surface $S$ thus generated, as shown in the adjoining figure.

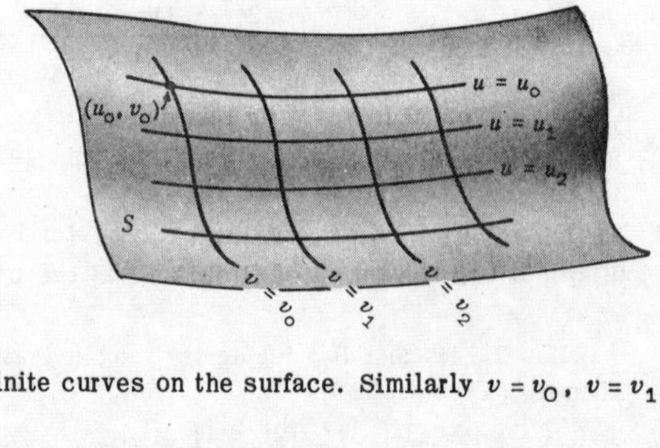


The curves $u = u_0$, $u = u_1$, ... represent definite curves on the surface. Similarly $v = v_0$, $v = v_1$, ... represent curves on the surface.

By assigning definite values to $u$ and $v$, we obtain a point on the surface. Thus curves $u = u_0$ and $v = v_0$, for example, intersect and define the point $(u_0, v_0)$ on the surface. We speak of the pair of numbers $(u,v)$ as defining the *curvilinear coordinates* on the surface. If all the curves $u$ = constant and $v$ = constant are perpendicular at each point of intersection, we call the curvilinear coordinate system *orthogonal*. For further discussion of curvilinear coordinates see Chapter 7.

(*b*) Consider point $P$ having coordinates $(u_0, v_0)$ on a surface $S$, as shown in the adjacent diagram. The vector $\partial\mathbf{r}/\partial u$ at $P$ is obtained by differentiating $\mathbf{r}$ with respect to $u$, keeping $v$ = constant = $v_0$. From the theory of space curves, it follows that $\partial\mathbf{r}/\partial u$ at $P$ represents a vector tangent to the curve $v = v_0$ at $P$, as shown in the adjoining figure. Similarly, $\partial\mathbf{r}/\partial v$ at $P$ represents a vector tangent to the curve $u$ = constant = $u_0$. Since $\partial\mathbf{r}/\partial u$ and $\partial\mathbf{r}/\partial v$ represent vectors at $P$ tangent to curves which lie on the surface $S$ at $P$, it follows that these vectors are tangent to the surface at $P$. Hence it follows that $\dfrac{\partial\mathbf{r}}{\partial u}\times\dfrac{\partial\mathbf{r}}{\partial v}$ is a vector normal to $S$ at $P$.

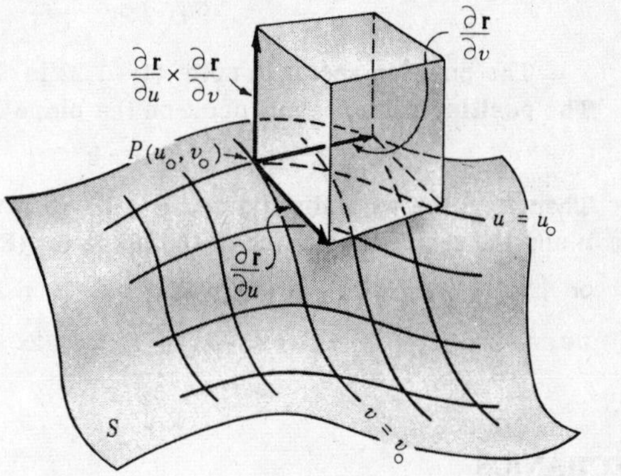


(*c*) $\dfrac{\partial\mathbf{r}}{\partial u} = -a\sin u\,\sin v\,\mathbf{i} + a\cos u\,\sin v\,\mathbf{j}$

$\dfrac{\partial\mathbf{r}}{\partial v} = a\cos u\,\cos v\,\mathbf{i} + a\sin u\,\cos v\,\mathbf{j} - a\sin v\,\mathbf{k}$

Then
$$\frac{\partial\mathbf{r}}{\partial u}\times\frac{\partial\mathbf{r}}{\partial v} = \begin{vmatrix} \mathbf{i} & \mathbf{j} & \mathbf{k} \\ -a\sin u\,\sin v & a\cos u\,\sin v & 0 \\ a\cos u\,\cos v & a\sin u\,\cos v & -a\sin v \end{vmatrix}$$
$$= -a^2\cos u\,\sin^2 v\,\mathbf{i} - a^2\sin u\,\sin^2 v\,\mathbf{j} - a^2\sin v\,\cos v\,\mathbf{k}$$

represents a vector normal to the surface at any point $(u,v)$.

A unit normal is obtained by dividing $\dfrac{\partial\mathbf{r}}{\partial u}\times\dfrac{\partial\mathbf{r}}{\partial v}$ by its magnitude, $\left|\dfrac{\partial\mathbf{r}}{\partial u}\times\dfrac{\partial\mathbf{r}}{\partial v}\right|$, given by

$$\sqrt{a^4\cos^2 u\,\sin^4 v + a^4\sin^2 u\,\sin^4 v + a^4\sin^2 v\,\cos^2 v}$$
$$= \sqrt{a^4(\cos^2 u + \sin^2 u)\sin^4 v + a^4\sin^2 v\,\cos^2 v}$$
$$= \sqrt{a^4\sin^2 v\,(\sin^2 v + \cos^2 v)} = \begin{cases} a^2\sin v & \text{if } \sin v > 0 \\ -a^2\sin v & \text{if } \sin v < 0 \end{cases}$$

Then there are two unit normals given by
$$\pm(\cos u\,\sin v\,\mathbf{i} + \sin u\,\sin v\,\mathbf{j} + \cos v\,\mathbf{k}) = \pm\mathbf{n}$$

It should be noted that the given surface is defined by $x = a\cos u\,\sin v$, $y = a\sin u\,\sin v$, $z = a\cos v$ from which it is seen that $x^2+y^2+z^2 = a^2$, which is a sphere of radius $a$. Since $\mathbf{r} = a\,\mathbf{n}$, it follows that

$$\mathbf{n} = \cos u\,\sin v\,\mathbf{i} + \sin u\,\sin v\,\mathbf{j} + \cos v\,\mathbf{k}$$

is the *outward drawn unit normal* to the sphere at the point $(u,v)$.

**26.** Find an equation for the tangent plane to the surface $z = x^2+y^2$ at the point $(1, -1, 2)$.

Let $x = u$, $y = v$, $z = u^2+v^2$ be parametric equations of the surface. The position vector to any point on the surface is
$$\mathbf{r} = u\,\mathbf{i} + v\,\mathbf{j} + (u^2+v^2)\mathbf{k}$$

Then $\dfrac{\partial \mathbf{r}}{\partial u} = \mathbf{i} + 2u\,\mathbf{k} = \mathbf{i} + 2\mathbf{k}$, $\dfrac{\partial \mathbf{r}}{\partial v} = \mathbf{j} + 2v\,\mathbf{k} = \mathbf{j} - 2\mathbf{k}$ at the point (1,–1,2), where $u = 1$ and $v = -1$.

By Problem 25, a normal $\mathbf{n}$ to the surface at this point is

$$\mathbf{n} = \frac{\partial \mathbf{r}}{\partial u} \times \frac{\partial \mathbf{r}}{\partial v} = (\mathbf{i} + 2\mathbf{k}) \times (\mathbf{j} - 2\mathbf{k}) = -2\mathbf{i} + 2\mathbf{j} + \mathbf{k}$$

The position vector to point (1,–1,2) is $\mathbf{R}_0 = \mathbf{i} - \mathbf{j} + 2\mathbf{k}$.
The position vector to any point on the plane is

$$\mathbf{R} = x\,\mathbf{i} + y\,\mathbf{j} + z\,\mathbf{k}$$

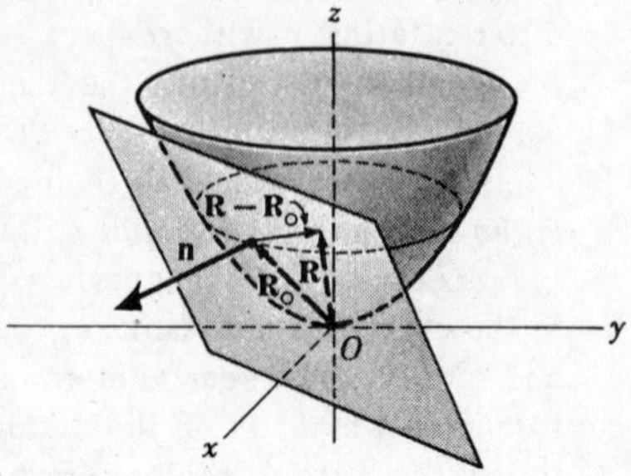


Then from the adjoining figure, $\mathbf{R} - \mathbf{R}_0$ is perpendicular to $\mathbf{n}$ and the required equation of the plane is $(\mathbf{R} - \mathbf{R}_0) \cdot \mathbf{n} = 0$
or $[(x\,\mathbf{i} + y\,\mathbf{j} + z\,\mathbf{k}) - (\mathbf{i} - \mathbf{j} + 2\mathbf{k})] \cdot [-2\mathbf{i} + 2\mathbf{j} + \mathbf{k}] = 0$
i.e. $-2(x-1) + 2(y+1) + (z-2) = 0$ or $2x - 2y - z = 2$.

## MECHANICS

**27.** Show that the acceleration $\mathbf{a}$ of a particle which travels along a space curve with velocity $\mathbf{v}$ is given by

$$\mathbf{a} = \frac{dv}{dt}\mathbf{T} + \frac{v^2}{\rho}\mathbf{N}$$

where $\mathbf{T}$ is the unit tangent vector to the space curve, $\mathbf{N}$ is its unit principal normal, and $\rho$ is the radius of curvature.

Velocity $\mathbf{v}$ = magnitude of $\mathbf{v}$ multiplied by unit tangent vector $\mathbf{T}$

or $\mathbf{v} = v\mathbf{T}$

Differentiating, $$\mathbf{a} = \frac{d\mathbf{v}}{dt} = \frac{d}{dt}(v\mathbf{T}) = \frac{dv}{dt}\mathbf{T} + v\frac{d\mathbf{T}}{dt}$$

But by Problem 18(*a*), $$\frac{d\mathbf{T}}{dt} = \frac{d\mathbf{T}}{ds}\frac{ds}{dt} = \kappa\mathbf{N}\frac{ds}{dt} = \kappa v\mathbf{N} = \frac{v\mathbf{N}}{\rho}$$

Then $$\mathbf{a} = \frac{dv}{dt}\mathbf{T} + v\left(\frac{v\mathbf{N}}{\rho}\right) = \frac{dv}{dt}\mathbf{T} + \frac{v^2}{\rho}\mathbf{N}$$

This shows that the component of the acceleration is $dv/dt$ in a direction tangent to the path and $v^2/\rho$ in a direction of the principal normal to the path. The latter acceleration is often called the *centripetal acceleration*. For a special case of this problem see Problem 12.

**28.** If $\mathbf{r}$ is the position vector of a particle of mass $m$ relative to point $O$ and $\mathbf{F}$ is the external force on the particle, then $\mathbf{r} \times \mathbf{F} = \mathbf{M}$ is the torque or moment of $\mathbf{F}$ about $O$. Show that $\mathbf{M} = d\mathbf{H}/dt$, where $\mathbf{H} = \mathbf{r} \times m\mathbf{v}$ and $\mathbf{v}$ is the velocity of the particle.

$$\mathbf{M} = \mathbf{r} \times \mathbf{F} = \mathbf{r} \times \frac{d}{dt}(m\mathbf{v}) \quad \text{by Newton's law.}$$

But $$\frac{d}{dt}(\mathbf{r} \times m\mathbf{v}) = \mathbf{r} \times \frac{d}{dt}(m\mathbf{v}) + \frac{d\mathbf{r}}{dt} \times m\mathbf{v}$$

$$= \mathbf{r} \times \frac{d}{dt}(m\mathbf{v}) + \mathbf{v} \times m\mathbf{v} = \mathbf{r} \times \frac{d}{dt}(m\mathbf{v}) + \mathbf{0}$$

i.e. $$\mathbf{M} = \frac{d}{dt}(\mathbf{r} \times m\mathbf{v}) = \frac{d\mathbf{H}}{dt}$$

Note that the result holds whether $m$ is constant or not. $\mathbf{H}$ is called the *angular momentum*. The result states that the torque is equal to the time rate of change of angular momentum.

This result is easily extended to a system of $n$ particles having respective masses $m_1, m_2, \ldots, m_n$ and position vectors $\mathbf{r}_1, \mathbf{r}_2, \ldots, \mathbf{r}_n$ with external forces $\mathbf{F}_1, \mathbf{F}_2, \ldots, \mathbf{F}_n$. For this case, $\mathbf{H} = \sum_{k=1}^{n} m_k \mathbf{r}_k \times \mathbf{v}_k$ is the total angular momentum, $\mathbf{M} = \sum_{k=1}^{n} \mathbf{r}_k \times \mathbf{F}_k$ is the total torque, and the result is $\mathbf{M} = \dfrac{d\mathbf{H}}{dt}$ as before.

**29.** An observer stationed at a point which is fixed relative to an $xyz$ coordinate system with origin $O$, as shown in the adjoining diagram, observes a vector $\mathbf{A} = A_1\mathbf{i} + A_2\mathbf{j} + A_3\mathbf{k}$ and calculates its time derivative to be $\dfrac{dA_1}{dt}\mathbf{i} + \dfrac{dA_2}{dt}\mathbf{j} + \dfrac{dA_3}{dt}\mathbf{k}$. Later, he finds out that he and his coordinate system are actually rotating with respect to an $XYZ$ coordinate system taken as fixed in space and having origin also at $O$. He asks, 'What would be the time derivative of $\mathbf{A}$ for an observer who is fixed relative to the $XYZ$ coordinate system?'

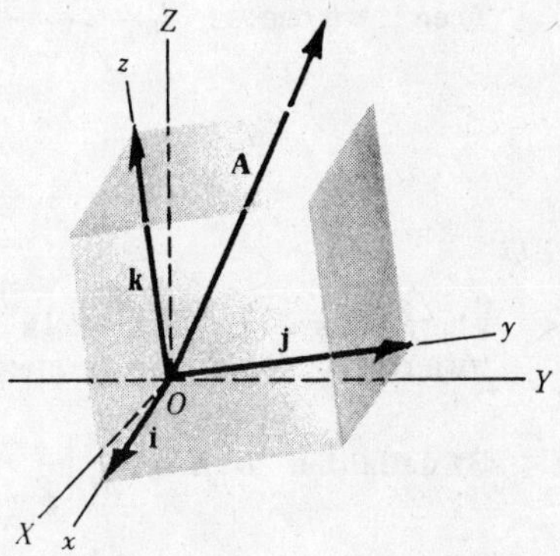


(*a*) If $\left.\dfrac{d\mathbf{A}}{dt}\right|_f$ and $\left.\dfrac{d\mathbf{A}}{dt}\right|_m$ denote respectively the time derivatives of $\mathbf{A}$ with respect to the fixed and moving systems, show that there exists a vector quantity $\boldsymbol{\omega}$ such that

$$\left.\frac{d\mathbf{A}}{dt}\right|_f = \left.\frac{d\mathbf{A}}{dt}\right|_m + \boldsymbol{\omega} \times \mathbf{A}$$

(*b*) Let $D_f$ and $D_m$ be symbolic time derivative operators in the fixed and moving systems respectively. Demonstrate the operator equivalence

$$D_f \equiv D_m + \boldsymbol{\omega} \times$$

(*a*) To the fixed observer the unit vectors $\mathbf{i}, \mathbf{j}, \mathbf{k}$ actually change with time. Hence such an observer would compute the time derivative of $\mathbf{A}$ as

$$(1) \qquad \frac{d\mathbf{A}}{dt} = \frac{dA_1}{dt}\mathbf{i} + \frac{dA_2}{dt}\mathbf{j} + \frac{dA_3}{dt}\mathbf{k} + A_1\frac{d\mathbf{i}}{dt} + A_2\frac{d\mathbf{j}}{dt} + A_3\frac{d\mathbf{k}}{dt} \qquad \text{i.e.}$$

$$(2) \qquad \left.\frac{d\mathbf{A}}{dt}\right|_f = \left.\frac{d\mathbf{A}}{dt}\right|_m + A_1\frac{d\mathbf{i}}{dt} + A_2\frac{d\mathbf{j}}{dt} + A_3\frac{d\mathbf{k}}{dt}$$

Since $\mathbf{i}$ is a unit vector, $d\mathbf{i}/dt$ is perpendicular to $\mathbf{i}$ (see Problem 9) and must therefore lie in the plane of $\mathbf{j}$ and $\mathbf{k}$. Then

$$(3) \qquad \frac{d\mathbf{i}}{dt} = \alpha_1\mathbf{j} + \alpha_2\mathbf{k}$$

Similarly,

$$(4) \qquad \frac{d\mathbf{j}}{dt} = \alpha_3\mathbf{k} + \alpha_4\mathbf{i}$$

$$(5) \qquad \frac{d\mathbf{k}}{dt} = \alpha_5\mathbf{i} + \alpha_6\mathbf{j}$$

From $\mathbf{i}\cdot\mathbf{j} = 0$, differentiation yields $\mathbf{i}\cdot\dfrac{d\mathbf{j}}{dt} + \dfrac{d\mathbf{i}}{dt}\cdot\mathbf{j} = 0$. But $\mathbf{i}\cdot\dfrac{d\mathbf{j}}{dt} = \alpha_4$ from (4), and $\dfrac{d\mathbf{i}}{dt}\cdot\mathbf{j} = \alpha_1$ from (3); then $\alpha_4 = -\alpha_1$.

Similarly from $\mathbf{i}\cdot\mathbf{k} = 0$, $\mathbf{i}\cdot\dfrac{d\mathbf{k}}{dt} + \dfrac{d\mathbf{i}}{dt}\cdot\mathbf{k} = 0$ and $\alpha_5 = -\alpha_2$; from $\mathbf{j}\cdot\mathbf{k} = 0$, $\mathbf{j}\cdot\dfrac{d\mathbf{k}}{dt} + \dfrac{d\mathbf{j}}{dt}\cdot\mathbf{k} = 0$ and $\alpha_6 = -\alpha_3$.

Then $\dfrac{d\mathbf{i}}{dt} = \alpha_1\mathbf{j} + \alpha_2\mathbf{k}$, $\dfrac{d\mathbf{j}}{dt} = \alpha_3\mathbf{k} - \alpha_1\mathbf{i}$, $\dfrac{d\mathbf{k}}{dt} = -\alpha_2\mathbf{i} - \alpha_3\mathbf{j}$ and

$$A_1\frac{d\mathbf{i}}{dt} + A_2\frac{d\mathbf{j}}{dt} + A_3\frac{d\mathbf{k}}{dt} = (-\alpha_1 A_2 - \alpha_2 A_3)\mathbf{i} + (\alpha_1 A_1 - \alpha_3 A_3)\mathbf{j} + (\alpha_2 A_1 + \alpha_3 A_2)\mathbf{k}$$

which can be written as

$$\begin{vmatrix} \mathbf{i} & \mathbf{j} & \mathbf{k} \\ \alpha_3 & -\alpha_2 & \alpha_1 \\ A_1 & A_2 & A_3 \end{vmatrix}$$

Then if we choose $\alpha_3 = \omega_1$, $-\alpha_2 = \omega_2$, $\alpha_1 = \omega_3$ the determinant becomes

$$\begin{vmatrix} \mathbf{i} & \mathbf{j} & \mathbf{k} \\ \omega_1 & \omega_2 & \omega_3 \\ A_1 & A_2 & A_3 \end{vmatrix} = \boldsymbol{\omega} \times \mathbf{A}$$

where $\boldsymbol{\omega} = \omega_1\mathbf{i} + \omega_2\mathbf{j} + \omega_3\mathbf{k}$. The quantity $\boldsymbol{\omega}$ is the angular velocity vector of the moving system with respect to the fixed system.

(*b*) By definition $D_f\mathbf{A} = \left.\frac{d\mathbf{A}}{dt}\right|_f$ = derivative in fixed system

$D_m\mathbf{A} = \left.\frac{d\mathbf{A}}{dt}\right|_m$ = derivative in moving system.

From (*a*), $$D_f\mathbf{A} = D_m\mathbf{A} + \boldsymbol{\omega}\times\mathbf{A} = (D_m + \boldsymbol{\omega}\times)\mathbf{A}$$

and shows the equivalence of the operators $D_f \equiv D_m + \boldsymbol{\omega}\times$ .

**30.** Determine the (*a*) velocity and (*b*) acceleration of a moving particle as seen by the two observers in Problem 29.

(*a*) Let vector **A** in Problem 29 be the position vector **r** of the particle. Using the operator notation of Problem 29(*b*), we have

(*1*) $$D_f\mathbf{r} = (D_m + \boldsymbol{\omega}\times)\mathbf{r} = D_m\mathbf{r} + \boldsymbol{\omega}\times\mathbf{r}$$

But $D_f\mathbf{r} = \mathbf{v}_{p|f}$ = velocity of particle relative to fixed system

$D_m\mathbf{r} = \mathbf{v}_{p|m}$ = velocity of particle relative to moving system

$\boldsymbol{\omega}\times\mathbf{r} = \mathbf{v}_{m|f}$ = velocity of moving system relative to fixed system.

Then (*1*) can be written as

(*2*) $$\mathbf{v}_{p|f} = \mathbf{v}_{p|m} + \boldsymbol{\omega}\times\mathbf{r}$$

or in the suggestive notation

(*3*) $$\mathbf{v}_{p|f} = \mathbf{v}_{p|m} + \mathbf{v}_{m|f}$$

Note that the roles of fixed and moving observers can, of course, be interchanged. Thus the fixed observer can think of himself as really moving with respect to the other. For this case we must interchange subscripts *m* and *f* and also change $\boldsymbol{\omega}$ to $-\boldsymbol{\omega}$ since the relative rotation is reversed. If this is done, (*2*) becomes

$$\mathbf{v}_{p|m} = \mathbf{v}_{p|f} - \boldsymbol{\omega}\times\mathbf{r} \quad \text{or} \quad \mathbf{v}_{p|f} = \mathbf{v}_{p|m} + \boldsymbol{\omega}\times\mathbf{r}$$

so that the result is valid for each observer.

(b) The acceleration of the particle as determined by the fixed observer at $O$ is $D_f^2\mathbf{r} = D_f(D_f\mathbf{r})$. Take $D_f$ of both sides of (1), using the operator equivalence established in Problem 29(b). Then

$$\begin{aligned} D_f(D_f\mathbf{r}) &= D_f(D_m\mathbf{r} + \boldsymbol{\omega}\times\mathbf{r}) \\ &= (D_m + \boldsymbol{\omega}\times)(D_m\mathbf{r} + \boldsymbol{\omega}\times\mathbf{r}) \\ &= D_m(D_m\mathbf{r} + \boldsymbol{\omega}\times\mathbf{r}) + \boldsymbol{\omega}\times(D_m\mathbf{r} + \boldsymbol{\omega}\times\mathbf{r}) \\ &= D_m^2\mathbf{r} + D_m(\boldsymbol{\omega}\times\mathbf{r}) + \boldsymbol{\omega}\times D_m\mathbf{r} + \boldsymbol{\omega}\times(\boldsymbol{\omega}\times\mathbf{r}) \end{aligned}$$

or

$$D_f^2\mathbf{r} = D_m^2\mathbf{r} + 2\boldsymbol{\omega}\times D_m\mathbf{r} + (D_m\boldsymbol{\omega})\times\mathbf{r} + \boldsymbol{\omega}\times(\boldsymbol{\omega}\times\mathbf{r})$$

Let $\mathbf{a}_{p|f} = D_f^2\mathbf{r}$ = acceleration of particle relative to fixed system

$\mathbf{a}_{p|m} = D_m^2\mathbf{r}$ = acceleration of particle relative to moving system.

Then $\mathbf{a}_{m|f} = 2\boldsymbol{\omega}\times D_m\mathbf{r} + (D_m\boldsymbol{\omega})\times\mathbf{r} + \boldsymbol{\omega}\times(\boldsymbol{\omega}\times\mathbf{r})$

= acceleration of moving system relative to fixed system

and we can write

$$\mathbf{a}_{p|f} = \mathbf{a}_{p|m} + \mathbf{a}_{m|f}\,.$$

For many cases of importance $\boldsymbol{\omega}$ is a constant vector, i.e. the rotation proceeds with constant angular velocity. Then $D_m\boldsymbol{\omega} = \mathbf{0}$ and

$$\mathbf{a}_{m|f} = 2\boldsymbol{\omega}\times D_m\mathbf{r} + \boldsymbol{\omega}\times(\boldsymbol{\omega}\times\mathbf{r}) = 2\boldsymbol{\omega}\times\mathbf{v}_m + \boldsymbol{\omega}\times(\boldsymbol{\omega}\times\mathbf{r})$$

The quantity $2\boldsymbol{\omega}\times\mathbf{v}_m$ is called the *Coriolis acceleration* and $\boldsymbol{\omega}\times(\boldsymbol{\omega}\times\mathbf{r})$ is called the *centripetal acceleration*.

Newton's laws are strictly valid only in *inertial systems*, i.e. systems which are either fixed or which move with constant velocity relative to a fixed system. The earth is not exactly an inertial system and this accounts for the presence of the so called 'fictitious' extra forces (Coriolis, etc.) which must be considered. If the mass of a particle is a constant $M$, then Newton's second law becomes

$$(4)\qquad M D_m^2\mathbf{r} = \mathbf{F} - 2M(\boldsymbol{\omega}\times D_m\mathbf{r}) - M[\boldsymbol{\omega}\times(\boldsymbol{\omega}\times\mathbf{r})]$$

where $D_m$ denotes $d/dt$ as computed by an observer on the earth, and $\mathbf{F}$ is the resultant of all real forces as measured by this observer. The last two terms on the right of (4) are negligible in most cases and are not used in practice.

The *theory of relativity* due to Einstein has modified quite radically the concepts of absolute motion which are implied by Newtonian concepts and has led to revision of Newton's laws.

## SUPPLEMENTARY PROBLEMS

**31.** If $\mathbf{R} = e^{-t}\mathbf{i} + \ln(t^2+1)\mathbf{j} - \tan t\,\mathbf{k}$, find (a) $\dfrac{d\mathbf{R}}{dt}$, (b) $\dfrac{d^2\mathbf{R}}{dt^2}$, (c) $\left|\dfrac{d\mathbf{R}}{dt}\right|$, (d) $\left|\dfrac{d^2\mathbf{R}}{dt^2}\right|$ at $t=0$.
*Ans.* (a) $-\mathbf{i}-\mathbf{k}$, (b) $\mathbf{i}+2\mathbf{j}$, (c) $\sqrt{2}$, (d) $\sqrt{5}$

**32.** Find the velocity and acceleration of a particle which moves along the curve $x = 2\sin 3t$, $y = 2\cos 3t$, $z = 8t$ at any time $t > 0$. Find the magnitude of the velocity and acceleration.
*Ans.* $\mathbf{v} = 6\cos 3t\,\mathbf{i} - 6\sin 3t\,\mathbf{j} + 8\mathbf{k}$, $\mathbf{a} = -18\sin 3t\,\mathbf{i} - 18\cos 3t\,\mathbf{j}$, $|\mathbf{v}| = 10$, $|\mathbf{a}| = 18$

**33.** Find a unit tangent vector to any point on the curve $x = a\cos\omega t$, $y = a\sin\omega t$, $z = bt$ where $a, b, \omega$ are constants. *Ans.* $\dfrac{-a\omega\sin\omega t\,\mathbf{i} + a\omega\cos\omega t\,\mathbf{j} + b\mathbf{k}}{\sqrt{a^2\omega^2 + b^2}}$

**34.** If $\mathbf{A} = t^2\mathbf{i} - t\mathbf{j} + (2t+1)\mathbf{k}$ and $\mathbf{B} = (2t-3)\mathbf{i} + \mathbf{j} - t\mathbf{k}$, find
(a) $\dfrac{d}{dt}(\mathbf{A}\cdot\mathbf{B})$, (b) $\dfrac{d}{dt}(\mathbf{A}\times\mathbf{B})$, (c) $\dfrac{d}{dt}|\mathbf{A}+\mathbf{B}|$, (d) $\dfrac{d}{dt}(\mathbf{A}\times\dfrac{d\mathbf{B}}{dt})$ at $t=1$. *Ans.* (a) $-6$, (b) $7\mathbf{j}+3\mathbf{k}$, (c) 1, (d) $\mathbf{i}+6\mathbf{j}+2\mathbf{k}$

35. If $\mathbf{A} = \sin u\ \mathbf{i} + \cos u\ \mathbf{j} + u\mathbf{k}$, $\mathbf{B} = \cos u\ \mathbf{i} - \sin u\ \mathbf{j} - 3\mathbf{k}$, and $\mathbf{C} = 2\mathbf{i} + 3\mathbf{j} - \mathbf{k}$, find $\dfrac{d}{du}(\mathbf{A}\times(\mathbf{B}\times\mathbf{C}))$ at $u=0$.
*Ans.* $7\mathbf{i} + 6\mathbf{j} - 6\mathbf{k}$

36. Find $\dfrac{d}{ds}\left(\mathbf{A}\cdot\dfrac{d\mathbf{B}}{ds} - \dfrac{d\mathbf{A}}{ds}\cdot\mathbf{B}\right)$ if $\mathbf{A}$ and $\mathbf{B}$ are differentiable functions of $s$. *Ans.* $\mathbf{A}\cdot\dfrac{d^2\mathbf{B}}{ds^2} - \dfrac{d^2\mathbf{A}}{ds^2}\cdot\mathbf{B}$

37. If $\mathbf{A}(t) = 3t^2\,\mathbf{i} - (t+4)\mathbf{j} + (t^2-2t)\mathbf{k}$ and $\mathbf{B}(t) = \sin t\ \mathbf{i} + 3e^{-t}\mathbf{j} - 3\cos t\ \mathbf{k}$, find $\dfrac{d^2}{dt^2}(\mathbf{A}\times\mathbf{B})$ at $t=0$.
*Ans.* $-30\mathbf{i} + 14\mathbf{j} + 20\mathbf{k}$

38. If $\dfrac{d^2\mathbf{A}}{dt^2} = 6t\,\mathbf{i} - 24t^2\,\mathbf{j} + 4\sin t\ \mathbf{k}$, find $\mathbf{A}$ given that $\mathbf{A} = 2\mathbf{i} + \mathbf{j}$ and $\dfrac{d\mathbf{A}}{dt} = -\mathbf{i} - 3\mathbf{k}$ at $t=0$.
*Ans.* $\mathbf{A} = (t^3 - t + 2)\mathbf{i} + (1 - 2t^4)\mathbf{j} + (t - 4\sin t)\mathbf{k}$

39. Show that $\mathbf{r} = e^{-t}(\mathbf{C}_1 \cos 2t + \mathbf{C}_2 \sin 2t)$, where $\mathbf{C}_1$ and $\mathbf{C}_2$ are constant vectors, is a solution of the differential equation $\dfrac{d^2\mathbf{r}}{dt^2} + 2\dfrac{d\mathbf{r}}{dt} + 5\mathbf{r} = \mathbf{0}$.

40. Show that the general solution of the differential equation $\dfrac{d^2\mathbf{r}}{dt^2} + 2\alpha\dfrac{d\mathbf{r}}{dt} + \omega^2\mathbf{r} = \mathbf{0}$, where $\alpha$ and $\omega$ are constants, is
(*a*) $\mathbf{r} = e^{-\alpha t}(\mathbf{C}_1 e^{\sqrt{\alpha^2-\omega^2}\,t} + \mathbf{C}_2 e^{-\sqrt{\alpha^2-\omega^2}\,t})$ if $\alpha^2 - \omega^2 > 0$
(*b*) $\mathbf{r} = e^{-\alpha t}(\mathbf{C}_1 \sin\sqrt{\omega^2-\alpha^2}\,t + \mathbf{C}_2 \cos\sqrt{\omega^2-\alpha^2}\,t)$ if $\alpha^2 - \omega^2 < 0$.
(*c*) $\mathbf{r} = e^{-\alpha t}(\mathbf{C}_1 + \mathbf{C}_2 t)$ if $\alpha^2 - \omega^2 = 0$,
where $\mathbf{C}_1$ and $\mathbf{C}_2$ are arbitrary constant vectors.

41. Solve (*a*) $\dfrac{d^2\mathbf{r}}{dt^2} - 4\dfrac{d\mathbf{r}}{dt} - 5\mathbf{r} = \mathbf{0}$, (*b*) $\dfrac{d^2\mathbf{r}}{dt^2} + 2\dfrac{d\mathbf{r}}{dt} + \mathbf{r} = \mathbf{0}$, (*c*) $\dfrac{d^2\mathbf{r}}{dt^2} + 4\mathbf{r} = \mathbf{0}$.
*Ans.* (*a*) $\mathbf{r} = \mathbf{C}_1 e^{5t} + \mathbf{C}_2 e^{-t}$, (*b*) $\mathbf{r} = e^{-t}(\mathbf{C}_1 + \mathbf{C}_2 t)$, (*c*) $\mathbf{r} = \mathbf{C}_1 \cos 2t + \mathbf{C}_2 \sin 2t$

42. Solve $\dfrac{d\mathbf{Y}}{dt} = \mathbf{X}$, $\dfrac{d\mathbf{X}}{dt} = -\mathbf{Y}$. *Ans.* $\mathbf{X} = \mathbf{C}_1 \cos t + \mathbf{C}_2 \sin t$, $\mathbf{Y} = \mathbf{C}_1 \sin t - \mathbf{C}_2 \cos t$

43. If $\mathbf{A} = \cos xy\ \mathbf{i} + (3xy - 2x^2)\mathbf{j} - (3x + 2y)\mathbf{k}$, find $\dfrac{\partial\mathbf{A}}{\partial x}, \dfrac{\partial\mathbf{A}}{\partial y}, \dfrac{\partial^2\mathbf{A}}{\partial x^2}, \dfrac{\partial^2\mathbf{A}}{\partial y^2}, \dfrac{\partial^2\mathbf{A}}{\partial x\,\partial y}, \dfrac{\partial^2\mathbf{A}}{\partial y\,\partial x}$.
*Ans.* $\dfrac{\partial\mathbf{A}}{\partial x} = -y\sin xy\ \mathbf{i} + (3y - 4x)\mathbf{j} - 3\mathbf{k}$, $\dfrac{\partial\mathbf{A}}{\partial y} = -x\sin xy\ \mathbf{i} + 3x\,\mathbf{j} - 2\mathbf{k}$,
$\dfrac{\partial^2\mathbf{A}}{\partial x^2} = -y^2\cos xy\ \mathbf{i} - 4\mathbf{j}$, $\dfrac{\partial^2\mathbf{A}}{\partial y^2} = -x^2\cos xy\ \mathbf{i}$, $\dfrac{\partial^2\mathbf{A}}{\partial x\,\partial y} = \dfrac{\partial^2\mathbf{A}}{\partial y\,\partial x} = -(xy\cos xy + \sin xy)\mathbf{i} + 3\mathbf{j}$

44. If $\mathbf{A} = x^2yz\,\mathbf{i} - 2xz^3\,\mathbf{j} + xz^2\,\mathbf{k}$ and $\mathbf{B} = 2z\,\mathbf{i} + y\,\mathbf{j} - x^2\,\mathbf{k}$, find $\dfrac{\partial^2}{\partial x\,\partial y}(\mathbf{A}\times\mathbf{B})$ at $(1,0,-2)$.
*Ans.* $-4\mathbf{i} - 8\mathbf{j}$

45. If $\mathbf{C}_1$ and $\mathbf{C}_2$ are constant vectors and $\lambda$ is a constant scalar, show that $\mathbf{H} = e^{-\lambda x}(\mathbf{C}_1 \sin\lambda y + \mathbf{C}_2 \cos\lambda y)$ satisfies the partial differential equation $\dfrac{\partial^2\mathbf{H}}{\partial x^2} + \dfrac{\partial^2\mathbf{H}}{\partial y^2} = \mathbf{0}$.

46. Prove that $\mathbf{A} = \dfrac{\mathbf{p}_0\, e^{i\omega(t-r/c)}}{r}$, where $\mathbf{p}_0$ is a constant vector, $\omega$ and $c$ are constant scalars and $i = \sqrt{-1}$, satisfies the equation $\dfrac{\partial^2\mathbf{A}}{\partial r^2} + \dfrac{2}{r}\dfrac{\partial\mathbf{A}}{\partial r} = \dfrac{1}{c^2}\dfrac{\partial^2\mathbf{A}}{\partial t^2}$. This result is of importance in *electromagnetic theory.*

## DIFFERENTIAL GEOMETRY

47. Find (*a*) the unit tangent $\mathbf{T}$, (*b*) the curvature $\kappa$, (*c*) the principal normal $\mathbf{N}$, (*d*) the binormal $\mathbf{B}$, and (*e*) the torsion $\tau$ for the space curve $x = t - t^3/3$, $y = t^2$, $z = t + t^3/3$.
*Ans.* (*a*) $\mathbf{T} = \dfrac{(1-t^2)\mathbf{i} + 2t\,\mathbf{j} + (1+t^2)\mathbf{k}}{\sqrt{2}(1+t^2)}$ (*c*) $\mathbf{N} = -\dfrac{2t}{1+t^2}\mathbf{i} + \dfrac{1-t^2}{1+t^2}\mathbf{j}$ (*e*) $\tau = \dfrac{1}{(1+t^2)^2}$
(*b*) $\kappa = \dfrac{1}{(1+t^2)^2}$ (*d*) $\mathbf{B} = \dfrac{(t^2-1)\mathbf{i} - 2t\,\mathbf{j} + (t^2+1)\mathbf{k}}{\sqrt{2}(1+t^2)}$

**48.** A space curve is defined in terms of the arc length parameter $s$ by the equations

$$x = \arc\tan s, \quad y = \tfrac{1}{2}\sqrt{2}\,\ln(s^2+1), \quad z = s - \arc\tan s$$

Find (*a*) **T**, (*b*) **N**, (*c*) **B**, (*d*) $\kappa$, (*e*) $\tau$, (*f*) $\rho$, (*g*) $\sigma$.

*Ans.* (*a*) $\mathbf{T} = \dfrac{\mathbf{i} + \sqrt{2}s\,\mathbf{j} + s^2\mathbf{k}}{s^2+1}$ (*d*) $\kappa = \dfrac{\sqrt{2}}{s^2+1}$

(*b*) $\mathbf{N} = \dfrac{-\sqrt{2}s\,\mathbf{i} + (1-s^2)\mathbf{j} + \sqrt{2}s\,\mathbf{k}}{s^2+1}$ (*e*) $\tau = \dfrac{\sqrt{2}}{s^2+1}$ (*g*) $\sigma = \dfrac{s^2+1}{\sqrt{2}}$

(*c*) $\mathbf{B} = \dfrac{s^2\mathbf{i} - \sqrt{2}s\,\mathbf{j} + \mathbf{k}}{s^2+1}$ (*f*) $\rho = \dfrac{s^2+1}{\sqrt{2}}$

**49.** Find $\kappa$ and $\tau$ for the space curve $x = t,\ y = t^2,\ z = t^3$ called the twisted cubic.

*Ans.* $\kappa = \dfrac{2\sqrt{9t^4 + 9t^2 + 1}}{(9t^4 + 4t^2 + 1)^{3/2}}, \quad \tau = \dfrac{3}{9t^4 + 9t^2 + 1}$

**50.** Show that for a plane curve the torsion $\tau = 0$.

**51.** Show that the radius of curvature of a plane curve with equations $y = f(x)$, $z = 0$, i.e. a curve in the $xy$ plane is given by $\rho = \dfrac{[1+(y')^2]^{3/2}}{|y''|}$.

**52.** Find the curvature and radius of curvature of the curve with position vector $\mathbf{r} = a\cos u\,\mathbf{i} + b\sin u\,\mathbf{j}$, where $a$ and $b$ are positive constants. Interpret the case where $a = b$.

*Ans.* $\kappa = \dfrac{ab}{(a^2\sin^2 u + b^2\cos^2 u)^{3/2}} = \dfrac{1}{\rho}$; if $a = b$, the given curve which is an ellipse, becomes a circle of radius $a$ and its radius of curvature $\rho = a$.

**53.** Show that the Frenet-Serret formulae can be written in the form $\dfrac{d\mathbf{T}}{ds} = \boldsymbol{\omega}\times\mathbf{T}$, $\dfrac{d\mathbf{N}}{ds} = \boldsymbol{\omega}\times\mathbf{N}$, $\dfrac{d\mathbf{B}}{ds} = \boldsymbol{\omega}\times\mathbf{B}$ and determine $\boldsymbol{\omega}$. *Ans.* $\boldsymbol{\omega} = \tau\mathbf{T} + \kappa\mathbf{B}$

**54.** Prove that the curvature of the space curve $\mathbf{r} = \mathbf{r}(t)$ is given numerically by $\kappa = \dfrac{|\dot{\mathbf{r}}\times\ddot{\mathbf{r}}|}{|\dot{\mathbf{r}}|^3}$, where dots denote differentiation with respect to $t$.

**55.** (*a*) Prove that $\tau = \dfrac{\dot{\mathbf{r}}\cdot\ddot{\mathbf{r}}\times\dddot{\mathbf{r}}}{|\dot{\mathbf{r}}\times\ddot{\mathbf{r}}|^2}$ for the space curve $\mathbf{r} = \mathbf{r}(t)$.

(*b*) If the parameter $t$ is the arc length $s$ show that $\tau = \dfrac{\dfrac{d\mathbf{r}}{ds}\cdot\dfrac{d^2\mathbf{r}}{ds^2}\times\dfrac{d^3\mathbf{r}}{ds^3}}{(d^2\mathbf{r}/ds^2)^2}$.

**56.** If $\mathbf{Q} = \dot{\mathbf{r}}\times\ddot{\mathbf{r}}$, show that $\kappa = \dfrac{Q}{|\dot{\mathbf{r}}|^3}$, $\tau = \dfrac{\mathbf{Q}\cdot\dddot{\mathbf{r}}}{Q^2}$.

**57.** Find $\kappa$ and $\tau$ for the space curve $x = \theta - \sin\theta,\ y = 1 - \cos\theta,\ z = 4\sin(\theta/2)$.

*Ans.* $\kappa = \dfrac{1}{8}\sqrt{6 - 2\cos\theta}, \quad \tau = \dfrac{(3+\cos\theta)\cos\theta/2 + 2\sin\theta\,\sin\theta/2}{12\cos\theta - 4}$

**58.** Find the torsion of the curve $x = \dfrac{2t+1}{t-1}$, $y = \dfrac{t^2}{t-1}$, $z = t + 2$. Explain your answer.

*Ans.* $\tau = 0$. The curve lies on the plane $x - 3y + 3z = 5$.

**59.** Show that the equations of the tangent line, principal normal and binormal to the space curve $\mathbf{r} = \mathbf{r}(t)$ at the point $t = t_o$ can be written respectively $\mathbf{r} = \mathbf{r}_o + t\mathbf{T}_o$, $\mathbf{r} = \mathbf{r}_o + t\mathbf{N}_o$, $\mathbf{r} = \mathbf{r}_o + t\mathbf{B}_o$, where $t$ is a parameter.

**60.** Find equations for the (*a*) tangent, (*b*) principal normal and (*c*) binormal to the curve $x = 3\cos t$, $y = 3\sin t$, $z = 4t$ at the point where $t = \pi$.

*Ans.* (*a*) Tangent: $\mathbf{r} = -3\mathbf{i} + 4\pi\mathbf{k} + t(-\frac{3}{5}\mathbf{j} + \frac{4}{5}\mathbf{k})$ or $x = -3,\ y = -\frac{3}{5}t,\ z = 4\pi + \frac{4}{5}t$.

(*b*) Normal: $\mathbf{r} = -3\mathbf{i} + 4\pi\mathbf{j} + t\mathbf{i}$ or $x = -3 + t,\ y = 4\pi,\ z = 0.$

(*c*) Binormal: $\mathbf{r} = -3\mathbf{i} + 4\pi\mathbf{j} + t(\frac{4}{5}\mathbf{j} + \frac{3}{5}\mathbf{k})$ or $x = -3,\ y = 4\pi + \frac{4}{5}t,\ z = \frac{3}{5}t.$

**61.** Find equations for the (*a*) osculating plane, (*b*) normal plane and (*c*) rectifying plane to the curve $x = 3t - t^3$, $y = 3t^2$, $z = 3t + t^3$ at the point where $t = 1$. *Ans.* (*a*) $y - z + 1 = 0$, (*b*) $y + z - 7 = 0$, (*c*) $x = 2$

**62.** (*a*) Show that the differential of arc length on the surface $\mathbf{r} = \mathbf{r}(u, v)$ is given by

$$ds^2 = E\,du^2 + 2F\,du\,dv + G\,dv^2$$

where $E = \frac{\partial \mathbf{r}}{\partial u}\cdot\frac{\partial \mathbf{r}}{\partial u} = (\frac{\partial \mathbf{r}}{\partial u})^2$, $F = \frac{\partial \mathbf{r}}{\partial u}\cdot\frac{\partial \mathbf{r}}{\partial v}$, $G = \frac{\partial \mathbf{r}}{\partial v}\cdot\frac{\partial \mathbf{r}}{\partial v} = (\frac{\partial \mathbf{r}}{\partial v})^2$.

(*b*) Prove that a necessary and sufficient condition that the $u, v$ curvilinear coordinate system be orthogonal is $F \equiv 0$.

**63.** Find an equation for the tangent plane to the surface $z = xy$ at the point (2,3,6). *Ans.* $3x + 2y - z = 6$

**64.** Find equations for the tangent plane and normal line to the surface $4z = x^2 - y^2$ at the point (3,1,2).
*Ans.* $3x - y - 2z = 4$; $x = 3t + 3,\ y = 1 - t,\ z = 2 - 2t$

**65.** Prove that a unit normal to the surface $\mathbf{r} = \mathbf{r}(u, v)$ is $\mathbf{n} = \pm \dfrac{\frac{\partial \mathbf{r}}{\partial u} \times \frac{\partial \mathbf{r}}{\partial v}}{\sqrt{EG - F^2}}$, where $E, F$, and $G$ are defined as in Problem 62.

## MECHANICS

**66.** A particle moves along the curve $\mathbf{r} = (t^3 - 4t)\mathbf{i} + (t^2 + 4t)\mathbf{j} + (8t^2 - 3t^3)\mathbf{k}$, where $t$ is the time. Find the magnitudes of the tangential and normal components of its acceleration when $t = 2$.
*Ans.* Tangential, 16 ; normal, $2\sqrt{73}$

**67.** If a particle has velocity $\mathbf{v}$ and acceleration $\mathbf{a}$ along a space curve, prove that the radius of curvature of its path is given numerically by $\rho = \dfrac{v^3}{|\mathbf{v} \times \mathbf{a}|}$.

**68.** An object is attracted to a fixed point $O$ with a force $\mathbf{F} = f(r)\mathbf{r}$, called a *central force*, where $\mathbf{r}$ is the position vector of the object relative to $O$. Show that $\mathbf{r} \times \mathbf{v} = \mathbf{h}$ where $\mathbf{h}$ is a constant vector. Prove that the angular momentum is constant.

**69.** Prove that the acceleration vector of a particle moving along a space curve always lies in the osculating plane.

**70.** (*a*) Find the acceleration of a particle moving in the $xy$ plane in terms of polar coordinates $(\rho, \phi)$.
(*b*) What are the components of the acceleration parallel and perpendicular to $\rho$ ?

*Ans.* (*a*) $$\ddot{\mathbf{r}} = [(\ddot{\rho} - \rho\dot{\phi}^2)\cos\phi - (\rho\ddot{\phi} + 2\dot{\rho}\dot{\phi})\sin\phi]\,\mathbf{i} + [(\ddot{\rho} - \rho\dot{\phi}^2)\sin\phi + (\rho\ddot{\phi} + 2\dot{\rho}\dot{\phi})\cos\phi]\,\mathbf{j}$$

(*b*) $\ddot{\rho} - \rho\dot{\phi}^2,\ \rho\ddot{\phi} + 2\dot{\rho}\dot{\phi}$

# Chapter 4

# GRADIENT, DIVERGENCE and CURL

**THE VECTOR DIFFERENTIAL OPERATOR DEL**, written $\nabla$, is defined by

$$\nabla \equiv \frac{\partial}{\partial x}\mathbf{i} + \frac{\partial}{\partial y}\mathbf{j} + \frac{\partial}{\partial z}\mathbf{k} \equiv \mathbf{i}\frac{\partial}{\partial x} + \mathbf{j}\frac{\partial}{\partial y} + \mathbf{k}\frac{\partial}{\partial z}$$

This vector operator possesses properties analogous to those of ordinary vectors. It is useful in defining three quantities which arise in practical applications and are known as the *gradient*, the *divergence* and the *curl*. The operator $\nabla$ is also known as *nabla*.

**THE GRADIENT.** Let $\phi(x,y,z)$ be defined and differentiable at each point $(x,y,z)$ in a certain region of space (i.e. $\phi$ defines a differentiable scalar field). Then the *gradient* of $\phi$, written $\nabla\phi$ or grad $\phi$, is defined by

$$\nabla\phi = \left(\frac{\partial}{\partial x}\mathbf{i} + \frac{\partial}{\partial y}\mathbf{j} + \frac{\partial}{\partial z}\mathbf{k}\right)\phi = \frac{\partial\phi}{\partial x}\mathbf{i} + \frac{\partial\phi}{\partial y}\mathbf{j} + \frac{\partial\phi}{\partial z}\mathbf{k}$$

Note that $\nabla\phi$ defines a vector field.

The component of $\nabla\phi$ in the direction of a unit vector $\mathbf{a}$ is given by $\nabla\phi\cdot\mathbf{a}$ and is called the *directional derivative* of $\phi$ in the direction $\mathbf{a}$. Physically, this is the rate of change of $\phi$ at $(x,y,z)$ in the direction $\mathbf{a}$.

**THE DIVERGENCE.** Let $\mathbf{V}(x,y,z) = V_1\mathbf{i} + V_2\mathbf{j} + V_3\mathbf{k}$ be defined and differentiable at each point $(x,y,z)$ in a certain region of space (i.e. $\mathbf{V}$ defines a differentiable vector field). Then the *divergence* of $\mathbf{V}$, written $\nabla\cdot\mathbf{V}$ or div $\mathbf{V}$, is defined by

$$\nabla\cdot\mathbf{V} = \left(\frac{\partial}{\partial x}\mathbf{i} + \frac{\partial}{\partial y}\mathbf{j} + \frac{\partial}{\partial z}\mathbf{k}\right)\cdot(V_1\mathbf{i} + V_2\mathbf{j} + V_3\mathbf{k})$$

$$= \frac{\partial V_1}{\partial x} + \frac{\partial V_2}{\partial y} + \frac{\partial V_3}{\partial z}$$

Note the analogy with $\mathbf{A}\cdot\mathbf{B} = A_1B_1 + A_2B_2 + A_3B_3$. Also note that $\nabla\cdot\mathbf{V} \neq \mathbf{V}\cdot\nabla$.

**THE CURL.** If $\mathbf{V}(x,y,z)$ is a differentiable vector field then the *curl* or *rotation* of $\mathbf{V}$, written $\nabla\times\mathbf{V}$, curl $\mathbf{V}$ or rot $\mathbf{V}$, is defined by

$$\nabla\times\mathbf{V} = \left(\frac{\partial}{\partial x}\mathbf{i} + \frac{\partial}{\partial y}\mathbf{j} + \frac{\partial}{\partial z}\mathbf{k}\right)\times(V_1\mathbf{i} + V_2\mathbf{j} + V_3\mathbf{k})$$

$$= \begin{vmatrix} \mathbf{i} & \mathbf{j} & \mathbf{k} \\ \frac{\partial}{\partial x} & \frac{\partial}{\partial y} & \frac{\partial}{\partial z} \\ V_1 & V_2 & V_3 \end{vmatrix}$$

$$= \begin{vmatrix} \frac{\partial}{\partial y} & \frac{\partial}{\partial z} \\ V_2 & V_3 \end{vmatrix} \mathbf{i} - \begin{vmatrix} \frac{\partial}{\partial x} & \frac{\partial}{\partial z} \\ V_1 & V_3 \end{vmatrix} \mathbf{j} + \begin{vmatrix} \frac{\partial}{\partial x} & \frac{\partial}{\partial y} \\ V_1 & V_2 \end{vmatrix} \mathbf{k}$$

$$= \left(\frac{\partial V_3}{\partial y} - \frac{\partial V_2}{\partial z}\right)\mathbf{i} + \left(\frac{\partial V_1}{\partial z} - \frac{\partial V_3}{\partial x}\right)\mathbf{j} + \left(\frac{\partial V_2}{\partial x} - \frac{\partial V_1}{\partial y}\right)\mathbf{k}$$

Note that in the expansion of the determinant the operators $\frac{\partial}{\partial x}, \frac{\partial}{\partial y}, \frac{\partial}{\partial z}$ must *precede* $V_1, V_2, V_3$.

**FORMULAE INVOLVING $\nabla$.** If **A** and **B** are differentiable vector functions, and $\phi$ and $\psi$ are differentiable scalar functions of position $(x,y,z)$, then

1. $\nabla(\phi+\psi) = \nabla\phi + \nabla\psi$ or $\text{grad}(\phi+\psi) = \text{grad}\,\phi + \text{grad}\,\psi$
2. $\nabla\cdot(\mathbf{A}+\mathbf{B}) = \nabla\cdot\mathbf{A} + \nabla\cdot\mathbf{B}$ or $\text{div}(\mathbf{A}+\mathbf{B}) = \text{div}\,\mathbf{A} + \text{div}\,\mathbf{B}$
3. $\nabla\times(\mathbf{A}+\mathbf{B}) = \nabla\times\mathbf{A} + \nabla\times\mathbf{B}$ or $\text{curl}(\mathbf{A}+\mathbf{B}) = \text{curl}\,\mathbf{A} + \text{curl}\,\mathbf{B}$
4. $\nabla\cdot(\phi\mathbf{A}) = (\nabla\phi)\cdot\mathbf{A} + \phi(\nabla\cdot\mathbf{A})$
5. $\nabla\times(\phi\mathbf{A}) = (\nabla\phi)\times\mathbf{A} + \phi(\nabla\times\mathbf{A})$
6. $\nabla\cdot(\mathbf{A}\times\mathbf{B}) = \mathbf{B}\cdot(\nabla\times\mathbf{A}) - \mathbf{A}\cdot(\nabla\times\mathbf{B})$
7. $\nabla\times(\mathbf{A}\times\mathbf{B}) = (\mathbf{B}\cdot\nabla)\mathbf{A} - \mathbf{B}(\nabla\cdot\mathbf{A}) - (\mathbf{A}\cdot\nabla)\mathbf{B} + \mathbf{A}(\nabla\cdot\mathbf{B})$
8. $\nabla(\mathbf{A}\cdot\mathbf{B}) = (\mathbf{B}\cdot\nabla)\mathbf{A} + (\mathbf{A}\cdot\nabla)\mathbf{B} + \mathbf{B}\times(\nabla\times\mathbf{A}) + \mathbf{A}\times(\nabla\times\mathbf{B})$
9. $\nabla\cdot(\nabla\phi) \equiv \nabla^2\phi \equiv \frac{\partial^2\phi}{\partial x^2} + \frac{\partial^2\phi}{\partial y^2} + \frac{\partial^2\phi}{\partial z^2}$

   where $\nabla^2 \equiv \frac{\partial^2}{\partial x^2} + \frac{\partial^2}{\partial y^2} + \frac{\partial^2}{\partial z^2}$ is called the *Laplacian operator.*
10. $\nabla\times(\nabla\phi) = \mathbf{0}$. The curl of the gradient of $\phi$ is zero.
11. $\nabla\cdot(\nabla\times\mathbf{A}) = 0$. The divergence of the curl of **A** is zero.
12. $\nabla\times(\nabla\times\mathbf{A}) = \nabla(\nabla\cdot\mathbf{A}) - \nabla^2\mathbf{A}$.

In Formulae 9-12, it is supposed that $\phi$ and **A** have continuous second partial derivatives.

**INVARIANCE.** Consider two rectangular coordinate systems or frames of reference $xyz$ and $x'y'z'$ (see figure below) having the same origin $O$ but with axes rotated with respect to each other.

A point $P$ in space has coordinates $(x,y,z)$ or $(x',y',z')$ relative to these coordinate systems. The equations of transformation between coordinates or the *coordinate transformations* are given by

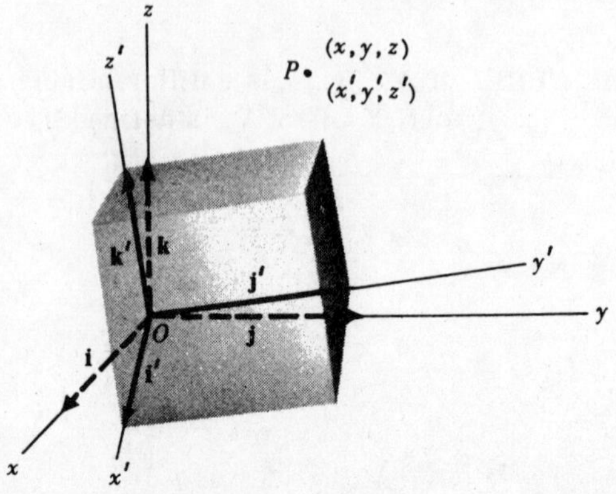


$$(1) \qquad \begin{aligned} x' &= l_{11}x + l_{12}y + l_{13}z \\ y' &= l_{21}x + l_{22}y + l_{23}z \\ z' &= l_{31}x + l_{32}y + l_{33}z \end{aligned}$$

where $l_{jk}$, $j,k = 1,2,3$, represent direction cosines of the $x'$, $y'$ and $z'$ axes with respect to the $x$, $y$, and

$z$ axes (see Problem 38). In case the origins of the two coordinate systems are not coincident the equations of transformation become

$$(2) \quad \begin{cases} x' = l_{11}x + l_{12}y + l_{13}z + a_1' \\ y' = l_{21}x + l_{22}y + l_{23}z + a_2' \\ z' = l_{31}x + l_{32}y + l_{33}z + a_3' \end{cases}$$

where origin $O$ of the $xyz$ coordinate system is located at $(a_1', a_2', a_3')$ relative to the $x'y'z'$ coordinate system.

The transformation equations (*1*) define a *pure rotation*, while equations (*2*) define a *rotation plus translation*. Any rigid body motion has the effect of a translation followed by a rotation. The transformation (*1*) is also called an *orthogonal transformation*. A general linear transformation is called an *affine transformation*.

Physically a scalar point function or scalar field $\phi(x,y,z)$ evaluated at a particular point should be independent of the coordinates of the point. Thus the temperature at a point is not dependent on whether coordinates $(x,y,z)$ or $(x',y',z')$ are used. Then if $\phi(x,y,z)$ is the temperature at point $P$ with coordinates $(x,y,z)$ while $\phi'(x',y',z')$ is the temperature at the same point $P$ with coordinates $(x',y',z')$, we must have $\phi(x,y,z) = \phi'(x',y',z')$. If $\phi(x,y,z) = \phi'(x',y',z')$, where $x,y,z$ and $x',y',z'$ are related by the transformation equations (*1*) or (*2*), we call $\phi(x,y,z)$ an *invariant* with respect to the transformation. For example, $x^2+y^2+z^2$ is invariant under the transformation of rotation (*1*), since $x^2+y^2+z^2 = x'^2 + y'^2 + z'^2$.

Similarly, a vector point function or vector field $\mathbf{A}(x,y,z)$ is called an *invariant* if $\mathbf{A}(x,y,z) = \mathbf{A}'(x',y',z')$. This will be true if

$$A_1(x,y,z)\mathbf{i} + A_2(x,y,z)\mathbf{j} + A_3(x,y,z)\mathbf{k} = A_1'(x',y',z')\mathbf{i}' + A_2'(x',y',z')\mathbf{j}' + A_3'(x',y',z')\mathbf{k}'$$

In Chap. 7 and 8, more general transformations are considered and the above concepts are extended.

It can be shown (see Problem 41) that the gradient of an invariant scalar field is an invariant vector field with respect to the transformations (*1*) or (*2*). Similarly, the divergence and curl of an invariant vector field are invariant under this transformation.

## SOLVED PROBLEMS

### THE GRADIENT

**1.** If $\phi(x,y,z) = 3x^2y - y^3z^2$, find $\nabla\phi$ (or grad $\phi$) at the point $(1,-2,-1)$.

$$\begin{aligned}
\nabla\phi &= \left(\frac{\partial}{\partial x}\mathbf{i} + \frac{\partial}{\partial y}\mathbf{j} + \frac{\partial}{\partial z}\mathbf{k}\right)(3x^2y - y^3z^2) \\
&= \mathbf{i}\frac{\partial}{\partial x}(3x^2y - y^3z^2) + \mathbf{j}\frac{\partial}{\partial y}(3x^2y - y^3z^2) + \mathbf{k}\frac{\partial}{\partial z}(3x^2y - y^3z^2) \\
&= 6xy\,\mathbf{i} + (3x^2 - 3y^2z^2)\mathbf{j} - 2y^3z\,\mathbf{k} \\
&= 6(1)(-2)\mathbf{i} + \{3(1)^2 - 3(-2)^2(-1)^2\}\mathbf{j} - 2(-2)^3(-1)\mathbf{k} \\
&= -12\mathbf{i} - 9\mathbf{j} - 16\mathbf{k}
\end{aligned}$$

2. Prove (*a*) $\nabla(F+G) = \nabla F + \nabla G$, (*b*) $\nabla(FG) = F\,\nabla G + G\,\nabla F$ where $F$ and $G$ are differentiable scalar functions of $x, y$ and $z$.

(*a*)
$$\nabla(F+G) = \left(\frac{\partial}{\partial x}\mathbf{i} + \frac{\partial}{\partial y}\mathbf{j} + \frac{\partial}{\partial z}\mathbf{k}\right)(F+G)$$
$$= \mathbf{i}\frac{\partial}{\partial x}(F+G) + \mathbf{j}\frac{\partial}{\partial y}(F+G) + \mathbf{k}\frac{\partial}{\partial z}(F+G)$$
$$= \mathbf{i}\frac{\partial F}{\partial x} + \mathbf{i}\frac{\partial G}{\partial x} + \mathbf{j}\frac{\partial F}{\partial y} + \mathbf{j}\frac{\partial G}{\partial y} + \mathbf{k}\frac{\partial F}{\partial z} + \mathbf{k}\frac{\partial G}{\partial z}$$
$$= \mathbf{i}\frac{\partial F}{\partial x} + \mathbf{j}\frac{\partial F}{\partial y} + \mathbf{k}\frac{\partial F}{\partial z} + \mathbf{i}\frac{\partial G}{\partial x} + \mathbf{j}\frac{\partial G}{\partial y} + \mathbf{k}\frac{\partial G}{\partial z}$$
$$= \left(\mathbf{i}\frac{\partial}{\partial x} + \mathbf{j}\frac{\partial}{\partial y} + \mathbf{k}\frac{\partial}{\partial z}\right)F + \left(\mathbf{i}\frac{\partial}{\partial x} + \mathbf{j}\frac{\partial}{\partial y} + \mathbf{k}\frac{\partial}{\partial z}\right)G = \nabla F + \nabla G$$

(*b*)
$$\nabla(FG) = \left(\frac{\partial}{\partial x}\mathbf{i} + \frac{\partial}{\partial y}\mathbf{j} + \frac{\partial}{\partial z}\mathbf{k}\right)(FG)$$
$$= \frac{\partial}{\partial x}(FG)\mathbf{i} + \frac{\partial}{\partial y}(FG)\mathbf{j} + \frac{\partial}{\partial z}(FG)\mathbf{k}$$
$$= \left(F\frac{\partial G}{\partial x} + G\frac{\partial F}{\partial x}\right)\mathbf{i} + \left(F\frac{\partial G}{\partial y} + G\frac{\partial F}{\partial y}\right)\mathbf{j} + \left(F\frac{\partial G}{\partial z} + G\frac{\partial F}{\partial z}\right)\mathbf{k}$$
$$= F\left(\frac{\partial G}{\partial x}\mathbf{i} + \frac{\partial G}{\partial y}\mathbf{j} + \frac{\partial G}{\partial z}\mathbf{k}\right) + G\left(\frac{\partial F}{\partial x}\mathbf{i} + \frac{\partial F}{\partial y}\mathbf{j} + \frac{\partial F}{\partial z}\mathbf{k}\right) = F\nabla G + G\nabla F$$

3. Find $\nabla\phi$ if (*a*) $\phi = \ln|\mathbf{r}|$, (*b*) $\phi = \dfrac{1}{r}$.

(*a*) $\mathbf{r} = x\mathbf{i} + y\mathbf{j} + z\mathbf{k}$. Then $|\mathbf{r}| = \sqrt{x^2+y^2+z^2}$ and $\phi = \ln|\mathbf{r}| = \frac{1}{2}\ln(x^2+y^2+z^2)$.

$$\nabla\phi = \tfrac{1}{2}\nabla\ln(x^2+y^2+z^2)$$
$$= \tfrac{1}{2}\left\{\mathbf{i}\frac{\partial}{\partial x}\ln(x^2+y^2+z^2) + \mathbf{j}\frac{\partial}{\partial y}\ln(x^2+y^2+z^2) + \mathbf{k}\frac{\partial}{\partial z}\ln(x^2+y^2+z^2)\right\}$$
$$= \tfrac{1}{2}\left\{\mathbf{i}\frac{2x}{x^2+y^2+z^2} + \mathbf{j}\frac{2y}{x^2+y^2+z^2} + \mathbf{k}\frac{2z}{x^2+y^2+z^2}\right\} = \frac{x\mathbf{i}+y\mathbf{j}+z\mathbf{k}}{x^2+y^2+z^2} = \frac{\mathbf{r}}{r^2}$$

(*b*)
$$\nabla\phi = \nabla\left(\frac{1}{r}\right) = \nabla\left(\frac{1}{\sqrt{x^2+y^2+z^2}}\right) = \nabla\{(x^2+y^2+z^2)^{-1/2}\}$$
$$= \mathbf{i}\frac{\partial}{\partial x}(x^2+y^2+z^2)^{-1/2} + \mathbf{j}\frac{\partial}{\partial y}(x^2+y^2+z^2)^{-1/2} + \mathbf{k}\frac{\partial}{\partial z}(x^2+y^2+z^2)^{-1/2}$$
$$= \mathbf{i}\left\{-\tfrac{1}{2}(x^2+y^2+z^2)^{-3/2}2x\right\} + \mathbf{j}\left\{-\tfrac{1}{2}(x^2+y^2+z^2)^{-3/2}2y\right\} + \mathbf{k}\left\{-\tfrac{1}{2}(x^2+y^2+z^2)^{-3/2}2z\right\}$$
$$= \frac{-x\mathbf{i}-y\mathbf{j}-z\mathbf{k}}{(x^2+y^2+z^2)^{3/2}} = -\frac{\mathbf{r}}{r^3}$$

4. Show that $\nabla r^n = nr^{n-2}\mathbf{r}$.

$$\nabla r^n = \nabla(\sqrt{x^2+y^2+z^2})^n = \nabla(x^2+y^2+z^2)^{n/2}$$
$$= \mathbf{i}\frac{\partial}{\partial x}\{(x^2+y^2+z^2)^{n/2}\} + \mathbf{j}\frac{\partial}{\partial y}\{(x^2+y^2+z^2)^{n/2}\} + \mathbf{k}\frac{\partial}{\partial z}\{(x^2+y^2+z^2)^{n/2}\}$$

$$= \mathbf{i}\{\frac{n}{2}(x^2+y^2+z^2)^{n/2-1}\,2x\} + \mathbf{j}\{\frac{n}{2}(x^2+y^2+z^2)^{n/2-1}\,2y\} + \mathbf{k}\{\frac{n}{2}(x^2+y^2+z^2)^{n/2-1}\,2z\}$$

$$= n(x^2+y^2+z^2)^{n/2-1}(x\mathbf{i} + y\mathbf{j} + z\mathbf{k})$$

$$= n(r^2)^{n/2-1}\mathbf{r} = nr^{n-2}\mathbf{r}$$

Note that if $\mathbf{r} = r\mathbf{r}_1$ where $\mathbf{r}_1$ is a unit vector in the direction $\mathbf{r}$, then $\nabla r^n = nr^{n-1}\mathbf{r}_1$.

5. Show that $\nabla\phi$ is a vector perpendicular to the surface $\phi(x,y,z) = c$ where $c$ is a constant.

Let $\mathbf{r} = x\mathbf{i} + y\mathbf{j} + z\mathbf{k}$ be the position vector to any point $P(x,y,z)$ on the surface. Then $d\mathbf{r} = dx\,\mathbf{i} + dy\,\mathbf{j} + dz\,\mathbf{k}$ lies in the tangent plane to the surface at $P$.

But $d\phi = \frac{\partial\phi}{\partial x}dx + \frac{\partial\phi}{\partial y}dy + \frac{\partial\phi}{\partial z}dz = 0$ or $(\frac{\partial\phi}{\partial x}\mathbf{i} + \frac{\partial\phi}{\partial y}\mathbf{j} + \frac{\partial\phi}{\partial z}\mathbf{k})\cdot(dx\,\mathbf{i} + dy\,\mathbf{j} + dz\,\mathbf{k}) = 0$

i.e. $\nabla\phi\cdot d\mathbf{r} = 0$ so that $\nabla\phi$ is perpendicular to $d\mathbf{r}$ and therefore to the surface.

6. Find a unit normal to the surface $x^2y + 2xz = 4$ at the point (2,–2,3).

$\nabla(x^2y + 2xz) = (2xy + 2z)\mathbf{i} + x^2\mathbf{j} + 2x\mathbf{k} = -2\mathbf{i} + 4\mathbf{j} + 4\mathbf{k}$ at the point (2,–2,3).

Then a unit normal to the surface $= \dfrac{-2\mathbf{i} + 4\mathbf{j} + 4\mathbf{k}}{\sqrt{(-2)^2+(4)^2+(4)^2}} = -\frac{1}{3}\mathbf{i} + \frac{2}{3}\mathbf{j} + \frac{2}{3}\mathbf{k}$.

Another unit normal is $\frac{1}{3}\mathbf{i} - \frac{2}{3}\mathbf{j} - \frac{2}{3}\mathbf{k}$ having direction opposite to that above.

7. Find an equation for the tangent plane to the surface $2xz^2 - 3xy - 4x = 7$ at the point (1,–1,2).

$$\nabla(2xz^2 - 3xy - 4x) = (2z^2 - 3y - 4)\mathbf{i} - 3x\mathbf{j} + 4xz\mathbf{k}$$

Then a normal to the surface at the point (1,–1,2) is $7\mathbf{i} - 3\mathbf{j} + 8\mathbf{k}$.

The equation of a plane passing through a point whose position vector is $\mathbf{r}_0$ and which is perpendicular to the normal $\mathbf{N}$ is $(\mathbf{r} - \mathbf{r}_0)\cdot\mathbf{N} = 0$. (See Chap.2, Prob.18.) Then the required equation is

$$[(x\mathbf{i} + y\mathbf{j} + z\mathbf{k}) - (\mathbf{i} - \mathbf{j} + 2\mathbf{k})]\cdot(7\mathbf{i} - 3\mathbf{j} + 8\mathbf{k}) = 0$$

or

$$7(x-1) - 3(y+1) + 8(z-2) = 0.$$

8. Let $\phi(x,y,z)$ and $\phi(x+\Delta x,\, y+\Delta y,\, z+\Delta z)$ be the temperatures at two neighbouring points $P(x,y,z)$ and $Q(x+\Delta x,\, y+\Delta y,\, z+\Delta z)$ of a certain region.

(a) Interpret physically the quantity $\dfrac{\Delta\phi}{\Delta s} = \dfrac{\phi(x+\Delta x,\, y+\Delta y,\, z+\Delta z) - \phi(x,y,z)}{\Delta s}$ where $\Delta s$ is the distance between points $P$ and $Q$.

(b) Evaluate $\lim\limits_{\Delta s\to 0}\dfrac{\Delta\phi}{\Delta s} = \dfrac{d\phi}{ds}$ and interpret physically.

(c) Show that $\dfrac{d\phi}{ds} = \nabla\phi\cdot\dfrac{d\mathbf{r}}{ds}$.

(a) Since $\Delta\phi$ is the change in temperature between points $P$ and $Q$ and $\Delta s$ is the distance between these points, $\dfrac{\Delta\phi}{\Delta s}$ represents the average rate of change in temperature per unit distance in the direction from $P$ to $Q$.

(b) From the calculus,

$$\Delta\phi = \frac{\partial\phi}{\partial x}\Delta x + \frac{\partial\phi}{\partial y}\Delta y + \frac{\partial\phi}{\partial z}\Delta z + \text{infinitesimals of order higher than } \Delta x, \Delta y \text{ and } \Delta z$$

Then
$$\lim_{\Delta s\to 0}\frac{\Delta\phi}{\Delta s} = \lim_{\Delta s\to 0}\frac{\partial\phi}{\partial x}\frac{\Delta x}{\Delta s} + \frac{\partial\phi}{\partial y}\frac{\Delta y}{\Delta s} + \frac{\partial\phi}{\partial z}\frac{\Delta z}{\Delta s}$$

or
$$\frac{d\phi}{ds} = \frac{\partial\phi}{\partial x}\frac{dx}{ds} + \frac{\partial\phi}{\partial y}\frac{dy}{ds} + \frac{\partial\phi}{\partial z}\frac{dz}{ds}$$

$\frac{d\phi}{ds}$ represents the rate of change of temperature with respect to distance at point $P$ in a direction toward $Q$. This is also called the *directional derivative* of $\phi$.

(c)
$$\frac{d\phi}{ds} = \frac{\partial\phi}{\partial x}\frac{dx}{ds} + \frac{\partial\phi}{\partial y}\frac{dy}{ds} + \frac{\partial\phi}{\partial z}\frac{dz}{ds} = (\frac{\partial\phi}{\partial x}\mathbf{i} + \frac{\partial\phi}{\partial y}\mathbf{j} + \frac{\partial\phi}{\partial z}\mathbf{k})\cdot(\frac{dx}{ds}\mathbf{i} + \frac{dy}{ds}\mathbf{j} + \frac{dz}{ds}\mathbf{k})$$
$$= \nabla\phi\cdot\frac{d\mathbf{r}}{ds}.$$

Note that since $\frac{d\mathbf{r}}{ds}$ is a unit vector, $\nabla\phi\cdot\frac{d\mathbf{r}}{ds}$ is the component of $\nabla\phi$ in the direction of this unit vector.

**9.** Show that the greatest rate of change of $\phi$, i.e. the maximum directional derivative, takes place in the direction of, and has the magnitude of, the vector $\nabla\phi$.

By Problem 8(c), $\frac{d\phi}{ds} = \nabla\phi\cdot\frac{d\mathbf{r}}{ds}$ is the projection of $\nabla\phi$ in the direction $\frac{d\mathbf{r}}{ds}$. This projection will be a maximum when $\nabla\phi$ and $\frac{d\mathbf{r}}{ds}$ have the same direction. Then the maximum value of $\frac{d\phi}{ds}$ takes place in the direction of $\nabla\phi$ and its magnitude is $|\nabla\phi|$.

**10.** Find the directional derivative of $\phi = x^2yz + 4xz^2$ at $(1,-2,-1)$ in the direction $2\mathbf{i} - \mathbf{j} - 2\mathbf{k}$.

$$\nabla\phi = \nabla(x^2yz + 4xz^2) = (2xyz + 4z^2)\mathbf{i} + x^2z\,\mathbf{j} + (x^2y + 8xz)\mathbf{k}$$
$$= 8\mathbf{i} - \mathbf{j} - 10\mathbf{k} \quad \text{at} \quad (1,-2,-1).$$

The unit vector in the direction of $2\mathbf{i} - \mathbf{j} - 2\mathbf{k}$ is

$$\mathbf{a} = \frac{2\mathbf{i} - \mathbf{j} - 2\mathbf{k}}{\sqrt{(2)^2 + (-1)^2 + (-2)^2}} = \frac{2}{3}\mathbf{i} - \frac{1}{3}\mathbf{j} - \frac{2}{3}\mathbf{k}$$

Then the required directional derivative is

$$\nabla\phi\cdot\mathbf{a} = (8\mathbf{i} - \mathbf{j} - 10\mathbf{k})\cdot(\frac{2}{3}\mathbf{i} - \frac{1}{3}\mathbf{j} - \frac{2}{3}\mathbf{k}) = \frac{16}{3} + \frac{1}{3} + \frac{20}{3} = \frac{37}{3}$$

Since this is positive, $\phi$ is increasing in this direction.

**11.** (a) In what direction from the point $(2,1,-1)$ is the directional derivative of $\phi = x^2yz^3$ a maximum? (b) What is the magnitude of this maximum?

$$\nabla\phi = \nabla(x^2yz^3) = 2xyz^3\,\mathbf{i} + x^2z^3\,\mathbf{j} + 3x^2yz^2\,\mathbf{k}$$
$$= -4\mathbf{i} - 4\mathbf{j} + 12\mathbf{k} \quad \text{at} \quad (2,1,-1).$$

Then by Problem 9,

(*a*) the directional derivative is a maximum in the direction $\nabla\phi = -4\mathbf{i} - 4\mathbf{j} + 12\mathbf{k}$,

(*b*) the magnitude of this maximum is $|\nabla\phi| = \sqrt{(-4)^2 + (-4)^2 + (12)^2} = \sqrt{176} = 4\sqrt{11}$.

**12.** Find the angle between the surfaces $x^2+y^2+z^2 = 9$ and $z = x^2+y^2-3$ at the point (2,–1,2).

The angle between the surfaces at the point is the angle between the normals to the surfaces at the point.

A normal to $x^2+y^2+z^2 = 9$ at (2,–1,2) is

$$\nabla\phi_1 = \nabla(x^2+y^2+z^2) = 2x\,\mathbf{i} + 2y\,\mathbf{j} + 2z\,\mathbf{k} = 4\mathbf{i} - 2\mathbf{j} + 4\mathbf{k}$$

A normal to $z = x^2+y^2-3$ or $x^2+y^2-z = 3$ at (2,–1,2) is

$$\nabla\phi_2 = \nabla(x^2+y^2-z) = 2x\,\mathbf{i} + 2y\,\mathbf{j} - \mathbf{k} = 4\mathbf{i} - 2\mathbf{j} - \mathbf{k}$$

$(\nabla\phi_1)\cdot(\nabla\phi_2) = |\nabla\phi_1|\,|\nabla\phi_2| \cos\theta$, where $\theta$ is the required angle. Then

$$(4\mathbf{i} - 2\mathbf{j} + 4\mathbf{k})\cdot(4\mathbf{i} - 2\mathbf{j} - \mathbf{k}) = |4\mathbf{i} - 2\mathbf{j} + 4\mathbf{k}|\,|4\mathbf{i} - 2\mathbf{j} - \mathbf{k}| \cos\theta$$

$$16 + 4 - 4 = \sqrt{(4)^2+(-2)^2+(4)^2}\,\sqrt{(4)^2+(-2)^2+(-1)^2}\,\cos\theta$$

and $\cos\theta = \dfrac{16}{6\sqrt{21}} = \dfrac{8\sqrt{21}}{63} = 0.5819$; thus the acute angle is $\theta = \text{arc cos } 0.5819 = 54^\circ 25'$.

**13.** Let $R$ be the distance from a fixed point $A(a,b,c)$ to any point $P(x,y,z)$. Show that $\nabla R$ is a unit vector in the direction $\mathbf{AP} = \mathbf{R}$.

If $\mathbf{r}_A$ and $\mathbf{r}_P$ are the position vectors $a\,\mathbf{i} + b\,\mathbf{j} + c\,\mathbf{k}$ and $x\,\mathbf{i} + y\,\mathbf{j} + z\,\mathbf{k}$ of $A$ and $P$ respectively, then $\mathbf{R} = \mathbf{r}_P - \mathbf{r}_A = (x-a)\,\mathbf{i} + (y-b)\,\mathbf{j} + (z-c)\,\mathbf{k}$, so that $R = \sqrt{(x-a)^2+(y-b)^2+(z-c)^2}$. Then

$$\nabla R = \nabla\left(\sqrt{(x-a)^2 + (y-b)^2 + (z-c)^2}\right) = \frac{(x-a)\,\mathbf{i} + (y-b)\,\mathbf{j} + (z-c)\,\mathbf{k}}{\sqrt{(x-a)^2 + (y-b)^2 + (z-c)^2}} = \frac{\mathbf{R}}{R}$$

is a unit vector in the direction $\mathbf{R}$.

**14.** Let $P$ be any point on an ellipse whose foci are at points $A$ and $B$, as shown in the figure below. Prove that lines $AP$ and $BP$ make equal angles with the tangent to the ellipse at $P$.

Let $\mathbf{R}_1 = \mathbf{AP}$ and $\mathbf{R}_2 = \mathbf{BP}$ denote vectors drawn respectively from foci $A$ and $B$ to point $P$ on the ellipse, and let $\mathbf{T}$ be a unit tangent to the ellipse at $P$.

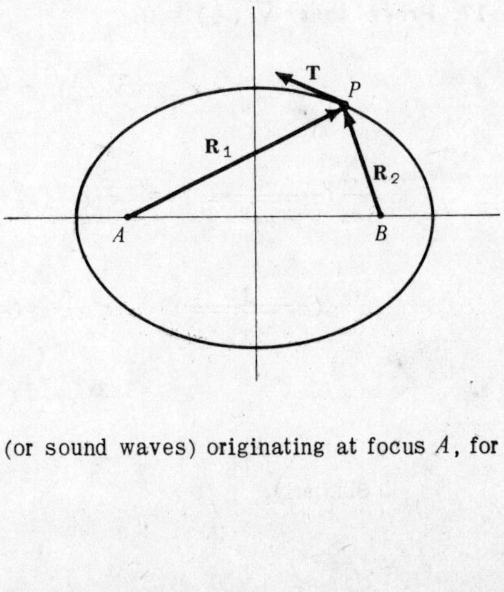


Since an ellipse is the locus of all points $P$ the sum of whose distances from two fixed points $A$ and $B$ is a constant $p$, it is seen that the equation of the ellipse is $R_1 + R_2 = p$.

By Problem 5, $\nabla(R_1+R_2)$ is a normal to the ellipse; hence $[\nabla(R_1+R_2)]\cdot\mathbf{T} = 0$ or $(\nabla R_2)\cdot\mathbf{T} = -(\nabla R_1)\cdot\mathbf{T}$.

Since $\nabla R_1$ and $\nabla R_2$ are unit vectors in direction $\mathbf{R}_1$ and $\mathbf{R}_2$ respectively (Problem 13), the cosine of the angle between $\nabla R_2$ and $\mathbf{T}$ is equal to the cosine of the angle between $\nabla R_1$ and $-\mathbf{T}$; hence the angles themselves are equal.

The problem has a physical interpretation. Light rays (or sound waves) originating at focus $A$, for example, will be reflected from the ellipse to focus $B$.

**THE DIVERGENCE**

**15.** If $\mathbf{A} = x^2z\,\mathbf{i} - 2y^3z^2\,\mathbf{j} + xy^2z\,\mathbf{k}$, find $\nabla\cdot\mathbf{A}$ (or div **A**) at the point (1,−1,1).

$$\nabla\cdot\mathbf{A} = \left(\frac{\partial}{\partial x}\mathbf{i} + \frac{\partial}{\partial y}\mathbf{j} + \frac{\partial}{\partial z}\mathbf{k}\right)\cdot(x^2z\,\mathbf{i} - 2y^3z^2\,\mathbf{j} + xy^2z\,\mathbf{k})$$

$$= \frac{\partial}{\partial x}(x^2z) + \frac{\partial}{\partial y}(-2y^3z^2) + \frac{\partial}{\partial z}(xy^2z)$$

$$= 2xz - 6y^2z^2 + xy^2 = 2(1)(1) - 6(-1)^2(1)^2 + (1)(-1)^2 = -3 \quad \text{at } (1,-1,1).$$

**16.** Given $\phi = 2x^3y^2z^4$. (*a*) Find $\nabla\cdot\nabla\phi$ (or div grad $\phi$).

(*b*) Show that $\nabla\cdot\nabla\phi = \nabla^2\phi$, where $\nabla^2 \equiv \dfrac{\partial^2}{\partial x^2} + \dfrac{\partial^2}{\partial y^2} + \dfrac{\partial^2}{\partial z^2}$ denotes the Laplacian operator.

(*a*) $$\nabla\phi = \mathbf{i}\frac{\partial}{\partial x}(2x^3y^2z^4) + \mathbf{j}\frac{\partial}{\partial y}(2x^3y^2z^4) + \mathbf{k}\frac{\partial}{\partial z}(2x^3y^2z^4)$$

$$= 6x^2y^2z^4\,\mathbf{i} + 4x^3yz^4\,\mathbf{j} + 8x^3y^2z^3\,\mathbf{k}$$

Then $$\nabla\cdot\nabla\phi = \left(\frac{\partial}{\partial x}\mathbf{i} + \frac{\partial}{\partial y}\mathbf{j} + \frac{\partial}{\partial z}\mathbf{k}\right)\cdot(6x^2y^2z^4\,\mathbf{i} + 4x^3yz^4\,\mathbf{j} + 8x^3y^2z^3\,\mathbf{k})$$

$$= \frac{\partial}{\partial x}(6x^2y^2z^4) + \frac{\partial}{\partial y}(4x^3yz^4) + \frac{\partial}{\partial z}(8x^3y^2z^3)$$

$$= 12xy^2z^4 + 4x^3z^4 + 24x^3y^2z^2$$

(*b*) $$\nabla\cdot\nabla\phi = \left(\frac{\partial}{\partial x}\mathbf{i} + \frac{\partial}{\partial y}\mathbf{j} + \frac{\partial}{\partial z}\mathbf{k}\right)\cdot\left(\frac{\partial\phi}{\partial x}\mathbf{i} + \frac{\partial\phi}{\partial y}\mathbf{j} + \frac{\partial\phi}{\partial z}\mathbf{k}\right)$$

$$= \frac{\partial}{\partial x}\left(\frac{\partial\phi}{\partial x}\right) + \frac{\partial}{\partial y}\left(\frac{\partial\phi}{\partial y}\right) + \frac{\partial}{\partial z}\left(\frac{\partial\phi}{\partial z}\right) = \frac{\partial^2\phi}{\partial x^2} + \frac{\partial^2\phi}{\partial y^2} + \frac{\partial^2\phi}{\partial z^2}$$

$$= \left(\frac{\partial^2}{\partial x^2} + \frac{\partial^2}{\partial y^2} + \frac{\partial^2}{\partial z^2}\right)\phi = \nabla^2\phi$$

**17.** Prove that $\nabla^2\left(\frac{1}{r}\right) = 0$.

$$\nabla^2\left(\frac{1}{r}\right) = \left(\frac{\partial^2}{\partial x^2} + \frac{\partial^2}{\partial y^2} + \frac{\partial^2}{\partial z^2}\right)\left(\frac{1}{\sqrt{x^2+y^2+z^2}}\right)$$

$$\frac{\partial}{\partial x}\left(\frac{1}{\sqrt{x^2+y^2+z^2}}\right) = \frac{\partial}{\partial x}(x^2+y^2+z^2)^{-1/2} = -x(x^2+y^2+z^2)^{-3/2}$$

$$\frac{\partial^2}{\partial x^2}\left(\frac{1}{\sqrt{x^2+y^2+z^2}}\right) = \frac{\partial}{\partial x}\left[-x(x^2+y^2+z^2)^{-3/2}\right]$$

$$= 3x^2(x^2+y^2+z^2)^{-5/2} - (x^2+y^2+z^2)^{-3/2} = \frac{2x^2-y^2-z^2}{(x^2+y^2+z^2)^{5/2}}$$

Similarly,

$$\frac{\partial^2}{\partial y^2}\left(\frac{1}{\sqrt{x^2+y^2+z^2}}\right) = \frac{2y^2 - z^2 - x^2}{(x^2+y^2+z^2)^{5/2}} \quad \text{and} \quad \frac{\partial^2}{\partial z^2}\left(\frac{1}{\sqrt{x^2+y^2+z^2}}\right) = \frac{2z^2 - x^2 - y^2}{(x^2+y^2+z^2)^{5/2}}$$

Then by addition, $\left(\frac{\partial^2}{\partial x^2} + \frac{\partial^2}{\partial y^2} + \frac{\partial^2}{\partial z^2}\right)\left(\frac{1}{\sqrt{x^2+y^2+z^2}}\right) = 0.$

The equation $\nabla^2\phi = 0$ is called *Laplace's equation.* It follows that $\phi = 1/r$ is a solution of this equation.

**18.** Prove: (*a*) $\nabla\cdot(\mathbf{A}+\mathbf{B}) = \nabla\cdot\mathbf{A} + \nabla\cdot\mathbf{B}$
(*b*) $\nabla\cdot(\phi\mathbf{A}) = (\nabla\phi)\cdot\mathbf{A} + \phi(\nabla\cdot\mathbf{A})$.

(*a*) Let $\mathbf{A} = A_1\mathbf{i} + A_2\mathbf{j} + A_3\mathbf{k}$, $\mathbf{B} = B_1\mathbf{i} + B_2\mathbf{j} + B_3\mathbf{k}$.

Then
$$\begin{aligned}
\nabla\cdot(\mathbf{A}+\mathbf{B}) &= \left(\frac{\partial}{\partial x}\mathbf{i} + \frac{\partial}{\partial y}\mathbf{j} + \frac{\partial}{\partial z}\mathbf{k}\right)\cdot\left[(A_1+B_1)\mathbf{i} + (A_2+B_2)\mathbf{j} + (A_3+B_3)\mathbf{k}\right] \\
&= \frac{\partial}{\partial x}(A_1+B_1) + \frac{\partial}{\partial y}(A_2+B_2) + \frac{\partial}{\partial z}(A_3+B_3) \\
&= \frac{\partial A_1}{\partial x} + \frac{\partial A_2}{\partial y} + \frac{\partial A_3}{\partial z} + \frac{\partial B_1}{\partial x} + \frac{\partial B_2}{\partial y} + \frac{\partial B_3}{\partial z} \\
&= \left(\frac{\partial}{\partial x}\mathbf{i} + \frac{\partial}{\partial y}\mathbf{j} + \frac{\partial}{\partial z}\mathbf{k}\right)\cdot(A_1\mathbf{i} + A_2\mathbf{j} + A_3\mathbf{k}) \\
&\qquad + \left(\frac{\partial}{\partial x}\mathbf{i} + \frac{\partial}{\partial y}\mathbf{j} + \frac{\partial}{\partial z}\mathbf{k}\right)\cdot(B_1\mathbf{i} + B_2\mathbf{j} + B_3\mathbf{k}) \\
&= \nabla\cdot\mathbf{A} + \nabla\cdot\mathbf{B}
\end{aligned}$$

(*b*)
$$\begin{aligned}
\nabla\cdot(\phi\mathbf{A}) &= \nabla\cdot(\phi A_1\mathbf{i} + \phi A_2\mathbf{j} + \phi A_3\mathbf{k}) \\
&= \frac{\partial}{\partial x}(\phi A_1) + \frac{\partial}{\partial y}(\phi A_2) + \frac{\partial}{\partial z}(\phi A_3) \\
&= \frac{\partial\phi}{\partial x}A_1 + \phi\frac{\partial A_1}{\partial x} + \frac{\partial\phi}{\partial y}A_2 + \phi\frac{\partial A_2}{\partial y} + \frac{\partial\phi}{\partial z}A_3 + \phi\frac{\partial A_3}{\partial z} \\
&= \frac{\partial\phi}{\partial x}A_1 + \frac{\partial\phi}{\partial y}A_2 + \frac{\partial\phi}{\partial z}A_3 + \phi\left(\frac{\partial A_1}{\partial x} + \frac{\partial A_2}{\partial y} + \frac{\partial A_3}{\partial z}\right) \\
&= \left(\frac{\partial\phi}{\partial x}\mathbf{i} + \frac{\partial\phi}{\partial y}\mathbf{j} + \frac{\partial\phi}{\partial z}\mathbf{k}\right)\cdot(A_1\mathbf{i} + A_2\mathbf{j} + A_3\mathbf{k}) + \phi\left(\frac{\partial}{\partial x}\mathbf{i} + \frac{\partial}{\partial y}\mathbf{j} + \frac{\partial}{\partial z}\mathbf{k}\right)\cdot(A_1\mathbf{i} + A_2\mathbf{j} + A_3\mathbf{k}) \\
&= (\nabla\phi)\cdot\mathbf{A} + \phi(\nabla\cdot\mathbf{A})
\end{aligned}$$

**19.** Prove $\nabla\cdot\left(\frac{\mathbf{r}}{r^3}\right) = 0$.

Let $\phi = r^{-3}$ and $\mathbf{A} = \mathbf{r}$ in the result of Problem 18(*b*).

Then
$$\begin{aligned}
\nabla\cdot(r^{-3}\mathbf{r}) &= (\nabla r^{-3})\cdot\mathbf{r} + (r^{-3})\nabla\cdot\mathbf{r} \\
&= -3r^{-5}\mathbf{r}\cdot\mathbf{r} + 3r^{-3} = 0, \quad \text{using Problem 4.}
\end{aligned}$$

**20.** Prove $\nabla\cdot(U\,\nabla V - V\,\nabla U) = U\,\nabla^2 V - V\,\nabla^2 U$.

From Problem 18(*b*), with $\phi = U$ and $\mathbf{A} = \nabla V$,

$$\nabla\cdot(U\,\nabla V) = (\nabla U)\cdot(\nabla V) + U(\nabla\cdot\nabla V) = (\nabla U)\cdot(\nabla V) + U\,\nabla^2 V$$

Interchanging $U$ and $V$ yields $\nabla\cdot(V\,\nabla U) = (\nabla V)\cdot(\nabla U) + V\,\nabla^2 U$.

Then subtracting,
$$\begin{aligned}\nabla\cdot(U\,\nabla V) - \nabla\cdot(V\,\nabla U) &= \nabla\cdot(U\,\nabla V - V\,\nabla U)\\ &= (\nabla U)\cdot(\nabla V) + U\,\nabla^2 V - [(\nabla V)\cdot(\nabla U) + V\,\nabla^2 U]\\ &= U\,\nabla^2 V - V\,\nabla^2 U\end{aligned}$$

**21.** A fluid moves so that its velocity at any point is $\mathbf{v}(x,y,z)$. Show that the gain of fluid per unit volume per unit time in a small parallelepiped having centre at $P(x,y,z)$ and edges parallel to the coordinate axes and having magnitude $\Delta x, \Delta y, \Delta z$ respectively, is given approximately by div $\mathbf{v} = \nabla\cdot\mathbf{v}$.

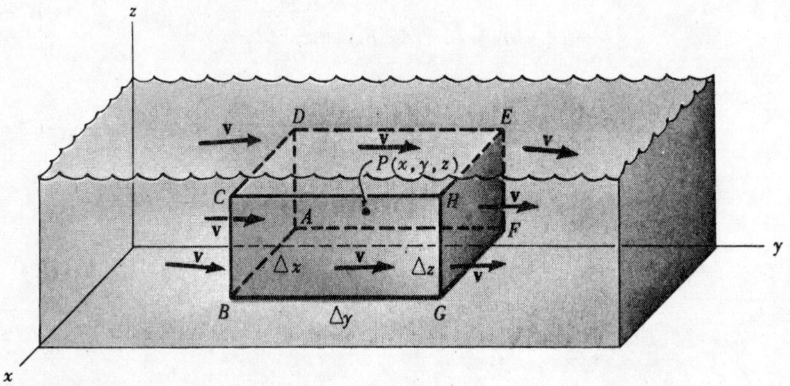


Referring to the figure above,

$x$ component of velocity $\mathbf{v}$ at $P$ $= v_1$

$x$ component of $\mathbf{v}$ at centre of face $AFED$ $= v_1 - \dfrac{1}{2}\dfrac{\partial v_1}{\partial x}\Delta x$ approx.

$x$ component of $\mathbf{v}$ at centre of face $GHCB$ $= v_1 + \dfrac{1}{2}\dfrac{\partial v_1}{\partial x}\Delta x$ approx.

Then (*1*) volume of fluid crossing $AFED$ per unit time $= (v_1 - \dfrac{1}{2}\dfrac{\partial v_1}{\partial x}\Delta x)\,\Delta y\,\Delta z$

(*2*) volume of fluid crossing $GHCB$ per unit time $= (v_1 + \dfrac{1}{2}\dfrac{\partial v_1}{\partial x}\Delta x)\,\Delta y\,\Delta z$.

Gain in volume per unit time in $x$ direction $= (2) - (1) = \dfrac{\partial v_1}{\partial x}\Delta x\,\Delta y\,\Delta z$.

Similarly, gain in volume per unit time in $y$ direction $= \dfrac{\partial v_2}{\partial y}\Delta x\,\Delta y\,\Delta z$

gain in volume per unit time in $z$ direction $= \dfrac{\partial v_3}{\partial z}\Delta x\,\Delta y\,\Delta z$.

Then, total gain in volume per unit volume per unit time

$$= \frac{\left(\dfrac{\partial v_1}{\partial x} + \dfrac{\partial v_2}{\partial y} + \dfrac{\partial v_3}{\partial z}\right)\Delta x\,\Delta y\,\Delta z}{\Delta x\,\Delta y\,\Delta z} = \text{div}\,\mathbf{v} = \nabla\cdot\mathbf{v}$$

This is true exactly only in the limit as the parallelepiped shrinks to $P$, i.e. as $\Delta x, \Delta y$ and $\Delta z$ approach zero. If there is no gain of fluid anywhere, then $\nabla\cdot\mathbf{v} = 0$. This is called the *continuity equation* for an incompressible fluid. Since fluid is neither created nor destroyed at any point, it is said to have no sources or sinks. A vector such as $\mathbf{v}$ whose divergence is zero is sometimes called *solenoidal.*

**22.** Determine the constant $a$ so that the vector $\mathbf{V} = (x+3y)\mathbf{i} + (y-2z)\mathbf{j} + (x+az)\mathbf{k}$ is solenoidal.

A vector $\mathbf{V}$ is solenoidal if its divergence is zero (Problem 21).

$$\nabla\cdot\mathbf{V} = \frac{\partial}{\partial x}(x+3y) + \frac{\partial}{\partial y}(y-2z) + \frac{\partial}{\partial z}(x+az) = 1 + 1 + a$$

Then $\nabla\cdot\mathbf{V} = a+2 = 0$ when $a = -2$.

## THE CURL

**23.** If $\mathbf{A} = xz^3\,\mathbf{i} - 2x^2yz\,\mathbf{j} + 2yz^4\,\mathbf{k}$, find $\nabla\times\mathbf{A}$ (or curl $\mathbf{A}$) at the point $(1,-1,1)$.

$$\nabla\times\mathbf{A} = \left(\frac{\partial}{\partial x}\mathbf{i} + \frac{\partial}{\partial y}\mathbf{j} + \frac{\partial}{\partial z}\mathbf{k}\right)\times(xz^3\,\mathbf{i} - 2x^2yz\,\mathbf{j} + 2yz^4\,\mathbf{k})$$

$$= \begin{vmatrix} \mathbf{i} & \mathbf{j} & \mathbf{k} \\ \frac{\partial}{\partial x} & \frac{\partial}{\partial y} & \frac{\partial}{\partial z} \\ xz^3 & -2x^2yz & 2yz^4 \end{vmatrix}$$

$$= \left[\frac{\partial}{\partial y}(2yz^4) - \frac{\partial}{\partial z}(-2x^2yz)\right]\mathbf{i} + \left[\frac{\partial}{\partial z}(xz^3) - \frac{\partial}{\partial x}(2yz^4)\right]\mathbf{j} + \left[\frac{\partial}{\partial x}(-2x^2yz) - \frac{\partial}{\partial y}(xz^3)\right]\mathbf{k}$$

$$= (2z^4 + 2x^2y)\,\mathbf{i} + 3xz^2\,\mathbf{j} - 4xyz\,\mathbf{k} = 3\mathbf{j} + 4\mathbf{k} \quad \text{at } (1,-1,1).$$

**24.** If $\mathbf{A} = x^2y\,\mathbf{i} - 2xz\,\mathbf{j} + 2yz\,\mathbf{k}$, find curl curl $\mathbf{A}$.

$$\text{curl curl } \mathbf{A} = \nabla\times(\nabla\times\mathbf{A})$$

$$= \nabla\times\begin{vmatrix} \mathbf{i} & \mathbf{j} & \mathbf{k} \\ \frac{\partial}{\partial x} & \frac{\partial}{\partial y} & \frac{\partial}{\partial z} \\ x^2y & -2xz & 2yz \end{vmatrix} = \nabla\times\left[(2x+2z)\,\mathbf{i} - (x^2+2z)\,\mathbf{k}\right]$$

$$= \begin{vmatrix} \mathbf{i} & \mathbf{j} & \mathbf{k} \\ \frac{\partial}{\partial x} & \frac{\partial}{\partial y} & \frac{\partial}{\partial z} \\ 2x+2z & 0 & -x^2-2z \end{vmatrix} = (2x+2)\,\mathbf{j}$$

**25.** Prove: (*a*) $\nabla\times(\mathbf{A}+\mathbf{B}) = \nabla\times\mathbf{A} + \nabla\times\mathbf{B}$
(*b*) $\nabla\times(\phi\mathbf{A}) = (\nabla\phi)\times\mathbf{A} + \phi(\nabla\times\mathbf{A})$.

(a) Let $\mathbf{A} = A_1\mathbf{i} + A_2\mathbf{j} + A_3\mathbf{k}$, $\mathbf{B} = B_1\mathbf{i} + B_2\mathbf{j} + B_3\mathbf{k}$. Then:

$$\nabla\times(\mathbf{A}+\mathbf{B}) = \left(\frac{\partial}{\partial x}\mathbf{i} + \frac{\partial}{\partial y}\mathbf{j} + \frac{\partial}{\partial z}\mathbf{k}\right) \times \left[(A_1+B_1)\mathbf{i} + (A_2+B_2)\mathbf{j} + (A_3+B_3)\mathbf{k}\right]$$

$$= \begin{vmatrix} \mathbf{i} & \mathbf{j} & \mathbf{k} \\ \frac{\partial}{\partial x} & \frac{\partial}{\partial y} & \frac{\partial}{\partial z} \\ A_1+B_1 & A_2+B_2 & A_3+B_3 \end{vmatrix}$$

$$= \left[\frac{\partial}{\partial y}(A_3+B_3) - \frac{\partial}{\partial z}(A_2+B_2)\right]\mathbf{i} + \left[\frac{\partial}{\partial z}(A_1+B_1) - \frac{\partial}{\partial x}(A_3+B_3)\right]\mathbf{j} + \left[\frac{\partial}{\partial x}(A_2+B_2) - \frac{\partial}{\partial y}(A_1+B_1)\right]\mathbf{k}$$

$$= \left[\frac{\partial A_3}{\partial y} - \frac{\partial A_2}{\partial z}\right]\mathbf{i} + \left[\frac{\partial A_1}{\partial z} - \frac{\partial A_3}{\partial x}\right]\mathbf{j} + \left[\frac{\partial A_2}{\partial x} - \frac{\partial A_1}{\partial y}\right]\mathbf{k} + \left[\frac{\partial B_3}{\partial y} - \frac{\partial B_2}{\partial z}\right]\mathbf{i} + \left[\frac{\partial B_1}{\partial z} - \frac{\partial B_3}{\partial x}\right]\mathbf{j} + \left[\frac{\partial B_2}{\partial x} - \frac{\partial B_1}{\partial y}\right]\mathbf{k}$$

$$= \nabla\times\mathbf{A} + \nabla\times\mathbf{B}$$

(b) $$\nabla\times(\phi\mathbf{A}) = \nabla\times(\phi A_1\mathbf{i} + \phi A_2\mathbf{j} + \phi A_3\mathbf{k})$$

$$= \begin{vmatrix} \mathbf{i} & \mathbf{j} & \mathbf{k} \\ \frac{\partial}{\partial x} & \frac{\partial}{\partial y} & \frac{\partial}{\partial z} \\ \phi A_1 & \phi A_2 & \phi A_3 \end{vmatrix}$$

$$= \left[\frac{\partial}{\partial y}(\phi A_3) - \frac{\partial}{\partial z}(\phi A_2)\right]\mathbf{i} + \left[\frac{\partial}{\partial z}(\phi A_1) - \frac{\partial}{\partial x}(\phi A_3)\right]\mathbf{j} + \left[\frac{\partial}{\partial x}(\phi A_2) - \frac{\partial}{\partial y}(\phi A_1)\right]\mathbf{k}$$

$$= \left[\phi\frac{\partial A_3}{\partial y} + \frac{\partial\phi}{\partial y}A_3 - \phi\frac{\partial A_2}{\partial z} - \frac{\partial\phi}{\partial z}A_2\right]\mathbf{i} + \left[\phi\frac{\partial A_1}{\partial z} + \frac{\partial\phi}{\partial z}A_1 - \phi\frac{\partial A_3}{\partial x} - \frac{\partial\phi}{\partial x}A_3\right]\mathbf{j} + \left[\phi\frac{\partial A_2}{\partial x} + \frac{\partial\phi}{\partial x}A_2 - \phi\frac{\partial A_1}{\partial y} - \frac{\partial\phi}{\partial y}A_1\right]\mathbf{k}$$

$$= \phi\left[\left(\frac{\partial A_3}{\partial y} - \frac{\partial A_2}{\partial z}\right)\mathbf{i} + \left(\frac{\partial A_1}{\partial z} - \frac{\partial A_3}{\partial x}\right)\mathbf{j} + \left(\frac{\partial A_2}{\partial x} - \frac{\partial A_1}{\partial y}\right)\mathbf{k}\right] + \left[\left(\frac{\partial\phi}{\partial y}A_3 - \frac{\partial\phi}{\partial z}A_2\right)\mathbf{i} + \left(\frac{\partial\phi}{\partial z}A_1 - \frac{\partial\phi}{\partial x}A_3\right)\mathbf{j} + \left(\frac{\partial\phi}{\partial x}A_2 - \frac{\partial\phi}{\partial y}A_1\right)\mathbf{k}\right]$$

$$= \phi(\nabla\times\mathbf{A}) + \begin{vmatrix} \mathbf{i} & \mathbf{j} & \mathbf{k} \\ \frac{\partial\phi}{\partial x} & \frac{\partial\phi}{\partial y} & \frac{\partial\phi}{\partial z} \\ A_1 & A_2 & A_3 \end{vmatrix}$$

$$= \phi(\nabla\times\mathbf{A}) + (\nabla\phi)\times\mathbf{A}.$$

26. Evaluate $\nabla\cdot(\mathbf{A}\times\mathbf{r})$ if $\nabla\times\mathbf{A}=\mathbf{0}$.

Let $\mathbf{A} = A_1\mathbf{i} + A_2\mathbf{j} + A_3\mathbf{k}$, $\mathbf{r} = x\mathbf{i} + y\mathbf{j} + z\mathbf{k}$.

Then $\mathbf{A}\times\mathbf{r} = \begin{vmatrix} \mathbf{i} & \mathbf{j} & \mathbf{k} \\ A_1 & A_2 & A_3 \\ x & y & z \end{vmatrix}$

$$= (zA_2 - yA_3)\mathbf{i} + (xA_3 - zA_1)\mathbf{j} + (yA_1 - xA_2)\mathbf{k}$$

and $\nabla\cdot(\mathbf{A}\times\mathbf{r}) = \frac{\partial}{\partial x}(zA_2 - yA_3) + \frac{\partial}{\partial y}(xA_3 - zA_1) + \frac{\partial}{\partial z}(yA_1 - xA_2)$

$$= z\frac{\partial A_2}{\partial x} - y\frac{\partial A_3}{\partial x} + x\frac{\partial A_3}{\partial y} - z\frac{\partial A_1}{\partial y} + y\frac{\partial A_1}{\partial z} - x\frac{\partial A_2}{\partial z}$$

$$= x(\frac{\partial A_3}{\partial y} - \frac{\partial A_2}{\partial z}) + y(\frac{\partial A_1}{\partial z} - \frac{\partial A_3}{\partial x}) + z(\frac{\partial A_2}{\partial x} - \frac{\partial A_1}{\partial y})$$

$$= [x\mathbf{i} + y\mathbf{j} + z\mathbf{k}]\cdot[(\frac{\partial A_3}{\partial y} - \frac{\partial A_2}{\partial z})\mathbf{i} + (\frac{\partial A_1}{\partial z} - \frac{\partial A_3}{\partial x})\mathbf{j} + (\frac{\partial A_2}{\partial x} - \frac{\partial A_1}{\partial y})\mathbf{k}]$$

$= \mathbf{r}\cdot(\nabla\times\mathbf{A}) = \mathbf{r}\cdot\text{curl }\mathbf{A}$. If $\nabla\times\mathbf{A}=\mathbf{0}$ this reduces to zero.

27. Prove: (*a*) $\nabla\times(\nabla\phi) = \mathbf{0}$ (curl grad $\phi = \mathbf{0}$), (*b*) $\nabla\cdot(\nabla\times\mathbf{A}) = 0$ (div curl $\mathbf{A} = 0$).

(*a*) $\nabla\times(\nabla\phi) = \nabla\times(\frac{\partial\phi}{\partial x}\mathbf{i} + \frac{\partial\phi}{\partial y}\mathbf{j} + \frac{\partial\phi}{\partial z}\mathbf{k})$

$$= \begin{vmatrix} \mathbf{i} & \mathbf{j} & \mathbf{k} \\ \frac{\partial}{\partial x} & \frac{\partial}{\partial y} & \frac{\partial}{\partial z} \\ \frac{\partial\phi}{\partial x} & \frac{\partial\phi}{\partial y} & \frac{\partial\phi}{\partial z} \end{vmatrix}$$

$$= [\frac{\partial}{\partial y}(\frac{\partial\phi}{\partial z}) - \frac{\partial}{\partial z}(\frac{\partial\phi}{\partial y})]\mathbf{i} + [\frac{\partial}{\partial z}(\frac{\partial\phi}{\partial x}) - \frac{\partial}{\partial x}(\frac{\partial\phi}{\partial z})]\mathbf{j} + [\frac{\partial}{\partial x}(\frac{\partial\phi}{\partial y}) - \frac{\partial}{\partial y}(\frac{\partial\phi}{\partial x})]\mathbf{k}$$

$$= (\frac{\partial^2\phi}{\partial y\,\partial z} - \frac{\partial^2\phi}{\partial z\,\partial y})\mathbf{i} + (\frac{\partial^2\phi}{\partial z\,\partial x} - \frac{\partial^2\phi}{\partial x\,\partial z})\mathbf{j} + (\frac{\partial^2\phi}{\partial x\,\partial y} - \frac{\partial^2\phi}{\partial y\,\partial x})\mathbf{k} = \mathbf{0}$$

provided we assume that $\phi$ has continuous second partial derivatives so that the order of differentiation is immaterial.

(*b*) $\nabla\cdot(\nabla\times\mathbf{A}) = \nabla\cdot\begin{vmatrix} \mathbf{i} & \mathbf{j} & \mathbf{k} \\ \frac{\partial}{\partial x} & \frac{\partial}{\partial y} & \frac{\partial}{\partial z} \\ A_1 & A_2 & A_3 \end{vmatrix}$

$$= \nabla\cdot[(\frac{\partial A_3}{\partial y} - \frac{\partial A_2}{\partial z})\mathbf{i} + (\frac{\partial A_1}{\partial z} - \frac{\partial A_3}{\partial x})\mathbf{j} + (\frac{\partial A_2}{\partial x} - \frac{\partial A_1}{\partial y})\mathbf{k}]$$

$$= \frac{\partial}{\partial x}(\frac{\partial A_3}{\partial y} - \frac{\partial A_2}{\partial z}) + \frac{\partial}{\partial y}(\frac{\partial A_1}{\partial z} - \frac{\partial A_3}{\partial x}) + \frac{\partial}{\partial z}(\frac{\partial A_2}{\partial x} - \frac{\partial A_1}{\partial y})$$

$$= \frac{\partial^2 A_3}{\partial x\, \partial y} - \frac{\partial^2 A_2}{\partial x\, \partial z} + \frac{\partial^2 A_1}{\partial y\, \partial z} - \frac{\partial^2 A_3}{\partial y\, \partial x} + \frac{\partial^2 A_2}{\partial z\, \partial x} - \frac{\partial^2 A_1}{\partial z\, \partial y} = 0$$

assuming that **A** has continuous second partial derivatives.

Note the similarity between the above results and the results $(\mathbf{C}\times\mathbf{C}m) = (\mathbf{C}\times\mathbf{C})m = \mathbf{0}$, where $m$ is a scalar and $\mathbf{C}\cdot(\mathbf{C}\times\mathbf{A}) = (\mathbf{C}\times\mathbf{C})\cdot\mathbf{A} = 0$.

**28.** Find $\text{curl}\,(\mathbf{r}\,f(r))$ where $f(r)$ is differentiable.

$$\text{curl}\,(\mathbf{r}\,f(r)) = \nabla\times(\mathbf{r}\,f(r))$$

$$= \nabla\times(x\,f(r)\mathbf{i} + y\,f(r)\mathbf{j} + z\,f(r)\mathbf{k})$$

$$= \begin{vmatrix} \mathbf{i} & \mathbf{j} & \mathbf{k} \\ \frac{\partial}{\partial x} & \frac{\partial}{\partial y} & \frac{\partial}{\partial z} \\ x\,f(r) & y\,f(r) & z\,f(r) \end{vmatrix}$$

$$= (z\frac{\partial f}{\partial y} - y\frac{\partial f}{\partial z})\mathbf{i} + (x\frac{\partial f}{\partial z} - z\frac{\partial f}{\partial x})\mathbf{j} + (y\frac{\partial f}{\partial x} - x\frac{\partial f}{\partial y})\mathbf{k}$$

But $\frac{\partial f}{\partial x} = (\frac{\partial f}{\partial r})(\frac{\partial r}{\partial x}) = \frac{\partial f}{\partial r}\frac{\partial}{\partial x}(\sqrt{x^2+y^2+z^2}) = \frac{f'(r)\,x}{\sqrt{x^2+y^2+z^2}} = \frac{f'\,x}{r}$. Similarly, $\frac{\partial f}{\partial y} = \frac{f'\,y}{r}$ and $\frac{\partial f}{\partial z} = \frac{f'\,z}{r}$.

Then the result $= (z\frac{f'\,y}{r} - y\frac{f'\,z}{r})\mathbf{i} + (x\frac{f'\,z}{r} - z\frac{f'\,x}{r})\mathbf{j} + (y\frac{f'\,x}{r} - x\frac{f'\,y}{r})\mathbf{k} = \mathbf{0}$.

**29.** Prove $\nabla\times(\nabla\times\mathbf{A}) = -\nabla^2\mathbf{A} + \nabla(\nabla\cdot\mathbf{A})$.

$$\nabla\times(\nabla\times\mathbf{A}) = \nabla\times\begin{vmatrix} \mathbf{i} & \mathbf{j} & \mathbf{k} \\ \frac{\partial}{\partial x} & \frac{\partial}{\partial y} & \frac{\partial}{\partial z} \\ A_1 & A_2 & A_3 \end{vmatrix}$$

$$= \nabla\times\left[(\frac{\partial A_3}{\partial y} - \frac{\partial A_2}{\partial z})\mathbf{i} + (\frac{\partial A_1}{\partial z} - \frac{\partial A_3}{\partial x})\mathbf{j} + (\frac{\partial A_2}{\partial x} - \frac{\partial A_1}{\partial y})\mathbf{k}\right]$$

$$= \begin{vmatrix} \mathbf{i} & \mathbf{j} & \mathbf{k} \\ \frac{\partial}{\partial x} & \frac{\partial}{\partial y} & \frac{\partial}{\partial z} \\ \frac{\partial A_3}{\partial y} - \frac{\partial A_2}{\partial z} & \frac{\partial A_1}{\partial z} - \frac{\partial A_3}{\partial x} & \frac{\partial A_2}{\partial x} - \frac{\partial A_1}{\partial y} \end{vmatrix}$$

$$= \left[\frac{\partial}{\partial y}(\frac{\partial A_2}{\partial x} - \frac{\partial A_1}{\partial y}) - \frac{\partial}{\partial z}(\frac{\partial A_1}{\partial z} - \frac{\partial A_3}{\partial x})\right]\mathbf{i}$$

$$+ \left[\frac{\partial}{\partial z}(\frac{\partial A_3}{\partial y} - \frac{\partial A_2}{\partial z}) - \frac{\partial}{\partial x}(\frac{\partial A_2}{\partial x} - \frac{\partial A_1}{\partial y})\right]\mathbf{j}$$

$$+ \left[\frac{\partial}{\partial x}(\frac{\partial A_1}{\partial z} - \frac{\partial A_3}{\partial x}) - \frac{\partial}{\partial y}(\frac{\partial A_3}{\partial y} - \frac{\partial A_2}{\partial z})\right]\mathbf{k}$$

$$= (-\frac{\partial^2 A_1}{\partial y^2} - \frac{\partial^2 A_1}{\partial z^2})\mathbf{i} + (-\frac{\partial^2 A_2}{\partial z^2} - \frac{\partial^2 A_2}{\partial x^2})\mathbf{j} + (-\frac{\partial^2 A_3}{\partial x^2} - \frac{\partial^2 A_3}{\partial y^2})\mathbf{k}$$

$$+ (\frac{\partial^2 A_2}{\partial y\,\partial x} + \frac{\partial^2 A_3}{\partial z\,\partial x})\mathbf{i} + (\frac{\partial^2 A_3}{\partial z\,\partial y} + \frac{\partial^2 A_1}{\partial x\,\partial y})\mathbf{j} + (\frac{\partial^2 A_1}{\partial x\,\partial z} + \frac{\partial^2 A_2}{\partial y\,\partial z})\mathbf{k}$$

$$= (-\frac{\partial^2 A_1}{\partial x^2} - \frac{\partial^2 A_1}{\partial y^2} - \frac{\partial^2 A_1}{\partial z^2})\mathbf{i} + (-\frac{\partial^2 A_2}{\partial x^2} - \frac{\partial^2 A_2}{\partial y^2} - \frac{\partial^2 A_2}{\partial z^2})\mathbf{j} + (-\frac{\partial^2 A_3}{\partial x^2} - \frac{\partial^2 A_3}{\partial y^2} - \frac{\partial^2 A_3}{\partial z^2})\mathbf{k}$$

$$+ (\frac{\partial^2 A_1}{\partial x^2} + \frac{\partial^2 A_2}{\partial y\,\partial x} + \frac{\partial^2 A_3}{\partial z\,\partial x})\mathbf{i} + (\frac{\partial^2 A_1}{\partial x\,\partial y} + \frac{\partial^2 A_2}{\partial y^2} + \frac{\partial^2 A_3}{\partial z\,\partial y})\mathbf{j} + (\frac{\partial^2 A_1}{\partial x\,\partial z} + \frac{\partial^2 A_2}{\partial y\,\partial z} + \frac{\partial^2 A_3}{\partial z^2})\mathbf{k}$$

$$= -(\frac{\partial^2}{\partial x^2} + \frac{\partial^2}{\partial y^2} + \frac{\partial^2}{\partial z^2})(A_1\mathbf{i} + A_2\mathbf{j} + A_3\mathbf{k})$$

$$+ \mathbf{i}\frac{\partial}{\partial x}(\frac{\partial A_1}{\partial x} + \frac{\partial A_2}{\partial y} + \frac{\partial A_3}{\partial z}) + \mathbf{j}\frac{\partial}{\partial y}(\frac{\partial A_1}{\partial x} + \frac{\partial A_2}{\partial y} + \frac{\partial A_3}{\partial z}) + \mathbf{k}\frac{\partial}{\partial z}(\frac{\partial A_1}{\partial x} + \frac{\partial A_2}{\partial y} + \frac{\partial A_3}{\partial z})$$

$$= -\nabla^2\mathbf{A} + \nabla(\frac{\partial A_1}{\partial x} + \frac{\partial A_2}{\partial y} + \frac{\partial A_3}{\partial z})$$

$$= -\nabla^2\mathbf{A} + \nabla(\nabla\cdot\mathbf{A})$$

If desired, the labour of writing can be shortened in this as well as other derivations by writing only the $\mathbf{i}$ components since the others can be obtained by symmetry.

The result can also be established formally as follows. From Problem 47(*a*), Chapter 2,

$$(1) \qquad \mathbf{A}\times(\mathbf{B}\times\mathbf{C}) = \mathbf{B}(\mathbf{A}\cdot\mathbf{C}) - (\mathbf{A}\cdot\mathbf{B})\mathbf{C}$$

Placing $\mathbf{A} = \mathbf{B} = \nabla$ and $\mathbf{C} = \mathbf{F}$,

$$\nabla\times(\nabla\times\mathbf{F}) = \nabla(\nabla\cdot\mathbf{F}) - (\nabla\cdot\nabla)\mathbf{F} = \nabla(\nabla\cdot\mathbf{F}) - \nabla^2\mathbf{F}$$

Note that the formula (*1*) must be written so that the operators **A** and **B** precede the operand **C**, otherwise the formalism fails to apply.

**30.** If $\mathbf{v} = \boldsymbol{\omega}\times\mathbf{r}$, prove $\boldsymbol{\omega} = \frac{1}{2}\,\text{curl}\,\mathbf{v}$ where $\boldsymbol{\omega}$ is a constant vector.

$$\text{curl}\,\mathbf{v} = \nabla\times\mathbf{v} = \nabla\times(\boldsymbol{\omega}\times\mathbf{r}) = \nabla\times\begin{vmatrix}\mathbf{i} & \mathbf{j} & \mathbf{k}\\ \omega_1 & \omega_2 & \omega_3\\ x & y & z\end{vmatrix}$$

$$= \nabla\times\left[(\omega_2 z - \omega_3 y)\mathbf{i} + (\omega_3 x - \omega_1 z)\mathbf{j} + (\omega_1 y - \omega_2 x)\mathbf{k}\right]$$

$$= \begin{vmatrix}\mathbf{i} & \mathbf{j} & \mathbf{k}\\ \frac{\partial}{\partial x} & \frac{\partial}{\partial y} & \frac{\partial}{\partial z}\\ \omega_2 z - \omega_3 y & \omega_3 x - \omega_1 z & \omega_1 y - \omega_2 x\end{vmatrix} = 2(\omega_1\mathbf{i} + \omega_2\mathbf{j} + \omega_3\mathbf{k}) = 2\boldsymbol{\omega}.$$

Then $\boldsymbol{\omega} = \frac{1}{2}\nabla \times \mathbf{v} = \frac{1}{2}\operatorname{curl}\mathbf{v}$.

This problem indicates that the curl of a vector field has something to do with rotational properties of the field. This is confirmed in Chapter 6. If the field $\mathbf{F}$ is that due to a moving fluid, for example, then a paddle wheel placed at various points in the field would tend to rotate in regions where $\operatorname{curl}\mathbf{F} \neq \mathbf{0}$, while if $\operatorname{curl}\mathbf{F} = \mathbf{0}$ in the region there would be no rotation and the field $\mathbf{F}$ is then called *irrotational*. A field which is not irrotational is sometimes called a *vortex field*.

**31.** If $\nabla\cdot\mathbf{E} = 0$, $\nabla\cdot\mathbf{H} = 0$, $\nabla\times\mathbf{E} = -\frac{\partial\mathbf{H}}{\partial t}$, $\nabla\times\mathbf{H} = \frac{\partial\mathbf{E}}{\partial t}$, show that $\mathbf{E}$ and $\mathbf{H}$ satisfy $\nabla^2 u = \frac{\partial^2 u}{\partial t^2}$.

$$\nabla\times(\nabla\times\mathbf{E}) = \nabla\times\left(-\frac{\partial\mathbf{H}}{\partial t}\right) = -\frac{\partial}{\partial t}(\nabla\times\mathbf{H}) = -\frac{\partial}{\partial t}\left(\frac{\partial\mathbf{E}}{\partial t}\right) = -\frac{\partial^2\mathbf{E}}{\partial t^2}$$

By Problem 29, $\nabla\times(\nabla\times\mathbf{E}) = -\nabla^2\mathbf{E} + \nabla(\nabla\cdot\mathbf{E}) = -\nabla^2\mathbf{E}$. Then $\nabla^2\mathbf{E} = \frac{\partial^2\mathbf{E}}{\partial t^2}$.

Similarly, $\nabla\times(\nabla\times\mathbf{H}) = \nabla\times\left(\frac{\partial\mathbf{E}}{\partial t}\right) = \frac{\partial}{\partial t}(\nabla\times\mathbf{E}) = \frac{\partial}{\partial t}\left(-\frac{\partial\mathbf{H}}{\partial t}\right) = -\frac{\partial^2\mathbf{H}}{\partial t^2}$.

But $\nabla\times(\nabla\times\mathbf{H}) = -\nabla^2\mathbf{H} + \nabla(\nabla\cdot\mathbf{H}) = -\nabla^2\mathbf{H}$. Then $\nabla^2\mathbf{H} = \frac{\partial^2\mathbf{H}}{\partial t^2}$.

The given equations are related to *Maxwell's equations of electromagnetic theory*. The equation $\frac{\partial^2 u}{\partial x^2} + \frac{\partial^2 u}{\partial y^2} + \frac{\partial^2 u}{\partial z^2} = \frac{\partial^2 u}{\partial t^2}$ is called the *wave equation*.

## MISCELLANEOUS PROBLEMS.

**32.** (*a*) A vector $\mathbf{V}$ is called irrotational if $\operatorname{curl}\mathbf{V} = \mathbf{0}$ (see Problem 30). Find constants $a,b,c$ so that

$$\mathbf{V} = (x + 2y + az)\mathbf{i} + (bx - 3y - z)\mathbf{j} + (4x + cy + 2z)\mathbf{k}$$

is irrotational.

(*b*) Show that $\mathbf{V}$ can be expressed as the gradient of a scalar function.

(*a*) $$\operatorname{curl}\mathbf{V} = \nabla\times\mathbf{V} = \begin{vmatrix} \mathbf{i} & \mathbf{j} & \mathbf{k} \\ \frac{\partial}{\partial x} & \frac{\partial}{\partial y} & \frac{\partial}{\partial z} \\ x+2y+az & bx-3y-z & 4x+cy+2z \end{vmatrix} = (c+1)\mathbf{i} + (a-4)\mathbf{j} + (b-2)\mathbf{k}$$

This equals zero when $a = 4$, $b = 2$, $c = -1$ and

$$\mathbf{V} = (x + 2y + 4z)\mathbf{i} + (2x - 3y - z)\mathbf{j} + (4x - y + 2z)\mathbf{k}$$

(*b*) Assume $\mathbf{V} = \nabla\phi = \frac{\partial\phi}{\partial x}\mathbf{i} + \frac{\partial\phi}{\partial y}\mathbf{j} + \frac{\partial\phi}{\partial z}\mathbf{k}$

Then (*1*) $\frac{\partial\phi}{\partial x} = x + 2y + 4z$, (*2*) $\frac{\partial\phi}{\partial y} = 2x - 3y - z$, (*3*) $\frac{\partial\phi}{\partial z} = 4x - y + 2z$.

Integrating (*1*) partially with respect to $x$, keeping $y$ and $z$ constant,

(*4*) $$\phi = \frac{x^2}{2} + 2xy + 4xz + f(y,z)$$

where $f(y,z)$ is an arbitrary function of $y$ and $z$. Similarly from (*2*) and (*3*),

(5) $$\phi = 2xy - \frac{3y^2}{2} - yz + g(x,z)$$

(6) $$\phi = 4xz - yz + z^2 + h(x,y).$$

Comparison of (4), (5) and (6) shows that there will be a common value of $\phi$ if we choose

$$f(y,z) = -\frac{3y^2}{2} + z^2, \qquad g(x,z) = \frac{x^2}{2} + z^2, \qquad h(x,y) = \frac{x^2}{2} - \frac{3y^2}{2}$$

so that

$$\phi = \frac{x^2}{2} - \frac{3y^2}{2} + z^2 + 2xy + 4xz - yz$$

Note that we can also add any constant to $\phi$. In general if $\nabla \times \mathbf{V} = \mathbf{0}$, then we can find $\phi$ so that $\mathbf{V} = \nabla\phi$. A vector field $\mathbf{V}$ which can be derived from a scalar field $\phi$ so that $\mathbf{V} = \nabla\phi$ is called a *conservative vector field* and $\phi$ is called the *scalar potential*. Note that conversely if $\mathbf{V} = \nabla\phi$, then $\nabla \times \mathbf{V} = \mathbf{0}$ (see Prob.27*a*).

**33.** Show that if $\phi(x,y,z)$ is any solution of Laplace's equation, then $\nabla\phi$ is a vector which is both solenoidal and irrotational.

By hypothesis, $\phi$ satisfies Laplace's equation $\nabla^2\phi = 0$, i.e. $\nabla\cdot(\nabla\phi) = 0$. Then $\nabla\phi$ is solenoidal (see Problems 21 and 22).

From Problem 27*a*, $\nabla \times (\nabla\phi) = \mathbf{0}$ so that $\nabla\phi$ is also irrotational.

**34.** Give a possible definition of grad **B**.

Assume $\mathbf{B} = B_1\mathbf{i} + B_2\mathbf{j} + B_3\mathbf{k}$. Formally, we can define grad **B** as

$$\begin{aligned} \nabla\mathbf{B} &= (\frac{\partial}{\partial x}\mathbf{i} + \frac{\partial}{\partial y}\mathbf{j} + \frac{\partial}{\partial z}\mathbf{k})(B_1\mathbf{i} + B_2\mathbf{j} + B_3\mathbf{k}) \\ &= \frac{\partial B_1}{\partial x}\mathbf{ii} + \frac{\partial B_2}{\partial x}\mathbf{ij} + \frac{\partial B_3}{\partial x}\mathbf{ik} \\ &\quad + \frac{\partial B_1}{\partial y}\mathbf{ji} + \frac{\partial B_2}{\partial y}\mathbf{jj} + \frac{\partial B_3}{\partial y}\mathbf{jk} \\ &\quad + \frac{\partial B_1}{\partial z}\mathbf{ki} + \frac{\partial B_2}{\partial z}\mathbf{kj} + \frac{\partial B_3}{\partial z}\mathbf{kk} \end{aligned}$$

The quantities **ii**, **ij**, etc., are called *unit dyads*. (Note that **ij**, for example, is not the same as **ji**.) A quantity of the form

$$a_{11}\mathbf{ii} + a_{12}\mathbf{ij} + a_{13}\mathbf{ik} + a_{21}\mathbf{ji} + a_{22}\mathbf{jj} + a_{23}\mathbf{jk} + a_{31}\mathbf{ki} + a_{32}\mathbf{kj} + a_{33}\mathbf{kk}$$

is called a *dyadic* and the coefficients $a_{11}, a_{12}, \ldots$ are its *components*. An array of these nine components in the form

$$\begin{pmatrix} a_{11} & a_{12} & a_{13} \\ a_{21} & a_{22} & a_{23} \\ a_{31} & a_{32} & a_{33} \end{pmatrix}$$

is called a 3 by 3 *matrix*. A dyadic is a generalization of a vector. Still further generalization leads to *triadics* which are quantities consisting of 27 terms of the form $a_{111}\mathbf{iii} + a_{211}\mathbf{jii} + \ldots$. A study of how the components of a dyadic or triadic transform from one system of coordinates to another leads to the subject of *tensor analysis* which is taken up in Chapter 8.

**35.** Let a vector **A** be defined by $\mathbf{A} = A_1\mathbf{i} + A_2\mathbf{j} + A_3\mathbf{k}$ and a dyadic $\mathbf{\Phi}$ by

$$\mathbf{\Phi} = a_{11}\mathbf{ii} + a_{12}\mathbf{ij} + a_{13}\mathbf{ik} + a_{21}\mathbf{ji} + a_{22}\mathbf{jj} + a_{23}\mathbf{jk} + a_{31}\mathbf{ki} + a_{32}\mathbf{kj} + a_{33}\mathbf{kk}$$

Give a possible definition of $\mathbf{A}\cdot\mathbf{\Phi}$.

Formally, assuming the distributive law to hold,

$$\mathbf{A}\cdot\mathbf{\Phi} = (A_1\mathbf{i} + A_2\mathbf{j} + A_3\mathbf{k})\cdot\mathbf{\Phi} = A_1\mathbf{i}\cdot\mathbf{\Phi} + A_2\mathbf{j}\cdot\mathbf{\Phi} + A_3\mathbf{k}\cdot\mathbf{\Phi}$$

As an example, consider $\mathbf{i}\cdot\mathbf{\Phi}$. This product is formed by taking the dot product of **i** with each term of $\mathbf{\Phi}$ and adding results. Typical examples are $\mathbf{i}\cdot a_{11}\mathbf{ii}$, $\mathbf{i}\cdot a_{12}\mathbf{ij}$, $\mathbf{i}\cdot a_{21}\mathbf{ji}$, $\mathbf{i}\cdot a_{32}\mathbf{kj}$, etc. If we give meaning to these as follows

$$\mathbf{i}\cdot a_{11}\mathbf{ii} = a_{11}(\mathbf{i}\cdot\mathbf{i})\mathbf{i} = a_{11}\mathbf{i} \qquad \text{since } \mathbf{i}\cdot\mathbf{i} = 1$$

$$\mathbf{i}\cdot a_{12}\mathbf{ij} = a_{12}(\mathbf{i}\cdot\mathbf{i})\mathbf{j} = a_{12}\mathbf{j} \qquad \text{since } \mathbf{i}\cdot\mathbf{i} = 1$$

$$\mathbf{i}\cdot a_{21}\mathbf{ji} = a_{21}(\mathbf{i}\cdot\mathbf{j})\mathbf{i} = \mathbf{0} \qquad \text{since } \mathbf{i}\cdot\mathbf{j} = 0$$

$$\mathbf{i}\cdot a_{32}\mathbf{kj} = a_{32}(\mathbf{i}\cdot\mathbf{k})\mathbf{j} = \mathbf{0} \qquad \text{since } \mathbf{i}\cdot\mathbf{k} = 0$$

and give analogous interpretation to the terms of $\mathbf{j}\cdot\mathbf{\Phi}$ and $\mathbf{k}\cdot\mathbf{\Phi}$, then

$$\begin{aligned}\mathbf{A}\cdot\mathbf{\Phi} &= A_1(a_{11}\mathbf{i} + a_{12}\mathbf{j} + a_{13}\mathbf{k}) + A_2(a_{21}\mathbf{i} + a_{22}\mathbf{j} + a_{23}\mathbf{k}) + A_3(a_{31}\mathbf{i} + a_{32}\mathbf{j} + a_{33}\mathbf{k}) \\ &= (A_1a_{11} + A_2a_{21} + A_3a_{31})\mathbf{i} + (A_1a_{12} + A_2a_{22} + A_3a_{32})\mathbf{j} + (A_1a_{13} + A_2a_{23} + A_3a_{33})\mathbf{k}\end{aligned}$$

which is a vector.

**36.** (*a*) Interpret the symbol $\mathbf{A}\cdot\nabla$. (*b*) Give a possible meaning to $(\mathbf{A}\cdot\nabla)\mathbf{B}$. (*c*) Is it possible to write this as $\mathbf{A}\cdot\nabla\mathbf{B}$ without ambiguity?

(*a*) Let $\mathbf{A} = A_1\mathbf{i} + A_2\mathbf{j} + A_3\mathbf{k}$. Then, formally,

$$\begin{aligned}\mathbf{A}\cdot\nabla &= (A_1\mathbf{i} + A_2\mathbf{j} + A_3\mathbf{k})\cdot(\frac{\partial}{\partial x}\mathbf{i} + \frac{\partial}{\partial y}\mathbf{j} + \frac{\partial}{\partial z}\mathbf{k}) \\ &= A_1\frac{\partial}{\partial x} + A_2\frac{\partial}{\partial y} + A_3\frac{\partial}{\partial z}\end{aligned}$$

is an operator. For example,

$$(\mathbf{A}\cdot\nabla)\phi = (A_1\frac{\partial}{\partial x} + A_2\frac{\partial}{\partial y} + A_3\frac{\partial}{\partial z})\phi = A_1\frac{\partial\phi}{\partial x} + A_2\frac{\partial\phi}{\partial y} + A_3\frac{\partial\phi}{\partial z}$$

Note that this is the same as $\mathbf{A}\cdot\nabla\phi$.

(*b*) Formally, using (*a*) with $\phi$ replaced by $\mathbf{B} = B_1\mathbf{i} + B_2\mathbf{j} + B_3\mathbf{k}$,

$$\begin{aligned}(\mathbf{A}\cdot\nabla)\mathbf{B} &= (A_1\frac{\partial}{\partial x} + A_2\frac{\partial}{\partial y} + A_3\frac{\partial}{\partial z})\mathbf{B} = A_1\frac{\partial\mathbf{B}}{\partial x} + A_2\frac{\partial\mathbf{B}}{\partial y} + A_3\frac{\partial\mathbf{B}}{\partial z} \\ &= (A_1\frac{\partial B_1}{\partial x} + A_2\frac{\partial B_1}{\partial y} + A_3\frac{\partial B_1}{\partial z})\mathbf{i} + (A_1\frac{\partial B_2}{\partial x} + A_2\frac{\partial B_2}{\partial y} + A_3\frac{\partial B_2}{\partial z})\mathbf{j} + (A_1\frac{\partial B_3}{\partial x} + A_2\frac{\partial B_3}{\partial y} + A_3\frac{\partial B_3}{\partial z})\mathbf{k}\end{aligned}$$

(*c*) Use the interpretation of $\nabla\mathbf{B}$ as given in Problem 34. Then, according to the symbolism established in Problem 35,

$$\begin{aligned}\mathbf{A}\cdot\nabla\mathbf{B} &= (A_1\mathbf{i} + A_2\mathbf{j} + A_3\mathbf{k})\cdot\nabla\mathbf{B} = A_1\mathbf{i}\cdot\nabla\mathbf{B} + A_2\mathbf{j}\cdot\nabla\mathbf{B} + A_3\mathbf{k}\cdot\nabla\mathbf{B} \\ &= A_1(\frac{\partial B_1}{\partial x}\mathbf{i} + \frac{\partial B_2}{\partial x}\mathbf{j} + \frac{\partial B_3}{\partial x}\mathbf{k}) + A_2(\frac{\partial B_1}{\partial y}\mathbf{i} + \frac{\partial B_2}{\partial y}\mathbf{j} + \frac{\partial B_3}{\partial y}\mathbf{k}) + A_3(\frac{\partial B_1}{\partial z}\mathbf{i} + \frac{\partial B_2}{\partial z}\mathbf{j} + \frac{\partial B_3}{\partial z}\mathbf{k})\end{aligned}$$

which gives the same result as that given in part (*b*). It follows that $(\mathbf{A}\cdot\nabla)\mathbf{B} = \mathbf{A}\cdot\nabla\mathbf{B}$ without ambiguity provided the concept of dyadics is introduced with properties as indicated.

**37.** If $\mathbf{A} = 2yz\,\mathbf{i} - x^2y\,\mathbf{j} + xz^2\,\mathbf{k}$, $\mathbf{B} = x^2\mathbf{i} + yz\,\mathbf{j} - xy\,\mathbf{k}$ and $\phi = 2x^2yz^3$, find (*a*) $(\mathbf{A}\cdot\nabla)\phi$, (*b*) $\mathbf{A}\cdot\nabla\phi$, (*c*) $(\mathbf{B}\cdot\nabla)\mathbf{A}$, (*d*) $(\mathbf{A}\times\nabla)\phi$, (*e*) $\mathbf{A}\times\nabla\phi$.

(*a*)
$$(\mathbf{A}\cdot\nabla)\phi = \left[(2yz\,\mathbf{i} - x^2y\,\mathbf{j} + xz^2\mathbf{k})\cdot\left(\frac{\partial}{\partial x}\mathbf{i} + \frac{\partial}{\partial y}\mathbf{j} + \frac{\partial}{\partial z}\mathbf{k}\right)\right]\phi$$
$$= \left(2yz\frac{\partial}{\partial x} - x^2y\frac{\partial}{\partial y} + xz^2\frac{\partial}{\partial z}\right)(2x^2yz^3)$$
$$= 2yz\frac{\partial}{\partial x}(2x^2yz^3) - x^2y\frac{\partial}{\partial y}(2x^2yz^3) + xz^2\frac{\partial}{\partial z}(2x^2yz^3)$$
$$= (2yz)(4xyz^3) - (x^2y)(2x^2z^3) + (xz^2)(6x^2yz^2)$$
$$= 8xy^2z^4 - 2x^4yz^3 + 6x^3yz^4$$

(*b*)
$$\mathbf{A}\cdot\nabla\phi = (2yz\,\mathbf{i} - x^2y\,\mathbf{j} + xz^2\,\mathbf{k})\cdot\left(\frac{\partial\phi}{\partial x}\mathbf{i} + \frac{\partial\phi}{\partial y}\mathbf{j} + \frac{\partial\phi}{\partial z}\mathbf{k}\right)$$
$$= (2yz\,\mathbf{i} - x^2y\,\mathbf{j} + xz^2\mathbf{k})\cdot(4xyz^3\,\mathbf{i} + 2x^2z^3\,\mathbf{j} + 6x^2yz^2\,\mathbf{k})$$
$$= 8xy^2z^4 - 2x^4yz^3 + 6x^3yz^4$$

Comparison with (*a*) illustrates the result $(\mathbf{A}\cdot\nabla)\phi = \mathbf{A}\cdot\nabla\phi$.

(*c*)
$$(\mathbf{B}\cdot\nabla)\mathbf{A} = \left[(x^2\mathbf{i} + yz\,\mathbf{j} - xy\,\mathbf{k})\cdot\left(\frac{\partial}{\partial x}\mathbf{i} + \frac{\partial}{\partial y}\mathbf{j} + \frac{\partial}{\partial z}\mathbf{k}\right)\right]\mathbf{A}$$
$$= \left(x^2\frac{\partial}{\partial x} + yz\frac{\partial}{\partial y} - xy\frac{\partial}{\partial z}\right)\mathbf{A} = x^2\frac{\partial\mathbf{A}}{\partial x} + yz\frac{\partial\mathbf{A}}{\partial y} - xy\frac{\partial\mathbf{A}}{\partial z}$$
$$= x^2(-2xy\,\mathbf{j} + z^2\mathbf{k}) + yz(2z\,\mathbf{i} - x^2\mathbf{j}) - xy(2y\,\mathbf{i} + 2xz\,\mathbf{k})$$
$$= (2yz^2 - 2xy^2)\mathbf{i} - (2x^3y + x^2yz)\mathbf{j} + (x^2z^2 - 2x^2yz)\mathbf{k}$$

For comparison of this with $\mathbf{B}\cdot\nabla\mathbf{A}$, see Problem 36(*c*).

(*d*)
$$(\mathbf{A}\times\nabla)\phi = \left[(2yz\,\mathbf{i} - x^2y\,\mathbf{j} + xz^2\mathbf{k})\times\left(\frac{\partial}{\partial x}\mathbf{i} + \frac{\partial}{\partial y}\mathbf{j} + \frac{\partial}{\partial z}\mathbf{k}\right)\right]\phi$$
$$= \begin{vmatrix} \mathbf{i} & \mathbf{j} & \mathbf{k} \\ 2yz & -x^2y & xz^2 \\ \frac{\partial}{\partial x} & \frac{\partial}{\partial y} & \frac{\partial}{\partial z} \end{vmatrix}\phi$$
$$= \left[\mathbf{i}\left(-x^2y\frac{\partial}{\partial z} - xz^2\frac{\partial}{\partial y}\right) + \mathbf{j}\left(xz^2\frac{\partial}{\partial x} - 2yz\frac{\partial}{\partial z}\right) + \mathbf{k}\left(2yz\frac{\partial}{\partial y} + x^2y\frac{\partial}{\partial x}\right)\right]\phi$$
$$= -\left(x^2y\frac{\partial\phi}{\partial z} + xz^2\frac{\partial\phi}{\partial y}\right)\mathbf{i} + \left(xz^2\frac{\partial\phi}{\partial x} - 2yz\frac{\partial\phi}{\partial z}\right)\mathbf{j} + \left(2yz\frac{\partial\phi}{\partial y} + x^2y\frac{\partial\phi}{\partial x}\right)\mathbf{k}$$
$$= -(6x^4y^2z^2 + 2x^3z^5)\mathbf{i} + (4x^2yz^5 - 12x^2y^2z^3)\mathbf{j} + (4x^2yz^4 + 4x^3y^2z^3)\mathbf{k}$$

(e) $\mathbf{A} \times \nabla\phi = (2yz\,\mathbf{i} - x^2y\,\mathbf{j} + xz^2\,\mathbf{k}) \times (\frac{\partial\phi}{\partial x}\mathbf{i} + \frac{\partial\phi}{\partial y}\mathbf{j} + \frac{\partial\phi}{\partial z}\mathbf{k})$

$$= \begin{vmatrix} \mathbf{i} & \mathbf{j} & \mathbf{k} \\ 2yz & -x^2y & xz^2 \\ \frac{\partial\phi}{\partial x} & \frac{\partial\phi}{\partial y} & \frac{\partial\phi}{\partial z} \end{vmatrix}$$

$$= (-x^2y\frac{\partial\phi}{\partial z} - xz^2\frac{\partial\phi}{\partial y})\mathbf{i} + (xz^2\frac{\partial\phi}{\partial x} - 2yz\frac{\partial\phi}{\partial z})\mathbf{j} + (2yz\frac{\partial\phi}{\partial y} + x^2y\frac{\partial\phi}{\partial x})\mathbf{k}$$

$$= -(6x^4y^2z^2 + 2x^3z^5)\mathbf{i} + (4x^2yz^5 - 12x^2y^2z^3)\mathbf{j} + (4x^2yz^4 + 4x^3y^2z^3)\mathbf{k}$$

Comparison with (d) illustrates the result $(\mathbf{A}\times\nabla)\phi = \mathbf{A}\times\nabla\phi$.

## INVARIANCE

**38.** Two rectangular $xyz$ and $x'y'z'$ coordinate systems having the same origin are rotated with respect to each other. Derive the transformation equations between the coordinates of a point in the two systems.

Let $\mathbf{r}$ and $\mathbf{r}'$ be the position vectors of any point $P$ in the two systems (see figure on page 58). Then since $\mathbf{r} = \mathbf{r}'$,

$$(1) \qquad x'\mathbf{i}' + y'\mathbf{j}' + z'\mathbf{k}' = x\mathbf{i} + y\mathbf{j} + z\mathbf{k}$$

Now for any vector $\mathbf{A}$ we have (Problem 20, Chapter 2),

$$\mathbf{A} = (\mathbf{A}\cdot\mathbf{i}')\mathbf{i}' + (\mathbf{A}\cdot\mathbf{j}')\mathbf{j}' + (\mathbf{A}\cdot\mathbf{k}')\mathbf{k}'$$

Then letting $\mathbf{A} = \mathbf{i}, \mathbf{j}, \mathbf{k}$ in succession,

$$(2) \qquad \begin{cases} \mathbf{i} = (\mathbf{i}\cdot\mathbf{i}')\mathbf{i}' + (\mathbf{i}\cdot\mathbf{j}')\mathbf{j}' + (\mathbf{i}\cdot\mathbf{k}')\mathbf{k}' = l_{11}\mathbf{i}' + l_{21}\mathbf{j}' + l_{31}\mathbf{k}' \\ \mathbf{j} = (\mathbf{j}\cdot\mathbf{i}')\mathbf{i}' + (\mathbf{j}\cdot\mathbf{j}')\mathbf{j}' + (\mathbf{j}\cdot\mathbf{k}')\mathbf{k}' = l_{12}\mathbf{i}' + l_{22}\mathbf{j}' + l_{32}\mathbf{k}' \\ \mathbf{k} = (\mathbf{k}\cdot\mathbf{i}')\mathbf{i}' + (\mathbf{k}\cdot\mathbf{j}')\mathbf{j}' + (\mathbf{k}\cdot\mathbf{k}')\mathbf{k}' = l_{13}\mathbf{i}' + l_{23}\mathbf{j}' + l_{33}\mathbf{k}' \end{cases}$$

Substituting equations (2) in (1) and equating coefficients of $\mathbf{i}', \mathbf{j}', \mathbf{k}'$ we find

$$(3) \qquad x' = l_{11}x + l_{12}y + l_{13}z, \qquad y' = l_{21}x + l_{22}y + l_{23}z, \qquad z' = l_{31}x + l_{32}y + l_{33}z$$

the required transformation equations.

**39.** Prove
$$\mathbf{i}' = l_{11}\mathbf{i} + l_{12}\mathbf{j} + l_{13}\mathbf{k}$$
$$\mathbf{j}' = l_{21}\mathbf{i} + l_{22}\mathbf{j} + l_{23}\mathbf{k}$$
$$\mathbf{k}' = l_{31}\mathbf{i} + l_{32}\mathbf{j} + l_{33}\mathbf{k}$$

For any vector $\mathbf{A}$ we have $\mathbf{A} = (\mathbf{A}\cdot\mathbf{i})\mathbf{i} + (\mathbf{A}\cdot\mathbf{j})\mathbf{j} + (\mathbf{A}\cdot\mathbf{k})\mathbf{k}$.

Then letting $\mathbf{A} = \mathbf{i}', \mathbf{j}', \mathbf{k}'$ in succession,

$$\mathbf{i}' = (\mathbf{i}'\cdot\mathbf{i})\mathbf{i} + (\mathbf{i}'\cdot\mathbf{j})\mathbf{j} + (\mathbf{i}'\cdot\mathbf{k})\mathbf{k} = l_{11}\mathbf{i} + l_{12}\mathbf{j} + l_{13}\mathbf{k}$$
$$\mathbf{j}' = (\mathbf{j}'\cdot\mathbf{i})\mathbf{i} + (\mathbf{j}'\cdot\mathbf{j})\mathbf{j} + (\mathbf{j}'\cdot\mathbf{k})\mathbf{k} = l_{21}\mathbf{i} + l_{22}\mathbf{j} + l_{23}\mathbf{k}$$
$$\mathbf{k}' = (\mathbf{k}'\cdot\mathbf{i})\mathbf{i} + (\mathbf{k}'\cdot\mathbf{j})\mathbf{j} + (\mathbf{k}'\cdot\mathbf{k})\mathbf{k} = l_{31}\mathbf{i} + l_{32}\mathbf{j} + l_{33}\mathbf{k}$$

**40.** Prove that $\sum_{p=1}^{3} l_{pm}\, l_{pn} = 1$ if $m = n$, and 0 if $m \neq n$, where $m$ and $n$ can assume any of the values 1, 2, 3.

From equations (*2*) of Problem 38,

$$\begin{aligned}
\mathbf{i}\cdot\mathbf{i} &= 1 = (l_{11}\mathbf{i}' + l_{21}\mathbf{j}' + l_{31}\mathbf{k}')\cdot(l_{11}\mathbf{i}' + l_{21}\mathbf{j}' + l_{31}\mathbf{k}') \\
&= l_{11}^2 + l_{21}^2 + l_{31}^2 \\
\mathbf{i}\cdot\mathbf{j} &= 0 = (l_{11}\mathbf{i}' + l_{21}\mathbf{j}' + l_{31}\mathbf{k}')\cdot(l_{12}\mathbf{i}' + l_{22}\mathbf{j}' + l_{32}\mathbf{k}') \\
&= l_{11}l_{12} + l_{21}l_{22} + l_{31}l_{32} \\
\mathbf{i}\cdot\mathbf{k} &= 0 = (l_{11}\mathbf{i}' + l_{21}\mathbf{j}' + l_{31}\mathbf{k}')\cdot(l_{13}\mathbf{i}' + l_{23}\mathbf{j}' + l_{33}\mathbf{k}') \\
&= l_{11}l_{13} + l_{21}l_{23} + l_{31}l_{33}
\end{aligned}$$

These establish the required result where $m$=1. By considering $\mathbf{j}\cdot\mathbf{i}, \mathbf{j}\cdot\mathbf{j}, \mathbf{j}\cdot\mathbf{k}, \mathbf{k}\cdot\mathbf{i}, \mathbf{k}\cdot\mathbf{j}$ and $\mathbf{k}\cdot\mathbf{k}$ the result can be proved for $m$=2 and $m$=3 .

By writing $\delta_{mn} = \begin{cases} 1 \text{ if } m=n \\ 0 \text{ if } m \neq n \end{cases}$ the result can be written $\sum_{p=1}^{3} l_{pm}\, l_{pn} = \delta_{mn}$.

The symbol $\delta_{mn}$ is called *Kronecker's symbol.*

**41.** If $\phi(x,y,z)$ is a scalar invariant with respect to a rotation of axes, prove that grad $\phi$ is a vector invariant under this transformation.

By hypothesis $\phi(x,y,z) = \phi'(x',y',z')$. To establish the desired result we must prove that

$$\frac{\partial\phi}{\partial x}\mathbf{i} + \frac{\partial\phi}{\partial y}\mathbf{j} + \frac{\partial\phi}{\partial z}\mathbf{k} = \frac{\partial\phi'}{\partial x'}\mathbf{i}' + \frac{\partial\phi'}{\partial y'}\mathbf{j}' + \frac{\partial\phi'}{\partial z'}\mathbf{k}'$$

Using the chain rule and the transformation equations (*3*) of Problem 38, we have

$$\begin{aligned}
\frac{\partial\phi}{\partial x} &= \frac{\partial\phi'}{\partial x'}\frac{\partial x'}{\partial x} + \frac{\partial\phi'}{\partial y'}\frac{\partial y'}{\partial x} + \frac{\partial\phi'}{\partial z'}\frac{\partial z'}{\partial x} = \frac{\partial\phi'}{\partial x'}l_{11} + \frac{\partial\phi'}{\partial y'}l_{21} + \frac{\partial\phi'}{\partial z'}l_{31} \\
\frac{\partial\phi}{\partial y} &= \frac{\partial\phi'}{\partial x'}\frac{\partial x'}{\partial y} + \frac{\partial\phi'}{\partial y'}\frac{\partial y'}{\partial y} + \frac{\partial\phi'}{\partial z'}\frac{\partial z'}{\partial y} = \frac{\partial\phi'}{\partial x'}l_{12} + \frac{\partial\phi'}{\partial y'}l_{22} + \frac{\partial\phi'}{\partial z'}l_{32} \\
\frac{\partial\phi}{\partial z} &= \frac{\partial\phi'}{\partial x'}\frac{\partial x'}{\partial z} + \frac{\partial\phi'}{\partial y'}\frac{\partial y'}{\partial z} + \frac{\partial\phi'}{\partial z'}\frac{\partial z'}{\partial z} = \frac{\partial\phi'}{\partial x'}l_{13} + \frac{\partial\phi'}{\partial y'}l_{23} + \frac{\partial\phi'}{\partial z'}l_{33}
\end{aligned}$$

Multiplying these equations by $\mathbf{i}, \mathbf{j}, \mathbf{k}$ respectively, adding and using Problem 39, the required result follows.

## SUPPLEMENTARY PROBLEMS

**42.** If $\phi = 2xz^4 - x^2 y$, find $\nabla\phi$ and $|\nabla\phi|$ at the point $(2,-2,-1)$. *Ans.* $10\mathbf{i} - 4\mathbf{j} - 16\mathbf{k}$, $2\sqrt{93}$

**43.** If $\mathbf{A} = 2x^2\,\mathbf{i} - 3yz\,\mathbf{j} + xz^2\,\mathbf{k}$ and $\phi = 2z - x^3 y$, find $\mathbf{A}\cdot\nabla\phi$ and $\mathbf{A}\times\nabla\phi$ at the point $(1,-1,1)$.
*Ans.* $5$, $7\mathbf{i} - \mathbf{j} - 11\mathbf{k}$

**44.** If $F = x^2 z + e^{y/x}$ and $G = 2z^2 y - xy^2$, find (a) $\nabla(F+G)$ and (b) $\nabla(FG)$ at the point $(1,0,-2)$.
*Ans.* (a) $-4\mathbf{i} + 9\mathbf{j} + \mathbf{k}$, (b) $-8\mathbf{j}$

**45.** Find $\nabla|\mathbf{r}|^3$. *Ans.* $3r\,\mathbf{r}$

**46.** Prove $\nabla f(r) = \dfrac{f'(r)\,\mathbf{r}}{r}$.

**47.** Evaluate $\nabla(3r^2 - 4\sqrt{r} + \dfrac{6}{\sqrt[3]{r}})$. *Ans.* $(6 - 2r^{-3/2} - 2r^{-7/3})\,\mathbf{r}$

**48.** If $\nabla U = 2r^4\,\mathbf{r}$, find $U$. *Ans.* $r^6/3$ + constant

**49.** Find $\phi(r)$ such that $\nabla\phi = \dfrac{\mathbf{r}}{r^5}$ and $\phi(1) = 0$. *Ans.* $\phi(r) = \dfrac{1}{3}(1 - \dfrac{1}{r^3})$

**50.** Find $\nabla\psi$ where $\psi = (x^2 + y^2 + z^2)\,e^{-\sqrt{x^2+y^2+z^2}}$. *Ans.* $(2 - r)\,e^{-r}\,\mathbf{r}$

**51.** If $\nabla\phi = 2xyz^3\,\mathbf{i} + x^2 z^3\,\mathbf{j} + 3x^2 yz^2\,\mathbf{k}$, find $\phi(x,y,z)$ if $\phi(1,-2,2) = 4$. *Ans.* $\phi = x^2 yz^3 + 20$

**52.** If $\nabla\psi = (y^2 - 2xyz^3)\,\mathbf{i} + (3 + 2xy - x^2 z^3)\,\mathbf{j} + (6z^3 - 3x^2 yz^2)\,\mathbf{k}$, find $\psi$.
*Ans.* $\psi = xy^2 - x^2 yz^3 + 3y + (3/2)\,z^4$ + constant

**53.** If $U$ is a differentiable function of $x,y,z$, prove $\nabla U\cdot d\mathbf{r} = dU$.

**54.** If $F$ is a differentiable function of $x,y,z,t$ where $x,y,z$ are differentiable functions of $t$, prove that

$$\frac{dF}{dt} = \frac{\partial F}{\partial t} + \nabla F\cdot\frac{d\mathbf{r}}{dt}$$

**55.** If $\mathbf{A}$ is a constant vector, prove $\nabla(\mathbf{r}\cdot\mathbf{A}) = \mathbf{A}$.

**56.** If $\mathbf{A}(x,y,z) = A_1\mathbf{i} + A_2\mathbf{j} + A_3\mathbf{k}$, show that $d\mathbf{A} = (\nabla A_1\cdot d\mathbf{r})\,\mathbf{i} + (\nabla A_2\cdot d\mathbf{r})\,\mathbf{j} + (\nabla A_3\cdot d\mathbf{r})\,\mathbf{k}$.

**57.** Prove $\nabla(\dfrac{F}{G}) = \dfrac{G\nabla F - F\nabla G}{G^2}$ if $G \neq 0$.

**58.** Find a unit vector which is perpendicular to the surface of the paraboloid of revolution $z = x^2 + y^2$ at the point $(1,2,5)$. *Ans.* $\dfrac{2\mathbf{i} + 4\mathbf{j} - \mathbf{k}}{\pm\sqrt{21}}$

**59.** Find the unit outward drawn normal to the surface $(x-1)^2 + y^2 + (z+2)^2 = 9$ at the point $(3,1,-4)$.
*Ans.* $(2\mathbf{i} + \mathbf{j} - 2\mathbf{k})/3$

**60.** Find an equation for the tangent plane to the surface $xz^2 + x^2 y = z - 1$ at the point $(1,-3,2)$.
*Ans.* $2x - y - 3z + 1 = 0$

**61.** Find equations for the tangent plane and normal line to the surface $z = x^2 + y^2$ at the point $(2,-1,5)$.
*Ans.* $4x - 2y - z = 5$, $\dfrac{x-2}{4} = \dfrac{y+1}{-2} = \dfrac{z-5}{-1}$ or $x = 4t+2$, $y = -2t-1$, $z = -t+5$

**62.** Find the directional derivative of $\phi = 4xz^3 - 3x^2 y^2 z$ at $(2,-1,2)$ in the direction $2\mathbf{i} - 3\mathbf{j} + 6\mathbf{k}$.
*Ans.* $376/7$

**63.** Find the directional derivative of $P = 4e^{2x-y+z}$ at the point $(1,1,-1)$ in a direction toward the point $(-3,5,6)$. *Ans.* $-20/9$

**64.** In what direction from the point (1,3,2) is the directional derivative of $\phi = 2xz - y^2$ a maximum? What is the magnitude of this maximum? *Ans.* In the direction of the vector $4\mathbf{i} - 6\mathbf{j} + 2\mathbf{k}$, $2\sqrt{14}$

**65.** Find the values of the constants $a,b,c$ so that the directional derivative of $\phi = axy^2 + byz + cz^2x^3$ at (1,2,−1) has a maximum of magnitude 64 in a direction parallel to the $z$ axis. *Ans.* $a = 6,\ b = 24,\ c = -8$

**66.** Find the acute angle between the surfaces $xy^2z = 3x + z^2$ and $3x^2 - y^2 + 2z = 1$ at the point (1,−2,1).

*Ans.* $\text{arc cos}\ \dfrac{3}{\sqrt{14}\sqrt{21}} = \text{arc cos}\ \dfrac{\sqrt{6}}{14} = 79°55'$

**67.** Find the constants $a$ and $b$ so that the surface $ax^2 - byz = (a+2)x$ will be orthogonal to the surface $4x^2y + z^3 = 4$ at the point (1,−1,2). *Ans.* $a = 5/2,\ b = 1$

**68.** (*a*) Let $u$ and $v$ be differentiable functions of $x,y$ and $z$. Show that a necessary and sufficient condition that $u$ and $v$ are functionally related by the equation $F(u,v) = 0$ is that $\nabla u \times \nabla v = \mathbf{0}$.

(*b*) Determine whether $u = \text{arc tan}\, x + \text{arc tan}\, y$ and $v = \dfrac{x+y}{1-xy}$ are functionally related.

*Ans.* (*b*) Yes ($v = \tan u$)

**69.** (*a*) Show that a necessary and sufficient condition that $u(x,y,z)$, $v(x,y,z)$ and $w(x,y,z)$ be functionally related through the equation $F(u,v,w) = 0$ is $\nabla u \cdot \nabla v \times \nabla w = 0$.

(*b*) Express $\nabla u \cdot \nabla v \times \nabla w$ in determinant form. This determinant is called the Jacobian of $u,v,w$ with respect to $x,y,z$ and is written $\dfrac{\partial(u,v,w)}{\partial(x,y,z)}$ or $J(\dfrac{u,v,w}{x,y,z})$.

(*c*) Determine whether $u = x+y+z$, $v = x^2+y^2+z^2$ and $w = xy+yz+zx$ are functionally related.

*Ans.* (*b*) $\begin{vmatrix} \dfrac{\partial u}{\partial x} & \dfrac{\partial u}{\partial y} & \dfrac{\partial u}{\partial z} \\ \dfrac{\partial v}{\partial x} & \dfrac{\partial v}{\partial y} & \dfrac{\partial v}{\partial z} \\ \dfrac{\partial w}{\partial x} & \dfrac{\partial w}{\partial y} & \dfrac{\partial w}{\partial z} \end{vmatrix}$ (*c*) Yes ($u^2 - v - 2w = 0$)

**70.** If $\mathbf{A} = 3xyz^2\,\mathbf{i} + 2xy^3\,\mathbf{j} - x^2yz\,\mathbf{k}$ and $\phi = 3x^2 - yz$, find (*a*) $\nabla\cdot\mathbf{A}$, (*b*) $\mathbf{A}\cdot\nabla\phi$, (*c*) $\nabla\cdot(\phi\mathbf{A})$, (*d*) $\nabla\cdot(\nabla\phi)$, at the point (1,−1,1). *Ans.* (*a*) 4, (*b*) −15, (*c*) 1, (*d*) 6

**71.** Evaluate $\text{div}\,(2x^2z\,\mathbf{i} - xy^2z\,\mathbf{j} + 3yz^2\,\mathbf{k})$. *Ans.* $4xz - 2xyz + 6yz$

**72.** If $\phi = 3x^2z - y^2z^3 + 4x^3y + 2x - 3y - 5$, find $\nabla^2\phi$. *Ans.* $6z + 24xy - 2z^3 - 6y^2z$

**73.** Evaluate $\nabla^2(\ln r)$. *Ans.* $1/r^2$

**74.** Prove $\nabla^2 r^n = n(n+1)r^{n-2}$ where $n$ is a constant.

**75.** If $\mathbf{F} = (3x^2y - z)\mathbf{i} + (xz^3 + y^4)\mathbf{j} - 2x^3z^2\,\mathbf{k}$, find $\nabla(\nabla\cdot\mathbf{F})$ at the point (2,−1,0). *Ans.* $-6\mathbf{i} + 24\mathbf{j} - 32\mathbf{k}$

**76.** If $\boldsymbol{\omega}$ is a constant vector and $\mathbf{v} = \boldsymbol{\omega}\times\mathbf{r}$, prove that $\text{div}\,\mathbf{v} = 0$.

**77.** Prove $\nabla^2(\phi\psi) = \phi\nabla^2\psi + 2\nabla\phi\cdot\nabla\psi + \psi\nabla^2\phi$.

**78.** If $U = 3x^2y$, $V = xz^2 - 2y$ evaluate $\text{grad}\,[(\text{grad}\,U)\cdot(\text{grad}\,V)]$. *Ans.* $(6yz^2 - 12x)\mathbf{i} + 6xz^2\,\mathbf{j} + 12xyz\,\mathbf{k}$

**79.** Evaluate $\nabla\cdot(r^3\mathbf{r})$. *Ans.* $6r^3$

**80.** Evaluate $\nabla\cdot[r\nabla(1/r^3)]$. *Ans.* $3r^{-4}$

**81.** Evaluate $\nabla^2[\nabla\cdot(\mathbf{r}/r^2)]$. *Ans.* $2r^{-4}$

**82.** If $\mathbf{A} = \mathbf{r}/r$, find grad div $\mathbf{A}$. *Ans.* $-2r^{-3}\,\mathbf{r}$

**83.** (*a*) Prove $\nabla^2 f(r) = \dfrac{d^2f}{dr^2} + \dfrac{2}{r}\dfrac{df}{dr}$. (*b*) Find $f(r)$ such that $\nabla^2 f(r) = 0$.

*Ans.* $f(r) = A + B/r$ where $A$ and $B$ are arbitrary constants.

**84.** Prove that the vector $\mathbf{A} = 3y^4z^2\,\mathbf{i} + 4x^3z^2\,\mathbf{j} - 3x^2y^2\,\mathbf{k}$ is solenoidal.

**85.** Show that $\mathbf{A} = (2x^2 + 8xy^2z)\mathbf{i} + (3x^3y - 3xy)\mathbf{j} - (4y^2z^2 + 2x^3z)\mathbf{k}$ is not solenoidal but $\mathbf{B} = xyz^2\,\mathbf{A}$ is solenoidal.

**86.** Find the most general differentiable function $f(r)$ so that $f(r)\mathbf{r}$ is solenoidal.
*Ans.* $f(r) = C/r^3$ where $C$ is an arbitrary constant.

**87.** Show that the vector field $\mathbf{V} = \dfrac{-x\,\mathbf{i} - y\,\mathbf{j}}{\sqrt{x^2+y^2}}$ is a "sink field". Plot and give a physical interpretation.

**88.** If $U$ and $V$ are differentiable scalar fields, prove that $\nabla U \times \nabla V$ is solenoidal.

**89.** If $\mathbf{A} = 2xz^2\,\mathbf{i} - yz\,\mathbf{j} + 3xz^3\,\mathbf{k}$ and $\phi = x^2yz$, find
(*a*) $\nabla\times\mathbf{A}$, (*b*) curl $(\phi\mathbf{A})$, (*c*) $\nabla\times(\nabla\times\mathbf{A})$, (*d*) $\nabla[\mathbf{A}\cdot\text{curl}\,\mathbf{A}]$, (*e*) curl grad $(\phi\mathbf{A})$ at the point (1,1,1).
*Ans.* (*a*) $\mathbf{i}+\mathbf{j}$, (*b*) $5\mathbf{i} - 3\mathbf{j} - 4\mathbf{k}$, (*c*) $5\mathbf{i} + 3\mathbf{k}$, (*d*) $-2\mathbf{i} + \mathbf{j} + 8\mathbf{k}$, (*e*) $\mathbf{0}$

**90.** If $F = x^2yz$, $G = xy - 3z^2$, find (*a*) $\nabla[(\nabla F)\cdot(\nabla G)]$, (*b*) $\nabla\cdot[(\nabla F)\times(\nabla G)]$, (*c*) $\nabla\times[(\nabla F)\times(\nabla G)]$.
*Ans.* (*a*) $(2y^2z + 3x^2z - 12xyz)\mathbf{i} + (4xyz - 6x^2z)\mathbf{j} + (2xy^2 + x^3 - 6x^2y)\mathbf{k}$
(*b*) 0
(*c*) $(x^2z - 24xyz)\mathbf{i} - (12x^2z + 2xyz)\mathbf{j} + (2xy^2 + 12yz^2 + x^3)\mathbf{k}$

**91.** Evaluate $\nabla\times(\mathbf{r}/r^2)$. *Ans.* $\mathbf{0}$

**92.** For what value of the constant $a$ will the vector $\mathbf{A} = (axy - z^3)\,\mathbf{i} + (a-2)x^2\,\mathbf{j} + (1-a)xz^2\,\mathbf{k}$ have its curl identically equal to zero? *Ans.* $a = 4$

**93.** Prove curl $(\phi \text{ grad } \phi) = \mathbf{0}$.

**94.** Graph the vector fields $\mathbf{A} = x\,\mathbf{i} + y\,\mathbf{j}$ and $\mathbf{B} = y\,\mathbf{i} - x\,\mathbf{j}$. Compute the divergence and curl of each vector field and explain the physical significance of the results obtained.

**95.** If $\mathbf{A} = x^2z\,\mathbf{i} + yz^3\,\mathbf{j} - 3xy\,\mathbf{k}$, $\mathbf{B} = y^2\,\mathbf{i} - yz\,\mathbf{j} + 2x\,\mathbf{k}$ and $\phi = 2x^2 + yz$, find
(*a*) $\mathbf{A}\cdot(\nabla\phi)$, (*b*) $(\mathbf{A}\cdot\nabla)\phi$, (*c*) $(\mathbf{A}\cdot\nabla)\mathbf{B}$, (*d*) $\mathbf{B}(\mathbf{A}\cdot\nabla)$, (*e*) $(\nabla\cdot\mathbf{A})\mathbf{B}$.
*Ans.* (*a*) $4x^3z + yz^4 - 3xy^2$, (*b*) $4x^3z + yz^4 - 3xy^2$ (same as (*a*)),
(*c*) $2y^2z^3\,\mathbf{i} + (3xy^2 - yz^4)\mathbf{j} + 2x^2z\,\mathbf{k}$,
(*d*) the operator $(x^2y^2z\,\mathbf{i} - x^2yz^2\,\mathbf{j} + 2x^3z\,\mathbf{k})\dfrac{\partial}{\partial x} + (y^3z^3\,\mathbf{i} - y^2z^4\,\mathbf{j} + 2xyz^3\,\mathbf{k})\dfrac{\partial}{\partial y}$
$+ (-3xy^3\,\mathbf{i} + 3xy^2z\,\mathbf{j} - 6x^2y\,\mathbf{k})\dfrac{\partial}{\partial z}$
(*e*) $(2xy^2z + y^2z^3)\mathbf{i} - (2xyz^2 + yz^4)\mathbf{j} + (4x^2z + 2xz^3)\mathbf{k}$

**96.** If $\mathbf{A} = yz^2\,\mathbf{i} - 3xz^2\,\mathbf{j} + 2xyz\,\mathbf{k}$, $\mathbf{B} = 3x\,\mathbf{i} + 4z\,\mathbf{j} - xy\,\mathbf{k}$ and $\phi = xyz$, find
(*a*) $\mathbf{A}\times(\nabla\phi)$, (*b*) $(\mathbf{A}\times\nabla)\phi$, (*c*) $(\nabla\times\mathbf{A})\times\mathbf{B}$, (*d*) $\mathbf{B}\cdot\nabla\times\mathbf{A}$.
*Ans.* (*a*) $-5x^2yz^2\,\mathbf{i} + xy^2z^2\,\mathbf{j} + 4xyz^3\,\mathbf{k}$
(*b*) $-5x^2yz^2\,\mathbf{i} + xy^2z^2\,\mathbf{j} + 4xyz^3\,\mathbf{k}$ (same as (*a*))
(*c*) $16z^3\,\mathbf{i} + (8x^2yz - 12xz^2)\mathbf{j} + 32xz^2\,\mathbf{k}$ (*d*) $24x^2z + 4xyz^2$

**97.** Find $\mathbf{A}\times(\nabla\times\mathbf{B})$ and $(\mathbf{A}\times\nabla)\times\mathbf{B}$ at the point (1,–1,2), if $\mathbf{A} = xz^2\,\mathbf{i} + 2y\,\mathbf{j} - 3xz\,\mathbf{k}$ and $\mathbf{B} = 3xz\,\mathbf{i} + 2yz\,\mathbf{j} - z^2\,\mathbf{k}$.
*Ans.* $\mathbf{A}\times(\nabla\times\mathbf{B}) = 18\mathbf{i} - 12\mathbf{j} + 16\mathbf{k}$, $(\mathbf{A}\times\nabla)\times\mathbf{B} = 4\mathbf{j} + 76\mathbf{k}$

**98.** Prove $(\mathbf{v}\cdot\nabla)\mathbf{v} = \frac{1}{2}\nabla v^2 - \mathbf{v}\times(\nabla\times\mathbf{v})$.

**99.** Prove $\nabla\cdot(\mathbf{A}\times\mathbf{B}) = \mathbf{B}\cdot(\nabla\times\mathbf{A}) - \mathbf{A}\cdot(\nabla\times\mathbf{B})$.

**100.** Prove $\nabla\times(\mathbf{A}\times\mathbf{B}) = (\mathbf{B}\cdot\nabla)\mathbf{A} - \mathbf{B}(\nabla\cdot\mathbf{A}) - (\mathbf{A}\cdot\nabla)\mathbf{B} + \mathbf{A}(\nabla\cdot\mathbf{B})$.

**101.** Prove $\nabla(\mathbf{A}\cdot\mathbf{B}) = (\mathbf{B}\cdot\nabla)\mathbf{A} + (\mathbf{A}\cdot\nabla)\mathbf{B} + \mathbf{B}\times(\nabla\times\mathbf{A}) + \mathbf{A}\times(\nabla\times\mathbf{B})$.

**102.** Show that $\mathbf{A} = (6xy + z^3)\mathbf{i} + (3x^2 - z)\mathbf{j} + (3xz^2 - y)\mathbf{k}$ is irrotational. Find $\phi$ such that $\mathbf{A} = \nabla\phi$.
*Ans.* $\phi = 3x^2y + xz^3 - yz$ + constant

**103.** Show that $\mathbf{E} = \mathbf{r}/r^2$ is irrotational. Find $\phi$ such that $\mathbf{E} = -\nabla\phi$ and such that $\phi(a) = 0$ where $a > 0$.
*Ans.* $\phi = \ln(a/r)$

**104.** If $\mathbf{A}$ and $\mathbf{B}$ are irrotational, prove that $\mathbf{A}\times\mathbf{B}$ is solenoidal.

**105.** If $f(r)$ is differentiable, prove that $f(r)\mathbf{r}$ is irrotational.

**106.** Is there a differentiable vector function $\mathbf{V}$ such that (*a*) curl $\mathbf{V} = \mathbf{r}$, (*b*) curl $\mathbf{V} = 2\mathbf{i} + \mathbf{j} + 3\mathbf{k}$? If so, find $\mathbf{V}$.
*Ans.* (*a*) No, (*b*) $\mathbf{V} = 3x\,\mathbf{j} + (2y - x)\mathbf{k} + \nabla\phi$, where $\phi$ is an arbitrary twice differentiable function.

**107.** Show that solutions to Maxwell's equations

$$\nabla\times\mathbf{H} = \frac{1}{c}\frac{\partial\mathbf{E}}{\partial t}, \quad \nabla\times\mathbf{E} = -\frac{1}{c}\frac{\partial\mathbf{H}}{\partial t}, \quad \nabla\cdot\mathbf{H} = 0, \quad \nabla\cdot\mathbf{E} = 4\pi\rho$$

where $\rho$ is a function of $x,y,z$ and $c$ is the velocity of light, assumed constant, are given by

$$\mathbf{E} = -\nabla\phi - \frac{1}{c}\frac{\partial\mathbf{A}}{\partial t}, \quad \mathbf{H} = \nabla\times\mathbf{A}$$

where $\mathbf{A}$ and $\phi$, called the *vector and scalar potentials* respectively, satisfy the equations

$$(1)\ \nabla\cdot\mathbf{A} + \frac{1}{c}\frac{\partial\phi}{\partial t} = 0, \quad (2)\ \nabla^2\phi - \frac{1}{c^2}\frac{\partial^2\phi}{\partial t^2} = -4\pi\rho, \quad (3)\ \nabla^2\mathbf{A} = \frac{1}{c^2}\frac{\partial^2\mathbf{A}}{\partial t^2}$$

**108.** (*a*) Given the dyadic $\mathbf{\Phi} = \mathbf{ii} + \mathbf{jj} + \mathbf{kk}$, evaluate $\mathbf{r}\cdot(\mathbf{\Phi}\cdot\mathbf{r})$ and $(\mathbf{r}\cdot\mathbf{\Phi})\cdot\mathbf{r}$. (*b*) Is there any ambiguity in writing $\mathbf{r}\cdot\mathbf{\Phi}\cdot\mathbf{r}$? (*c*) What does $\mathbf{r}\cdot\mathbf{\Phi}\cdot\mathbf{r} = 1$ represent geometrically?
*Ans.* (*a*) $\mathbf{r}\cdot(\mathbf{\Phi}\cdot\mathbf{r}) = (\mathbf{r}\cdot\mathbf{\Phi})\cdot\mathbf{r} = x^2+y^2+z^2$, (*b*) No, (*c*) Sphere of radius one with centre at the origin.

**109.** (*a*) If $\mathbf{A} = xz\,\mathbf{i} - y^2\,\mathbf{j} + yz^2\,\mathbf{k}$ and $\mathbf{B} = 2z^2\,\mathbf{i} - xy\,\mathbf{j} + y^3\,\mathbf{k}$, give a possible significance to $(\mathbf{A}\times\nabla)\mathbf{B}$ at the point $(1,-1,1)$.
(*b*) Is it possible to write the result as $\mathbf{A}\times(\nabla\mathbf{B})$ by use of dyadics?
*Ans.* (*a*) $-4\mathbf{ii} - \mathbf{ij} + 3\mathbf{ik} - \mathbf{jj} - 4\mathbf{ji} + 3\mathbf{kk}$
(*b*) Yes, if the operations are suitably performed.

**110.** Prove that $\phi(x,y,z) = x^2 + y^2 + z^2$ is a scalar invariant under a rotation of axes.

**111.** If $\mathbf{A}(x,y,z)$ is an invariant differentiable vector field with respect to a rotation of axes, prove that (*a*) div $\mathbf{A}$ and (*b*) curl $\mathbf{A}$ are invariant scalar and vector fields respectively under the transformation.

**112.** Solve equations (*3*) of Solved Problem 38 for $x,y,z$ in terms of $x',y',z'$.
*Ans.* $x = l_{11}x' + l_{21}y' + l_{31}z', \quad y = l_{12}x' + l_{22}y' + l_{32}z', \quad z = l_{13}x' + l_{23}y' + l_{33}z'$

**113.** If $\mathbf{A}$ and $\mathbf{B}$ are invariant under rotation show that $\mathbf{A}\cdot\mathbf{B}$ and $\mathbf{A}\times\mathbf{B}$ are also invariant.

**114.** Show that under a rotation

$$\nabla = \mathbf{i}\frac{\partial}{\partial x} + \mathbf{j}\frac{\partial}{\partial y} + \mathbf{k}\frac{\partial}{\partial z} = \mathbf{i}'\frac{\partial}{\partial x'} + \mathbf{j}'\frac{\partial}{\partial y'} + \mathbf{k}'\frac{\partial}{\partial z'} = \nabla'$$

**115.** Show that the Laplacian operator is invariant under a rotation.

# Chapter 5

# VECTOR INTEGRATION

**ORDINARY INTEGRALS OF VECTORS.** Let $\mathbf{R}(u) = R_1(u)\mathbf{i} + R_2(u)\mathbf{j} + R_3(u)\mathbf{k}$ be a vector depending on a single scalar variable $u$, where $R_1(u)$, $R_2(u)$, $R_3(u)$ are supposed continuous in a specified interval. Then

$$\int \mathbf{R}(u)\,du = \mathbf{i}\int R_1(u)\,du + \mathbf{j}\int R_2(u)\,du + \mathbf{k}\int R_3(u)\,du$$

is called an *indefinite integral* of $\mathbf{R}(u)$. If there exists a vector $\mathbf{S}(u)$ such that $\mathbf{R}(u) = \frac{d}{du}(\mathbf{S}(u))$, then

$$\int \mathbf{R}(u)\,du = \int \frac{d}{du}(\mathbf{S}(u))\,du = \mathbf{S}(u) + \mathbf{c}$$

where $\mathbf{c}$ is an *arbitrary constant vector* independent of $u$. The *definite integral* between limits $u=a$ and $u=b$ can in such case be written

$$\int_a^b \mathbf{R}(u)\,du = \int_a^b \frac{d}{du}(\mathbf{S}(u))\,du = \mathbf{S}(u) + \mathbf{c}\,\Big|_a^b = \mathbf{S}(b) - \mathbf{S}(a)$$

This integral can also be defined as a limit of a sum in a manner analogous to that of elementary integral calculus.

**LINE INTEGRALS.** Let $\mathbf{r}(u) = x(u)\mathbf{i} + y(u)\mathbf{j} + z(u)\mathbf{k}$, where $\mathbf{r}(u)$ is the position vector of $(x,y,z)$, define a curve $C$ joining points $P_1$ and $P_2$, where $u=u_1$ and $u=u_2$ respectively.

We assume that $C$ is composed of a finite number of curves for each of which $\mathbf{r}(u)$ has a continuous derivative. Let $\mathbf{A}(x,y,z) = A_1\mathbf{i} + A_2\mathbf{j} + A_3\mathbf{k}$ be a vector function of position defined and continuous along $C$. Then the integral of the tangential component of $\mathbf{A}$ along $C$ from $P_1$ to $P_2$, written as

$$\int_{P_1}^{P_2} \mathbf{A}\cdot d\mathbf{r} = \int_C \mathbf{A}\cdot d\mathbf{r} = \int_C A_1\,dx + A_2\,dy + A_3\,dz$$

is an example of a *line integral*. If $\mathbf{A}$ is the force $\mathbf{F}$ on a particle moving along $C$, this line integral represents the work done by the force. If $C$ is a closed curve (which we shall suppose is a *simple closed curve*, i.e. a curve which does not intersect itself anywhere) the integral around $C$ is often denoted by

$$\oint \mathbf{A}\cdot d\mathbf{r} = \oint A_1\,dx + A_2\,dy + A_3\,dz$$

In aerodynamics and fluid mechanics this integral is called the *circulation* of $\mathbf{A}$ about $C$, where $\mathbf{A}$ represents the velocity of a fluid.

In general, any integral which is to be evaluated along a curve is called a line integral. Such integrals can be defined in terms of limits of sums as are the integrals of elementary calculus.

For methods of evaluation of line integrals, see the Solved Problems.

The following theorem is important.

**THEOREM.** If $\mathbf{A} = \nabla\phi$ everywhere in a region $R$ of space, defined by $a_1 \leqq x \leqq a_2$, $b_1 \leqq y \leqq b_2$, $c_1 \leqq z \leqq c_2$, where $\phi(x,y,z)$ is single-valued and has continuous derivatives in $R$, then

1. $\displaystyle\int_{P_1}^{P_2} \mathbf{A}\cdot d\mathbf{r}$ is independent of the path $C$ in $R$ joining $P_1$ and $P_2$.

2. $\displaystyle\oint_C \mathbf{A}\cdot d\mathbf{r} = 0$ around any closed curve $C$ in $R$.

In such case $\mathbf{A}$ is called a *conservative vector field* and $\phi$ is its *scalar potential.*

A vector field $\mathbf{A}$ is conservative if and only if $\nabla\times\mathbf{A} = \mathbf{0}$, or equivalently $\mathbf{A} = \nabla\phi$. In such case $\mathbf{A}\cdot d\mathbf{r} = A_1\,dx + A_2\,dy + A_3\,dz = d\phi$, an exact differential. See Problems 10-14.

**SURFACE INTEGRALS.** Let $S$ be a two-sided surface, such as shown in the figure below. Let one side of $S$ be considered arbitrarily as the positive side (if $S$ is a closed surface this is taken as the outer side). A unit normal $\mathbf{n}$ to any point of the positive side of $S$ is called a *positive* or *outward drawn* unit normal.

Associate with the differential of surface area $dS$ a vector $d\mathbf{S}$ whose magnitude is $dS$ and whose direction is that of $\mathbf{n}$. Then $d\mathbf{S} = \mathbf{n}\,dS$. The integral

$$\iint_S \mathbf{A}\cdot d\mathbf{S} \quad = \quad \iint_S \mathbf{A}\cdot\mathbf{n}\,dS$$

is an example of a surface integral called the *flux* of $\mathbf{A}$ over $S$. Other surface integrals are

$$\iint_S \phi\,dS, \quad \iint_S \phi\,\mathbf{n}\,dS, \quad \iint_S \mathbf{A}\times d\mathbf{S}$$

where $\phi$ is a scalar function. Such integrals can be defined in terms of limits of sums as in elementary calculus (see Problem 17).

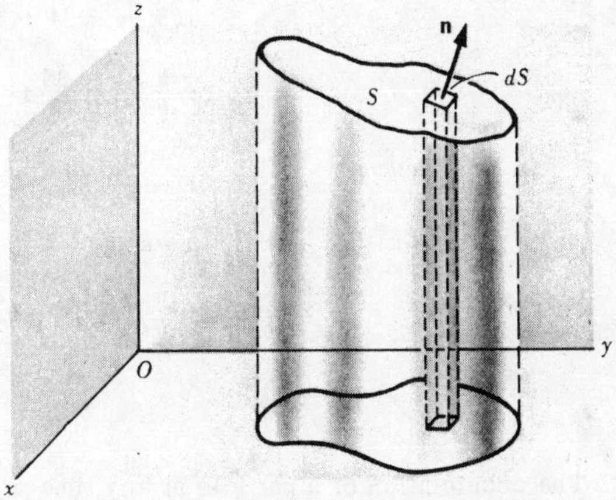


The notation $\oiint_S$ is sometimes used to indicate integration over the closed surface $S$. Where no confusion can arise the notation $\oint_S$ may also be used.

To evaluate surface integrals, it is convenient to express them as double integrals taken over the projected area of the surface $S$ on one of the coordinate planes. This is possible if any line perpendicular to the coordinate plane chosen meets the surface in no more than one point. However, this does not pose any real problem since we can generally subdivide $S$ into surfaces which do satisfy this restriction.

**VOLUME INTEGRALS.** Consider a closed surface in space enclosing a volume $V$. Then

$$\iiint_V \mathbf{A}\,dV \quad \text{and} \quad \iiint_V \phi\,dV$$

are examples of *volume integrals* or *space integrals* as they are sometimes called. For evaluation of such integrals, see the Solved Problems.

## SOLVED PROBLEMS

1. If $\mathbf{R}(u) = (u-u^2)\mathbf{i} + 2u^3\,\mathbf{j} - 3\mathbf{k}$, find (a) $\int \mathbf{R}(u)\,du$ and (b) $\int_1^2 \mathbf{R}(u)\,du$.

$$(a)\ \int \mathbf{R}(u)\,du = \int [(u-u^2)\mathbf{i} + 2u^3\,\mathbf{j} - 3\mathbf{k}]\,du$$
$$= \mathbf{i}\int (u-u^2)\,du + \mathbf{j}\int 2u^3\,du + \mathbf{k}\int -3\,du$$
$$= \mathbf{i}(\frac{u^2}{2} - \frac{u^3}{3} + c_1) + \mathbf{j}(\frac{u^4}{2} + c_2) + \mathbf{k}(-3u + c_3)$$
$$= (\frac{u^2}{2} - \frac{u^3}{3})\mathbf{i} + \frac{u^4}{2}\mathbf{j} - 3u\,\mathbf{k} + c_1\,\mathbf{i} + c_2\,\mathbf{j} + c_3\,\mathbf{k}$$
$$= (\frac{u^2}{2} - \frac{u^3}{3})\mathbf{i} + \frac{u^4}{2}\mathbf{j} - 3u\,\mathbf{k} + \mathbf{c}$$

where $\mathbf{c}$ is the constant vector $c_1\,\mathbf{i} + c_2\,\mathbf{j} + c_3\,\mathbf{k}$.

(b) From (a), $$\int_1^2 \mathbf{R}(u)\,du = (\frac{u^2}{2} - \frac{u^3}{3})\mathbf{i} + \frac{u^4}{2}\mathbf{j} - 3u\,\mathbf{k} + \mathbf{c}\,\Big|_1^2$$
$$= [(\frac{2^2}{2} - \frac{2^3}{3})\mathbf{i} + \frac{2^4}{2}\mathbf{j} - 3(2)\mathbf{k} + \mathbf{c}] - [(\frac{1^2}{2} - \frac{1^3}{3})\mathbf{i} + \frac{1^4}{2}\mathbf{j} - 3(1)\mathbf{k} + \mathbf{c}]$$
$$= -\frac{5}{6}\mathbf{i} + \frac{15}{2}\mathbf{j} - 3\mathbf{k}$$

*Another Method.*

$$\int_1^2 \mathbf{R}(u)\,du = \mathbf{i}\int_1^2 (u-u^2)du + \mathbf{j}\int_1^2 2u^3\,du + \mathbf{k}\int_1^2 -3\,du$$
$$= \mathbf{i}(\frac{u^2}{2} - \frac{u^3}{3})\Big|_1^2 + \mathbf{j}(\frac{u^4}{2})\Big|_1^2 + \mathbf{k}(-3u)\Big|_1^2 = -\frac{5}{6}\mathbf{i} + \frac{15}{2}\mathbf{j} - 3\mathbf{k}$$

2. The acceleration of a particle at any time $t \geq 0$ is given by
$$\mathbf{a} = \frac{d\mathbf{v}}{dt} = 12\cos 2t\,\mathbf{i} - 8\sin 2t\,\mathbf{j} + 16t\,\mathbf{k}$$
If the velocity $\mathbf{v}$ and displacement $\mathbf{r}$ are zero at $t=0$, find $\mathbf{v}$ and $\mathbf{r}$ at any time.

Integrating, $$\mathbf{v} = \mathbf{i}\int 12\cos 2t\,dt + \mathbf{j}\int -8\sin 2t\,dt + \mathbf{k}\int 16t\,dt$$
$$= 6\sin 2t\,\mathbf{i} + 4\cos 2t\,\mathbf{j} + 8t^2\,\mathbf{k} + \mathbf{c}_1$$

Putting $\mathbf{v}=\mathbf{0}$ when $t=0$, we find $\mathbf{0} = 0\mathbf{i} + 4\mathbf{j} + 0\mathbf{k} + \mathbf{c}_1$ and $\mathbf{c}_1 = -4\mathbf{j}$.

Then $$\mathbf{v} = 6\sin 2t\,\mathbf{i} + (4\cos 2t - 4)\,\mathbf{j} + 8t^2\,\mathbf{k}$$
so that $$\frac{d\mathbf{r}}{dt} = 6\sin 2t\,\mathbf{i} + (4\cos 2t - 4)\,\mathbf{j} + 8t^2\,\mathbf{k}.$$

Integrating, $$\mathbf{r} = \mathbf{i}\int 6\sin 2t\,dt + \mathbf{j}\int (4\cos 2t - 4)\,dt + \mathbf{k}\int 8t^2\,dt$$
$$= -3\cos 2t\,\mathbf{i} + (2\sin 2t - 4t)\,\mathbf{j} + \frac{8}{3}t^3\,\mathbf{k} + \mathbf{c}_2$$

Putting $\mathbf{r}=\mathbf{0}$ when $t=0$, $\mathbf{0} = -3\mathbf{i} + 0\mathbf{j} + 0\mathbf{k} + \mathbf{c}_2$ and $\mathbf{c}_2 = 3\mathbf{i}$.

Then $\mathbf{r} = (3 - 3\cos 2t)\mathbf{i} + (2\sin 2t - 4t)\mathbf{j} + \frac{8}{3}t^3\,\mathbf{k}$.

3. Evaluate $\displaystyle\int \mathbf{A} \times \frac{d^2\mathbf{A}}{dt^2}\,dt$ .

$$\frac{d}{dt}(\mathbf{A} \times \frac{d\mathbf{A}}{dt}) = \mathbf{A} \times \frac{d^2\mathbf{A}}{dt^2} + \frac{d\mathbf{A}}{dt} \times \frac{d\mathbf{A}}{dt} = \mathbf{A} \times \frac{d^2\mathbf{A}}{dt^2}$$

Integrating, $\displaystyle\int \mathbf{A} \times \frac{d^2\mathbf{A}}{dt^2}\,dt = \int \frac{d}{dt}(\mathbf{A} \times \frac{d\mathbf{A}}{dt})\,dt = \mathbf{A} \times \frac{d\mathbf{A}}{dt} + \mathbf{c}$.

4. The equation of motion of a particle $P$ of mass $m$ is given by

$$m\frac{d^2\mathbf{r}}{dt^2} = f(r)\,\mathbf{r}_1$$

where $\mathbf{r}$ is the position vector of $P$ measured from an origin $O$, $\mathbf{r}_1$ is a unit vector in the direction $\mathbf{r}$, and $f(r)$ is a function of the distance of $P$ from $O$.

(a) Show that $\mathbf{r} \times \frac{d\mathbf{r}}{dt} = \mathbf{c}$ where $\mathbf{c}$ is a constant vector.

(b) Interpret physically the cases $f(r) < 0$ and $f(r) > 0$.

(c) Interpret the result in (a) geometrically.

(d) Describe how the results obtained relate to the motion of the planets in our solar system.

(a) Multiply both sides of $m\frac{d^2\mathbf{r}}{dt^2} = f(r)\,\mathbf{r}_1$ by $\mathbf{r}\times$. Then

$$m\,\mathbf{r} \times \frac{d^2\mathbf{r}}{dt^2} = f(r)\,\mathbf{r} \times \mathbf{r}_1 = \mathbf{0}$$

since $\mathbf{r}$ and $\mathbf{r}_1$ are collinear and so $\mathbf{r} \times \mathbf{r}_1 = \mathbf{0}$. Thus

$$\mathbf{r} \times \frac{d^2\mathbf{r}}{dt^2} = \mathbf{0} \quad \text{and} \quad \frac{d}{dt}(\mathbf{r} \times \frac{d\mathbf{r}}{dt}) = \mathbf{0}$$

Integrating, $\mathbf{r} \times \frac{d\mathbf{r}}{dt} = \mathbf{c}$, where $\mathbf{c}$ is a constant vector. (Compare with Problem 3).

(b) If $f(r) < 0$ the acceleration $\frac{d^2\mathbf{r}}{dt^2}$ has direction opposite to $\mathbf{r}_1$; hence the force is directed toward $O$ and the particle is always *attracted* toward $O$.

If $f(r) > 0$ the force is directed away from $O$ and the particle is under the influence of a *repulsive* force at $O$.

A force directed toward or away from a fixed point $O$ and having magnitude depending only on the distance $r$ from $O$ is called a *central force*.

(c) In time $\Delta t$ the particle moves from $M$ to $N$ (see adjoining figure). The area swept out by the position vector in this time is approximately half the area of a parallelogram with sides $\mathbf{r}$ and $\Delta\mathbf{r}$, or $\frac{1}{2}\mathbf{r} \times \Delta\mathbf{r}$. Then the approximate area swept out by the radius vector per unit time is $\frac{1}{2}\mathbf{r} \times \frac{\Delta\mathbf{r}}{\Delta t}$; hence the instantaneous time rate of change in area is

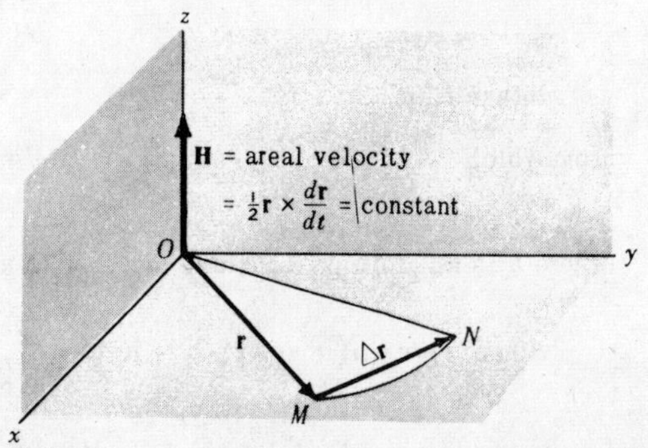


$$\lim_{\Delta t \to 0} \frac{1}{2}\mathbf{r} \times \frac{\Delta\mathbf{r}}{\Delta t} = \frac{1}{2}\mathbf{r} \times \frac{d\mathbf{r}}{dt} = \frac{1}{2}\mathbf{r} \times \mathbf{v}$$

where $\mathbf{v}$ is the instantaneous velocity of the particle. The quantity $\mathbf{H} = \frac{1}{2}\mathbf{r} \times \frac{d\mathbf{r}}{dt} = \frac{1}{2}\mathbf{r} \times \mathbf{v}$ is called the *areal velocity*. From part (*a*),

$$\text{Areal Velocity} = \mathbf{H} = \tfrac{1}{2}\mathbf{r} \times \frac{d\mathbf{r}}{dt} = \text{constant}$$

Since $\mathbf{r} \cdot \mathbf{H} = 0$, the motion takes place in a plane, which we take as the $xy$ plane in the figure above.

(*d*) A planet (such as the earth) is attracted toward the sun according to Newton's universal law of gravitation, which states that any two objects of mass $m$ and $M$ respectively are attracted toward each other with a force of magnitude $F = \dfrac{GMm}{r^2}$, where $r$ is the distance between objects and $G$ is a universal constant. Let $m$ and $M$ be the masses of the planet and sun respectively and choose a set of coordinate axes with the origin $O$ at the sun. Then the equation of motion of the planet is

$$m\frac{d^2\mathbf{r}}{dt^2} = -\frac{GMm}{r^2}\mathbf{r}_1 \qquad \text{or} \qquad \frac{d^2\mathbf{r}}{dt^2} = -\frac{GM}{r^2}\mathbf{r}_1$$

assuming the influence of the other planets to be negligible.

According to part (*c*), a planet moves around the sun so that its position vector sweeps out equal areas in equal times. This result and that of Problem 5 are two of Kepler's famous three laws which he deduced empirically from volumes of data compiled by the astronomer Tycho Brahe. These laws enabled Newton to formulate his universal law of gravitation. For Kepler's third law see Problem 36.

5. Show that the path of a planet around the sun is an ellipse with the sun at one focus.

From Problems 4(*c*) and 4(*d*),

(*1*) $$\frac{d\mathbf{v}}{dt} = -\frac{GM}{r^2}\mathbf{r}_1$$

(*2*) $$\mathbf{r} \times \mathbf{v} = 2\mathbf{H} = \mathbf{h}$$

Now $\mathbf{r} = r\,\mathbf{r}_1$, $\dfrac{d\mathbf{r}}{dt} = r\dfrac{d\mathbf{r}_1}{dt} + \dfrac{dr}{dt}\mathbf{r}_1$ so that

(*3*) $$\mathbf{h} = \mathbf{r} \times \mathbf{v} = r\,\mathbf{r}_1 \times \left(r\frac{d\mathbf{r}_1}{dt} + \frac{dr}{dt}\mathbf{r}_1\right) = r^2\mathbf{r}_1 \times \frac{d\mathbf{r}_1}{dt}$$

From (*1*), $$\frac{d\mathbf{v}}{dt} \times \mathbf{h} = -\frac{GM}{r^2}\mathbf{r}_1 \times \mathbf{h} = -GM\,\mathbf{r}_1 \times \left(\mathbf{r}_1 \times \frac{d\mathbf{r}_1}{dt}\right)$$

$$= -GM\left[\left(\mathbf{r}_1 \cdot \frac{d\mathbf{r}_1}{dt}\right)\mathbf{r}_1 - (\mathbf{r}_1 \cdot \mathbf{r}_1)\frac{d\mathbf{r}_1}{dt}\right] = GM\frac{d\mathbf{r}_1}{dt}$$

using equation (*3*) and the fact that $\mathbf{r}_1 \cdot \dfrac{d\mathbf{r}_1}{dt} = 0$ (Problem 9, Chapter 3).

But since $\mathbf{h}$ is a constant vector, $\dfrac{d\mathbf{v}}{dt} \times \mathbf{h} = \dfrac{d}{dt}(\mathbf{v} \times \mathbf{h})$ so that

$$\frac{d}{dt}(\mathbf{v} \times \mathbf{h}) = GM\frac{d\mathbf{r}_1}{dt}$$

Integrating, $$\mathbf{v} \times \mathbf{h} = GM\,\mathbf{r}_1 + \mathbf{p}$$

from which $$\mathbf{r} \cdot (\mathbf{v} \times \mathbf{h}) = GM\,\mathbf{r} \cdot \mathbf{r}_1 + \mathbf{r} \cdot \mathbf{p}$$
$$= GMr + r\,\mathbf{r}_1 \cdot \mathbf{p} = GMr + rp\cos\theta$$

where $\mathbf{p}$ is an arbitrary constant vector with magnitude $p$, and $\theta$ is the angle between $\mathbf{p}$ and $\mathbf{r}_1$.

Since $\mathbf{r} \cdot (\mathbf{v} \times \mathbf{h}) = (\mathbf{r} \times \mathbf{v}) \cdot \mathbf{h} = \mathbf{h} \cdot \mathbf{h} = h^2$, we have $h^2 = GMr + rp\cos\theta$ and

$$r = \frac{h^2}{GM + p\cos\theta} = \frac{h^2/GM}{1 + (p/GM)\cos\theta}$$

From analytical geometry, the polar equation of a conic section with focus at the origin and eccentricity $\epsilon$ is $r = \dfrac{a}{1+\epsilon\cos\theta}$ where $a$ is a constant. Comparing this with the equation derived, it is seen that the required orbit is a conic section with eccentricity $\epsilon = p/GM$. The orbit is an ellipse, parabola or hyperbola according as $\epsilon$ is less than, equal to or greater than one. Since orbits of planets are closed curves it follows that they must be ellipses.

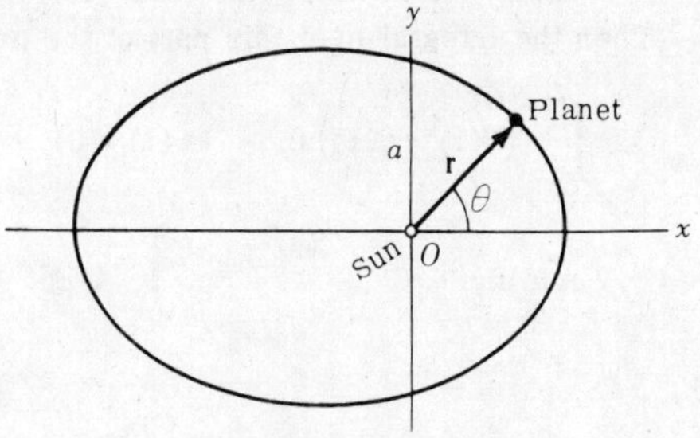


Ellipse $r = \dfrac{a}{1+\epsilon\cos\theta}$

## LINE INTEGRALS

6. If $\mathbf{A} = (3x^2+6y)\mathbf{i} - 14yz\,\mathbf{j} + 20xz^2\,\mathbf{k}$, evaluate $\int_C \mathbf{A}\cdot d\mathbf{r}$ from (0,0,0) to (1,1,1) along the following paths $C$:
(a) $x = t,\ y = t^2,\ z = t^3$.
(b) the straight lines from (0,0,0) to (1,0,0), then to (1,1,0), and then to (1,1,1).
(c) the straight line joining (0,0,0) and (1,1,1).

$$\int_C \mathbf{A}\cdot d\mathbf{r} = \int_C \left[(3x^2+6y)\mathbf{i} - 14yz\,\mathbf{j} + 20xz^2\mathbf{k}\right]\cdot(dx\,\mathbf{i} + dy\,\mathbf{j} + dz\,\mathbf{k})$$
$$= \int_C (3x^2+6y)\,dx - 14yz\,dy + 20xz^2\,dz$$

(a) If $x=t,\ y=t^2,\ z=t^3$, points (0,0,0) and (1,1,1) correspond to $t=0$ and $t=1$ respectively. Then

$$\int_C \mathbf{A}\cdot d\mathbf{r} = \int_{t=0}^{1} (3t^2+6t^2)\,dt - 14(t^2)(t^3)\,d(t^2) + 20(t)(t^3)^2\,d(t^3)$$
$$= \int_{t=0}^{1} 9t^2\,dt - 28t^6\,dt + 60t^9\,dt$$
$$= \int_{t=0}^{1} (9t^2 - 28t^6 + 60t^9)\,dt = 3t^3 - 4t^7 + 6t^{10}\Big|_0^1 = 5$$

*Another Method.*

Along $C$, $\mathbf{A} = 9t^2\mathbf{i} - 14t^5\mathbf{j} + 20t^7\mathbf{k}$ and $\mathbf{r} = x\mathbf{i} + y\mathbf{j} + z\mathbf{k} = t\mathbf{i} + t^2\mathbf{j} + t^3\mathbf{k}$ and $d\mathbf{r} = (\mathbf{i} + 2t\mathbf{j} + 3t^2\mathbf{k})\,dt$.

Then $$\int_C \mathbf{A}\cdot d\mathbf{r} = \int_{t=0}^{1} (9t^2\,\mathbf{i} - 14t^5\,\mathbf{j} + 20t^7\,\mathbf{k})\cdot(\mathbf{i} + 2t\,\mathbf{j} + 3t^2\,\mathbf{k})\,dt$$
$$= \int_0^1 (9t^2 - 28t^6 + 60t^9)\,dt = 5$$

(b) Along the straight line from (0,0,0) to (1,0,0) $y = 0,\ z = 0,\ dy = 0,\ dz = 0$ while $x$ varies from 0 to 1. Then the integral over this part of the path is

$$\int_{x=0}^{1} \big(3x^2 + 6(0)\big)\,dx - 14(0)(0)(0) + 20x(0)^2(0) = \int_{x=0}^{1} 3x^2\,dx = x^3\Big|_0^1 = 1$$

Along the straight line from (1,0,0) to (1,1,0) $x = 1,\ z = 0,\ dx = 0,\ dz = 0$ while $y$ varies from 0 to 1. Then the integral over this part of the path is

$$\int_{y=0}^{1} (3(1)^2 + 6y)\,0 - 14y(0)\,dy + 20(1)(0)^2\,0 = 0$$

Along the straight line from (1,1,0) to (1,1,1) $x=1$, $y=1$, $dx=0$, $dy=0$ while $z$ varies from 0 to 1. Then the integral over this part of the path is

$$\int_{z=0}^{1} \left(3(1)^2+6(1)\right)0 \;-\; 14(1)\,z(0) \;+\; 20(1)\,z^2\,dz \;=\; \int_{z=0}^{1} 20\,z^2\,dz \;=\; \frac{20\,z^3}{3}\Big|_0^1 \;=\; \frac{20}{3}$$

Adding, $$\int_C \mathbf{A}\cdot d\mathbf{r} \;=\; 1 + 0 + \frac{20}{3} \;=\; \frac{23}{3}$$

(c) The straight line joining (0,0,0) and (1,1,1) is given in parametric form by $x=t$, $y=t$, $z=t$. Then

$$\int_C \mathbf{A}\cdot d\mathbf{r} \;=\; \int_{t=0}^{1} (3t^2+6t)\,dt \;-\; 14(t)(t)\,dt \;+\; 20(t)(t)^2\,dt$$

$$= \int_{t=0}^{1} (3t^2+6t-14t^2+20t^3)\,dt \;=\; \int_{t=0}^{1} (6t-11t^2+20t^3)\,dt \;=\; \frac{13}{3}$$

7. Find the total work done in moving a particle in a force field given by $\mathbf{F} = 3xy\,\mathbf{i} - 5z\,\mathbf{j} + 10x\,\mathbf{k}$ along the curve $x=t^2+1$, $y=2t^2$, $z=t^3$ from $t=1$ to $t=2$.

$$\text{Total work} \;=\; \int_C \mathbf{F}\cdot d\mathbf{r} \;=\; \int_C (3xy\,\mathbf{i} - 5z\,\mathbf{j} + 10x\,\mathbf{k})\cdot(dx\,\mathbf{i} + dy\,\mathbf{j} + dz\,\mathbf{k})$$

$$= \int_C 3xy\,dx \;-\; 5z\,dy \;+\; 10x\,dz$$

$$= \int_{t=1}^{2} 3(t^2+1)(2t^2)\,d(t^2+1) \;-\; 5(t^3)\,d(2t^2) \;+\; 10(t^2+1)\,d(t^3)$$

$$= \int_1^2 (12t^5 + 10t^4 + 12t^3 + 30t^2)\,dt \;=\; 303$$

8. If $\mathbf{F} = 3xy\,\mathbf{i} - y^2\,\mathbf{j}$, evaluate $\int_C \mathbf{F}\cdot d\mathbf{r}$ where $C$ is the curve in the $xy$ plane, $y=2x^2$, from (0,0) to (1,2).

Since the integration is performed in the $xy$ plane ($z=0$), we can take $\mathbf{r} = x\,\mathbf{i} + y\,\mathbf{j}$. Then

$$\int_C \mathbf{F}\cdot d\mathbf{r} \;=\; \int_C (3xy\,\mathbf{i} - y^2\,\mathbf{j})\cdot(dx\,\mathbf{i} + dy\,\mathbf{j})$$

$$= \int_C 3xy\,dx \;-\; y^2\,dy$$

*First Method.* Let $x=t$ in $y=2x^2$. Then the parametric equations of $C$ are $x=t$, $y=2t^2$. Points (0,0) and (1,2) correspond to $t=0$ and $t=1$ respectively. Then

$$\int_C \mathbf{F}\cdot d\mathbf{r} \;=\; \int_{t=0}^{1} 3(t)(2t^2)\,dt \;-\; (2t^2)^2\,d(2t^2) \;=\; \int_{t=0}^{1} (6t^3-16t^5)\,dt \;=\; -\frac{7}{6}$$

*Second Method.* Substitute $y=2x^2$ directly, where $x$ goes from 0 to 1. Then

$$\int_C \mathbf{F}\cdot d\mathbf{r} \;=\; \int_{x=0}^{1} 3x(2x^2)\,dx \;-\; (2x^2)^2\,d(2x^2) \;=\; \int_{x=0}^{1} (6x^3-16x^5)\,dx \;=\; -\frac{7}{6}$$

Note that if the curve were traversed in the opposite sense, i.e. from (1,2) to (0,0), the value of the integral would have been $7/6$ instead of $-7/6$.

**9.** Find the work done in moving a particle once around a circle $C$ in the $xy$ plane, if the circle has centre at the origin and radius 3 and if the force field is given by

$$\mathbf{F} = (2x - y + z)\mathbf{i} + (x + y - z^2)\mathbf{j} + (3x - 2y + 4z)\mathbf{k}$$

In the plane $z=0$, $\mathbf{F} = (2x-y)\mathbf{i} + (x+y)\mathbf{j} + (3x-2y)\mathbf{k}$ and $d\mathbf{r} = dx\,\mathbf{i} + dy\,\mathbf{j}$ so that the work done is

$$\int_C \mathbf{F}\cdot d\mathbf{r} = \int_C [(2x-y)\mathbf{i} + (x+y)\mathbf{j} + (3x-2y)\mathbf{k}]\cdot[dx\,\mathbf{i} + dy\,\mathbf{j}]$$

$$= \int_C (2x-y)\,dx + (x+y)\,dy$$

Choose the parametric equations of the circle as $x = 3\cos t$, $y = 3\sin t$ where $t$ varies from 0 to $2\pi$ (see adjoining figure). Then the line integral equals

$$\int_{t=0}^{2\pi} [2(3\cos t) - 3\sin t][-3\sin t]\,dt + [3\cos t + 3\sin t][3\cos t]\,dt$$

$$= \int_0^{2\pi} (9 - 9\sin t\cos t)\,dt = 9t - \frac{9}{2}\sin^2 t\,\Big|_0^{2\pi} = 18\pi$$

In traversing $C$ we have chosen the counterclockwise direction indicated in the adjoining figure. We call this the *positive* direction, or say that $C$ has been traversed in the positive sense. If $C$ were traversed in the clockwise (negative) direction the value of the integral would be $-18\pi$.

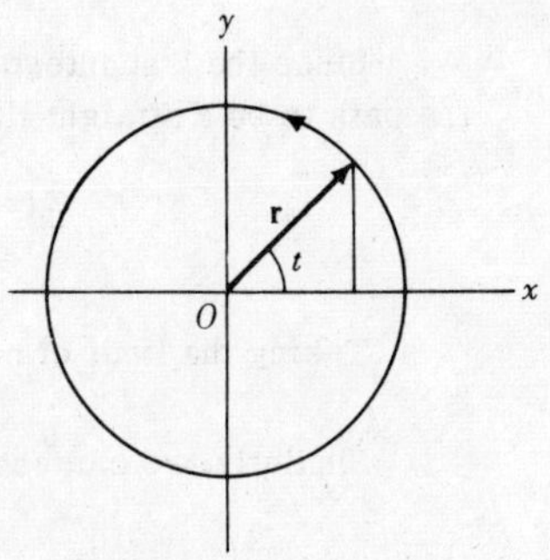

$\mathbf{r} = x\,\mathbf{i} + y\,\mathbf{j}$
$= 3\cos t\,\mathbf{i} + 3\sin t\,\mathbf{j}$

**10.** (*a*) If $\mathbf{F} = \nabla\phi$, where $\phi$ is single-valued and has continuous partial derivatives, show that the work done in moving a particle from one point $P_1 \equiv (x_1, y_1, z_1)$ in this field to another point $P_2 \equiv (x_2, y_2, z_2)$ is independent of the path joining the two points.

(*b*) Conversely, if $\int_C \mathbf{F}\cdot d\mathbf{r}$ is independent of the path $C$ joining any two points, show that there exists a function $\phi$ such that $\mathbf{F} = \nabla\phi$.

(*a*) Work done $= \int_{P_1}^{P_2} \mathbf{F}\cdot d\mathbf{r} = \int_{P_1}^{P_2} \nabla\phi\cdot d\mathbf{r}$

$$= \int_{P_1}^{P_2} (\frac{\partial\phi}{\partial x}\mathbf{i} + \frac{\partial\phi}{\partial y}\mathbf{j} + \frac{\partial\phi}{\partial z}\mathbf{k})\cdot(dx\,\mathbf{i} + dy\,\mathbf{j} + dz\,\mathbf{k})$$

$$= \int_{P_1}^{P_2} \frac{\partial\phi}{\partial x}dx + \frac{\partial\phi}{\partial y}dy + \frac{\partial\phi}{\partial z}dz$$

$$= \int_{P_1}^{P_2} d\phi = \phi(P_2) - \phi(P_1) = \phi(x_2, y_2, z_2) - \phi(x_1, y_1, z_1)$$

Then the integral depends only on points $P_1$ and $P_2$ and not on the path joining them. This is true of course only if $\phi(x,y,z)$ is single-valued at all points $P_1$ and $P_2$.

(*b*) Let $\mathbf{F} = F_1\mathbf{i} + F_2\mathbf{j} + F_3\mathbf{k}$. By hypothesis, $\int_C \mathbf{F}\cdot d\mathbf{r}$ is independent of the path $C$ joining any two points, which we take as $(x_1, y_1, z_1)$ and $(x,y,z)$ respectively. Then

$$\phi(x,y,z) = \int_{(x_1,y_1,z_1)}^{(x,y,z)} \mathbf{F}\cdot d\mathbf{r} = \int_{(x_1,y_1,z_1)}^{(x,y,z)} F_1dx + F_2dy + F_3dz$$

is independent of the path joining $(x_1, y_1, z_1)$ and $(x,y,z)$. Thus

$$\phi(x+\Delta x, y, z) - \phi(x,y,z) = \int_{(x_1,y_1,z_1)}^{(x+\Delta x,y,z)} \mathbf{F}\cdot d\mathbf{r} - \int_{(x_1,y_1,z_1)}^{(x,y,z)} \mathbf{F}\cdot d\mathbf{r}$$

$$= \int_{(x,y,z)}^{(x_1,y_1,z_1)} \mathbf{F}\cdot d\mathbf{r} + \int_{(x_1,y_1,z_1)}^{(x+\Delta x,y,z)} \mathbf{F}\cdot d\mathbf{r}$$

$$= \int_{(x,y,z)}^{(x+\Delta x,y,z)} \mathbf{F}\cdot d\mathbf{r} = \int_{(x,y,z)}^{(x+\Delta x,y,z)} F_1\,dx + F_2\,dy + F_3\,dz$$

Since the last integral must be independent of the path joining $(x,y,z)$ and $(x+\Delta x,y,z)$, we may choose the path to be a straight line joining these points so that $dy$ and $dz$ are zero. Then

$$\frac{\phi(x+\Delta x, y, z) - \phi(x,y,z)}{\Delta x} = \frac{1}{\Delta x}\int_{(x,y,z)}^{(x+\Delta x,y,z)} F_1\,dx$$

Taking the limit of both sides as $\Delta x \to 0$, we have $\dfrac{\partial\phi}{\partial x} = F_1$.

Similarly, we can show that $\dfrac{\partial\phi}{\partial y} = F_2$ and $\dfrac{\partial\phi}{\partial z} = F_3$.

Then $\mathbf{F} = F_1\mathbf{i} + F_2\mathbf{j} + F_3\mathbf{k} = \dfrac{\partial\phi}{\partial x}\mathbf{i} + \dfrac{\partial\phi}{\partial y}\mathbf{j} + \dfrac{\partial\phi}{\partial z}\mathbf{k} = \nabla\phi$.

If $\int_{P_1}^{P_2} \mathbf{F}\cdot d\mathbf{r}$ is independent of the path $C$ joining $P_1$ and $P_2$, then $\mathbf{F}$ is called a *conservative field.* It follows that if $\mathbf{F} = \nabla\phi$ then $\mathbf{F}$ is conservative, and conversely.

*Proof using vectors.* If the line integral is independent of the path, then

$$\phi(x,y,z) = \int_{(x_1,y_1,z_1)}^{(x,y,z)} \mathbf{F}\cdot d\mathbf{r} = \int_{(x_1,y_1,z_1)}^{(x,y,z)} \mathbf{F}\cdot\frac{d\mathbf{r}}{ds}\,ds$$

By differentiation, $\dfrac{d\phi}{ds} = \mathbf{F}\cdot\dfrac{d\mathbf{r}}{ds}$. But $\dfrac{d\phi}{ds} = \nabla\phi\cdot\dfrac{d\mathbf{r}}{ds}$ so that $(\nabla\phi - \mathbf{F})\cdot\dfrac{d\mathbf{r}}{ds} = 0$.

Since this must hold irrespective of $\dfrac{d\mathbf{r}}{ds}$, we have $\mathbf{F} = \nabla\phi$.

**11.** (*a*) If $\mathbf{F}$ is a conservative field, prove that $\operatorname{curl}\mathbf{F} = \nabla\times\mathbf{F} = \mathbf{0}$ (i.e. $\mathbf{F}$ is irrotational).
(*b*) Conversely, if $\nabla\times\mathbf{F} = \mathbf{0}$ (i.e. $\mathbf{F}$ is irrotational), prove that $\mathbf{F}$ is conservative.

(*a*) If $\mathbf{F}$ is a conservative field, then by Problem 10, $\mathbf{F} = \nabla\phi$.
Thus $\operatorname{curl}\mathbf{F} = \nabla\times\nabla\phi = \mathbf{0}$ (see Problem 27(*a*), Chapter 4).

(*b*) If $\nabla\times\mathbf{F} = \mathbf{0}$, then $\begin{vmatrix} \mathbf{i} & \mathbf{j} & \mathbf{k} \\ \dfrac{\partial}{\partial x} & \dfrac{\partial}{\partial y} & \dfrac{\partial}{\partial z} \\ F_1 & F_2 & F_3 \end{vmatrix} = \mathbf{0}$ and thus

$$\frac{\partial F_3}{\partial y} = \frac{\partial F_2}{\partial z}, \quad \frac{\partial F_1}{\partial z} = \frac{\partial F_3}{\partial x}, \quad \frac{\partial F_2}{\partial x} = \frac{\partial F_1}{\partial y}$$

We must prove that $\mathbf{F} = \nabla\phi$ follows as a consequence of this.

The work done in moving a particle from $(x_1,y_1,z_1)$ to $(x,y,z)$ in the force field $\mathbf{F}$ is

$$\int_C F_1(x,y,z)\,dx \;+\; F_2(x,y,z)\,dy \;+\; F_3(x,y,z)\,dz$$

where $C$ is a path joining $(x_1, y_1, z_1)$ and $(x,y,z)$. Let us choose as a particular path the straight line segments from $(x_1, y_1, z_1)$ to $(x, y_1, z_1)$ to $(x,y,z_1)$ to $(x,y,z)$ and call $\phi(x,y,z)$ the work done along this particular path. Then

$$\phi(x,y,z) \;=\; \int_{x_1}^{x} F_1(x, y_1, z_1)\,dx \;+\; \int_{y_1}^{y} F_2(x, y, z_1)\,dy \;+\; \int_{z_1}^{z} F_3(x,y,z)\,dz$$

It follows that

$$\frac{\partial\phi}{\partial z} \;=\; F_3(x,y,z)$$

$$\begin{aligned}
\frac{\partial\phi}{\partial y} &= F_2(x,y,z_1) + \int_{z_1}^{z} \frac{\partial F_3}{\partial y}(x,y,z)\,dz \\
&= F_2(x,y,z_1) + \int_{z_1}^{z} \frac{\partial F_2}{\partial z}(x,y,z)\,dz \\
&= F_2(x,y,z_1) + F_2(x,y,z)\Big|_{z_1}^{z} \;=\; F_2(x,y,z_1) + F_2(x,y,z) - F_2(x,y,z_1) \;=\; F_2(x,y,z)
\end{aligned}$$

$$\begin{aligned}
\frac{\partial\phi}{\partial x} &= F_1(x,y_1,z_1) + \int_{y_1}^{y} \frac{\partial F_2}{\partial x}(x,y,z_1)\,dy + \int_{z_1}^{z} \frac{\partial F_3}{\partial x}(x,y,z)\,dz \\
&= F_1(x,y_1,z_1) + \int_{y_1}^{y} \frac{\partial F_1}{\partial y}(x,y,z_1)\,dy + \int_{z_1}^{z} \frac{\partial F_1}{\partial z}(x,y,z)\,dz \\
&= F_1(x,y_1,z_1) + F_1(x,y,z_1)\Big|_{y_1}^{y} + F_1(x,y,z)\Big|_{z_1}^{z} \\
&= F_1(x,y_1,z_1) + F_1(x,y,z_1) - F(x,y_1,z_1) + F_1(x,y,z) - F(x,y,z_1) \;=\; F_1(x,y,z)
\end{aligned}$$

Then $\qquad \mathbf{F} \;=\; F_1\,\mathbf{i} + F_2\,\mathbf{j} + F_3\,\mathbf{k} \;=\; \dfrac{\partial\phi}{\partial x}\mathbf{i} + \dfrac{\partial\phi}{\partial y}\mathbf{j} + \dfrac{\partial\phi}{\partial z}\mathbf{k} \;=\; \nabla\phi.$

Thus a necessary and sufficient condition that a field $\mathbf{F}$ be conservative is that $\text{curl}\,\mathbf{F} = \nabla\times\mathbf{F} = \mathbf{0}$.

**12.** (*a*) Show that $\mathbf{F} = (2xy + z^3)\mathbf{i} + x^2\,\mathbf{j} + 3xz^2\,\mathbf{k}$ is a conservative force field. (*b*) Find the scalar potential. (*c*) Find the work done in moving an object in this field from $(1,-2,1)$ to $(3,1,4)$.

(*a*) From Problem 11, a necessary and sufficient condition that a force will be conservative is that $\text{curl}\,\mathbf{F} = \nabla\times\mathbf{F} = \mathbf{0}$.

Now $\quad \nabla\times\mathbf{F} \;=\; \begin{vmatrix} \mathbf{i} & \mathbf{j} & \mathbf{k} \\ \dfrac{\partial}{\partial x} & \dfrac{\partial}{\partial y} & \dfrac{\partial}{\partial z} \\ 2xy + z^3 & x^2 & 3xz^2 \end{vmatrix} \;=\; \mathbf{0}.$

Thus $\mathbf{F}$ is a conservative force field.

(*b*) *First Method.*

By Problem 10, $\mathbf{F} = \nabla\phi$ or $\frac{\partial\phi}{\partial x}\mathbf{i} + \frac{\partial\phi}{\partial y}\mathbf{j} + \frac{\partial\phi}{\partial z}\mathbf{k} = (2xy + z^3)\mathbf{i} + x^2\mathbf{j} + 3xz^2\mathbf{k}$. Then

$$(1)\ \frac{\partial\phi}{\partial x} = 2xy + z^3 \qquad (2)\ \frac{\partial\phi}{\partial y} = x^2 \qquad (3)\ \frac{\partial\phi}{\partial z} = 3xz^2$$

Integrating, we find from (*1*), (*2*) and (*3*) respectively,

$$\phi = x^2y + xz^3 + f(y,z)$$
$$\phi = x^2y + g(x,z)$$
$$\phi = xz^3 + h(x,y)$$

These agree if we choose $f(y,z) = 0$, $g(x,z) = xz^3$, $h(x,y) = x^2y$ so that $\phi = x^2y + xz^3$ to which may be added any constant.

*Second Method.*

Since $\mathbf{F}$ is conservative, $\int_C \mathbf{F}\cdot d\mathbf{r}$ is independent of the path $C$ joining $(x_1, y_1, z_1)$ and $(x,y,z)$. Using the method of Problem 11(*b*),

$$\phi(x,y,z) = \int_{x_1}^{x} (2xy_1 + z_1^3)\,dx + \int_{y_1}^{y} x^2\,dy + \int_{z_1}^{z} 3xz^2\,dz$$

$$= (x^2y_1 + xz_1^3)\Big|_{x_1}^{x} + x^2y\Big|_{y_1}^{y} + xz^3\Big|_{z_1}^{z}$$

$$= x^2y_1 + xz_1^3 - x_1^2y_1 - x_1z_1^3 + x^2y - x^2y_1 + xz^3 - xz_1^3$$

$$= x^2y + xz^3 - x_1^2y_1 - x_1z_1^3 = x^2y + xz^3 + \text{constant}$$

*Third Method.* $\mathbf{F}\cdot d\mathbf{r} = \nabla\phi\cdot d\mathbf{r} = \frac{\partial\phi}{\partial x}dx + \frac{\partial\phi}{\partial y}dy + \frac{\partial\phi}{\partial z}dz = d\phi$

Then $d\phi = \mathbf{F}\cdot d\mathbf{r} = (2xy + z^3)\,dx + x^2\,dy + 3xz^2\,dz$

$$= (2xy\,dx + x^2\,dy) + (z^3\,dx + 3xz^2\,dz)$$

$$= d(x^2y) + d(xz^3) = d(x^2y + xz^3)$$

and $\phi = x^2y + xz^3 + \text{constant}$.

(*c*) Work done $= \int_{P_1}^{P_2} \mathbf{F}\cdot d\mathbf{r}$

$$= \int_{P_1}^{P_2} (2xy + z^3)\,dx + x^2\,dy + 3xz^2\,dz$$

$$= \int_{P_1}^{P_2} d(x^2y + xz^3) = x^2y + xz^3\Big|_{P_1}^{P_2} = x^2y + xz^3\Big|_{(1,-2,1)}^{(3,1,4)} = 202$$

*Another Method.*

From part (*b*), $\phi(x,y,z) = x^2y + xz^3 + \text{constant}$.

Then work done $= \phi(3,1,4) - \phi(1,-2,1) = 202$.

**13.** Prove that if $\int_{P_1}^{P_2} \mathbf{F}\cdot d\mathbf{r}$ is independent of the path joining any two points $P_1$ and $P_2$ in a given region, then $\oint \mathbf{F}\cdot d\mathbf{r} = 0$ for all closed paths in the region and conversely.

Let $P_1AP_2BP_1$ (see adjacent figure) be a closed curve. Then

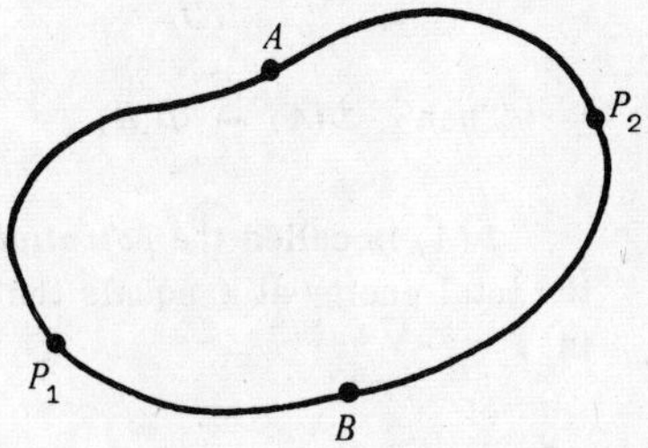


$$\oint \mathbf{F}\cdot d\mathbf{r} = \int_{P_1AP_2BP_1} \mathbf{F}\cdot d\mathbf{r} = \int_{P_1AP_2} \mathbf{F}\cdot d\mathbf{r} + \int_{P_2BP_1} \mathbf{F}\cdot d\mathbf{r}$$
$$= \int_{P_1AP_2} \mathbf{F}\cdot d\mathbf{r} - \int_{P_1BP_2} \mathbf{F}\cdot d\mathbf{r} = 0$$

since the integral from $P_1$ to $P_2$ along a path through $A$ is the same as that along a path through $B$, by hypothesis.

Conversely if $\oint \mathbf{F}\cdot d\mathbf{r} = 0$, then

$$\int_{P_1AP_2BP_1} \mathbf{F}\cdot d\mathbf{r} = \int_{P_1AP_2} \mathbf{F}\cdot d\mathbf{r} + \int_{P_2BP_1} \mathbf{F}\cdot d\mathbf{r} = \int_{P_1AP_2} \mathbf{F}\cdot d\mathbf{r} - \int_{P_1BP_2} \mathbf{F}\cdot d\mathbf{r} = 0$$

so that, $\int_{P_1AP_2} \mathbf{F}\cdot d\mathbf{r} = \int_{P_1BP_2} \mathbf{F}\cdot d\mathbf{r}$.

**14.** (*a*) Show that a necessary and sufficient condition that $F_1\,dx + F_2\,dy + F_3\,dz$ be an exact differential is that $\nabla\times\mathbf{F} = \mathbf{0}$ where $\mathbf{F} = F_1\mathbf{i} + F_2\mathbf{j} + F_3\mathbf{k}$.

(*b*) Show that $(y^2z^3\cos x - 4x^3z)\,dx + 2z^3y\sin x\,dy + (3y^2z^2\sin x - x^4)\,dz$ is an exact differential of a function $\phi$ and find $\phi$.

(*a*) Suppose $F_1\,dx + F_2\,dy + F_3\,dz = d\phi = \frac{\partial\phi}{\partial x}dx + \frac{\partial\phi}{\partial y}dy + \frac{\partial\phi}{\partial z}dz$, an exact differential. Then since $x, y$ and $z$ are independent variables,

$$F_1 = \frac{\partial\phi}{\partial x},\quad F_2 = \frac{\partial\phi}{\partial y},\quad F_3 = \frac{\partial\phi}{\partial z}$$

and so $\mathbf{F} = F_1\mathbf{i} + F_2\mathbf{j} + F_3\mathbf{k} = \frac{\partial\phi}{\partial x}\mathbf{i} + \frac{\partial\phi}{\partial y}\mathbf{j} + \frac{\partial\phi}{\partial z}\mathbf{k} = \nabla\phi$. Thus $\nabla\times\mathbf{F} = \nabla\times\nabla\phi = \mathbf{0}$.

Conversely if $\nabla\times\mathbf{F} = \mathbf{0}$ then by Problem 11, $\mathbf{F} = \nabla\phi$ and so $\mathbf{F}\cdot d\mathbf{r} = \nabla\phi\cdot d\mathbf{r} = d\phi$, i.e. $F_1\,dx + F_2\,dy + F_3\,dz = d\phi$, an exact differential.

(*b*) $\mathbf{F} = (y^2z^3\cos x - 4x^3z)\mathbf{i} + 2z^3y\sin x\,\mathbf{j} + (3y^2z^2\sin x - x^4)\mathbf{k}$ and $\nabla\times\mathbf{F}$ is computed to be zero, so that by part (*a*)

$$(y^2z^3\cos x - 4x^3z)\,dx + 2z^3y\sin x\,dy + (3y^2z^2\sin x - x^4)\,dz = d\phi$$

By any of the methods of Problem 12 we find $\phi = y^2z^3\sin x - x^4z + \text{constant}$.

**15.** Let $\mathbf{F}$ be a conservative force field such that $\mathbf{F} = -\nabla\phi$. Suppose a particle of constant mass $m$ to move in this field. If $A$ and $B$ are any two points in space, prove that

$$\phi(A) + \tfrac{1}{2}mv_A^2 = \phi(B) + \tfrac{1}{2}mv_B^2$$

where $v_A$ and $v_B$ are the magnitudes of the velocities of the particle at $A$ and $B$ respectively.

$$\mathbf{F} = m\mathbf{a} = m\frac{d^2\mathbf{r}}{dt^2}. \quad \text{Then} \quad \mathbf{F}\cdot\frac{d\mathbf{r}}{dt} = m\frac{d\mathbf{r}}{dt}\cdot\frac{d^2\mathbf{r}}{dt^2} = \frac{m}{2}\frac{d}{dt}\left(\frac{d\mathbf{r}}{dt}\right)^2.$$

Integrating, $$\int_A^B \mathbf{F}\cdot d\mathbf{r} = \frac{m}{2}v^2\Big|_A^B = \frac{1}{2}mv_B^2 - \frac{1}{2}mv_A^2.$$

If $\mathbf{F} = -\nabla\phi$, $$\int_A^B \mathbf{F}\cdot d\mathbf{r} = -\int_A^B \nabla\phi\cdot d\mathbf{r} = -\int_A^B d\phi = \phi(A) - \phi(B).$$

Then $\phi(A) - \phi(B) = \frac{1}{2}mv_B^2 - \frac{1}{2}mv_A^2$ and the result follows.

$\phi(A)$ is called the *potential energy* at $A$ and $\frac{1}{2}mv_A^2$ is the *kinetic energy* at $A$. The result states that the total energy at $A$ equals the total energy at $B$ (conservation of energy). Note the use of the minus sign in $\mathbf{F} = -\nabla\phi$.

**16.** If $\phi = 2xyz^2$, $\mathbf{F} = xy\,\mathbf{i} - z\,\mathbf{j} + x^2\,\mathbf{k}$ and $C$ is the curve $x=t^2$, $y=2t$, $z=t^3$ from $t=0$ to $t=1$, evaluate the line integrals (a) $\int_C \phi\, d\mathbf{r}$, (b) $\int_C \mathbf{F}\times d\mathbf{r}$.

(a) Along $C$, $\phi = 2xyz^2 = 2(t^2)(2t)(t^3)^2 = 4t^9$,

$\mathbf{r} = x\,\mathbf{i} + y\,\mathbf{j} + z\,\mathbf{k} = t^2\,\mathbf{i} + 2t\,\mathbf{j} + t^3\,\mathbf{k}$, and

$d\mathbf{r} = (2t\,\mathbf{i} + 2\mathbf{j} + 3t^2\,\mathbf{k})\,dt$. Then

$$\int_C \phi\, d\mathbf{r} = \int_{t=0}^1 4t^9(2t\,\mathbf{i} + 2\mathbf{j} + 3t^2\,\mathbf{k})\,dt$$

$$= \mathbf{i}\int_0^1 8t^{10}\,dt + \mathbf{j}\int_0^1 8t^9\,dt + \mathbf{k}\int_0^1 12t^{11}\,dt = \frac{8}{11}\mathbf{i} + \frac{4}{5}\mathbf{j} + \mathbf{k}$$

(b) Along $C$, $\mathbf{F} = xy\,\mathbf{i} - z\,\mathbf{j} + x^2\,\mathbf{k} = 2t^3\,\mathbf{i} - t^3\,\mathbf{j} + t^4\,\mathbf{k}$.

Then $\mathbf{F}\times d\mathbf{r} = (2t^3\,\mathbf{i} - t^3\,\mathbf{j} + t^4\,\mathbf{k}) \times (2t\,\mathbf{i} + 2\mathbf{j} + 3t^2\,\mathbf{k})\,dt$

$$= \begin{vmatrix} \mathbf{i} & \mathbf{j} & \mathbf{k} \\ 2t^3 & -t^3 & t^4 \\ 2t & 2 & 3t^2 \end{vmatrix} dt = \left[(-3t^5 - 2t^4)\mathbf{i} + (2t^5 - 6t^5)\mathbf{j} + (4t^3 + 2t^4)\mathbf{k}\right] dt$$

and $$\int_C \mathbf{F}\times d\mathbf{r} = \mathbf{i}\int_0^1 (-3t^5 - 2t^4)\,dt + \mathbf{j}\int_0^1 (-4t^5)\,dt + \mathbf{k}\int_0^1 (4t^3 + 2t^4)\,dt$$

$$= -\frac{9}{10}\mathbf{i} - \frac{2}{3}\mathbf{j} + \frac{7}{5}\mathbf{k}$$

## SURFACE INTEGRALS.

**17.** Give a definition of $\iint_S \mathbf{A}\cdot\mathbf{n}\, dS$ over a surface $S$ in terms of limit of a sum.

Subdivide the area $S$ into $M$ elements of area $\Delta S_p$ where $p = 1,2,3,\ldots, M$. Choose any point $P_p$ within $\Delta S_p$ whose coordinates are $(x_p, y_p, z_p)$. Define $\mathbf{A}(x_p, y_p, z_p) = \mathbf{A}_p$. Let $\mathbf{n}_p$ be the positive unit normal to $\Delta S_p$ at $P$. Form the sum

$$\sum_{p=1}^{M} \mathbf{A}_p \cdot \mathbf{n}_p \, \Delta S_p$$

where $\mathbf{A}_p \cdot \mathbf{n}_p$ is the normal component of $\mathbf{A}_p$ at $P_p$.

Now take the limit of this sum as $M \to \infty$ in such a way that the largest dimension of each $\Delta S_p$ approaches zero. This limit, if it exists, is called the surface integral of the normal component of $\mathbf{A}$ over $S$ and is denoted by

$$\iint_S \mathbf{A} \cdot \mathbf{n} \, dS$$

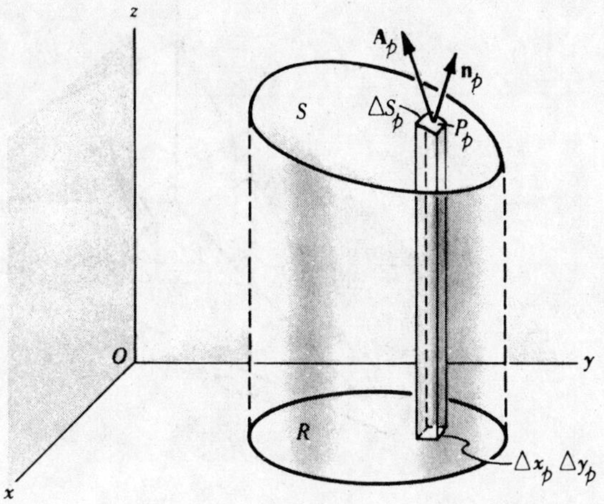


**18.** Suppose that the surface $S$ has projection $R$ on the $xy$ plane (see figure of Prob.17). Show that

$$\iint_S \mathbf{A} \cdot \mathbf{n} \, dS \quad = \quad \iint_R \mathbf{A} \cdot \mathbf{n} \, \frac{dx\, dy}{|\mathbf{n} \cdot \mathbf{k}|}$$

By Problem 17, the surface integral is the limit of the sum

$$(1) \qquad \sum_{p=1}^{M} \mathbf{A}_p \cdot \mathbf{n}_p \, \Delta S_p$$

The projection of $\Delta S_p$ on the $xy$ plane is $|(\mathbf{n}_p \, \Delta S_p) \cdot \mathbf{k}|$ or $|\mathbf{n}_p \cdot \mathbf{k}| \, \Delta S_p$ which is equal to $\Delta x_p \, \Delta y_p$ so that $\Delta S_p = \dfrac{\Delta x_p \, \Delta y_p}{|\mathbf{n}_p \cdot \mathbf{k}|}$. Thus the sum $(1)$ becomes

$$(2) \qquad \sum_{p=1}^{M} \mathbf{A}_p \cdot \mathbf{n}_p \frac{\Delta x_p \, \Delta y_p}{|\mathbf{n}_p \cdot \mathbf{k}|}$$

By the fundamental theorem of integral calculus the limit of this sum as $M \to \infty$ in such a manner that the largest $\Delta x_p$ and $\Delta y_p$ approach zero is

$$\iint_R \mathbf{A} \cdot \mathbf{n} \, \frac{dx\, dy}{|\mathbf{n} \cdot \mathbf{k}|}$$

and so the required result follows.

Strictly speaking, the result $\Delta S_p = \dfrac{\Delta x_p \, \Delta y_p}{|\mathbf{n}_p \cdot \mathbf{k}|}$ is only approximately true but it can be shown on closer examination that they differ from each other by infinitesimals of order higher than $\Delta x_p \, \Delta y_p$, and using this the *limits* of $(1)$ and $(2)$ can in fact be shown equal.

**19.** Evaluate $\displaystyle\iint_S \mathbf{A} \cdot \mathbf{n} \, dS$, where $\mathbf{A} = 18z\,\mathbf{i} - 12\mathbf{j} + 3y\,\mathbf{k}$ and $S$ is that part of the plane $2x + 3y + 6z = 12$ which is located in the first octant.

The surface $S$ and its projection $R$ on the $xy$ plane are shown in the figure below.

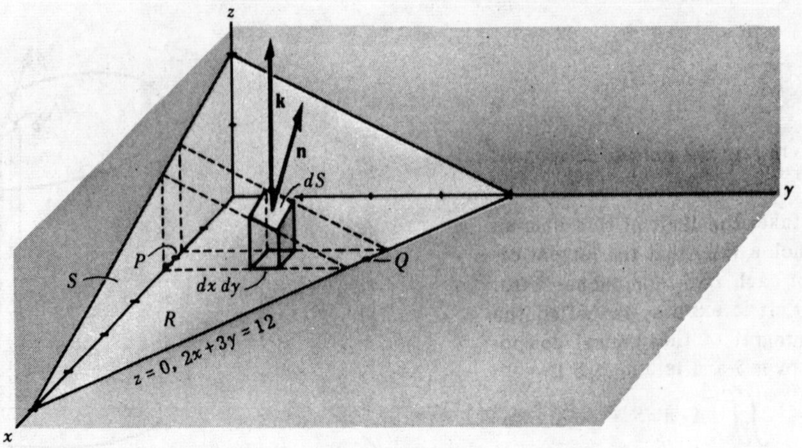


From Problem 17,

$$\iint_S \mathbf{A}\cdot\mathbf{n}\, dS \quad = \quad \iint_R \mathbf{A}\cdot\mathbf{n}\, \frac{dx\, dy}{|\mathbf{n}\cdot\mathbf{k}|}$$

To obtain **n** note that a vector perpendicular to the surface $2x+3y+6z = 12$ is given by $\nabla(2x+3y+6z) = 2\mathbf{i}+3\mathbf{j}+6\mathbf{k}$ (see Problem 5 of Chapter 4). Then a unit normal to any point of $S$ (see figure above) is

$$\mathbf{n} \;=\; \frac{2\mathbf{i}+3\mathbf{j}+6\mathbf{k}}{\sqrt{2^2+3^2+6^2}} \;=\; \tfrac{2}{7}\mathbf{i} + \tfrac{3}{7}\mathbf{j} + \tfrac{6}{7}\mathbf{k}$$

Thus $\quad \mathbf{n}\cdot\mathbf{k} \;=\; (\tfrac{2}{7}\mathbf{i}+\tfrac{3}{7}\mathbf{j}+\tfrac{6}{7}\mathbf{k})\cdot\mathbf{k} \;=\; \dfrac{6}{7}$ and so $\dfrac{dx\,dy}{|\mathbf{n}\cdot\mathbf{k}|} \;=\; \dfrac{7}{6}\,dx\,dy$.

Also $\quad \mathbf{A}\cdot\mathbf{n} \;=\; (18z\,\mathbf{i} - 12\mathbf{j} + 3y\,\mathbf{k})\cdot(\tfrac{2}{7}\mathbf{i}+\tfrac{3}{7}\mathbf{j}+\tfrac{6}{7}\mathbf{k}) \;=\; \dfrac{36z-36+18y}{7} \;=\; \dfrac{36-12x}{7}$,

using the fact that $z = \dfrac{12-2x-3y}{6}$ from the equation of $S$. Then

$$\iint_S \mathbf{A}\cdot\mathbf{n}\, dS \;=\; \iint_R \mathbf{A}\cdot\mathbf{n}\,\frac{dx\,dy}{|\mathbf{n}\cdot\mathbf{k}|} \;=\; \iint_R \left(\frac{36-12x}{7}\right)\frac{7}{6}\,dx\,dy \;=\; \iint_R (6-2x)\,dx\,dy$$

To evaluate this double integral over $R$, keep $x$ fixed and integrate with respect to $y$ from $y=0$ ($P$ in the figure above) to $y = \dfrac{12-2x}{3}$ ($Q$ in the figure above); then integrate with respect to $x$ from $x=0$ to $x=6$. In this manner $R$ is completely covered. The integral becomes

$$\int_{x=0}^{6}\int_{y=0}^{(12-2x)/3} (6-2x)\,dy\,dx \;=\; \int_{x=0}^{6} (24 - 12x + \frac{4x^2}{3})\,dx \;=\; 24$$

If we had chosen the positive unit normal **n** opposite to that in the figure above, we would have obtained the result $-24$.

**20.** Evaluate $\displaystyle\iint_S \mathbf{A}\cdot\mathbf{n}\,dS$, where $\mathbf{A} = z\,\mathbf{i} + x\,\mathbf{j} - 3y^2 z\,\mathbf{k}$ and $S$ is the surface of the cylinder $x^2+y^2=16$ included in the first octant between $z=0$ and $z=5$.

Project $S$ on the $xz$ plane as in the figure below and call the projection $R$. Note that the projection of $S$ on the $xy$ plane cannot be used here. Then

$$\iint_S \mathbf{A}\cdot\mathbf{n}\,dS = \iint_R \mathbf{A}\cdot\mathbf{n}\,\frac{dx\,dz}{|\mathbf{n}\cdot\mathbf{j}|}$$

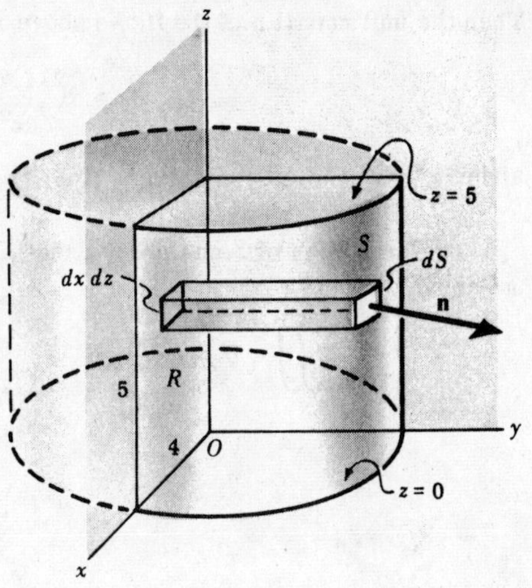


A normal to $x^2+y^2=16$ is $\nabla(x^2+y^2) = 2x\mathbf{i}+2y\mathbf{j}$. Thus the unit normal to $S$ as shown in the adjoining figure, is

$$\mathbf{n} = \frac{2x\,\mathbf{i}+2y\,\mathbf{j}}{\sqrt{(2x)^2+(2y)^2}} = \frac{x\,\mathbf{i}+y\,\mathbf{j}}{4}$$

since $x^2+y^2=16$ on $S$.

$$\mathbf{A}\cdot\mathbf{n} = (z\,\mathbf{i}+x\,\mathbf{j}-3y^2z\,\mathbf{k})\cdot\left(\frac{x\,\mathbf{i}+y\,\mathbf{j}}{4}\right) = \frac{1}{4}(xz+xy)$$

$$\mathbf{n}\cdot\mathbf{j} = \frac{x\,\mathbf{i}+y\,\mathbf{j}}{4}\cdot\mathbf{j} = \frac{y}{4}.$$

Then the surface integral equals

$$\iint_R \frac{xz+xy}{y}\,dx\,dz = \int_{z=0}^{5}\int_{x=0}^{4}\left(\frac{xz}{\sqrt{16-x^2}}+x\right)dx\,dz = \int_{z=0}^{5}(4z+8)\,dz = 90$$

**21.** Evaluate $\iint_S \phi\mathbf{n}\,dS$ where $\phi = \frac{3}{8}xyz$ and $S$ is the surface of Problem 20.

We have
$$\iint_S \phi\mathbf{n}\,dS = \iint_R \phi\mathbf{n}\,\frac{dx\,dz}{|\mathbf{n}\cdot\mathbf{j}|}$$

Using $\mathbf{n} = \frac{x\,\mathbf{i}+y\,\mathbf{j}}{4}$, $\mathbf{n}\cdot\mathbf{j} = \frac{y}{4}$ as in Problem 20, this last integral becomes

$$\iint_R \frac{3}{8}xz\,(x\,\mathbf{i}+y\,\mathbf{j})\,dx\,dz = \frac{3}{8}\int_{z=0}^{5}\int_{x=0}^{4}(x^2z\,\mathbf{i}+xz\sqrt{16-x^2}\,\mathbf{j})\,dx\,dz$$

$$= \frac{3}{8}\int_{z=0}^{5}\left(\frac{64}{3}z\,\mathbf{i}+\frac{64}{3}z\,\mathbf{j}\right)dz = 100\,\mathbf{i}+100\,\mathbf{j}$$

**22.** If $\mathbf{F} = y\,\mathbf{i}+(x-2xz)\mathbf{j}-xy\,\mathbf{k}$, evaluate $\iint_S (\nabla\times\mathbf{F})\cdot\mathbf{n}\,dS$ where $S$ is the surface of the sphere $x^2+y^2+z^2=a^2$ above the $xy$ plane.

$$\nabla\times\mathbf{F} = \begin{vmatrix} \mathbf{i} & \mathbf{j} & \mathbf{k} \\ \frac{\partial}{\partial x} & \frac{\partial}{\partial y} & \frac{\partial}{\partial z} \\ y & x-2xz & -xy \end{vmatrix} = x\,\mathbf{i}+y\,\mathbf{j}-2z\,\mathbf{k}$$

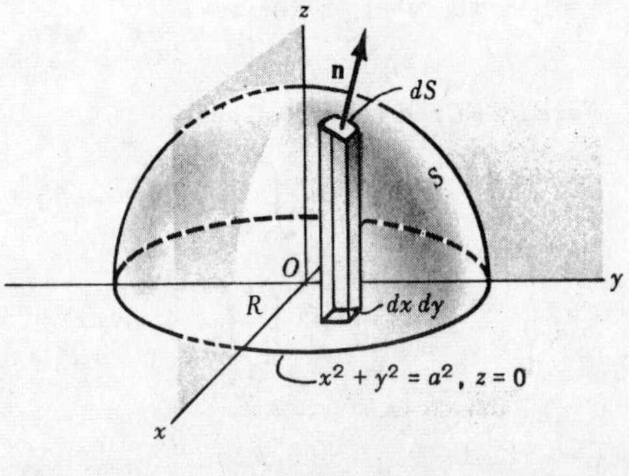


A normal to $x^2+y^2+z^2=a^2$ is

$$\nabla(x^2+y^2+z^2) = 2x\,\mathbf{i}+2y\,\mathbf{j}+2z\,\mathbf{k}$$

Then the unit normal **n** of the figure above is given by

$$\mathbf{n} = \frac{2x\,\mathbf{i} + 2y\,\mathbf{j} + 2z\,\mathbf{k}}{\sqrt{4x^2+4y^2+4z^2}} = \frac{x\,\mathbf{i} + y\,\mathbf{j} + z\,\mathbf{k}}{a}$$

since $x^2+y^2+z^2 = a^2$.

The projection of $S$ on the $xy$ plane is the region $R$ bounded by the circle $x^2+y^2 = a^2$, $z=0$ (see figure above). Then

$$\iint_S (\nabla\times\mathbf{F})\cdot\mathbf{n}\,dS = \iint_R (\nabla\times\mathbf{F})\cdot\mathbf{n}\,\frac{dx\,dy}{|\mathbf{n}\cdot\mathbf{k}|}$$

$$= \iint_R (x\,\mathbf{i} + y\,\mathbf{j} - 2z\,\mathbf{k})\cdot\left(\frac{x\,\mathbf{i}+y\,\mathbf{j}+z\,\mathbf{k}}{a}\right)\frac{dx\,dy}{z/a}$$

$$= \int_{x=-a}^{a}\int_{y=-\sqrt{a^2-x^2}}^{\sqrt{a^2-x^2}} \frac{3(x^2+y^2)-2a^2}{\sqrt{a^2-x^2-y^2}}\,dy\,dx$$

using the fact that $z = \sqrt{a^2-x^2-y^2}$. To evaluate the double integral, transform to polar coordinates $(\rho,\phi)$ where $x = \rho\cos\phi$, $y = \rho\sin\phi$ and $dy\,dx$ is replaced by $\rho\,d\rho\,d\phi$. The double integral becomes

$$\int_{\phi=0}^{2\pi}\int_{\rho=0}^{a} \frac{3\rho^2-2a^2}{\sqrt{a^2-\rho^2}}\,\rho\,d\rho\,d\phi = \int_{\phi=0}^{2\pi}\int_{\rho=0}^{a} \frac{3(\rho^2-a^2)+a^2}{\sqrt{a^2-\rho^2}}\,\rho\,d\rho\,d\phi$$

$$= \int_{\phi=0}^{2\pi}\int_{\rho=0}^{a} \left(-3\rho\sqrt{a^2-\rho^2} + \frac{a^2\rho}{\sqrt{a^2-\rho^2}}\right)d\rho\,d\phi$$

$$= \int_{\phi=0}^{2\pi} \left[(a^2-\rho^2)^{3/2} - a^2\sqrt{a^2-\rho^2}\,\Big|_{\rho=0}^{a}\right]d\phi$$

$$= \int_{\phi=0}^{2\pi} (a^3-a^3)\,d\phi = 0$$

**23.** If $\mathbf{F} = 4xz\,\mathbf{i} - y^2\,\mathbf{j} + yz\,\mathbf{k}$, evaluate $\iint_S \mathbf{F}\cdot\mathbf{n}\,dS$ where $S$ is the surface of the cube bounded by $x=0$, $x=1$, $y=0$, $y=1$, $z=0$, $z=1$.

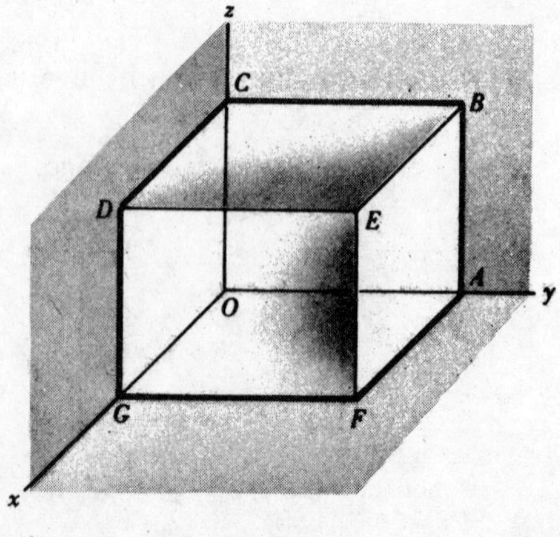


*Face DEFG:* $\mathbf{n}=\mathbf{i}$, $x=1$. Then

$$\iint_{DEFG} \mathbf{F}\cdot\mathbf{n}\,dS = \int_0^1\int_0^1 (4z\,\mathbf{i} - y^2\,\mathbf{j} + yz\,\mathbf{k})\cdot\mathbf{i}\,dy\,dz$$

$$= \int_0^1\int_0^1 4z\,dy\,dz = 2$$

*Face ABCO*: $\mathbf{n} = -\mathbf{i}$, $x = 0$. Then

$$\iint_{ABCO} \mathbf{F}\cdot\mathbf{n}\, dS \quad = \quad \int_0^1\int_0^1 (-y^2\,\mathbf{j} + yz\,\mathbf{k})\cdot(-\mathbf{i})\, dy\, dz \quad = \quad 0$$

*Face ABEF*: $\mathbf{n} = \mathbf{j}$, $y = 1$. Then

$$\iint_{ABEF} \mathbf{F}\cdot\mathbf{n}\, dS \quad = \quad \int_0^1\int_0^1 (4xz\,\mathbf{i} - \mathbf{j} + z\,\mathbf{k})\cdot\mathbf{j}\, dx\, dz \quad = \quad \int_0^1\int_0^1 -dx\, dz \quad = \quad -1$$

*Face OGDC*: $\mathbf{n} = -\mathbf{j}$, $y = 0$. Then

$$\iint_{OGDC} \mathbf{F}\cdot\mathbf{n}\, dS \quad = \quad \int_0^1\int_0^1 (4xz\,\mathbf{i})\cdot(-\mathbf{j})\, dx\, dz \quad = \quad 0$$

*Face BCDE*: $\mathbf{n} = \mathbf{k}$, $z = 1$. Then

$$\iint_{BCDE} \mathbf{F}\cdot\mathbf{n}\, dS \quad = \quad \int_0^1\int_0^1 (4x\,\mathbf{i} - y^2\,\mathbf{j} + y\,\mathbf{k})\cdot\mathbf{k}\, dx\, dy \quad = \quad \int_0^1\int_0^1 y\, dx\, dy \quad = \quad \tfrac{1}{2}$$

*Face AFGO*: $\mathbf{n} = -\mathbf{k}$, $z = 0$. Then

$$\iint_{AFGO} \mathbf{F}\cdot\mathbf{n}\, dS \quad = \quad \int_0^1\int_0^1 (-y^2\,\mathbf{j})\cdot(-\mathbf{k})\, dx\, dy \quad = \quad 0$$

Adding, $$\iint_S \mathbf{F}\cdot\mathbf{n}\, dS \quad = \quad 2 + 0 + (-1) + 0 + \tfrac{1}{2} + 0 \quad = \quad \frac{3}{2}.$$

**24.** In dealing with surface integrals we have restricted ourselves to surfaces which are two-sided. Give an example of a surface which is not two-sided.

Take a strip of paper such as $ABCD$ as shown in the adjoining figure. Twist the strip so that points $A$ and $B$ fall on $D$ and $C$ respectively, as in the adjoining figure. If $\mathbf{n}$ is the positive normal at point $P$ of the surface, we find that as $\mathbf{n}$ moves around the surface it reverses its original direction when it reaches $P$ again. If we tried to colour only one side of the surface we would find the whole thing coloured. This surface, called a *Moebius strip*, is an example of a one-sided surface. This is sometimes called a *non-orientable* surface. A two-sided surface is *orientable*.

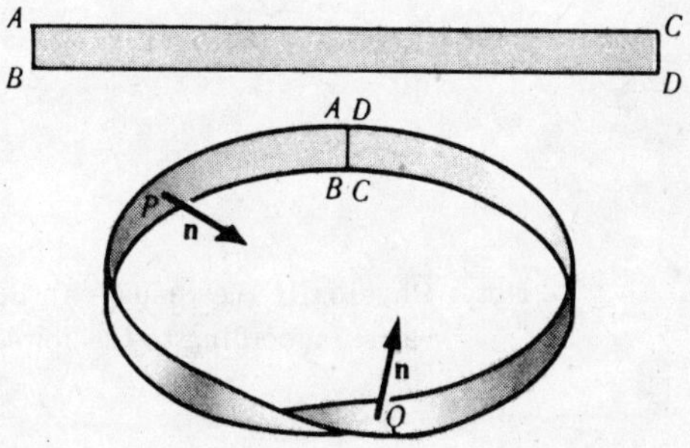


**VOLUME INTEGRALS**

**25.** Let $\phi = 45x^2y$ and let $V$ denote the closed region bounded by the planes $4x + 2y + z = 8$, $x = 0$, $y = 0$, $z = 0$. (*a*) Express $\iiint_V \phi\, dV$ as the limit of a sum. (*b*) Evaluate the integral in (*a*).

(a) Subdivide region $V$ into $M$ cubes having volume $\Delta V_k = \Delta x_k \Delta y_k \Delta z_k$ $k = 1, 2, \ldots, M$ as indicated in the adjoining figure and let $(x_k, y_k, z_k)$ be a point within this cube. Define $\phi(x_k, y_k, z_k) = \phi_k$. Consider the sum

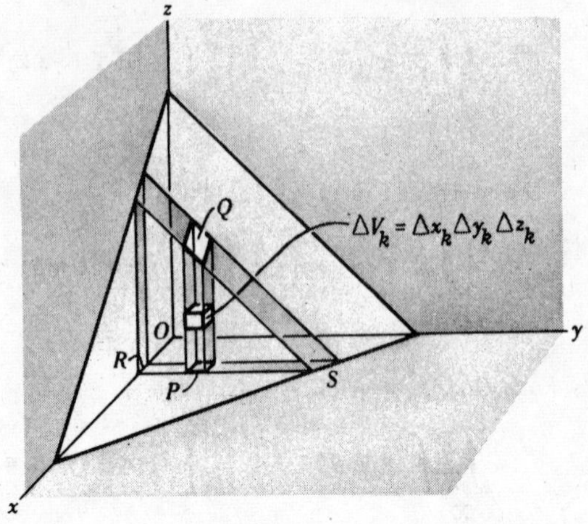


$$(1) \qquad \sum_{k=1}^{M} \phi_k \Delta V_k$$

taken over all possible cubes in the region. The limit of this sum, when $M \to \infty$ in such a manner that the largest of the quantities $\Delta V_k$ will approach zero, if it exists, is denoted by $\iiint_V \phi \, dV$. It can be shown that this limit is independent of the method of subdivision if $\phi$ is continuous throughout $V$.

In forming the sum (1) over all possible cubes in the region, it is advisable to proceed in an orderly fashion. One possibility is to add first all terms in (1) corresponding to volume elements contained in a column such as $PQ$ in the above figure. This amounts to keeping $x_k$ and $y_k$ fixed and adding over all $z_k$'s. Next, keep $x_k$ fixed but sum over all $y_k$'s. This amounts to adding all columns such as $PQ$ contained in a slab $RS$, and consequently amounts to summing over all cubes contained in such a slab. Finally, vary $x_k$. This amounts to addition of all slabs such as $RS$.

In the process outlined the summation is taken first over $z_k$'s then over $y_k$'s and finally over $x_k$'s. However, the summation can clearly be taken in any other order.

(b) The ideas involved in the method of summation outlined in (a) can be used in evaluating the integral. Keeping $x$ and $y$ constant, integrate from $z = 0$ (base of column $PQ$) to $z = 8 - 4x - 2y$ (top of column $PQ$). Next keep $x$ constant and integrate with respect to $y$. This amounts to addition of columns having bases in the $xy$ plane ($z = 0$) located anywhere from $R$ (where $y = 0$) to $S$ (where $4x + 2y = 8$ or $y = 4 - 2x$), and the integration is from $y = 0$ to $y = 4 - 2x$. Finally, we add all slabs parallel to the $yz$ plane, which amounts to integration from $x = 0$ to $x = 2$. The integration can be written

$$\int_{x=0}^{2} \int_{y=0}^{4-2x} \int_{z=0}^{8-4x-2y} 45x^2 y \, dz\, dy\, dx \;=\; 45 \int_{x=0}^{2} \int_{y=0}^{4-2x} x^2 y\,(8 - 4x - 2y)\, dy\, dx$$

$$= \; 45 \int_{x=0}^{2} \tfrac{1}{3} x^2 (4 - 2x)^3 \, dx \;=\; 128$$

Note: Physically the result can be interpreted as the mass of the region $V$ in which the density $\phi$ varies according to the formula $\phi = 45x^2 y$.

**26.** Let $\mathbf{F} = 2xz\,\mathbf{i} - x\,\mathbf{j} + y^2\,\mathbf{k}$. Evaluate $\iiint_V \mathbf{F}\, dV$ where $V$ is the region bounded by the surfaces $x = 0$, $y = 0$, $y = 6$, $z = x^2$, $z = 4$.

The region $V$ is covered (a) by keeping $x$ and $y$ fixed and integrating from $z = x^2$ to $z = 4$ (base to top of column $PQ$), (b) then by keeping $x$ fixed and integrating from $y = 0$ to $y = 6$ ($R$ to $S$ in the slab), (c) finally integrating from $x = 0$ to $x = 2$ (where $z = x^2$ meets $z = 4$). Then the required integral is

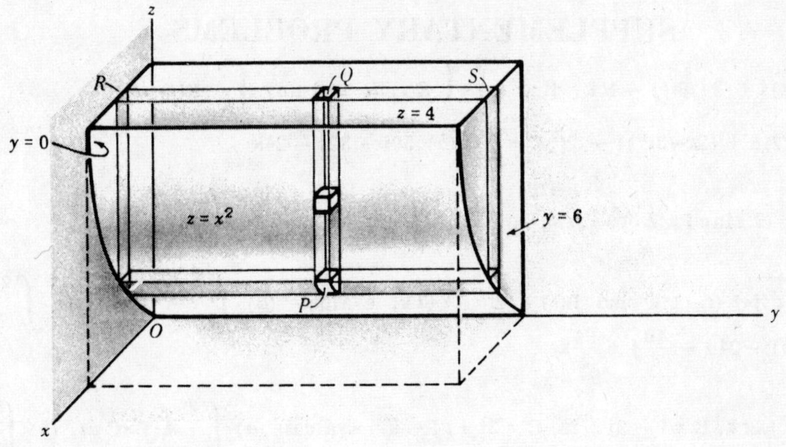


$$\int_{x=0}^{2}\int_{y=0}^{6}\int_{z=x^2}^{4}(2xz\,\mathbf{i} - x\,\mathbf{j} + y^2\,\mathbf{k})\,dz\,dy\,dx$$

$$= \mathbf{i}\int_0^2\int_0^6\int_{x^2}^4 2xz\,dz\,dy\,dx \;-\; \mathbf{j}\int_0^2\int_0^6\int_{x^2}^4 x\,dz\,dy\,dx \;+\; \mathbf{k}\int_0^2\int_0^6\int_{x^2}^4 y^2\,dz\,dy\,dx$$

$$= 128\,\mathbf{i} - 24\,\mathbf{j} + 384\,\mathbf{k}$$

**27.** Find the volume of the region common to the intersecting cylinders $x^2+y^2 = a^2$ and $x^2+z^2 = a^2$.

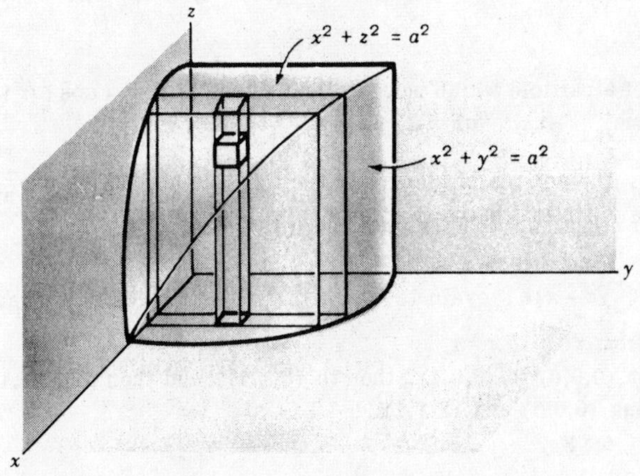


Required volume = 8 times volume of region shown in above figure

$$= 8\int_{x=0}^{a}\int_{y=0}^{\sqrt{a^2-x^2}}\int_{z=0}^{\sqrt{a^2-x^2}} dz\,dy\,dx$$

$$= 8\int_{x=0}^{a}\int_{y=0}^{\sqrt{a^2-x^2}}\sqrt{a^2-x^2}\,dy\,dx = 8\int_{x=0}^{a}(a^2-x^2)\,dx = \frac{16a^3}{3}$$

## SUPPLEMENTARY PROBLEMS

**28.** If $\mathbf{R}(t) = (3t^2-t)\mathbf{i} + (2-6t)\mathbf{j} - 4t\mathbf{k}$, find (a) $\int \mathbf{R}(t)\,dt$ and (b) $\int_2^4 \mathbf{R}(t)\,dt$.

*Ans.* (a) $(t^3-t^2/2)\mathbf{i} + (2t-3t^2)\mathbf{j} - 2t^2\mathbf{k} + \mathbf{c}$ (b) $50\mathbf{i} - 32\mathbf{j} - 24\mathbf{k}$

**29.** Evaluate $\int_0^{\pi/2} (3\sin u\,\mathbf{i} + 2\cos u\,\mathbf{j})\,du$ *Ans.* $3\mathbf{i} + 2\mathbf{j}$

**30.** If $\mathbf{A}(t) = t\,\mathbf{i} - t^2\mathbf{j} + (t-1)\mathbf{k}$ and $\mathbf{B}(t) = 2t^2\mathbf{i} + 6t\mathbf{k}$, evaluate (a) $\int_0^2 \mathbf{A}\cdot\mathbf{B}\,dt$, (b) $\int_0^2 \mathbf{A}\times\mathbf{B}\,dt$.

*Ans.* (a) 12 (b) $-24\mathbf{i} - \frac{40}{3}\mathbf{j} + \frac{64}{5}\mathbf{k}$

**31.** Let $\mathbf{A} = t\mathbf{i} - 3\mathbf{j} + 2t\mathbf{k}$, $\mathbf{B} = \mathbf{i} - 2\mathbf{j} + 2\mathbf{k}$, $\mathbf{C} = 3\mathbf{i} + t\mathbf{j} - \mathbf{k}$. Evaluate (a) $\int_1^2 \mathbf{A}\cdot\mathbf{B}\times\mathbf{C}\,dt$, (b) $\int_1^2 \mathbf{A}\times(\mathbf{B}\times\mathbf{C})\,dt$.

*Ans.* (a) 0 (b) $-\frac{87}{2}\mathbf{i} - \frac{44}{3}\mathbf{j} + \frac{15}{2}\mathbf{k}$

**32.** The acceleration $\mathbf{a}$ of a particle at any time $t \geqq 0$ is given by $\mathbf{a} = e^{-t}\mathbf{i} - 6(t+1)\mathbf{j} + 3\sin t\,\mathbf{k}$. If the velocity $\mathbf{v}$ and displacement $\mathbf{r}$ are zero at $t=0$, find $\mathbf{v}$ and $\mathbf{r}$ at any time.

*Ans.* $\mathbf{v} = (1-e^{-t})\mathbf{i} - (3t^2+6t)\mathbf{j} + (3-3\cos t)\mathbf{k}$, $\mathbf{r} = (t-1+e^{-t})\mathbf{i} - (t^3+3t^2)\mathbf{j} + (3t-3\sin t)\mathbf{k}$

**33.** The acceleration $\mathbf{a}$ of an object at any time $t$ is given by $\mathbf{a} = -g\,\mathbf{j}$, where $g$ is a constant. At $t=0$ the velocity is given by $\mathbf{v} = v_0\cos\theta_0\,\mathbf{i} + v_0\sin\theta_0\,\mathbf{j}$ and the displacement $\mathbf{r}=\mathbf{0}$. Find $\mathbf{v}$ and $\mathbf{r}$ at any time $t>0$. This describes the motion of a projectile fired from a cannon inclined at angle $\theta_0$ with the positive $x$-axis with initial velocity of magnitude $v_0$.

*Ans.* $\mathbf{v} = v_0\cos\theta_0\,\mathbf{i} + (v_0\sin\theta_0 - gt)\mathbf{j}$, $\mathbf{r} = (v_0\cos\theta_0)t\,\mathbf{i} + [(v_0\sin\theta_0)t - \frac{1}{2}gt^2]\mathbf{j}$

**34.** Evaluate $\int_2^3 \mathbf{A}\cdot\frac{d\mathbf{A}}{dt}\,dt$ if $\mathbf{A}(2) = 2\mathbf{i} - \mathbf{j} + 2\mathbf{k}$ and $\mathbf{A}(3) = 4\mathbf{i} - 2\mathbf{j} + 3\mathbf{k}$. *Ans.* 10

**35.** Find the areal velocity of a particle which moves along the path $\mathbf{r} = a\cos\omega t\,\mathbf{i} + b\sin\omega t\,\mathbf{j}$ where $a,b,\omega$ are constants and $t$ is time. *Ans.* $\frac{1}{2}ab\omega\,\mathbf{k}$

**36.** Prove that the squares of the periods of planets in their motion around the sun are proportional to the cubes of the major axes of their elliptical paths (Kepler's third law).

**37.** If $\mathbf{A} = (2y+3)\mathbf{i} + xz\,\mathbf{j} + (yz-x)\mathbf{k}$, evaluate $\int_C \mathbf{A}\cdot d\mathbf{r}$ along the following paths $C$:

(a) $x=2t^2$, $y=t$, $z=t^3$ from $t=0$ to $t=1$,
(b) the straight lines from (0,0,0) to (0,0,1), then to (0,1,1), and then to (2,1,1),
(c) the straight line joining (0,0,0) and (2,1,1).

*Ans.* (a) 288/35 (b) 10 (c) 8

**38.** If $\mathbf{F} = (5xy-6x^2)\mathbf{i} + (2y-4x)\mathbf{j}$, evaluate $\int_C \mathbf{F}\cdot d\mathbf{r}$ along the curve $C$ in the $xy$ plane, $y=x^3$ from the point (1,1) to (2,8). *Ans.* 35

**39.** If $\mathbf{F} = (2x+y)\mathbf{i} + (3y-x)\mathbf{j}$, evaluate $\int_C \mathbf{F}\cdot d\mathbf{r}$ where $C$ is the curve in the $xy$ plane consisting of the straight lines from (0,0) to (2,0) and then to (3,2). *Ans.* 11

**40.** Find the work done in moving a particle in the force field $\mathbf{F} = 3x^2\mathbf{i} + (2xz-y)\mathbf{j} + z\,\mathbf{k}$ along

(a) the straight line from (0,0,0) to (2,1,3).
(b) the space curve $x=2t^2$, $y=t$, $z=4t^2-t$ from $t=0$ to $t=1$.
(c) the curve defined by $x^2=4y$, $3x^3=8z$ from $x=0$ to $x=2$.

*Ans.* (a) 16 (b) 14.2 (c) 16

**41.** Evaluate $\oint_C \mathbf{F}\cdot d\mathbf{r}$ where $\mathbf{F} = (x-3y)\mathbf{i} + (y-2x)\mathbf{j}$ and $C$ is the closed curve in the $xy$ plane, $x = 2\cos t$, $y = 3\sin t$ from $t=0$ to $t=2\pi$. *Ans.* $6\pi$, if $C$ is traversed in the positive (counterclockwise) direction.

**42.** If $\mathbf{T}$ is a unit tangent vector to the curve $C$, $\mathbf{r}=\mathbf{r}(u)$, show that the work done in moving a particle in a force field $\mathbf{F}$ along $C$ is given by $\int_C \mathbf{F}\cdot\mathbf{T}\,ds$ where $s$ is the arc length.

**43.** If $\mathbf{F} = (2x+y^2)\mathbf{i} + (3y-4x)\mathbf{j}$, evaluate $\oint_C \mathbf{F}\cdot d\mathbf{r}$ around the triangle $C$ of Figure 1, $(a)$ in the indicated direction, $(b)$ opposite to the indicated direction. *Ans.* $(a)\ -14/3$ $(b)\ 14/3$

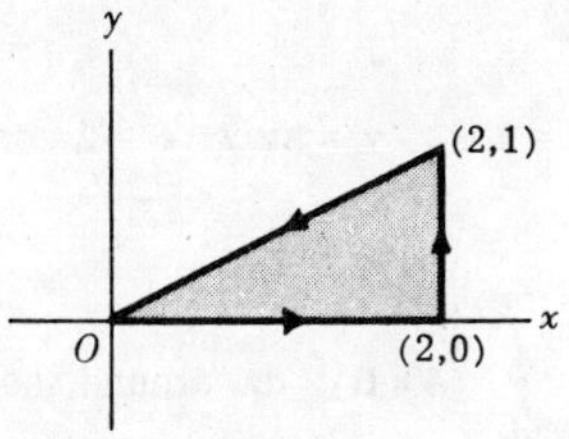


Fig. 1

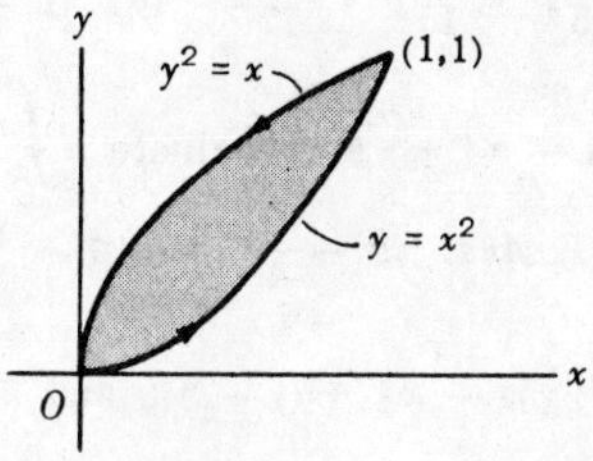


Fig. 2

**44.** Evaluate $\oint_C \mathbf{A}\cdot d\mathbf{r}$ around the closed curve $C$ of Fig. 2 above if $\mathbf{A} = (x-y)\mathbf{i} + (x+y)\mathbf{j}$. *Ans.* 2/3

**45.** If $\mathbf{A} = (y-2x)\mathbf{i} + (3x+2y)\mathbf{j}$, compute the circulation of $\mathbf{A}$ about a circle $C$ in the $xy$ plane with centre at the origin and radius 2, if $C$ is traversed in the positive direction. *Ans.* $8\pi$

**46.** $(a)$ If $\mathbf{A} = (4xy-3x^2z^2)\mathbf{i} + 2x^2\mathbf{j} - 2x^3z\,\mathbf{k}$, prove that $\int_C \mathbf{A}\cdot d\mathbf{r}$ is independent of the curve $C$ joining two given points. $(b)$ Show that there is a differentiable function $\phi$ such that $\mathbf{A} = \nabla\phi$ and find it.
*Ans.* $(b)\ \phi = 2x^2y - x^3z^2 + \text{constant}$

**47.** $(a)$ Prove that $\mathbf{F} = (y^2\cos x + z^3)\mathbf{i} + (2y\sin x - 4)\mathbf{j} + (3xz^2+2)\mathbf{k}$ is a conservative force field.
$(b)$ Find the scalar potential for $\mathbf{F}$.
$(c)$ Find the work done in moving an object in this field from $(0,1,-1)$ to $(\pi/2,-1,2)$.
*Ans.* $(b)\ \phi = y^2\sin x + xz^3 - 4y + 2z + \text{constant}$ $(c)\ 15 + 4\pi$

**48.** Prove that $\mathbf{F} = r^2\mathbf{r}$ is conservative and find the scalar potential. *Ans.* $\phi = \dfrac{r^4}{4} + \text{constant}$

**49.** Determine whether the force field $\mathbf{F} = 2xz\,\mathbf{i} + (x^2-y)\mathbf{j} + (2z-x^2)\mathbf{k}$ is conservative or non-conservative.
*Ans.* non-conservative

**50.** Show that the work done on a particle in moving it from $A$ to $B$ equals its change in kinetic energies at these points whether the force field is conservative or not.

**51.** Evaluate $\int_C \mathbf{A}\cdot d\mathbf{r}$ along the curve $x^2+y^2=1$, $z=1$ in the positive direction from $(0,1,1)$ to $(1,0,1)$ if $\mathbf{A} = (yz+2x)\mathbf{i} + xz\,\mathbf{j} + (xy+2z)\mathbf{k}$. *Ans.* 1

**52.** $(a)$ If $\mathbf{E} = r\mathbf{r}$, is there a function $\phi$ such that $\mathbf{E} = -\nabla\phi$? If so, find it. $(b)$ Evaluate $\oint_C \mathbf{E}\cdot d\mathbf{r}$ if $C$ is any simple closed curve. *Ans.* $(a)\ \phi = -\dfrac{r^3}{3} + \text{constant}$ $(b)\ 0$

**53.** Show that $(2x\cos y + z\sin y)\,dx + (xz\cos y - x^2\sin y)\,dy + x\sin y\,dz$ is an exact differential. Hence

solve the differential equation $(2x\cos y + z\sin y)\,dx + (xz\cos y - x^2\sin y)\,dy + x\sin y\,dz = 0$.
*Ans.* $x^2\cos y + xz\sin y = \text{constant}$

**54.** Solve (*a*) $(e^{-y} + 3x^2y^2)\,dx + (2x^3y - xe^{-y})\,dy = 0$,

(*b*) $(z - e^{-x}\sin y)\,dx + (1 + e^{-x}\cos y)\,dy + (x - 8z)\,dz = 0$.

*Ans.* (*a*) $xe^{-y} + x^3y^2 = \text{constant}$ (*b*) $xz + e^{-x}\sin y + y - 4z^2 = \text{constant}$

**55.** If $\phi = 2xy^2z + x^2y$, evaluate $\int_C \phi\,d\mathbf{r}$ where $C$

(*a*) is the curve $x = t,\ y = t^2,\ z = t^3$ from $t = 0$ to $t = 1$

(*b*) consists of the straight lines from (0,0,0) to (1,0,0), then to (1,1,0), and then to (1,1,1).

*Ans.* (*a*) $\frac{19}{45}\mathbf{i} + \frac{11}{15}\mathbf{j} + \frac{75}{77}\mathbf{k}$ (*b*) $\frac{1}{2}\mathbf{j} + 2\mathbf{k}$

**56.** If $\mathbf{F} = 2y\,\mathbf{i} - z\,\mathbf{j} + x\,\mathbf{k}$, evaluate $\int_C \mathbf{F}\times d\mathbf{r}$ along the curve $x = \cos t,\ y = \sin t,\ z = 2\cos t$ from $t = 0$ to $t = \pi/2$. *Ans.* $(2 - \frac{\pi}{4})\mathbf{i} + (\pi - \frac{1}{2})\mathbf{j}$

**57.** If $\mathbf{A} = (3x + y)\mathbf{i} - x\,\mathbf{j} + (y - 2)\mathbf{k}$ and $\mathbf{B} = 2\mathbf{i} - 3\mathbf{j} + \mathbf{k}$, evaluate $\oint_C (\mathbf{A}\times\mathbf{B})\times d\mathbf{r}$ around the circle in the $xy$ plane having centre at the origin and radius 2 traversed in the positive direction. *Ans.* $4\pi(7\mathbf{i} + 3\mathbf{j})$

**58.** Evaluate $\iint_S \mathbf{A}\cdot\mathbf{n}\,dS$ for each of the following cases.

(*a*) $\mathbf{A} = y\,\mathbf{i} + 2x\,\mathbf{j} - z\,\mathbf{k}$ and $S$ is the surface of the plane $2x + y = 6$ in the first octant cut off by the plane $z = 4$.

(*b*) $\mathbf{A} = (x + y^2)\mathbf{i} - 2x\,\mathbf{j} + 2yz\,\mathbf{k}$ and $S$ is the surface of the plane $2x + y + 2z = 6$ in the first octant.
*Ans.* (*a*) 108 (*b*) 81

**59.** If $\mathbf{F} = 2y\,\mathbf{i} - z\,\mathbf{j} + x^2\,\mathbf{k}$ and $S$ is the surface of the parabolic cylinder $y^2 = 8x$ in the first octant bounded by the planes $y = 4$ and $z = 6$, evaluate $\iint_S \mathbf{F}\cdot\mathbf{n}\,dS$. *Ans.* 132

**60.** Evaluate $\iint_S \mathbf{A}\cdot\mathbf{n}\,dS$ over the entire surface $S$ of the region bounded by the cylinder $x^2 + z^2 = 9$, $x = 0$, $y = 0$, $z = 0$ and $y = 8$, if $\mathbf{A} = 6z\,\mathbf{i} + (2x + y)\mathbf{j} - x\,\mathbf{k}$. *Ans.* $18\pi$

**61.** Evaluate $\iint_S \mathbf{r}\cdot\mathbf{n}\,dS$ over: (*a*) the surface $S$ of the unit cube bounded by the coordinate planes and the planes $x = 1,\ y = 1,\ z = 1$; (*b*) the surface of a sphere of radius $a$ with centre at (0,0,0).
*Ans.* (*a*) 3 (*b*) $4\pi a^3$

**62.** Evaluate $\iint_S \mathbf{A}\cdot\mathbf{n}\,dS$ over the entire surface of the region above the $xy$ plane bounded by the cone $z^2 = x^2 + y^2$ and the plane $z = 4$, if $\mathbf{A} = 4xz\,\mathbf{i} + xyz^2\,\mathbf{j} + 3z\,\mathbf{k}$. *Ans.* $320\pi$

**63.** (*a*) Let $R$ be the projection of a surface $S$ on the $xy$ plane. Prove that the surface area of $S$ is given by
$\iint_R \sqrt{1 + (\frac{\partial z}{\partial x})^2 + (\frac{\partial z}{\partial y})^2}\,dx\,dy$ if the equation for $S$ is $z = f(x,y)$.

(*b*) What is the surface area if $S$ has the equation $F(x,y,z)=0$? *Ans.* $\displaystyle\iint_R \frac{\sqrt{(\frac{\partial \mathbf{F}}{\partial x})^2+(\frac{\partial \mathbf{F}}{\partial y})^2+(\frac{\partial \mathbf{F}}{\partial z})^2}}{\left|\frac{\partial \mathbf{F}}{\partial z}\right|}\,dx\,dy$

**64.** Find the surface area of the plane $x+2y+2z=12$ cut off by: (*a*) $x=0, y=0, x=1, y=1$; (*b*) $x=0, y=0$, and $x^2+y^2=16$. *Ans.* (*a*) 3/2 (*b*) $6\pi$

**65.** Find the surface area of the region common to the intersecting cylinders $x^2+y^2=a^2$ and $x^2+z^2=a^2$. *Ans.* $16a^2$

**66.** Evaluate (*a*) $\displaystyle\iint_S (\nabla\times\mathbf{F})\cdot\mathbf{n}\,dS$ and (*b*) $\displaystyle\iint_S \phi\,\mathbf{n}\,dS$ if $\mathbf{F}=(x+2y)\mathbf{i}-3z\,\mathbf{j}+x\,\mathbf{k}$, $\phi=4x+3y-2z$, and $S$ is the surface of $2x+y+2z=6$ bounded by $x=0, x=1, y=0$ and $y=2$.
*Ans.* (*a*) 1 (*b*) $2\mathbf{i}+\mathbf{j}+2\mathbf{k}$

**67.** Solve the preceding problem if $S$ is the surface of $2x+y+2z=6$ bounded by $x=0$, $y=0$, and $z=0$.
*Ans.* (*a*) 9/2 (*b*) $72\mathbf{i}+36\mathbf{j}+72\mathbf{k}$

**68.** Evaluate $\displaystyle\iint_R \sqrt{x^2+y^2}\,dx\,dy$ over the region $R$ in the $xy$ plane bounded by $x^2+y^2=36$. *Ans.* $144\pi$

**69.** Evaluate $\displaystyle\iiint_V (2x+y)\,dV$, where $V$ is the closed region bounded by the cylinder $z=4-x^2$ and the planes $x=0, y=0, y=2$ and $z=0$. *Ans.* 80/3

**70.** If $\mathbf{F}=(2x^2-3z)\mathbf{i}-2xy\,\mathbf{j}-4x\,\mathbf{k}$, evaluate (*a*) $\displaystyle\iiint_V \nabla\cdot\mathbf{F}\,dV$ and (*b*) $\displaystyle\iiint_V \nabla\times\mathbf{F}\,dV$, where $V$ is the closed region bounded by the planes $x=0, y=0, z=0$ and $2x+2y+z=4$. *Ans.* (*a*) $\frac{8}{3}$ (*b*) $\frac{8}{3}(\mathbf{j}-\mathbf{k})$

# Chapter 6

# The DIVERGENCE THEOREM, STOKES' THEOREM, and RELATED INTEGRAL THEOREMS

**THE DIVERGENCE THEOREM OF GAUSS** states that if $V$ is the volume bounded by a closed surface $S$ and $\mathbf{A}$ is a vector function of position with continuous derivatives, then

$$\iiint_V \nabla\cdot\mathbf{A}\,dV \quad = \quad \iint_S \mathbf{A}\cdot\mathbf{n}\,dS \quad = \quad \oiint_S \mathbf{A}\cdot d\mathbf{S}$$

where $\mathbf{n}$ is the positive (outward drawn) normal to $S$.

**STOKES' THEOREM** states that if $S$ is an open, two-sided surface bounded by a closed, non-intersecting curve $C$ (simple closed curve) then if $\mathbf{A}$ has continuous derivatives

$$\oint_C \mathbf{A}\cdot d\mathbf{r} \quad = \quad \iint_S (\nabla\times\mathbf{A})\cdot\mathbf{n}\,dS \quad = \quad \iint_S (\nabla\times\mathbf{A})\cdot d\mathbf{S}$$

where $C$ is traversed in the positive direction. The direction of $C$ is called *positive* if an observer, walking on the boundary of $S$ in this direction, with his head pointing in the direction of the positive normal to $S$, has the surface on his left.

**GREEN'S THEOREM IN THE PLANE.** If $R$ is a closed region of the $xy$ plane bounded by a simple closed curve $C$ and if $M$ and $N$ are continuous functions of $x$ and $y$ having continuous derivatives in $R$, then

$$\oint_C M\,dx + N\,dy \quad = \quad \iint_R \left(\frac{\partial N}{\partial x} - \frac{\partial M}{\partial y}\right) dx\,dy$$

where $C$ is traversed in the positive (counterclockwise) direction. Unless otherwise stated we shall always assume $\oint$ to mean that the integral is described in the positive sense.

Green's theorem in the plane is a special case of Stokes' theorem (see Problem 4). Also, it is of interest to notice that Gauss' divergence theorem is a generalization of Green's theorem in the plane where the (plane) region $R$ and its closed boundary (curve) $C$ are replaced by a (space) region $V$ and its closed boundary (surface) $S$. For this reason the divergence theorem is often called *Green's theorem in space* (see Problem 4).

Green's theorem in the plane also holds for regions bounded by a finite number of simple closed curves which do not intersect (see Problems 10 and 11).

**RELATED INTEGRAL THEOREMS.**

1. $$\iiint_V [\phi\nabla^2\psi + (\nabla\phi)\cdot(\nabla\psi)]\, dV = \iint_S (\phi\nabla\psi)\cdot d\mathbf{S}$$

This is called *Green's first identity or theorem.*

2. $$\iiint_V (\phi\nabla^2\psi - \psi\nabla^2\phi)\, dV = \iint_S (\phi\nabla\psi - \psi\nabla\phi)\cdot d\mathbf{S}$$

This is called *Green's second identity or symmetrical theorem.* See Problem 21.

3. $$\iiint_V \nabla\times\mathbf{A}\, dV = \iint_S (\mathbf{n}\times\mathbf{A})\, dS = \iint_S d\mathbf{S}\times\mathbf{A}$$

Note that here the dot product of Gauss' divergence theorem is replaced by the cross product. See Problem 23.

4. $$\oint_C \phi\, d\mathbf{r} = \iint_S (\mathbf{n}\times\nabla\phi)\, dS = \iint_S d\mathbf{S}\times\nabla\phi$$

5. Let $\psi$ represent either a vector or scalar function according as the symbol $\circ$ denotes a dot or cross, or an ordinary multiplication. Then

$$\iiint_V \nabla\circ\psi\, dV = \iint_S \mathbf{n}\circ\psi\, dS = \iint_S d\mathbf{S}\circ\psi$$

$$\oint_C d\mathbf{r}\circ\psi = \iint_S (\mathbf{n}\times\nabla)\circ\psi\, dS = \iint_S (d\mathbf{S}\times\nabla)\circ\psi$$

Gauss' divergence theorem, Stokes' theorem and the results *3* and *4* are special cases of these. See Problems 22, 23, and 34.

**INTEGRAL OPERATOR FORM FOR $\nabla$.** It is of interest that, using the terminology of Problem 19, the operator $\nabla$ can be expressed symbolically in the form

$$\nabla\circ \equiv \lim_{\Delta V\to 0} \frac{1}{\Delta V} \oiint_{\Delta S} d\mathbf{S}\circ$$

where $\circ$ denotes a dot, cross or an ordinary multiplication (see Problem 25). The result proves useful in extending the concepts of gradient, divergence and curl to coordinate systems other than rectangular (see Problems 19, 24 and also Chapter 7).

## SOLVED PROBLEMS

### GREEN'S THEOREM IN THE PLANE

1. Prove Green's theorem in the plane if $C$ is a closed curve which has the property that any straight line parallel to the coordinate axes cuts $C$ in at most two points.

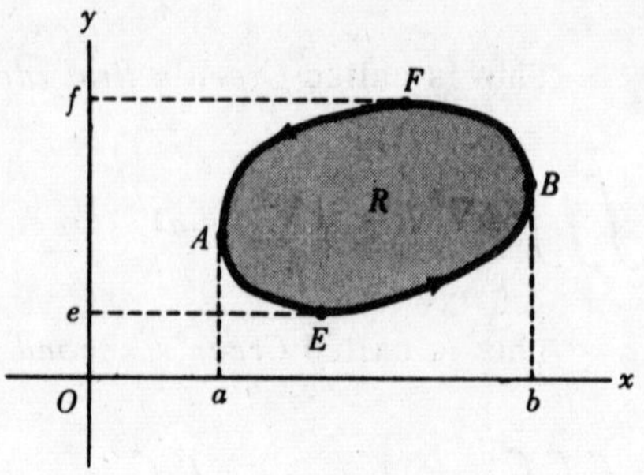


Let the equations of the curves $AEB$ and $AFB$ (see adjoining figure) be $y = Y_1(x)$ and $y = Y_2(x)$ respectively. If $R$ is the region bounded by $C$, we have

$$\iint_R \frac{\partial M}{\partial y}\,dx\,dy = \int_{x=a}^{b}\left[\int_{y=Y_1(x)}^{Y_2(x)} \frac{\partial M}{\partial y}\,dy\right]dx = \int_{x=a}^{b} M(x,y)\Big|_{y=Y_1(x)}^{Y_2(x)}\,dx = \int_a^b \left[M(x,Y_2) - M(x,Y_1)\right]dx$$

$$= -\int_a^b M(x,Y_1)\,dx - \int_b^a M(x,Y_2)\,dx = -\oint_C M\,dx$$

Then

$$(1) \qquad \oint_C M\,dx = -\iint_R \frac{\partial M}{\partial y}\,dx\,dy$$

Similarly let the equations of curves $EAF$ and $EBF$ be $x = X_1(y)$ and $x = X_2(y)$ respectively. Then

$$\iint_R \frac{\partial N}{\partial x}\,dx\,dy = \int_{y=e}^{f}\left[\int_{x=X_1(y)}^{X_2(y)} \frac{\partial N}{\partial x}\,dx\right]dy = \int_e^f \left[N(X_2,y) - N(X_1,y)\right]dy$$

$$= \int_f^e N(X_1,y)\,dy + \int_e^f N(X_2,y)\,dy = \oint_C N\,dy$$

Then

$$(2) \qquad \oint_C N\,dy = \iint_R \frac{\partial N}{\partial x}\,dx\,dy$$

Adding (1) and (2), $\displaystyle \oint_C M\,dx + N\,dy = \iint_R \left(\frac{\partial N}{\partial x} - \frac{\partial M}{\partial y}\right)dx\,dy.$

2. Verify Green's theorem in the plane for $\oint_C (xy+y^2)\,dx + x^2\,dy$ where $C$ is the closed curve of the region bounded by $y=x$ and $y=x^2$.

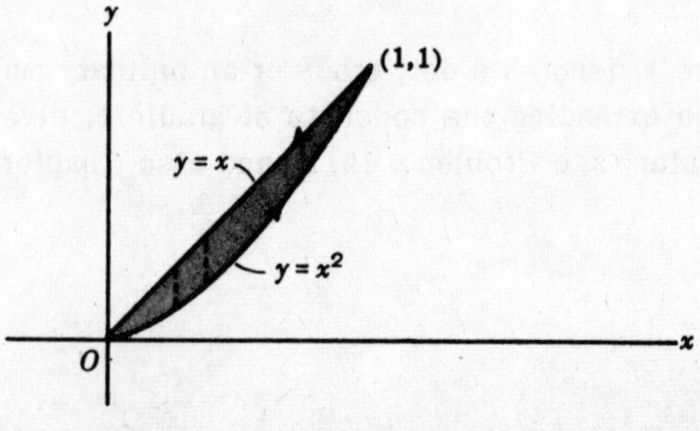


$y=x$ and $y=x^2$ intersect at (0,0) and (1,1). The positive direction in traversing $C$ is as shown in the adjacent diagram.

Along $y = x^2$, the line integral equals

$$\int_0^1 \left((x)(x^2) + x^4\right) dx \quad + \quad (x^2)(2x)\, dx \quad = \quad \int_0^1 (3x^3 + x^4)\, dx \quad = \quad \frac{19}{20}$$

Along $y = x$ from (1,1) to (0,0) the line integral equals

$$\int_1^0 \left((x)(x) + x^2\right) dx \quad + \quad x^2\, dx \quad = \quad \int_1^0 3x^2\, dx \quad = \quad -1$$

Then the required line integral $= \dfrac{19}{20} - 1 = -\dfrac{1}{20}$.

$$\iint_R \left(\frac{\partial N}{\partial x} - \frac{\partial M}{\partial y}\right) dx\, dy \quad = \quad \iint_R \left[\frac{\partial}{\partial x}(x^2) - \frac{\partial}{\partial y}(xy + y^2)\right] dx\, dy$$

$$= \quad \iint_R (x - 2y)\, dx\, dy \quad = \quad \int_{x=0}^1 \int_{y=x^2}^x (x - 2y)\, dy\, dx$$

$$= \quad \int_0^1 \left[\int_{x^2}^x (x - 2y)\, dy\right] dx \quad = \quad \int_0^1 (xy - y^2)\Big|_{x^2}^x\, dx$$

$$= \quad \int_0^1 (x^4 - x^3)\, dx \quad = \quad -\frac{1}{20}$$

so that the theorem is verified.

3. Extend the proof of Green's theorem in the plane given in Problem 1 to the curves $C$ for which lines parallel to the coordinate axes may cut $C$ in more than two points.

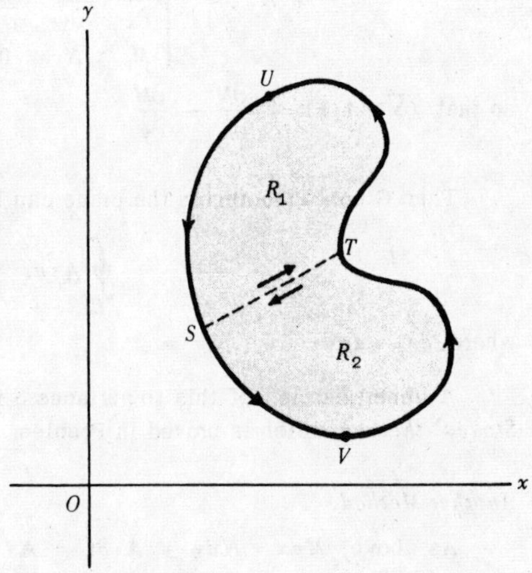


Consider a closed curve $C$ such as shown in the adjoining figure, in which lines parallel to the axes may meet $C$ in more than two points. By constructing line $ST$ the region is divided into two regions $R_1$ and $R_2$ which are of the type considered in Problem 1 and for which Green's theorem applies, i.e.,

$$(1) \qquad \int_{STUS} M\, dx + N\, dy \quad = \quad \iint_{R_1} \left(\frac{\partial N}{\partial x} - \frac{\partial M}{\partial y}\right) dx\, dy$$

$$(2) \qquad \int_{SVTS} M\, dx + N\, dy \quad = \quad \iint_{R_2} \left(\frac{\partial N}{\partial x} - \frac{\partial M}{\partial y}\right) dx\, dy$$

Adding the left hand sides of (1) and (2), we have, omitting the integrand $M\, dx + N\, dy$ in each case,

$$\int_{STUS} + \int_{SVTS} \quad = \quad \int_{ST} + \int_{TUS} + \int_{SVT} + \int_{TS} \quad = \quad \int_{TUS} + \int_{SVT} \quad = \quad \int_{TUSVT}$$

using the fact that $\displaystyle\int_{ST} = -\int_{TS}$

Adding the right hand sides of (1) and (2), omitting the integrand,

$$\iint_{R_1} + \iint_{R_2} = \iint_{R}$$

where $R$ consists of regions $R_1$ and $R_2$.

Then $\displaystyle\int_{TUSVT} M\,dx + N\,dy = \iint_R \left(\frac{\partial N}{\partial x} - \frac{\partial M}{\partial y}\right) dx\,dy$ and the theorem is proved.

A region $R$ such as considered here and in Problem 1, for which any closed curve lying in $R$ can be continuously shrunk to a point without leaving $R$, is called a *simply-connected region.* A region which is not simply-connected is called *multiply-connected.* We have shown here that Green's theorem in the plane applies to simply-connected regions bounded by closed curves. In Problem 10 the theorem is extended to multiply-connected regions.

For more complicated simply-connected regions it may be necessary to construct more lines, such as $ST$, to establish the theorem.

**4.** Express Green's theorem in the plane in vector notation.

We have $M\,dx + N\,dy = (M\mathbf{i}+N\mathbf{j})\cdot(dx\,\mathbf{i} + dy\,\mathbf{j}) = \mathbf{A}\cdot d\mathbf{r}$, where $\mathbf{A} = M\mathbf{i} + N\mathbf{j}$ and $\mathbf{r} = x\mathbf{i} + y\mathbf{j}$ so that $d\mathbf{r} = dx\,\mathbf{i} + dy\,\mathbf{j}$.

Also, if $\mathbf{A} = M\mathbf{i} + N\mathbf{j}$ then

$$\nabla\times\mathbf{A} = \begin{vmatrix} \mathbf{i} & \mathbf{j} & \mathbf{k} \\ \frac{\partial}{\partial x} & \frac{\partial}{\partial y} & \frac{\partial}{\partial z} \\ M & N & 0 \end{vmatrix} = -\frac{\partial N}{\partial z}\mathbf{i} + \frac{\partial M}{\partial z}\mathbf{j} + \left(\frac{\partial N}{\partial x} - \frac{\partial M}{\partial y}\right)\mathbf{k}$$

so that $(\nabla\times\mathbf{A})\cdot\mathbf{k} = \dfrac{\partial N}{\partial x} - \dfrac{\partial M}{\partial y}$.

Then Green's theorem in the plane can be written

$$\oint_C \mathbf{A}\cdot d\mathbf{r} = \iint_R (\nabla\times\mathbf{A})\cdot\mathbf{k}\,dR$$

where $dR = dx\,dy$.

A generalization of this to surfaces $S$ in space having a curve $C$ as boundary leads quite naturally to *Stokes' theorem* which is proved in Problem 31.

*Another Method.*

As above, $M\,dx + N\,dy = \mathbf{A}\cdot d\mathbf{r} = \mathbf{A}\cdot\dfrac{d\mathbf{r}}{ds}\,ds = \mathbf{A}\cdot\mathbf{T}\,ds$, where $\dfrac{d\mathbf{r}}{ds} = \mathbf{T}$ = unit tangent vector to $C$ (see adjacent figure). If $\mathbf{n}$ is the outward drawn unit normal to $C$, then $\mathbf{T} = \mathbf{k}\times\mathbf{n}$ so that

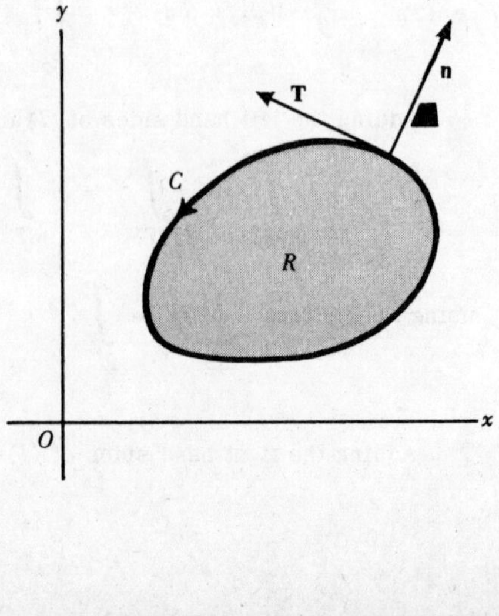


$$M\,dx + N\,dy = \mathbf{A}\cdot\mathbf{T}\,ds = \mathbf{A}\cdot(\mathbf{k}\times\mathbf{n})\,ds = (\mathbf{A}\times\mathbf{k})\cdot\mathbf{n}\,ds$$

Since $\mathbf{A} = M\mathbf{i} + N\mathbf{j}$, $\mathbf{B} = \mathbf{A}\times\mathbf{k} = (M\mathbf{i}+N\mathbf{j})\times\mathbf{k} = N\mathbf{i} - M\mathbf{j}$ and $\dfrac{\partial N}{\partial x} - \dfrac{\partial M}{\partial y} = \nabla\cdot\mathbf{B}$. Then Green's theorem in the plane becomes

$$\oint_C \mathbf{B}\cdot\mathbf{n}\,ds = \iint_R \nabla\cdot\mathbf{B}\,dR$$

where $dR = dx\,dy$.

Generalization of this to the case where the differential arc length $ds$ of a closed curve $C$ is replaced by the differential of surface area $dS$ of a closed surface $S$, and the corresponding plane region $R$ enclosed by $C$ is replaced by the volume $V$ enclosed by $S$, leads to *Gauss' divergence theorem* or *Green's theorem in space.*

$$\iint_S \mathbf{B}\cdot\mathbf{n}\,dS = \iiint_V \nabla\cdot\mathbf{B}\,dV$$

**5.** Interpret physically the first result of Problem 4.

If $\mathbf{A}$ denotes the force field acting on a particle, then $\oint_C \mathbf{A}\cdot d\mathbf{r}$ is the work done in moving the particle around a closed path $C$ and is determined by the value of $\nabla\times\mathbf{A}$. It follows in particular that if $\nabla\times\mathbf{A}=\mathbf{0}$ or equivalently if $\mathbf{A}=\nabla\phi$, then the integral around a closed path is zero. This amounts to saying that the work done in moving the particle from one point in the plane to another is independent of the path in the plane joining the points or that the force field is conservative. These results have already been demonstrated for force fields and curves in space (see Chapter 5).

Conversely, if the integral is independent of the path joining any two points of a region, i.e. if the integral around any closed path is zero, then $\nabla\times\mathbf{A}=\mathbf{0}$. In the plane, the condition $\nabla\times\mathbf{A}=\mathbf{0}$ is equivalent to the condition $\dfrac{\partial M}{\partial y}=\dfrac{\partial N}{\partial x}$ where $\mathbf{A}=M\mathbf{i}+N\mathbf{j}$.

**6.** Evaluate $\displaystyle\int_{(0,0)}^{(2,1)} (10x^4-2xy^3)\,dx - 3x^2y^2\,dy$ along the path $x^4-6xy^3=4y^2$.

A direct evaluation is difficult. However, noting that $M=10x^4-2xy^3$, $N=-3x^2y^2$ and $\dfrac{\partial M}{\partial y}=-6xy^2=\dfrac{\partial N}{\partial x}$, it follows that the integral is independent of the path. Then we can use any path, for example the path consisting of straight line segments from (0,0) to (2,0) and then from (2,0) to (2,1).

Along the straight line path from (0,0) to (2,0), $y=0$, $dy=0$ and the integral equals $\displaystyle\int_{x=0}^{2} 10x^4\,dx = 64$.

Along the straight line path from (2,0) to (2,1), $x=2$, $dx=0$ and the integral equals $\displaystyle\int_{y=0}^{1} -12y^2\,dy = -4$.

Then the required value of the line integral $= 64-4 = 60$.

*Another Method.*

Since $\dfrac{\partial M}{\partial y}=\dfrac{\partial N}{\partial x}$, $(10x^4-2xy^3)\,dx - 3x^2y^2\,dy$ is an exact differential (of $2x^5-x^2y^3$). Then

$$\int_{(0,0)}^{(2,1)} (10x^4-2xy^3)\,dx - 3x^2y^2\,dy = \int_{(0,0)}^{(2,1)} d(2x^5-x^2y^3) = 2x^5-x^2y^3\Big|_{(0,0)}^{(2,1)} = 60$$

**7.** Show that the area bounded by a simple closed curve $C$ is given by $\frac{1}{2}\oint_C x\,dy-y\,dx$.

In Green's theorem, put $M=-y$, $N=x$. Then

$$\oint_C x\,dy - y\,dx = \iint_R \left(\frac{\partial}{\partial x}(x)-\frac{\partial}{\partial y}(-y)\right)dx\,dy = 2\iint_R dx\,dy = 2A$$

where $A$ is the required area. Thus $A=\frac{1}{2}\oint_C x\,dy - y\,dx$.

**8.** Find the area of the ellipse $x = a\cos\theta,\ y = b\sin\theta$.

$$\text{Area} = \tfrac{1}{2}\oint_C x\,dy - y\,dx = \tfrac{1}{2}\int_0^{2\pi} (a\cos\theta)(b\cos\theta)\,d\theta - (b\sin\theta)(-a\sin\theta)\,d\theta$$

$$= \tfrac{1}{2}\int_0^{2\pi} ab\,(\cos^2\theta + \sin^2\theta)\,d\theta = \tfrac{1}{2}\int_0^{2\pi} ab\,d\theta = \pi ab$$

**9.** Evaluate $\oint_C (y - \sin x)\,dx + \cos x\,dy$, where $C$ is the triangle of the adjoining figure:
(*a*) directly,
(*b*) by using Green's theorem in the plane.

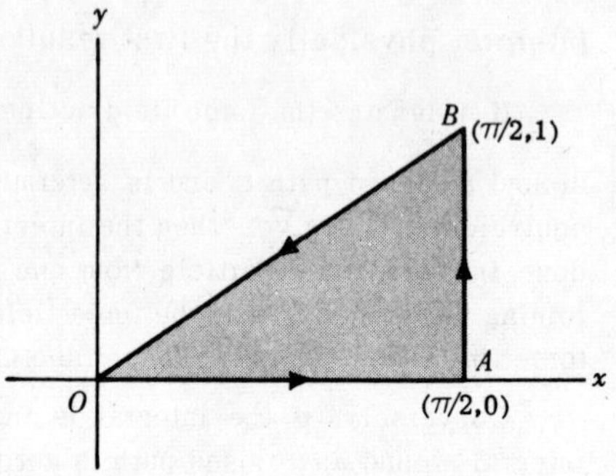


(*a*) Along $OA$, $y = 0$, $dy = 0$ and the integral equals

$$\int_0^{\pi/2} (0 - \sin x)\,dx + (\cos x)(0) = \int_0^{\pi/2} -\sin x\,dx$$

$$= \cos x\,\Big|_0^{\pi/2} = -1$$

Along $AB$, $x = \dfrac{\pi}{2}$, $dx = 0$ and the integral equals

$$\int_0^1 (y-1)0 + 0\,dy = 0$$

Along $BO$, $y = \dfrac{2x}{\pi}$, $dy = \dfrac{2}{\pi}dx$ and the integral equals

$$\int_{\pi/2}^{0} \left(\frac{2x}{\pi} - \sin x\right)dx + \frac{2}{\pi}\cos x\,dx = \left(\frac{x^2}{\pi} + \cos x + \frac{2}{\pi}\sin x\right)\Big|_{\pi/2}^{0} = 1 - \frac{\pi}{4} - \frac{2}{\pi}$$

Then the integral along $C$ $= -1 + 0 + 1 - \dfrac{\pi}{4} - \dfrac{2}{\pi} = -\dfrac{\pi}{4} - \dfrac{2}{\pi}$.

(*b*) $M = y - \sin x$, $N = \cos x$, $\dfrac{\partial N}{\partial x} = -\sin x$, $\dfrac{\partial M}{\partial y} = 1$ and

$$\oint_C M\,dx + N\,dy = \iint_R \left(\frac{\partial N}{\partial x} - \frac{\partial M}{\partial y}\right)dx\,dy = \iint_R (-\sin x - 1)\,dy\,dx$$

$$= \int_{x=0}^{\pi/2}\left[\int_{y=0}^{2x/\pi} (-\sin x - 1)\,dy\right]dx = \int_{x=0}^{\pi/2} (-y\sin x - y)\Big|_0^{2x/\pi}\,dx$$

$$= \int_0^{\pi/2} \left(-\frac{2x}{\pi}\sin x - \frac{2x}{\pi}\right)dx = -\frac{2}{\pi}(-x\cos x + \sin x) - \frac{x^2}{\pi}\Big|_0^{\pi/2} = -\frac{2}{\pi} - \frac{\pi}{4}$$

in agreement with part (*a*).

Note that although there exist lines parallel to the coordinate axes (coincident with the coordinate axes in this case) which meet $C$ in an *infinite* number of points, Green's theorem in the plane still holds. In general the theorem is valid when $C$ is composed of a finite number of straight line segments.

**10.** Show that Green's theorem in the plane is also valid for a multiply-connected region $R$ such as shown in the figure below.

The shaded region $R$, shown in the figure below, is multiply-connected since not every closed curve lying in $R$ can be shrunk to a point without leaving $R$, as is observed by considering a curve surrounding $DEFGD$ for example. The boundary of $R$, which consists of the exterior boundary $AHJKLA$ and the interior boundary $DEFGD$, is to be traversed in the positive direction, so that a person travelling in this direction always has the region on his left. It is seen that the positive directions are those indicated in the adjoining figure.

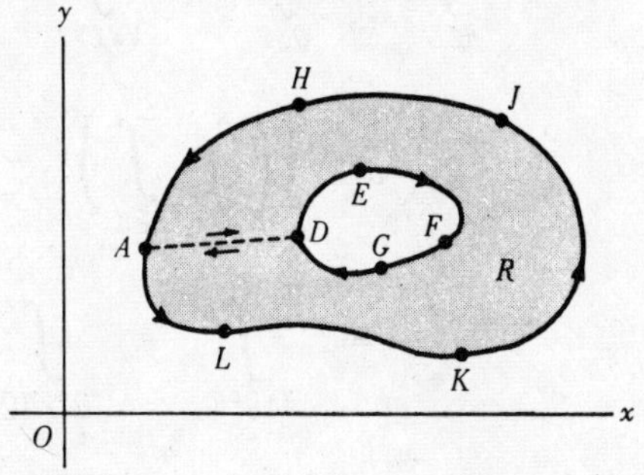


In order to establish the theorem, construct a line, such as $AD$, called a *cross-cut*, connecting the exterior and interior boundaries. The region bounded by $ADEFGDALKJHA$ is simply-connected, and so Green's theorem is valid. Then

$$\oint_{ADEFGDALKJHA} M\,dx + N\,dy \quad = \quad \iint_R \left(\frac{\partial N}{\partial x} - \frac{\partial M}{\partial y}\right) dx\,dy$$

But the integral on the left, leaving out the integrand, is equal to

$$\int_{AD} + \int_{DEFGD} + \int_{DA} + \int_{ALKJHA} \quad = \quad \int_{DEFGD} + \int_{ALKJHA}$$

since $\int_{AD} = -\int_{DA}$. Thus if $C_1$ is the curve $ALKJHA$, $C_2$ is the curve $DEFGD$ and $C$ is the boundary of $R$ consisting of $C_1$ and $C_2$ (traversed in the positive directions), then $\int_{C_1} + \int_{C_2} = \int_C$ and so

$$\oint_C M\,dx + N\,dy \quad = \quad \iint_R \left(\frac{\partial N}{\partial x} - \frac{\partial M}{\partial y}\right) dx\,dy$$

**11.** Show that Green's theorem in the plane holds for the region $R$, of the figure below, bounded by the simple closed curves $C_1(ABDEFGA)$, $C_2(HKLPH)$, $C_3(QSTUQ)$ and $C_4(VWXYV)$.

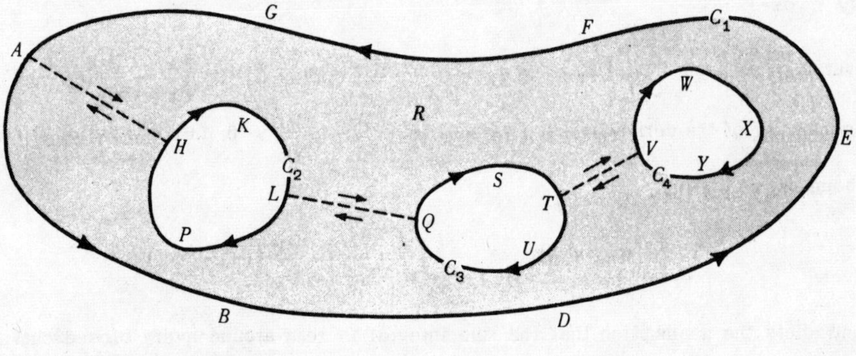


Construct the cross-cuts $AH$, $LQ$ and $TV$. Then the region bounded by $AHKLQSTVWXYVTUQLPHA$-$BDEFGA$ is simply-connected and Green's theorem applies. The integral over this boundary is equal to

$$\int_{AH} + \int_{HKL} + \int_{LQ} + \int_{QST} + \int_{TV} + \int_{VWXYV} + \int_{VT} + \int_{TUQ} + \int_{QL} + \int_{LPH} + \int_{HA} + \int_{ABDEFGA}$$

Since the integrals along $AH$ and $HA$, $LQ$ and $QL$, $TV$ and $VT$ cancel out in pairs, this becomes

$$\int_{HKL} + \int_{QST} + \int_{VWXYV} + \int_{TUQ} + \int_{LPH} + \int_{ABDEFGA}$$

$$= \left(\int_{HKL} + \int_{LPH}\right) + \left(\int_{QST} + \int_{TUQ}\right) + \int_{VWXYV} + \int_{ABDEFGA}$$

$$= \int_{HKLPH} + \int_{QSTUQ} + \int_{VWXYV} + \int_{ABDEFGA}$$

$$= \int_{C_2} + \int_{C_3} + \int_{C_4} + \int_{C_1} = \int_C$$

where $C$ is the boundary consisting of $C_1$, $C_2$, $C_3$ and $C_4$. Then

$$\oint_C M\,dx + N\,dy = \iint_R \left(\frac{\partial N}{\partial x} - \frac{\partial M}{\partial y}\right) dx\,dy$$

as required.

12. Prove that $\oint_C M\,dx + N\,dy = 0$ around every closed curve $C$ in a simply-connected region if and only if $\frac{\partial M}{\partial y} = \frac{\partial N}{\partial x}$ everywhere in the region.

Assume that $M$ and $N$ are continuous and have continuous partial derivatives everywhere in the region $R$ bounded by $C$, so that Green's theorem is applicable. Then

$$\oint_C M\,dx + N\,dy = \iint_R \left(\frac{\partial N}{\partial x} - \frac{\partial M}{\partial y}\right) dx\,dy$$

If $\frac{\partial M}{\partial y} = \frac{\partial N}{\partial x}$ in $R$, then clearly $\oint_C M\,dx + N\,dy = 0$.

Conversely, suppose $\oint_C M\,dx + N\,dy = 0$ for all curves $C$. If $\frac{\partial N}{\partial x} - \frac{\partial M}{\partial y} > 0$ at a point $P$, then from the continuity of the derivatives it follows that $\frac{\partial N}{\partial x} - \frac{\partial M}{\partial y} > 0$ in some region $A$ surrounding $P$. If $\Gamma$ is the boundary of $A$ then

$$\oint_\Gamma M\,dx + N\,dy = \iint_A \left(\frac{\partial N}{\partial x} - \frac{\partial M}{\partial y}\right) dx\,dy > 0$$

which contradicts the assumption that the line integral is zero around every closed curve. Similarly the assumption $\frac{\partial N}{\partial x} - \frac{\partial M}{\partial y} < 0$ leads to a contradiction. Thus $\frac{\partial N}{\partial x} - \frac{\partial M}{\partial y} = 0$ at all points.

Note that the condition $\frac{\partial M}{\partial y} = \frac{\partial N}{\partial x}$ is equivalent to the condition $\nabla \times \mathbf{A} = \mathbf{0}$ where $\mathbf{A} = M\mathbf{i} + N\mathbf{j}$ (see Problems 10 and 11, Chapter 5). For a generalization to space curves, see Problem 31.

**13.** Let $\mathbf{F} = \dfrac{-y\mathbf{i} + x\mathbf{j}}{x^2+y^2}$. (*a*) Calculate $\nabla \times \mathbf{F}$. (*b*) Evaluate $\oint \mathbf{F}\cdot d\mathbf{r}$ around any closed path and explain the results.

$$(a)\ \nabla\times\mathbf{F} = \begin{vmatrix} \mathbf{i} & \mathbf{j} & \mathbf{k} \\ \dfrac{\partial}{\partial x} & \dfrac{\partial}{\partial y} & \dfrac{\partial}{\partial z} \\ \dfrac{-y}{x^2+y^2} & \dfrac{x}{x^2+y^2} & 0 \end{vmatrix} = \mathbf{0} \quad \text{in any region excluding } (0,0).$$

(*b*) $\oint \mathbf{F}\cdot d\mathbf{r} = \oint \dfrac{-y\,dx + x\,dy}{x^2+y^2}$. Let $x = \rho\cos\phi$, $y = \rho\sin\phi$, where $(\rho,\phi)$ are polar coordinates.

Then

$$dx = -\rho\sin\phi\,d\phi + d\rho\cos\phi, \qquad dy = \rho\cos\phi\,d\phi + d\rho\sin\phi$$

and so

$$\frac{-y\,dx + x\,dy}{x^2+y^2} = d\phi = d\left(\arctan\frac{y}{x}\right)$$

For a closed curve $ABCDA$ (see Figure (*a*) below) surrounding the origin, $\phi = 0$ at $A$ and $\phi = 2\pi$ after a complete circuit back to $A$. In this case the line integral equals $\int_0^{2\pi} d\phi = 2\pi$.

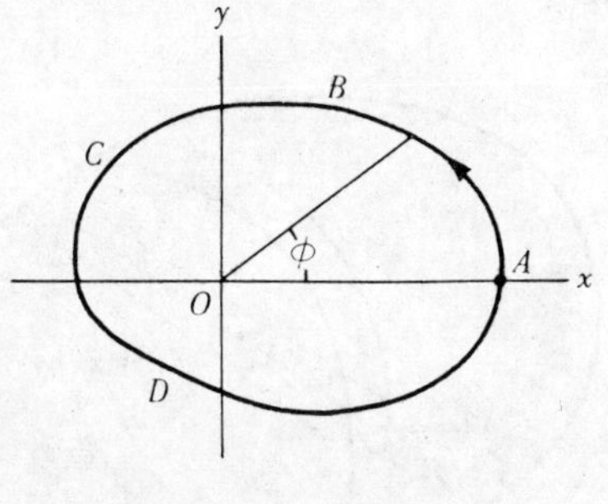


Fig. (*a*)

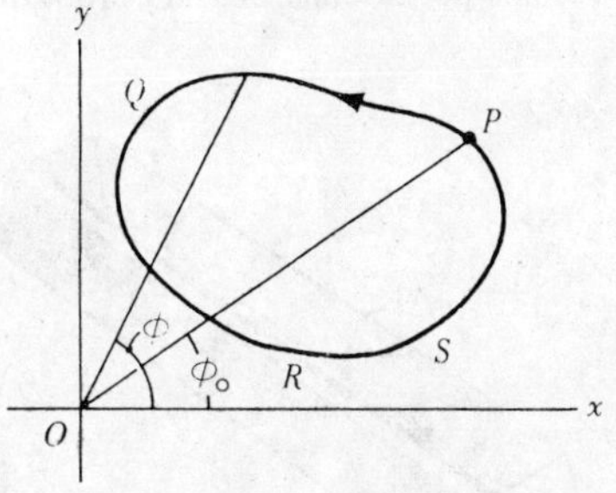


Fig. (*b*)

For a closed curve $PQRSP$ (see Figure (*b*) above) not surrounding the origin, $\phi = \phi_0$ at $P$ and $\phi = \phi_0$ after a complete circuit back to $P$. In this case the line integral equals $\int_{\phi_0}^{\phi_0} d\phi = 0$.

Since $\mathbf{F} = M\mathbf{i} + N\mathbf{j}$, $\nabla\times\mathbf{F} = \mathbf{0}$ is equivalent to $\dfrac{\partial M}{\partial y} = \dfrac{\partial N}{\partial x}$ and the results would seem to contradict those of Problem 12. However, no contradiction exists since $M = \dfrac{-y}{x^2+y^2}$ and $N = \dfrac{x}{x^2+y^2}$ do not have continuous derivatives throughout any region including (0,0), and this was assumed in Prob.12.

## THE DIVERGENCE THEOREM

**14.** (*a*) Express the divergence theorem in words and (*b*) write it in rectangular form.

(*a*) The surface integral of the normal component of a vector **A** taken over a closed surface is equal to the integral of the divergence of **A** taken over the volume enclosed by the surface.

(b) Let $\mathbf{A} = A_1\mathbf{i} + A_2\mathbf{j} + A_3\mathbf{k}$. Then $\text{div}\,\mathbf{A} = \nabla\cdot\mathbf{A} = \dfrac{\partial A_1}{\partial x} + \dfrac{\partial A_2}{\partial y} + \dfrac{\partial A_3}{\partial z}$.

The unit normal to $S$ is $\mathbf{n} = n_1\,\mathbf{i} + n_2\,\mathbf{j} + n_3\,\mathbf{k}$. Then $n_1 = \mathbf{n}\cdot\mathbf{i} = \cos\alpha$, $n_2 = \mathbf{n}\cdot\mathbf{j} = \cos\beta$ and $n_3 = \mathbf{n}\cdot\mathbf{k} = \cos\gamma$, where $\alpha, \beta, \gamma$ are the angles which $\mathbf{n}$ makes with the positive $x, y, z$ axes or $\mathbf{i}, \mathbf{j}, \mathbf{k}$ directions respectively. The quantities $\cos\alpha$, $\cos\beta$, $\cos\gamma$ are the direction cosines of $\mathbf{n}$. Then

$$\begin{aligned}\mathbf{A}\cdot\mathbf{n} &= (A_1\mathbf{i} + A_2\mathbf{j} + A_3\mathbf{k})\cdot(\cos\alpha\,\mathbf{i} + \cos\beta\,\mathbf{j} + \cos\gamma\,\mathbf{k}) \\ &= A_1\cos\alpha + A_2\cos\beta + A_3\cos\gamma\end{aligned}$$

and the divergence theorem can be written

$$\iiint_V \left(\frac{\partial A_1}{\partial x} + \frac{\partial A_2}{\partial y} + \frac{\partial A_3}{\partial z}\right) dx\,dy\,dz = \iint_S (A_1\cos\alpha + A_2\cos\beta + A_3\cos\gamma)\,dS$$

**15.** Demonstrate the divergence theorem physically.

Let $\mathbf{A}$ = velocity $\mathbf{v}$ at any point of a moving fluid. From Figure (a) below:

Volume of fluid crossing $dS$ in $\Delta t$ seconds

$$\begin{aligned}&= \text{volume contained in cylinder of base } dS \text{ and slant height } \mathbf{v}\Delta t \\ &= (\mathbf{v}\Delta t)\cdot\mathbf{n}\,dS = \mathbf{v}\cdot\mathbf{n}\,dS\,\Delta t\end{aligned}$$

Then, volume per second of fluid crossing $dS = \mathbf{v}\cdot\mathbf{n}\,dS$

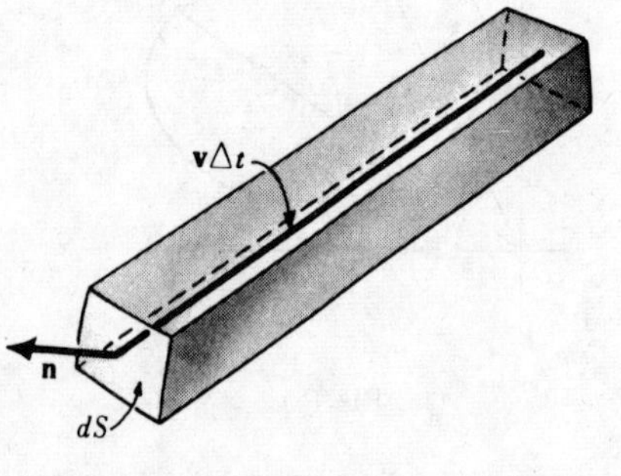


Fig. (a)

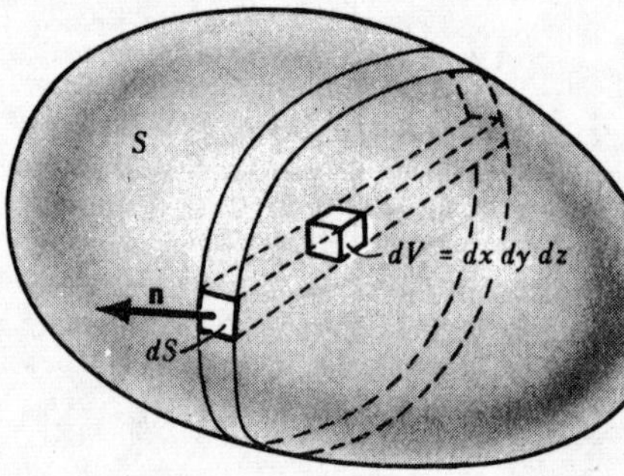


Fig. (b)

From Figure (b) above:

Total volume per second of fluid emerging from closed surface $S$

$$= \iint_S \mathbf{v}\cdot\mathbf{n}\,dS$$

From Problem 21 of Chapter 4, $\nabla\cdot\mathbf{v}\,dV$ is the volume per second of fluid emerging from a volume element $dV$. Then

Total volume per second of fluid emerging from all volume elements in $S$

$$= \iiint_V \nabla\cdot\mathbf{v}\,dV$$

Thus

$$\iint_S \mathbf{v}\cdot\mathbf{n}\,dS = \iiint_V \nabla\cdot\mathbf{v}\,dV$$

**16.** Prove the divergence theorem.

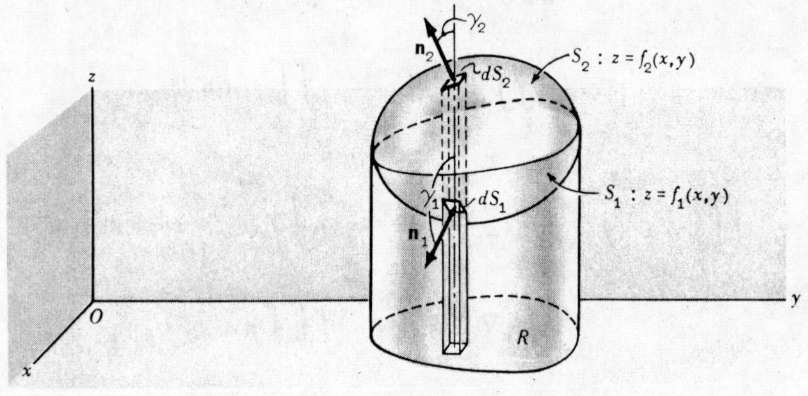


Let $S$ be a closed surface which is such that any line parallel to the coordinate axes cuts $S$ in at most two points. Assume the equations of the lower and upper portions, $S_1$ and $S_2$, to be $z = f_1(x,y)$ and $z = f_2(x,y)$ respectively. Denote the projection of the surface on the $xy$ plane by $R$. Consider

$$\iiint_V \frac{\partial A_3}{\partial z}\, dV = \iiint_V \frac{\partial A_3}{\partial z}\, dz\, dy\, dx = \iint_R \left[ \int_{z=f_1(x,y)}^{f_2(x,y)} \frac{\partial A_3}{\partial z}\, dz \right] dy\, dx$$

$$= \iint_R A_3(x,y,z) \Big|_{z=f_1}^{f_2} dy\, dx = \iint_R [A_3(x,y,f_2) - A_3(x,y,f_1)]\, dy\, dx$$

For the upper portion $S_2$, $dy\, dx = \cos \gamma_2\, dS_2 = \mathbf{k}\cdot\mathbf{n}_2\, dS_2$ since the normal $\mathbf{n}_2$ to $S_2$ makes an acute angle $\gamma_2$ with $\mathbf{k}$.

For the lower portion $S_1$, $dy\, dx = -\cos \gamma_1\, dS_1 = -\mathbf{k}\cdot\mathbf{n}_1\, dS_1$ since the normal $\mathbf{n}_1$ to $S_1$ makes an obtuse angle $\gamma_1$ with $\mathbf{k}$.

Then
$$\iint_R A_3(x,y,f_2)\, dy\, dx = \iint_{S_2} A_3\, \mathbf{k}\cdot\mathbf{n}_2\, dS_2$$

$$\iint_R A_3(x,y,f_1)\, dy\, dx = -\iint_{S_1} A_3\, \mathbf{k}\cdot\mathbf{n}_1\, dS_1$$

and

$$\iint_R A_3(x,y,f_2)\, dy\, dx - \iint_R A_3(x,y,f_1)\, dy\, dx = \iint_{S_2} A_3\, \mathbf{k}\cdot\mathbf{n}_2\, dS_2 + \iint_{S_1} A_3\, \mathbf{k}\cdot\mathbf{n}_1\, dS_1$$

$$= \iint_S A_3\, \mathbf{k}\cdot\mathbf{n}\, dS$$

so that

$$(1) \qquad \iiint_V \frac{\partial A_3}{\partial z}\, dV = \iint_S A_3\, \mathbf{k}\cdot\mathbf{n}\, dS$$

Similarly, by projecting $S$ on the other coordinate planes,

$$(2) \qquad \iiint_V \frac{\partial A_1}{\partial x}\, dV \;=\; \iint_S A_1\, \mathbf{i}\cdot\mathbf{n}\, dS$$

$$(3) \qquad \iiint_V \frac{\partial A_2}{\partial y}\, dV \;=\; \iint_S A_2\, \mathbf{j}\cdot\mathbf{n}\, dS$$

Adding (*1*), (*2*) and (*3*),

$$\iiint_V \left(\frac{\partial A_1}{\partial x} + \frac{\partial A_2}{\partial y} + \frac{\partial A_3}{\partial z}\right) dV \;=\; \iint_S (A_1\mathbf{i} + A_2\mathbf{j} + A_3\mathbf{k})\cdot\mathbf{n}\, dS$$

or

$$\iiint_V \nabla\cdot\mathbf{A}\, dV \;=\; \iint_S \mathbf{A}\cdot\mathbf{n}\, dS$$

The theorem can be extended to surfaces which are such that lines parallel to the coordinate axes meet them in more than two points. To establish this extension, subdivide the region bounded by $S$ into subregions whose surfaces do satisfy this condition. The procedure is analogous to that used in Green's theorem for the plane.

**17.** Evaluate $\iint_S \mathbf{F}\cdot\mathbf{n}\, dS$, where $\mathbf{F} = 4xz\,\mathbf{i} - y^2\,\mathbf{j} + yz\,\mathbf{k}$ and $S$ is the surface of the cube bounded by $x=0,\ x=1,\ y=0,\ y=1,\ z=0,\ z=1$.

By the divergence theorem, the required integral is equal to

$$\iiint_V \nabla\cdot\mathbf{F}\, dV \;=\; \iiint_V \left[\frac{\partial}{\partial x}(4xz) + \frac{\partial}{\partial y}(-y^2) + \frac{\partial}{\partial z}(yz)\right] dV$$

$$= \iiint_V (4z - y)\, dV \;=\; \int_{x=0}^{1}\int_{y=0}^{1}\int_{z=0}^{1} (4z-y)\, dz\, dy\, dx$$

$$= \int_{x=0}^{1}\int_{y=0}^{1} 2z^2 - yz\,\Big|_{z=0}^{1}\, dy\, dx \;=\; \int_{x=0}^{1}\int_{y=0}^{1} (2-y)\, dy\, dx \;=\; \frac{3}{2}$$

The surface integral may also be evaluated directly as in Problem 23, Chapter 5.

**18.** Verify the divergence theorem for $\mathbf{A} = 4x\,\mathbf{i} - 2y^2\,\mathbf{j} + z^2\,\mathbf{k}$ taken over the region bounded by $x^2 + y^2 = 4,\ z=0$ and $z=3$.

$$\text{Volume integral} \;=\; \iiint_V \nabla\cdot\mathbf{A}\, dV \;=\; \iiint_V \left[\frac{\partial}{\partial x}(4x) + \frac{\partial}{\partial y}(-2y^2) + \frac{\partial}{\partial z}(z^2)\right] dV$$

$$= \iiint_V (4 - 4y + 2z)\, dV \;=\; \int_{x=-2}^{2}\int_{y=-\sqrt{4-x^2}}^{\sqrt{4-x^2}}\int_{z=0}^{3} (4 - 4y + 2z)\, dz\, dy\, dx \;=\; 84\pi$$

The surface $S$ of the cylinder consists of a base $S_1$ $(z=0)$, the top $S_2$ $(z=3)$ and the convex portion $S_3$ $(x^2+y^2=4)$. Then

$$\text{Surface integral} \;=\; \iint_S \mathbf{A\cdot n}\, dS \;=\; \iint_{S_1} \mathbf{A\cdot n}\, dS_1 + \iint_{S_2} \mathbf{A\cdot n}\, dS_2 + \iint_{S_3} \mathbf{A\cdot n}\, dS_3$$

On $S_1$ $(z=0)$, $\mathbf{n}=-\mathbf{k}$, $\mathbf{A} = 4x\,\mathbf{i} - 2y^2\mathbf{j}$ and $\mathbf{A\cdot n} = 0$, so that $\iint_{S_1} \mathbf{A\cdot n}\, dS_1 = 0$.

On $S_2$ $(z=3)$, $\mathbf{n}=\mathbf{k}$, $\mathbf{A} = 4x\,\mathbf{i} - 2y^2\mathbf{j} + 9\mathbf{k}$ and $\mathbf{A\cdot n} = 9$, so that

$$\iint_{S_2} \mathbf{A\cdot n}\, dS_2 \;=\; 9\iint_{S_2} dS_2 \;=\; 36\pi, \qquad \text{since area of } S_2 = 4\pi$$

On $S_3$ $(x^2+y^2=4)$. A perpendicular to $x^2+y^2=4$ has the direction $\nabla(x^2+y^2) = 2x\,\mathbf{i} + 2y\,\mathbf{j}$.

Then a unit normal is $\mathbf{n} = \dfrac{2x\,\mathbf{i} + 2y\,\mathbf{j}}{\sqrt{4x^2+4y^2}} = \dfrac{x\,\mathbf{i} + y\,\mathbf{j}}{2}$ since $x^2+y^2=4$.

$$\mathbf{A\cdot n} \;=\; (4x\,\mathbf{i} - 2y^2\mathbf{j} + z^2\mathbf{k})\cdot\left(\frac{x\,\mathbf{i}+y\,\mathbf{j}}{2}\right) \;=\; 2x^2 - y^3$$

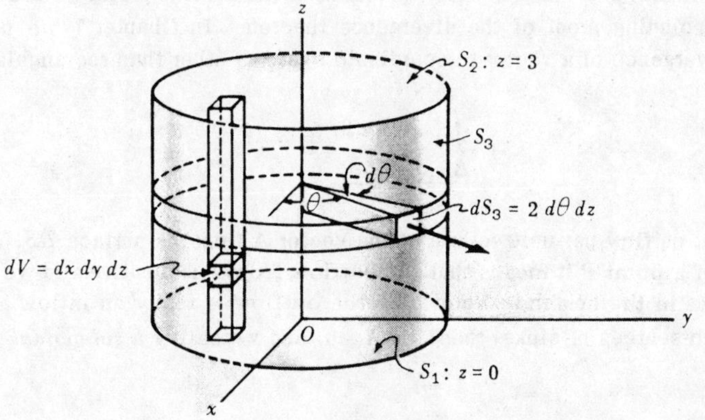


From the figure above, $x = 2\cos\theta$, $y = 2\sin\theta$, $dS_3 = 2\,d\theta\,dz$ and so

$$\iint_{S_3} \mathbf{A\cdot n}\, dS_3 \;=\; \int_{\theta=0}^{2\pi}\int_{z=0}^{3} \left[2(2\cos\theta)^2 - (2\sin\theta)^3\right] 2\,dz\,d\theta$$

$$=\; \int_{\theta=0}^{2\pi} (48\cos^2\theta - 48\sin^3\theta)\,d\theta \;=\; \int_{\theta=0}^{2\pi} 48\cos^2\theta\,d\theta \;=\; 48\pi$$

Then the surface integral $= 0 + 36\pi + 48\pi = 84\pi$, agreeing with the volume integral and verifying the divergence theorem.

Note that evaluation of the surface integral over $S_3$ could also have been done by projection of $S_3$ on the $xz$ or $yz$ coordinate planes.

**19.** If div **A** denotes the divergence of a vector field **A** at a point $P$, show that

$$\operatorname{div}\mathbf{A} \;=\; \lim_{\Delta V\to 0} \frac{\iint_{\Delta S} \mathbf{A\cdot n}\, dS}{\Delta V}$$

where $\Delta V$ is the volume enclosed by the surface $\Delta S$ and the limit is obtained by shrinking $\Delta V$ to the point $P$.

By the divergence theorem, $\iiint\limits_{\Delta V} \operatorname{div} \mathbf{A}\, dV = \iint\limits_{\Delta S} \mathbf{A}\cdot\mathbf{n}\, dS$

By the mean-value theorem for integrals, the left side can be written

$$\overline{\operatorname{div} \mathbf{A}} \iiint\limits_{\Delta V} dV = \overline{\operatorname{div} \mathbf{A}}\ \Delta V$$

where $\overline{\operatorname{div} \mathbf{A}}$ is some value intermediate between the maximum and minimum of $\operatorname{div} \mathbf{A}$ throughout $\Delta V$. Then

$$\overline{\operatorname{div} \mathbf{A}} = \frac{\iint\limits_{\Delta S} \mathbf{A}\cdot\mathbf{n}\, dS}{\Delta V}$$

Taking the limit as $\Delta V \to 0$ such that $P$ is always interior to $\Delta V$, $\overline{\operatorname{div} \mathbf{A}}$ approaches the value $\operatorname{div} \mathbf{A}$ at point $P$; hence

$$\operatorname{div} \mathbf{A} = \lim_{\Delta V \to 0} \frac{\iint\limits_{\Delta S} \mathbf{A}\cdot\mathbf{n}\, dS}{\Delta V}$$

This result can be taken as a starting point for defining the divergence of **A**, and from it all the properties may be derived including proof of the divergence theorem. In Chapter 7 we use this definition to extend the concept of divergence of a vector to coordinate systems other than rectangular. Physically,

$$\frac{\iint\limits_{\Delta S} \mathbf{A}\cdot\mathbf{n}\, dS}{\Delta V}$$

represents the flux or net outflow per unit volume of the vector **A** from the surface $\Delta S$. If $\operatorname{div} \mathbf{A}$ is positive in the neighbourhood of a point $P$ it means that the outflow from $P$ is positive and we call $P$ a *source*. Similarly, if $\operatorname{div} \mathbf{A}$ is negative in the neighbourhood of $P$ the outflow is really an inflow and $P$ is called a *sink*. If in a region there are no sources or sinks, then $\operatorname{div} \mathbf{A} = 0$ and we call **A** a *solenoidal* vector field.

**20.** Evaluate $\iint\limits_S \mathbf{r}\cdot\mathbf{n}\, dS$, where $S$ is a closed surface.

By the divergence theorem,

$$\iint\limits_S \mathbf{r}\cdot\mathbf{n}\, dS = \iiint\limits_V \nabla\cdot\mathbf{r}\, dV$$

$$= \iiint\limits_V \left(\frac{\partial}{\partial x}\mathbf{i} + \frac{\partial}{\partial y}\mathbf{j} + \frac{\partial}{\partial z}\mathbf{k}\right)\cdot(x\,\mathbf{i} + y\,\mathbf{j} + z\,\mathbf{k})\, dV$$

$$= \iiint\limits_V \left(\frac{\partial x}{\partial x} + \frac{\partial y}{\partial y} + \frac{\partial z}{\partial z}\right) dV = 3\iiint\limits_V dV = 3V$$

where $V$ is the volume enclosed by $S$.

**21.** Prove $\iiint\limits_V (\phi\nabla^2\psi - \psi\nabla^2\phi)\, dV = \iint\limits_S (\phi\nabla\psi - \psi\nabla\phi)\cdot d\mathbf{S}$.

Let $\mathbf{A} = \phi\nabla\psi$ in the divergence theorem. Then

$$\iiint_V \nabla\cdot(\phi\nabla\psi)\,dV = \iint_S (\phi\nabla\psi)\cdot\mathbf{n}\,dS = \iint_S (\phi\nabla\psi)\cdot d\mathbf{S}$$

But $\nabla\cdot(\phi\nabla\psi) = \phi(\nabla\cdot\nabla\psi) + (\nabla\phi)\cdot(\nabla\psi) = \phi\nabla^2\psi + (\nabla\phi)\cdot(\nabla\psi)$

Thus $\iiint_V \nabla\cdot(\phi\nabla\psi)\,dV = \iiint_V [\phi\nabla^2\psi + (\nabla\phi)\cdot(\nabla\psi)]\,dV$

or

$$(1) \qquad \iiint_V [\phi\nabla^2\psi + (\nabla\phi)\cdot(\nabla\psi)]\,dV = \iint_S (\phi\nabla\psi)\cdot d\mathbf{S}$$

which proves *Green's first identity*. Interchanging $\phi$ and $\psi$ in (1),

$$(2) \qquad \iiint_V [\psi\nabla^2\phi + (\nabla\psi)\cdot(\nabla\phi)]\,dV = \iint_S (\psi\nabla\phi)\cdot d\mathbf{S}$$

Subtracting (2) from (1), we have

$$(3) \qquad \iiint_V (\phi\nabla^2\psi - \psi\nabla^2\phi)\,dV = \iint_S (\phi\nabla\psi - \psi\nabla\phi)\cdot d\mathbf{S}$$

which is *Green's second identity* or *symmetrical theorem*. In the proof we have assumed that $\phi$ and $\psi$ are scalar functions of position with continuous derivatives of the second order at least.

22. Prove $\iiint_V \nabla\phi\,dV = \iint_S \phi\mathbf{n}\,dS$.

In the divergence theorem, let $\mathbf{A} = \phi\mathbf{C}$ where $\mathbf{C}$ is a constant vector. Then

$$\iiint_V \nabla\cdot(\phi\mathbf{C})\,dV = \iint_S \phi\mathbf{C}\cdot\mathbf{n}\,dS$$

Since $\nabla\cdot(\phi\mathbf{C}) = (\nabla\phi)\cdot\mathbf{C} = \mathbf{C}\cdot\nabla\phi$ and $\phi\mathbf{C}\cdot\mathbf{n} = \mathbf{C}\cdot(\phi\mathbf{n})$,

$$\iiint_V \mathbf{C}\cdot\nabla\phi\,dV = \iint_S \mathbf{C}\cdot(\phi\mathbf{n})\,dS$$

Taking $\mathbf{C}$ outside the integrals,

$$\mathbf{C}\cdot\iiint_V \nabla\phi\,dV = \mathbf{C}\cdot\iint_S \phi\mathbf{n}\,dS$$

and since $\mathbf{C}$ is an arbitrary constant vector,

$$\iiint_V \nabla\phi\,dV = \iint_S \phi\mathbf{n}\,dS$$

23. Prove $\iiint_V \nabla\times\mathbf{B}\,dV = \iint_S \mathbf{n}\times\mathbf{B}\,dS$.

In the divergence theorem, let $\mathbf{A} = \mathbf{B} \times \mathbf{C}$ where $\mathbf{C}$ is a constant vector. Then

$$\iiint_V \nabla \cdot (\mathbf{B} \times \mathbf{C})\, dV = \iint_S (\mathbf{B} \times \mathbf{C}) \cdot \mathbf{n}\, dS$$

Since $\nabla \cdot (\mathbf{B} \times \mathbf{C}) = \mathbf{C} \cdot (\nabla \times \mathbf{B})$ and $(\mathbf{B} \times \mathbf{C}) \cdot \mathbf{n} = \mathbf{B} \cdot (\mathbf{C} \times \mathbf{n}) = (\mathbf{C} \times \mathbf{n}) \cdot \mathbf{B} = \mathbf{C} \cdot (\mathbf{n} \times \mathbf{B})$,

$$\iiint_V \mathbf{C} \cdot (\nabla \times \mathbf{B})\, dV = \iint_S \mathbf{C} \cdot (\mathbf{n} \times \mathbf{B})\, dS$$

Taking $\mathbf{C}$ outside the integrals,

$$\mathbf{C} \cdot \iiint_V \nabla \times \mathbf{B}\, dV = \mathbf{C} \cdot \iint_S \mathbf{n} \times \mathbf{B}\, dS$$

and since $\mathbf{C}$ is an arbitrary constant vector,

$$\iiint_V \nabla \times \mathbf{B}\, dV = \iint_S \mathbf{n} \times \mathbf{B}\, dS$$

**24.** Show that at any point $P$

$$(a)\quad \nabla \phi = \lim_{\Delta V \to 0} \frac{\iint_{\Delta S} \phi \mathbf{n}\, dS}{\Delta V} \qquad \text{and} \qquad (b)\quad \nabla \times \mathbf{A} = \lim_{\Delta V \to 0} \frac{\iint_{\Delta S} \mathbf{n} \times \mathbf{A}\, dS}{\Delta V}$$

where $\Delta V$ is the volume enclosed by the surface $\Delta S$, and the limit is obtained by shrinking $\Delta V$ to the point $P$.

(a) From Problem 22, $\iiint_{\Delta V} \nabla \phi\, dV = \iint_{\Delta S} \phi \mathbf{n}\, dS$. Then $\iiint_{\Delta V} \nabla \phi \cdot \mathbf{i}\, dV = \iint_{\Delta S} \phi \mathbf{n} \cdot \mathbf{i}\, dS$.

Using the same principle employed in Problem 19, we have

$$\overline{\nabla \phi \cdot \mathbf{i}} = \frac{\iint_{\Delta S} \phi \mathbf{n} \cdot \mathbf{i}\, dS}{\Delta V}$$

where $\overline{\nabla \phi \cdot \mathbf{i}}$ is some value intermediate between the maximum and minimum of $\nabla \phi \cdot \mathbf{i}$ throughout $\Delta V$. Taking the limit as $\Delta V \to 0$ in such a way that $P$ is always interior to $\Delta V$, $\overline{\nabla \phi \cdot \mathbf{i}}$ approaches the value

$$(1)\qquad \nabla \phi \cdot \mathbf{i} = \lim_{\Delta V \to 0} \frac{\iint_S \phi \mathbf{n} \cdot \mathbf{i}\, dS}{\Delta V}$$

Similarly we find

$$(2)\qquad \nabla \phi \cdot \mathbf{j} = \lim_{\Delta V \to 0} \frac{\iint_S \phi \mathbf{n} \cdot \mathbf{j}\, dS}{\Delta V}$$

$$(3)\qquad \nabla \phi \cdot \mathbf{k} = \lim_{\Delta V \to 0} \frac{\iint_S \phi \mathbf{n} \cdot \mathbf{k}\, dS}{\Delta V}$$

Multiplying (*1*), (*2*), (*3*) by **i**, **j**, **k** respectively, and adding, using

$$\nabla\phi = (\nabla\phi\cdot\mathbf{i})\mathbf{i} + (\nabla\phi\cdot\mathbf{j})\mathbf{j} + (\nabla\phi\cdot\mathbf{k})\mathbf{k}, \qquad \mathbf{n} = (\mathbf{n}\cdot\mathbf{i})\mathbf{i} + (\mathbf{n}\cdot\mathbf{j})\mathbf{j} + (\mathbf{n}\cdot\mathbf{k})\mathbf{k}$$

(see Problem 20, Chapter 2) the result follows.

(*b*) From Problem 23, replacing **B** by **A**, $\iiint\limits_{\Delta V} \nabla\times\mathbf{A}\, dV = \iint\limits_{\Delta S} \mathbf{n}\times\mathbf{A}\, dS.$

Then as in part (*a*), we can show that

$$(\nabla\times\mathbf{A})\cdot\mathbf{i} = \lim_{\Delta V\to 0} \frac{\iint\limits_{\Delta S} (\mathbf{n}\times\mathbf{A})\cdot\mathbf{i}\, dS}{\Delta V}$$

and similar results with **j** and **k** replacing **i**. Multiplying by **i**, **j**, **k** and adding, the result follows.

The results obtained can be taken as starting points for definition of gradient and curl. Using these definitions, extensions can be made to coordinate systems other than rectangular.

**25.** Establish the operator equivalence

$$\nabla\circ \equiv \lim_{\Delta V\to 0} \frac{1}{\Delta V} \oiint\limits_{\Delta S} d\mathbf{S}\,\circ$$

where $\circ$ indicates a dot product, cross product or ordinary product.

To establish the equivalence, the results of the operation on a vector or scalar field must be consistent with already established results.

If $\circ$ is the dot product, then for a vector **A**,

$$\nabla\circ\mathbf{A} = \lim_{\Delta V\to 0} \frac{1}{\Delta V} \iint\limits_{\Delta S} d\mathbf{S}\circ\mathbf{A}$$

or

$$\operatorname{div}\mathbf{A} = \lim_{\Delta V\to 0} \frac{1}{\Delta V} \iint\limits_{\Delta S} d\mathbf{S}\cdot\mathbf{A}$$

$$= \lim_{\Delta V\to 0} \frac{1}{\Delta V} \iint\limits_{\Delta S} \mathbf{A}\cdot\mathbf{n}\, dS$$

established in Problem 19.

Similarly if $\circ$ is the cross product,

$$\operatorname{curl}\mathbf{A} = \nabla\times\mathbf{A} = \lim_{\Delta V\to 0} \frac{1}{\Delta V} \iint\limits_{\Delta S} d\mathbf{S}\times\mathbf{A}$$

$$= \lim_{\Delta V\to 0} \frac{1}{\Delta V} \iint\limits_{\Delta S} \mathbf{n}\times\mathbf{A}\, dS$$

established in Problem 24(*b*).

Also if $\circ$ is ordinary multiplication, then for a scalar $\phi$,

$$\nabla\circ\phi = \lim_{\Delta V\to 0} \frac{1}{\Delta V} \iint\limits_{\Delta S} d\mathbf{S}\circ\phi \qquad \text{or} \qquad \nabla\phi = \lim_{\Delta V\to 0} \frac{1}{\Delta V} \iint\limits_{\Delta S} \phi\, d\mathbf{S}$$

established in Problem 24(*a*).

**26.** Let $S$ be a closed surface and let $\mathbf{r}$ denote the position vector of any point $(x,y,z)$ measured from an origin $O$. Prove that

$$\iint_S \frac{\mathbf{n}\cdot\mathbf{r}}{r^3}\,dS$$

is equal to (*a*) zero if $O$ lies outside $S$; (*b*) $4\pi$ if $O$ lies inside $S$. This result is known as *Gauss' theorem.*

(*a*) By the divergence theorem, $\displaystyle\iint_S \frac{\mathbf{n}\cdot\mathbf{r}}{r^3}\,dS = \iiint_V \nabla\cdot\frac{\mathbf{r}}{r^3}\,dV.$

But $\nabla\cdot\frac{\mathbf{r}}{r^3} = 0$ (Problem 19, Chapter 4) everywhere within $V$ provided $r \neq 0$ in $V$, i.e. provided $O$ is outside of $V$ and thus outside of $S$. Then $\displaystyle\iint_S \frac{\mathbf{n}\cdot\mathbf{r}}{r^3}\,dS = 0.$

(*b*) If $O$ is inside $S$, surround $O$ by a small sphere $s$ of radius $a$. Let $\tau$ denote the region bounded by $S$ and $s$. Then by the divergence theorem

$$\iint_{S+s} \frac{\mathbf{n}\cdot\mathbf{r}}{r^3}\,dS = \iint_S \frac{\mathbf{n}\cdot\mathbf{r}}{r^3}\,dS + \iint_s \frac{\mathbf{n}\cdot\mathbf{r}}{r^3}\,dS = \iiint_\tau \nabla\cdot\frac{\mathbf{r}}{r^3}\,dV = 0$$

since $r \neq 0$ in $\tau$. Thus

$$\iint_S \frac{\mathbf{n}\cdot\mathbf{r}}{r^3}\,dS = -\iint_s \frac{\mathbf{n}\cdot\mathbf{r}}{r^3}\,dS$$

Now on $s$, $r = a$, $\mathbf{n} = -\frac{\mathbf{r}}{a}$ so that $\frac{\mathbf{n}\cdot\mathbf{r}}{r^3} = \frac{(-\mathbf{r}/a)\cdot\mathbf{r}}{a^3} = -\frac{\mathbf{r}\cdot\mathbf{r}}{a^4} = -\frac{a^2}{a^4} = -\frac{1}{a^2}$ and

$$\iint_S \frac{\mathbf{n}\cdot\mathbf{r}}{r^3}\,dS = -\iint_s \frac{\mathbf{n}\cdot\mathbf{r}}{r^3}\,dS = \iint_s \frac{1}{a^2}\,dS = \frac{1}{a^2}\iint_s dS = \frac{4\pi a^2}{a^2} = 4\pi$$

**27.** Interpret Gauss' theorem (Problem 26) geometrically.

Let $dS$ denote an element of surface area and connect all points on the boundary of $dS$ to $O$ (see adjoining figure), thereby forming a cone. Let $d\Omega$ be the area of that portion of a sphere with $O$ as centre and radius $r$ which is cut out by this cone; then the *solid angle* subtended by $dS$ at $O$ is defined as $d\omega = \frac{d\Omega}{r^2}$ and is numerically equal to the area of that portion of a sphere with centre $O$ and unit radius cut out by the cone. Let $\mathbf{n}$ be the positive unit normal to $dS$ and call $\theta$ the angle between $\mathbf{n}$ and $\mathbf{r}$; then $\cos\theta = \frac{\mathbf{n}\cdot\mathbf{r}}{r}$. Also, $d\Omega = \pm\,dS\cos\theta = \pm\frac{\mathbf{n}\cdot\mathbf{r}}{r}\,dS$ so that $d\omega = \pm\frac{\mathbf{n}\cdot\mathbf{r}}{r^3}\,dS$, the + or − being chosen according as $\mathbf{n}$ and $\mathbf{r}$ form an acute or an obtuse angle $\theta$ with each other.

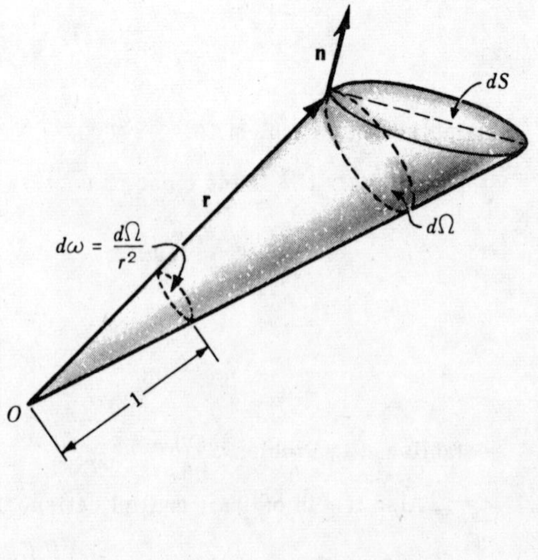


Let $S$ be a surface, as in Figure (*a*) below, such that any line meets $S$ in not more than two points. If $O$ lies outside $S$, then at a position such as 1, $\frac{\mathbf{n}\cdot\mathbf{r}}{r^3}\,dS = d\omega$; whereas at the corresponding position 2,

$\frac{\mathbf{n}\cdot\mathbf{r}}{r^3}\,dS = -d\omega$. An integration over these two regions gives zero, since the contributions to the solid angle cancel out. When the integration is performed over $S$ it thus follows that $\iint_S \frac{\mathbf{n}\cdot\mathbf{r}}{r^3}\,dS = 0$, since for every positive contribution there is a negative one.

In case $O$ is inside $S$, however, then at a position such as 3, $\frac{\mathbf{n}\cdot\mathbf{r}}{r^3}\,dS = d\omega$ and at 4, $\frac{\mathbf{n}\cdot\mathbf{r}}{r^3}\,dS = d\omega$ so that the contributions add instead of cancel. The total solid angle in this case is equal to the area of a unit sphere which is $4\pi$, so that $\iint_S \frac{\mathbf{n}\cdot\mathbf{r}}{r^3}\,dS = 4\pi$.

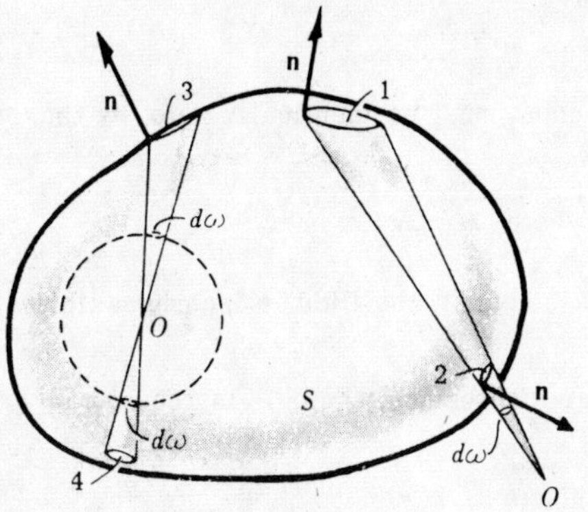


Fig. (a)

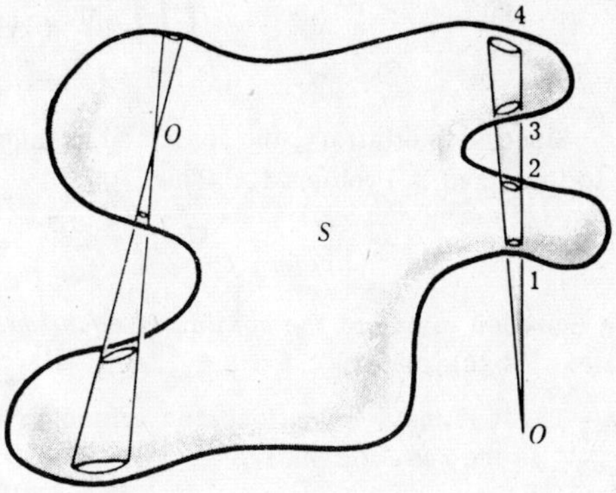


Fig. (b)

For surfaces $S$, such that a line may meet $S$ in more than two points, an exactly similar situation holds as is seen by reference to Figure (b) above. If $O$ is outside $S$, for example, then a cone with vertex at $O$ intersects $S$ at an even number of places and the contribution to the surface integral is zero since the solid angles subtended at $O$ cancel out in pairs. If $O$ is inside $S$, however, a cone having vertex at $O$ intersects $S$ at an odd number of places and since cancellation occurs only for an even number of these, there will always be a contribution of $4\pi$ for the entire surface $S$.

**28.** A fluid of density $\rho(x,y,z,t)$ moves with velocity $\mathbf{v}(x,y,z,t)$. If there are no sources or sinks, prove that

$$\nabla\cdot\mathbf{J} + \frac{\partial\rho}{\partial t} = 0 \qquad \text{where } \mathbf{J} = \rho\mathbf{v}$$

Consider an arbitrary surface enclosing a volume $V$ of the fluid. At any time the mass of fluid within $V$ is

$$M = \iiint_V \rho\,dV$$

The time rate of increase of this mass is

$$\frac{\partial M}{\partial t} = \frac{\partial}{\partial t}\iiint_V \rho\,dV = \iiint_V \frac{\partial\rho}{\partial t}\,dV$$

The mass of fluid per unit time leaving $V$ is

$$\iint_S \rho\mathbf{v}\cdot\mathbf{n}\,dS$$

(see Problem 15) and the time rate of increase in mass is therefore

$$-\iint_S \rho\mathbf{v}\cdot\mathbf{n}\,dS \quad = \quad -\iiint_V \nabla\cdot(\rho\mathbf{v})\,dV$$

by the divergence theorem. Then

$$\iiint_V \frac{\partial \rho}{\partial t}\,dV \quad = \quad -\iiint_V \nabla\cdot(\rho\mathbf{v})\,dV$$

or

$$\iiint_V (\nabla\cdot(\rho\mathbf{v}) + \frac{\partial \rho}{\partial t})\,dV \quad = \quad 0$$

Since $V$ is arbitrary, the integrand, assumed continuous, must be identically zero, by reasoning similar to that used in Problem 12. Then

$$\nabla\cdot\mathbf{J} + \frac{\partial \rho}{\partial t} \quad = \quad 0 \qquad \text{where } \mathbf{J} = \rho\mathbf{v}$$

The equation is called the *continuity equation.* If $\rho$ is a constant, the fluid is incompressible and $\nabla\cdot\mathbf{v} = 0$, i.e. $\mathbf{v}$ is solenoidal.

The continuity equation also arises in electromagnetic theory, where $\rho$ is the *charge density* and $\mathbf{J} = \rho\mathbf{v}$ is the *current density.*

**29.** If the temperature at any point $(x,y,z)$ of a solid at time $t$ is $U(x,y,z,t)$ and if $\kappa, \rho$ and $c$ are respectively the thermal conductivity, density and specific heat of the solid, assumed constant, show that

$$\frac{\partial U}{\partial t} \quad = \quad k\,\nabla^2 U \qquad \text{where } k = \kappa/\rho c$$

Let $V$ be an arbitrary volume lying within the solid, and let $S$ denote its surface. The total flux of heat across $S$, or the quantity of heat leaving $S$ per unit time, is

$$\iint_S (-\kappa\,\nabla U)\cdot\mathbf{n}\,dS$$

Thus the quantity of heat entering $S$ per unit time is

(1) $$\iint_S (\kappa\,\nabla U)\cdot\mathbf{n}\,dS \quad = \quad \iiint_V \nabla\cdot(\kappa\,\nabla U)\,dV$$

by the divergence theorem. The heat contained in a volume $V$ is given by

$$\iiint_V c\rho\,U\,dV$$

Then the time rate of increase of heat is

(2) $$\frac{\partial}{\partial t}\iiint_V c\rho\,U\,dV \quad = \quad \iiint_V c\rho\,\frac{\partial U}{\partial t}\,dV$$

Equating the right hand sides of (1) and (2),

$$\iiint_V [c\rho\,\frac{\partial U}{\partial t} - \nabla\cdot(\kappa\,\nabla U)]\,dV \quad = \quad 0$$

and since $V$ is arbitrary, the integrand, assumed continuous, must be identically zero so that

$$c\rho\,\frac{\partial U}{\partial t} = \nabla\cdot(\kappa\,\nabla U)$$

or if $\kappa, c, \rho$ are constants,

$$\frac{\partial U}{\partial t} = \frac{\kappa}{c\rho}\,\nabla\cdot\nabla U = k\,\nabla^2 U$$

The quantity $k$ is called the *diffusivity*. For steady-state heat flow (i.e. $\frac{\partial U}{\partial t} = 0$ or $U$ is independent of time) the equation reduces to Laplace's equation $\nabla^2 U = 0$.

## STOKES' THEOREM

**30.** (*a*) Express Stokes' theorem in words and (*b*) write it in rectangular form.

(*a*) The line integral of the tangential component of a vector **A** taken around a simple closed curve $C$ is equal to the surface integral of the normal component of the curl of **A** taken over any surface $S$ having $C$ as its boundary.

(*b*) As in Problem 14(*b*),

$$\mathbf{A} = A_1\mathbf{i} + A_2\mathbf{j} + A_3\mathbf{k}, \qquad \mathbf{n} = \cos\alpha\,\mathbf{i} + \cos\beta\,\mathbf{j} + \cos\gamma\,\mathbf{k}$$

Then

$$\nabla\times\mathbf{A} = \begin{vmatrix} \mathbf{i} & \mathbf{j} & \mathbf{k} \\ \frac{\partial}{\partial x} & \frac{\partial}{\partial y} & \frac{\partial}{\partial z} \\ A_1 & A_2 & A_3 \end{vmatrix} = \left(\frac{\partial A_3}{\partial y} - \frac{\partial A_2}{\partial z}\right)\mathbf{i} + \left(\frac{\partial A_1}{\partial z} - \frac{\partial A_3}{\partial x}\right)\mathbf{j} + \left(\frac{\partial A_2}{\partial x} - \frac{\partial A_1}{\partial y}\right)\mathbf{k}$$

$$(\nabla\times\mathbf{A})\cdot\mathbf{n} = \left(\frac{\partial A_3}{\partial y} - \frac{\partial A_2}{\partial z}\right)\cos\alpha + \left(\frac{\partial A_1}{\partial z} - \frac{\partial A_3}{\partial x}\right)\cos\beta + \left(\frac{\partial A_2}{\partial x} - \frac{\partial A_1}{\partial y}\right)\cos\gamma$$

$$\mathbf{A}\cdot d\mathbf{r} = (A_1\mathbf{i} + A_2\mathbf{j} + A_3\mathbf{k})\cdot(dx\,\mathbf{i} + dy\,\mathbf{j} + dz\,\mathbf{k}) = A_1dx + A_2dy + A_3dz$$

and Stokes' theorem becomes

$$\iint_S \left[\left(\frac{\partial A_3}{\partial y} - \frac{\partial A_2}{\partial z}\right)\cos\alpha + \left(\frac{\partial A_1}{\partial z} - \frac{\partial A_3}{\partial x}\right)\cos\beta + \left(\frac{\partial A_2}{\partial x} - \frac{\partial A_1}{\partial y}\right)\cos\gamma\right] dS = \oint_C A_1dx + A_2dy + A_3dz$$

**31.** Prove Stokes' theorem.

Let $S$ be a surface which is such that its projections on the $xy$, $yz$ and $xz$ planes are regions bounded by simple closed curves, as indicated in the adjoining figure. Assume $S$ to have representation $z = f(x,y)$ or $x = g(y,z)$ or $y = h(x,z)$, where $f, g, h$ are single-valued, continuous and differentiable functions. We must show that

$$\iint_S (\nabla\times\mathbf{A})\cdot\mathbf{n}\,dS = \iint_S [\nabla\times(A_1\mathbf{i} + A_2\mathbf{j} + A_3\mathbf{k})]\cdot\mathbf{n}\,dS = \oint_C \mathbf{A}\cdot d\mathbf{r}$$

where $C$ is the boundary of $S$.

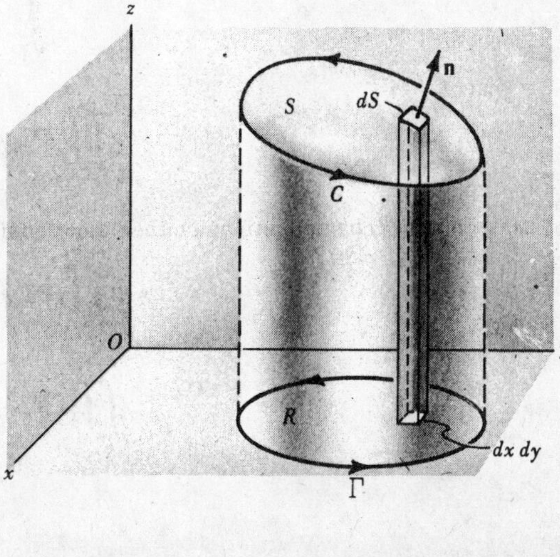

Consider first $\iint_S [\nabla\times(A_1\mathbf{i})]\cdot\mathbf{n}\,dS$.

Since $\nabla\times(A_1\mathbf{i}) = \begin{vmatrix} \mathbf{i} & \mathbf{j} & \mathbf{k} \\ \frac{\partial}{\partial x} & \frac{\partial}{\partial y} & \frac{\partial}{\partial z} \\ A_1 & 0 & 0 \end{vmatrix} = \frac{\partial A_1}{\partial z}\mathbf{j} - \frac{\partial A_1}{\partial y}\mathbf{k}$,

$$(1) \qquad [\nabla\times(A_1\mathbf{i})]\cdot\mathbf{n}\,dS = \left(\frac{\partial A_1}{\partial z}\mathbf{n}\cdot\mathbf{j} - \frac{\partial A_1}{\partial y}\mathbf{n}\cdot\mathbf{k}\right)dS$$

If $z = f(x,y)$ is taken as the equation of $S$, then the position vector to any point of $S$ is $\mathbf{r} = x\mathbf{i}+y\mathbf{j}+z\mathbf{k} = x\mathbf{i}+y\mathbf{j}+f(x,y)\mathbf{k}$ so that $\frac{\partial\mathbf{r}}{\partial y} = \mathbf{j}+\frac{\partial z}{\partial y}\mathbf{k} = \mathbf{j}+\frac{\partial f}{\partial y}\mathbf{k}$. But $\frac{\partial\mathbf{r}}{\partial y}$ is a vector tangent to $S$ (see Problem 25, Chapter 3) and thus perpendicular to $\mathbf{n}$, so that

$$\mathbf{n}\cdot\frac{\partial\mathbf{r}}{\partial y} = \mathbf{n}\cdot\mathbf{j}+\frac{\partial z}{\partial y}\mathbf{n}\cdot\mathbf{k} = 0 \qquad \text{or} \qquad \mathbf{n}\cdot\mathbf{j} = -\frac{\partial z}{\partial y}\mathbf{n}\cdot\mathbf{k}$$

Substitute in (1) to obtain

$$\left(\frac{\partial A_1}{\partial z}\mathbf{n}\cdot\mathbf{j} - \frac{\partial A_1}{\partial y}\mathbf{n}\cdot\mathbf{k}\right)dS = \left(-\frac{\partial A_1}{\partial z}\frac{\partial z}{\partial y}\mathbf{n}\cdot\mathbf{k} - \frac{\partial A_1}{\partial y}\mathbf{n}\cdot\mathbf{k}\right)dS$$

or

$$(2) \qquad [\nabla\times(A_1\mathbf{i})]\cdot\mathbf{n}\,dS = -\left(\frac{\partial A_1}{\partial y}+\frac{\partial A_1}{\partial z}\frac{\partial z}{\partial y}\right)\mathbf{n}\cdot\mathbf{k}\,dS$$

Now on $S$, $A_1(x,y,z) = A_1(x,y,f(x,y)) = F(x,y)$; hence $\frac{\partial A_1}{\partial y}+\frac{\partial A_1}{\partial z}\frac{\partial z}{\partial y} = \frac{\partial F}{\partial y}$ and (2) becomes

$$[\nabla\times(A_1\mathbf{i})]\cdot\mathbf{n}\,dS = -\frac{\partial F}{\partial y}\mathbf{n}\cdot\mathbf{k}\,dS = -\frac{\partial F}{\partial y}\,dx\,dy$$

Then

$$\iint_S [\nabla\times(A_1\mathbf{i})]\cdot\mathbf{n}\,dS = \iint_R -\frac{\partial F}{\partial y}\,dx\,dy$$

where $R$ is the projection of $S$ on the $xy$ plane. By Green's theorem for the plane the last integral equals $\oint_\Gamma F\,dx$ where $\Gamma$ is the boundary of $R$. Since at each point $(x,y)$ of $\Gamma$ the value of $F$ is the same as the value of $A_1$ at each point $(x,y,z)$ of $C$, and since $dx$ is the same for both curves, we must have

$$\oint_\Gamma F\,dx = \oint_C A_1\,dx$$

or

$$\iint_S [\nabla\times(A_1\mathbf{i})]\cdot\mathbf{n}\,dS = \oint_C A_1\,dx$$

Similarly, by projections on the other coordinate planes,

$$\iint_S [\nabla\times(A_2\mathbf{j})]\cdot\mathbf{n}\,dS = \oint_C A_2\,dy$$

$$\iint_S [\nabla\times(A_3\mathbf{k})]\cdot\mathbf{n}\,dS = \oint_C A_3\,dz$$

Thus by addition,

$$\iint_S (\nabla\times \mathbf{A})\cdot \mathbf{n}\, dS = \oint_C \mathbf{A}\cdot d\mathbf{r}$$

The theorem is also valid for surfaces $S$ which may not satisfy the restrictions imposed above. For assume that $S$ can be subdivided into surfaces $S_1, S_2, \ldots S_k$ with boundaries $C_1, C_2, \ldots C_k$ which do satisfy the restrictions. Then Stokes' theorem holds for each such surface. Adding these surface integrals, the total surface integral over $S$ is obtained. Adding the corresponding line integrals over $C_1, C_2, \ldots C_k$, the line integral over $C$ is obtained.

**32.** Verify Stokes' theorem for $\mathbf{A} = (2x-y)\mathbf{i} - yz^2\mathbf{j} - y^2z\mathbf{k}$, where $S$ is the upper half surface of the sphere $x^2+y^2+z^2 = 1$ and $C$ is its boundary.

The boundary $C$ of $S$ is a circle in the $xy$ plane of radius one and centre at the origin. Let $x = \cos t$, $y = \sin t$, $z = 0$, $0 \leqq t < 2\pi$ be parametric equations of $C$. Then

$$\oint_C \mathbf{A}\cdot d\mathbf{r} = \oint_C (2x-y)\,dx - yz^2\,dy - y^2z\,dz$$

$$= \int_0^{2\pi} (2\cos t - \sin t)(-\sin t)\,dt = \pi$$

Also,

$$\nabla\times\mathbf{A} = \begin{vmatrix} \mathbf{i} & \mathbf{j} & \mathbf{k} \\ \dfrac{\partial}{\partial x} & \dfrac{\partial}{\partial y} & \dfrac{\partial}{\partial z} \\ 2x-y & -yz^2 & -y^2z \end{vmatrix} = \mathbf{k}$$

Then

$$\iint_S (\nabla\times\mathbf{A})\cdot\mathbf{n}\,dS = \iint_S \mathbf{k}\cdot\mathbf{n}\,dS = \iint_R dx\,dy$$

since $\mathbf{n}\cdot\mathbf{k}\,dS = dx\,dy$ and $R$ is the projection of $S$ on the $xy$ plane. This last integral equals

$$\int_{x=-1}^{1}\int_{y=-\sqrt{1-x^2}}^{\sqrt{1-x^2}} dy\,dx = 4\int_0^1\int_0^{\sqrt{1-x^2}} dy\,dx = 4\int_0^1 \sqrt{1-x^2}\,dx = \pi$$

and Stokes' theorem is verified.

**33.** Prove that a necessary and sufficient condition that $\oint_C \mathbf{A}\cdot d\mathbf{r} = 0$ for every closed curve $C$ is that $\nabla\times\mathbf{A} = \mathbf{0}$ identically.

*Sufficiency.* Suppose $\nabla\times\mathbf{A} = \mathbf{0}$. Then by Stokes' theorem

$$\oint_C \mathbf{A}\cdot d\mathbf{r} = \iint_S (\nabla\times\mathbf{A})\cdot\mathbf{n}\,dS = 0$$

*Necessity.* Suppose $\oint_C \mathbf{A}\cdot d\mathbf{r} = 0$ around every closed path $C$, and assume $\nabla\times\mathbf{A} \neq \mathbf{0}$ at some point $P$. Then assuming $\nabla\times\mathbf{A}$ is continuous there will be a region with $P$ as an interior point, where $\nabla\times\mathbf{A} \neq \mathbf{0}$. Let $S$ be a surface contained in this region whose normal $\mathbf{n}$ at each point has the same direction as $\nabla\times\mathbf{A}$, i.e. $\nabla\times\mathbf{A} = \alpha\mathbf{n}$ where $\alpha$ is a positive constant. Let $C$ be the boundary of $S$. Then by Stokes' theorem

$$\oint_C \mathbf{A}\cdot d\mathbf{r} = \iint_S (\nabla\times\mathbf{A})\cdot\mathbf{n}\, dS = \alpha \iint_S \mathbf{n}\cdot\mathbf{n}\, dS > 0$$

which contradicts the hypothesis that $\oint_C \mathbf{A}\cdot d\mathbf{r} = 0$ and shows that $\nabla\times\mathbf{A} = \mathbf{0}$.

It follows that $\nabla\times\mathbf{A} = \mathbf{0}$ is also a necessary and sufficient condition for a line integral $\int_{P_1}^{P_2} \mathbf{A}\cdot d\mathbf{r}$ to be independent of the path joining points $P_1$ and $P_2$. (See Problems 10 and 11, Chapter 5.)

**34.** Prove $\oint d\mathbf{r}\times\mathbf{B} = \iint_S (\mathbf{n}\times\nabla)\times\mathbf{B}\, dS$.

In Stokes' theorem, let $\mathbf{A} = \mathbf{B}\times\mathbf{C}$ where $\mathbf{C}$ is a constant vector, Then

$$\oint d\mathbf{r}\cdot(\mathbf{B}\times\mathbf{C}) = \iint_S [\nabla\times(\mathbf{B}\times\mathbf{C})]\cdot\mathbf{n}\, dS$$

$$\oint \mathbf{C}\cdot(d\mathbf{r}\times\mathbf{B}) = \iint_S [(\mathbf{C}\cdot\nabla)\mathbf{B} - \mathbf{C}(\nabla\cdot\mathbf{B})]\cdot\mathbf{n}\, dS$$

$$\mathbf{C}\cdot\oint d\mathbf{r}\times\mathbf{B} = \iint_S [(\mathbf{C}\cdot\nabla)\mathbf{B}]\cdot\mathbf{n}\, dS - \iint_S [\mathbf{C}(\nabla\cdot\mathbf{B})]\cdot\mathbf{n}\, dS$$

$$= \iint_S \mathbf{C}\cdot[\nabla(\mathbf{B}\cdot\mathbf{n})]\, dS - \iint_S \mathbf{C}\cdot[\mathbf{n}(\nabla\cdot\mathbf{B})]\, dS$$

$$= \mathbf{C}\cdot\iint_S [\nabla(\mathbf{B}\cdot\mathbf{n}) - \mathbf{n}(\nabla\cdot\mathbf{B})]\, dS = \mathbf{C}\cdot\iint_S (\mathbf{n}\times\nabla)\times\mathbf{B}\, dS$$

Since $\mathbf{C}$ is an arbitrary constant vector $\oint d\mathbf{r}\times\mathbf{B} = \iint_S (\mathbf{n}\times\nabla)\times\mathbf{B}\, dS$

**35.** If $\Delta S$ is a surface bounded by a simple closed curve $C$, $P$ is any point of $\Delta S$ not on $C$ and $\mathbf{n}$ is a unit normal to $\Delta S$ at $P$, show that at $P$

$$(\text{curl}\,\mathbf{A})\cdot\mathbf{n} = \lim_{\Delta S\to 0} \frac{\oint_C \mathbf{A}\cdot d\mathbf{r}}{\Delta S}$$

where the limit is taken in such a way that $\Delta S$ shrinks to $P$.

By Stokes' theorem, $\iint_{\Delta S} (\text{curl}\,\mathbf{A})\cdot\mathbf{n}\, dS = \oint_C \mathbf{A}\cdot d\mathbf{r}$.

Using the mean value theorem for integrals as in Problems 19 and 24, this can be written

$$\overline{(\text{curl}\,\mathbf{A})\cdot\mathbf{n}} = \frac{\oint_C \mathbf{A}\cdot d\mathbf{r}}{\Delta S}$$

and the required result follows upon taking the limit as $\Delta S \to 0$.

This can be used as a starting point for defining curl **A** (see Problem 36) and is useful in obtaining curl **A** in coordinate systems other than rectangular. Since $\oint_C \mathbf{A} \cdot d\mathbf{r}$ is called the circulation of **A** about $C$, the normal component of the curl can be interpreted physically as the limit of the circulation per unit area, thus accounting for the synonym rotation of **A** (rot **A**) instead of curl of **A**.

**36.** If curl **A** is defined according to the limiting process of Problem 35, find the $z$ component of curl **A**.

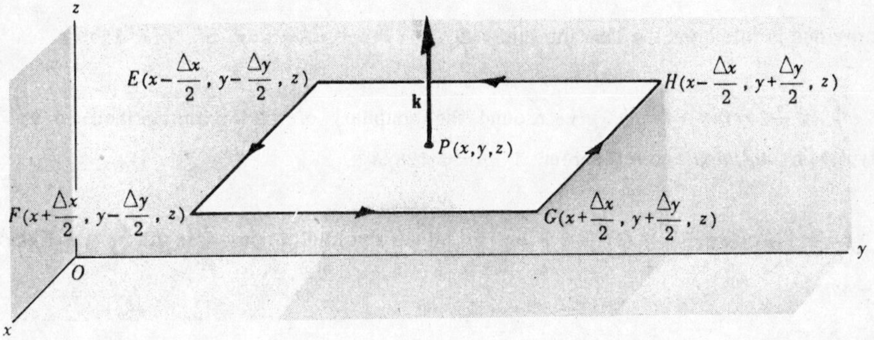


Let $EFGH$ be a rectangle parallel to the $xy$ plane with interior point $P(x,y,z)$ taken as midpoint, as shown in the figure above. Let $A_1$ and $A_2$ be the components of **A** at $P$ in the positive $x$ and $y$ directions respectively.

If $C$ is the boundary of the rectangle, then

$$\oint_C \mathbf{A} \cdot d\mathbf{r} = \int_{EF} \mathbf{A} \cdot d\mathbf{r} + \int_{FG} \mathbf{A} \cdot d\mathbf{r} + \int_{GH} \mathbf{A} \cdot d\mathbf{r} + \int_{HE} \mathbf{A} \cdot d\mathbf{r}$$

But
$$\int_{EF} \mathbf{A} \cdot d\mathbf{r} = (A_1 - \frac{1}{2}\frac{\partial A_1}{\partial y}\Delta y)\Delta x \qquad \int_{GH} \mathbf{A} \cdot d\mathbf{r} = -(A_1 + \frac{1}{2}\frac{\partial A_1}{\partial y}\Delta y)\Delta x$$

$$\int_{FG} \mathbf{A} \cdot d\mathbf{r} = (A_2 + \frac{1}{2}\frac{\partial A_2}{\partial x}\Delta x)\Delta y \qquad \int_{HE} \mathbf{A} \cdot d\mathbf{r} = -(A_2 - \frac{1}{2}\frac{\partial A_2}{\partial x}\Delta x)\Delta y$$

except for infinitesimals of higher order than $\Delta x\, \Delta y$.

Adding, we have approximately $\oint_C \mathbf{A} \cdot d\mathbf{r} = (\frac{\partial A_2}{\partial x} - \frac{\partial A_1}{\partial y})\Delta x\, \Delta y$.

Then, since $\Delta S = \Delta x\, \Delta y$,

$$z \text{ component of curl } \mathbf{A} = (\text{curl } \mathbf{A}) \cdot \mathbf{k} = \lim_{\Delta S \to 0} \frac{\oint \mathbf{A} \cdot d\mathbf{r}}{\Delta S}$$

$$= \lim_{\substack{\Delta x \to 0 \\ \Delta y \to 0}} \frac{(\frac{\partial A_2}{\partial x} - \frac{\partial A_1}{\partial y})\Delta x\, \Delta y}{\Delta x\, \Delta y}$$

$$= \frac{\partial A_2}{\partial x} - \frac{\partial A_1}{\partial y}$$

## SUPPLEMENTARY PROBLEMS

**37.** Verify Green's theorem in the plane for $\oint_C (3x^2 - 8y^2)\,dx + (4y - 6xy)\,dy$, where $C$ is the boundary of the region defined by: (a) $y = \sqrt{x}$, $y = x^2$; (b) $x = 0$, $y = 0$, $x + y = 1$.
*Ans.* (a) common value = 3/2 (b) common value = 5/3

**38.** Evaluate $\oint_C (3x + 4y)\,dx + (2x - 3y)\,dy$ where $C$, a circle of radius two with centre at the origin of the $xy$ plane, is traversed in the positive sense. *Ans.* $-8\pi$

**39.** Work the previous problem for the line integral $\oint_C (x^2 + y^2)\,dx + 3xy^2\,dy$. *Ans.* $12\pi$

**40.** Evaluate $\oint (x^2 - 2xy)\,dx + (x^2y + 3)\,dy$ around the boundary of the region defined by $y^2 = 8x$ and $x = 2$ (a) directly, (b) by using Green's theorem. *Ans.* 128/5

**41.** Evaluate $\int_{(0,0)}^{(\pi,2)} (6xy - y^2)\,dx + (3x^2 - 2xy)\,dy$ along the cycloid $x = \theta - \sin\theta$, $y = 1 - \cos\theta$.
*Ans.* $6\pi^2 - 4\pi$

**42.** Evaluate $\oint (3x^2 + 2y)\,dx - (x + 3\cos y)\,dy$ around the parallelogram having vertices at (0,0), (2,0), (3,1) and (1,1). *Ans.* $-6$

**43.** Find the area bounded by one arch of the cycloid $x = a(\theta - \sin\theta)$, $y = a(1 - \cos\theta)$, $a > 0$, and the $x$ axis.
*Ans.* $3\pi a^2$

**44.** Find the area bounded by the hypocycloid $x^{2/3} + y^{2/3} = a^{2/3}$, $a > 0$.
Hint: Parametric equations are $x = a\cos^3\theta$, $y = a\sin^3\theta$. *Ans.* $3\pi a^2/8$

**45.** Show that in polar coordinates $(\rho, \phi)$ the expression $x\,dy - y\,dx = \rho^2\,d\phi$. Interpret $\frac{1}{2}\int x\,dy - y\,dx$.

**46.** Find the area of a loop of the four-leafed rose $\rho = 3\sin 2\phi$. *Ans.* $9\pi/8$

**47.** Find the area of both loops of the lemniscate $\rho^2 = a^2\cos 2\phi$. *Ans.* $a^2$

**48.** Find the area of the loop of the folium of Descartes $x^3 + y^3 = 3axy$, $a > 0$ (see adjoining figure).
Hint: Let $y = tx$ and obtain the parametric equations of the curve. Then use the fact that

$$\text{Area} = \frac{1}{2}\oint x\,dy - y\,dx = \frac{1}{2}\oint x^2\,d\left(\frac{y}{x}\right) = \frac{1}{2}\oint x^2\,dt$$

*Ans.* $3a^2/2$

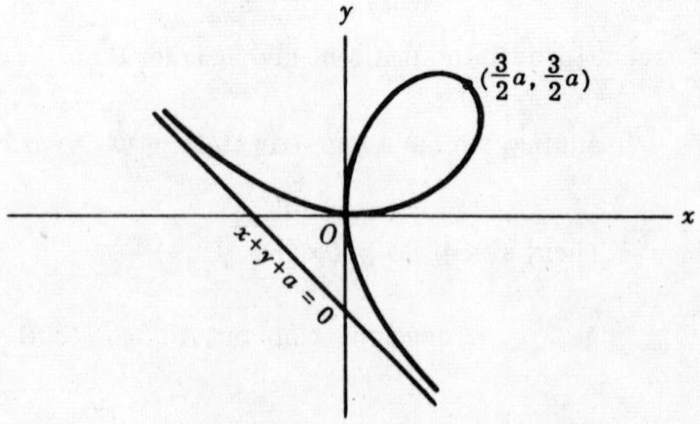


**49.** Verify Green's theorem in the plane for $\oint_C (2x - y^3)\,dx - xy\,dy$, where $C$ is the boundary of the region enclosed by the circles $x^2 + y^2 = 1$ and $x^2 + y^2 = 9$. *Ans.* common value = $60\pi$

**50.** Evaluate $\int_{(1,0)}^{(-1,0)} \frac{-y\,dx + x\,dy}{x^2 + y^2}$ along the following paths:

(*a*) straight line segments from (1,0) to (1,1), then to (−1,1), then to (−1,0).
(*b*) straight line segments from (1,0) to (1,−1), then to (−1,−1), then to (−1,0).

Show that although $\dfrac{\partial M}{\partial y} = \dfrac{\partial N}{\partial x}$, the line integral *is dependent* on the path joining (1,0) to (−1,0) and explain.

*Ans.* (*a*) $\pi$ (*b*) $-\pi$

**51.** By changing variables from $(x,y)$ to $(u,v)$ according to the transformation $x = x(u,v)$, $y = y(u,v)$, show that the area $A$ of a region $R$ bounded by a simple closed curve $C$ is given by

$$A = \iint_R \left| J\left(\frac{x,y}{u,v}\right) \right| du\,dv \qquad \text{where} \qquad J\left(\frac{x,y}{u,v}\right) \equiv \begin{vmatrix} \dfrac{\partial x}{\partial u} & \dfrac{\partial y}{\partial u} \\ \dfrac{\partial x}{\partial v} & \dfrac{\partial y}{\partial v} \end{vmatrix}$$

is the Jacobian of $x$ and $y$ with respect to $u$ and $v$. What restrictions should you make ? Illustrate the result where $u$ and $v$ are polar coordinates.

Hint: Use the result $A = \frac{1}{2}\int x\,dy - y\,dx$, transform to $u,v$ coordinates and then use Green's theorem.

**52.** Evaluate $\iint_S \mathbf{F}\cdot\mathbf{n}\,dS$, where $\mathbf{F} = 2xy\,\mathbf{i} + yz^2\,\mathbf{j} + xz\,\mathbf{k}$ and $S$ is:

(*a*) the surface of the parallelepiped bounded by $x=0$, $y=0$, $z=0$, $x=2$, $y=1$ and $z=3$,
(*b*) the surface of the region bounded by $x=0$, $y=0$, $y=3$, $z=0$ and $x+2z=6$.

*Ans.* (*a*) 30 (*b*) 351/2

**53.** Verify the divergence theorem or $\mathbf{A} = 2x^2y\,\mathbf{i} - y^2\,\mathbf{j} + 4xz^2\,\mathbf{k}$ taken over the region in the first octant bounded by $y^2+z^2=9$ and $x=2$. *Ans.* 180

**54.** Evaluate $\iint_S \mathbf{r}\cdot\mathbf{n}\,dS$ where (*a*) $S$ is the sphere of radius 2 with centre at (0,0,0), (*b*) $S$ is the surface of the cube bounded by $x=-1$, $y=-1$, $z=-1$, $x=1$, $y=1$, $z=1$, (*c*) $S$ is the surface bounded by the paraboloid $z = 4-(x^2+y^2)$ and the $xy$ plane. *Ans.* (*a*) $32\pi$ (*b*) 24 (*c*) $24\pi$

**55.** If $S$ is any closed surface enclosing a volume $V$ and $\mathbf{A} = ax\,\mathbf{i} + by\,\mathbf{j} + cz\,\mathbf{k}$, prove that $\iint_S \mathbf{A}\cdot\mathbf{n}\,dS = (a+b+c)V$.

**56.** If $\mathbf{H} = \text{curl}\,\mathbf{A}$, prove that $\iint_S \mathbf{H}\cdot\mathbf{n}\,dS = 0$ for any closed surface $S$.

**57.** If $\mathbf{n}$ is the unit outward drawn normal to any closed surface of area $S$, show that $\iiint_V \text{div}\,\mathbf{n}\,dV = S$.

**58.** Prove $\iiint_V \dfrac{dV}{r^2} = \iint_S \dfrac{\mathbf{r}\cdot\mathbf{n}}{r^2}\,dS$.

**59.** Prove $\iint_S r^5\mathbf{n}\,dS = \iiint_V 5r^3\mathbf{r}\,dV$.

**60.** Prove $\iint_S \mathbf{n}\,dS = \mathbf{0}$ for any closed surface $S$.

**61.** Show that Green's second identity can be written $\iiint_V (\phi\nabla^2\psi - \psi\nabla^2\phi)\,dV = \iint_S \left(\phi\dfrac{d\psi}{dn} - \psi\dfrac{d\phi}{dn}\right)dS$

**62.** Prove $\iint_S \mathbf{r}\times d\mathbf{S} = \mathbf{0}$ for any closed surface $S$.

**63.** Verify Stokes' theorem for $\mathbf{A} = (y-z+2)\mathbf{i} + (yz+4)\mathbf{j} - xz\,\mathbf{k}$, where $S$ is the surface of the cube $x=0$, $y=0$, $z=0$, $x=2$, $y=2$, $z=2$ above the $xy$ plane. *Ans.* common value $= -4$

**64.** Verify Stokes' theorem for $\mathbf{F} = xz\,\mathbf{i} - y\,\mathbf{j} + x^2y\,\mathbf{k}$, where $S$ is the surface of the region bounded by $x=0$, $y=0$, $z=0$, $2x+y+2z=8$ which is not included in the $xz$ plane. *Ans.* common value $= 32/3$

**65.** Evaluate $\iint_S (\nabla\times\mathbf{A})\cdot\mathbf{n}\,dS$, where $\mathbf{A} = (x^2+y-4)\mathbf{i} + 3xy\,\mathbf{j} + (2xz+z^2)\mathbf{k}$ and $S$ is the surface of (*a*) the hemisphere $x^2+y^2+z^2 = 16$ above the $xy$ plane, (*b*) the paraboloid $z = 4-(x^2+y^2)$ above the $xy$ plane.
*Ans.* (*a*) $-16\pi$, (*b*) $-4\pi$

**66.** If $\mathbf{A} = 2yz\,\mathbf{i} - (x+3y-2)\mathbf{j} + (x^2+z)\mathbf{k}$, evaluate $\iint_S (\nabla\times\mathbf{A})\cdot\mathbf{n}\,dS$ over the surface of intersection of the cylinders $x^2+y^2 = a^2$, $x^2+z^2 = a^2$ which is included in the first octant. *Ans.* $-\dfrac{a^2}{12}(3\pi+8a)$

**67.** A vector $\mathbf{B}$ is always normal to a given closed surface $S$. Show that $\iiint_V \operatorname{curl}\mathbf{B}\,dV = \mathbf{0}$, where $V$ is the region bounded by $S$.

**68.** If $\oint_C \mathbf{E}\cdot d\mathbf{r} = -\dfrac{1}{c}\dfrac{\partial}{\partial t}\iint_S \mathbf{H}\cdot d\mathbf{S}$, where $S$ is any surface bounded by the curve $C$, show that $\nabla\times\mathbf{E} = -\dfrac{1}{c}\dfrac{\partial\mathbf{H}}{\partial t}$.

**69.** Prove $\oint_C \phi\,d\mathbf{r} = \iint_S d\mathbf{S}\times\nabla\phi$.

**70.** Use the operator equivalence of Solved Problem 25 to arrive at (*a*) $\nabla\phi$, (*b*) $\nabla\cdot\mathbf{A}$, (*c*) $\nabla\times\mathbf{A}$ in rectangular coordinates.

**71.** Prove $\iiint_V \nabla\phi\cdot\mathbf{A}\,dV = \iint_S \phi\mathbf{A}\cdot\mathbf{n}\,dS - \iiint_V \phi\nabla\cdot\mathbf{A}\,dV$.

**72.** Let $\mathbf{r}$ be the position vector of any point relative to an origin $O$. Suppose $\phi$ has continuous derivatives of order two, at least, and let $S$ be a closed surface bounding a volume $V$. Denote $\phi$ at $O$ by $\phi_o$. Show that

$$\iint_S \left[\frac{1}{r}\nabla\phi - \phi\nabla\left(\frac{1}{r}\right)\right]\cdot d\mathbf{S} = \iiint_V \frac{\nabla^2\phi}{r}\,dV + \alpha$$

where $\alpha = 0$ or $4\pi\phi_o$ according as $O$ is outside or inside $S$.

**73.** The potential $\phi(P)$ at a point $P(x,y,z)$ due to a system of charges (or masses) $q_1, q_2, \ldots, q_n$ having position vectors $\mathbf{r}_1, \mathbf{r}_2, \ldots, \mathbf{r}_n$ with respect to $P$ is given by

$$\phi = \sum_{m=1}^{n}\frac{q_m}{r_m}$$

Prove *Gauss' law*

$$\iint_S \mathbf{E}\cdot d\mathbf{S} = 4\pi Q$$

where $\mathbf{E} = -\nabla\phi$ is the electric field intensity, $S$ is a surface enclosing all the charges and $Q = \sum_{m=1}^{n} q_m$ is the total charge within $S$.

**74.** If a region $V$ bounded by a surface $S$ has a continuous charge (or mass) distribution of density $\rho$, the potential $\phi(P)$ at a point $P$ is defined by $\phi = \iiint_V \dfrac{\rho\,dV}{r}$. Deduce the following under suitable assumptions:

(*a*) $\iint_S \mathbf{E}\cdot d\mathbf{S} = 4\pi\iiint_V \rho\,dV$, where $\mathbf{E} = -\nabla\phi$.

(*b*) $\nabla^2\phi = -4\pi\rho$ (Poisson's equation) at all points $P$ where charges exist, and $\nabla^2\phi = 0$ (Laplace's equation) where no charges exist.

# Chapter 7

# CURVILINEAR COORDINATES

**TRANSFORMATION OF COORDINATES.** Let the rectangular coordinates $(x,y,z)$ of any point be expressed as functions of $(u_1, u_2, u_3)$ so that

$$(1) \qquad x = x(u_1, u_2, u_3), \qquad y = y(u_1, u_2, u_3), \qquad z = z(u_1, u_2, u_3)$$

Suppose that *(1)* can be solved for $u_1, u_2, u_3$ in terms of $x, y, z$, i.e.,

$$(2) \qquad u_1 = u_1(x, y, z), \qquad u_2 = u_2(x, y, z), \qquad u_3 = u_3(x, y, z)$$

The functions in *(1)* and *(2)* are assumed to be single-valued and to have continuous derivatives so that the correspondence between $(x, y, z)$ and $(u_1, u_2, u_3)$ is unique. In practice this assumption may not apply at certain points and special consideration is required.

Given a point $P$ with rectangular coordinates $(x, y, z)$ we can, from *(2)* associate a unique set of coordinates $(u_1, u_2, u_3)$ called the *curvilinear coordinates* of $P$. The sets of equations *(1)* or *(2)* define a *transformation of coordinates*.

**ORTHOGONAL CURVILINEAR COORDINATES.**

The surfaces $u_1 = c_1$, $u_2 = c_2$, $u_3 = c_3$, where $c_1, c_2, c_3$ are constants, are called *coordinate surfaces* and each pair of these surfaces intersect in curves called *coordinate curves or lines* (see Fig. 1). If the coordinate surfaces intersect at right angles the curvilinear coordinate system is called *orthogonal*. The $u_1, u_2$ and $u_3$ coordinate curves of a curvilinear system are analogous to the $x, y$ and $z$ coordinate axes of a rectangular system.

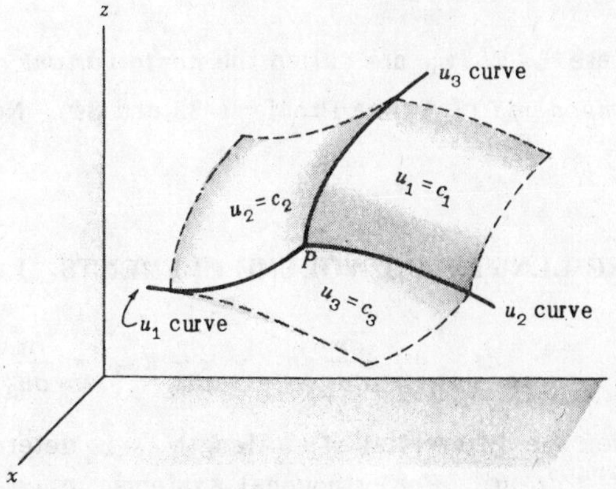

Fig. 1

**UNIT VECTORS IN CURVILINEAR SYSTEMS.** Let $\mathbf{r} = x\mathbf{i} + y\mathbf{j} + z\mathbf{k}$ be the position vector of a point $P$. Then *(1)* can be written $\mathbf{r} = \mathbf{r}(u_1, u_2, u_3)$. A tangent vector to the $u_1$ curve at $P$ (for which $u_2$ and $u_3$ are constants) is $\dfrac{\partial \mathbf{r}}{\partial u_1}$. Then a unit tangent vector in this direction is $\mathbf{e}_1 = \dfrac{\partial \mathbf{r}}{\partial u_1} \Big/ \left|\dfrac{\partial \mathbf{r}}{\partial u_1}\right|$ so that $\dfrac{\partial \mathbf{r}}{\partial u_1} = h_1 \mathbf{e}_1$ where $h_1 = \left|\dfrac{\partial \mathbf{r}}{\partial u_1}\right|$. Similarly, if $\mathbf{e}_2$ and $\mathbf{e}_3$ are unit tangent vectors to the $u_2$ and $u_3$ curves at $P$ respectively, then $\dfrac{\partial \mathbf{r}}{\partial u_2} = h_2 \mathbf{e}_2$ and $\dfrac{\partial \mathbf{r}}{\partial u_3} = h_3 \mathbf{e}_3$ where $h_2 = \left|\dfrac{\partial \mathbf{r}}{\partial u_2}\right|$ and $h_3 = \left|\dfrac{\partial \mathbf{r}}{\partial u_3}\right|$. The quantities $h_1, h_2, h_3$ are called *scale factors*. The unit vectors $\mathbf{e}_1, \mathbf{e}_2, \mathbf{e}_3$ are in the directions of increasing $u_1, u_2, u_3$, respectively.

Since $\nabla u_1$ is a vector at $P$ normal to the surface $u_1 = c_1$, a unit vector in this direction is given by $\mathbf{E}_1 = \nabla u_1 / |\nabla u_1|$. Similarly, the unit vectors $\mathbf{E}_2 = \nabla u_2 / |\nabla u_2|$ and $\mathbf{E}_3 = \nabla u_3 / |\nabla u_3|$ at $P$ are normal to the surfaces $u_2 = c_2$ and $u_3 = c_3$ respectively.

Thus at each point $P$ of a curvilinear system there exist, in general, two sets of unit vectors, $\mathbf{e}_1, \mathbf{e}_2, \mathbf{e}_3$ tangent to the coordinate curves and $\mathbf{E}_1, \mathbf{E}_2, \mathbf{E}_3$ normal to the coordinate surfaces (see Fig.2). The sets become identical if and only if the curvilinear coordinate system is orthogonal (see Problem 19). Both sets are analogous to the $\mathbf{i}, \mathbf{j}, \mathbf{k}$ unit vectors in rectangular coordinates but are unlike them in that they may change directions from point to point. It can be shown (see Problem 15) that the sets $\frac{\partial \mathbf{r}}{\partial u_1}, \frac{\partial \mathbf{r}}{\partial u_2}, \frac{\partial \mathbf{r}}{\partial u_3}$ and $\nabla u_1, \nabla u_2, \nabla u_3$ constitute reciprocal systems of vectors.

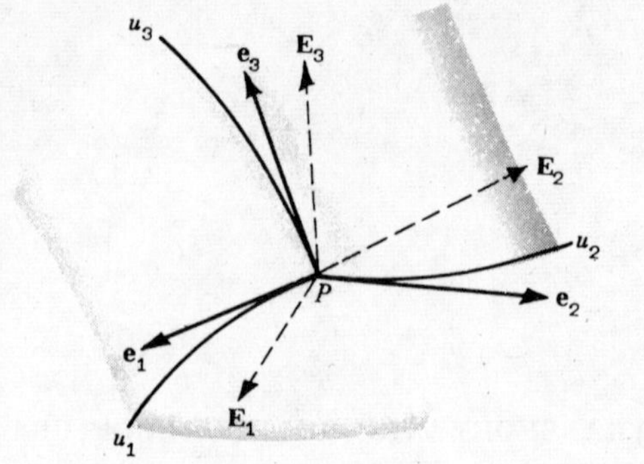

Fig. 2

A vector $\mathbf{A}$ can be represented in terms of the unit base vectors $\mathbf{e}_1, \mathbf{e}_2, \mathbf{e}_3$ or $\mathbf{E}_1, \mathbf{E}_2, \mathbf{E}_3$ in the form

$$\mathbf{A} = A_1 \mathbf{e}_1 + A_2 \mathbf{e}_2 + A_3 \mathbf{e}_3 = a_1 \mathbf{E}_1 + a_2 \mathbf{E}_2 + a_3 \mathbf{E}_3$$

where $A_1, A_2, A_3$ and $a_1, a_2, a_3$ are the respective *components* of $\mathbf{A}$ in each system.

We can also represent $\mathbf{A}$ in terms of the base vectors $\frac{\partial \mathbf{r}}{\partial u_1}, \frac{\partial \mathbf{r}}{\partial u_2}, \frac{\partial \mathbf{r}}{\partial u_3}$ or $\nabla u_1, \nabla u_2, \nabla u_3$ which are called *unitary base vectors* but *are not* unit vectors in general. In this case

$$\mathbf{A} = C_1 \frac{\partial \mathbf{r}}{\partial u_1} + C_2 \frac{\partial \mathbf{r}}{\partial u_2} + C_3 \frac{\partial \mathbf{r}}{\partial u_3} = C_1 \boldsymbol{\alpha}_1 + C_2 \boldsymbol{\alpha}_2 + C_3 \boldsymbol{\alpha}_3$$

and

$$\mathbf{A} = c_1 \nabla u_1 + c_2 \nabla u_2 + c_3 \nabla u_3 = c_1 \boldsymbol{\beta}_1 + c_2 \boldsymbol{\beta}_2 + c_3 \boldsymbol{\beta}_3$$

where $C_1, C_2, C_3$ are called the *contravariant components* of $\mathbf{A}$ and $c_1, c_2, c_3$ are called the *covariant components* of $\mathbf{A}$ (see Problems 33 and 34). Note that $\boldsymbol{\alpha}_p = \frac{\partial \mathbf{r}}{\partial u_p}$, $\boldsymbol{\beta}_p = \nabla u_p$, $p = 1, 2, 3$.

**ARC LENGTH AND VOLUME ELEMENTS.** From $\mathbf{r} = \mathbf{r}(u_1, u_2, u_3)$ we have

$$d\mathbf{r} = \frac{\partial \mathbf{r}}{\partial u_1} du_1 + \frac{\partial \mathbf{r}}{\partial u_2} du_2 + \frac{\partial \mathbf{r}}{\partial u_3} du_3 = h_1 \, du_1 \, \mathbf{e}_1 + h_2 \, du_2 \, \mathbf{e}_2 + h_3 \, du_3 \, \mathbf{e}_3$$

Then the differential of arc length $ds$ is determined from $ds^2 = d\mathbf{r} \cdot d\mathbf{r}$. For orthogonal systems, $\mathbf{e}_1 \cdot \mathbf{e}_2 = \mathbf{e}_2 \cdot \mathbf{e}_3 = \mathbf{e}_3 \cdot \mathbf{e}_1 = 0$ and

$$ds^2 = h_1^2 \, du_1^2 + h_2^2 \, du_2^2 + h_3^2 \, du_3^2$$

For non-orthogonal or general curvilinear systems see Problem 17.

Along a $u_1$ curve, $u_2$ and $u_3$ are constants so that $d\mathbf{r} = h_1 \, du_1 \, \mathbf{e}_1$. Then the differential of arc length $ds_1$ along $u_1$ at $P$ is $h_1 \, du_1$. Similarly the differential arc lengths along $u_2$ and $u_3$ at $P$ are $ds_2 = h_2 du_2$, $ds_3 = h_3 du_3$.

Referring to Fig.3 the volume element for an orthogonal curvilinear coordinate system is given by

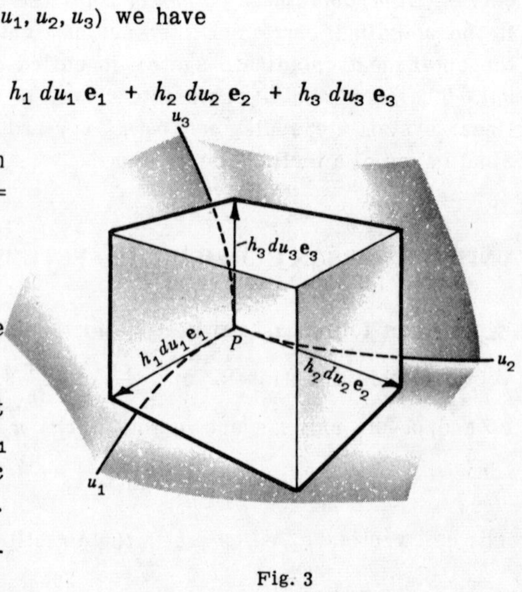

Fig. 3

$$dV = |(h_1 du_1 \mathbf{e}_1)\cdot(h_2 du_2 \mathbf{e}_2) \times (h_3 du_3 \mathbf{e}_3)| = h_1 h_2 h_3 \, du_1 du_2 du_3$$

since $|\mathbf{e}_1 \cdot \mathbf{e}_2 \times \mathbf{e}_3| = 1$.

**THE GRADIENT, DIVERGENCE AND CURL** can be expressed in terms of curvilinear coordinates. If $\Phi$ is a scalar function and $\mathbf{A} = A_1\mathbf{e}_1 + A_2\mathbf{e}_2 + A_3\mathbf{e}_3$ a vector function of orthogonal curvilinear coordinates $u_1, u_2, u_3$, then the following results are valid.

*1.* $$\nabla\Phi = \text{grad } \Phi = \frac{1}{h_1}\frac{\partial\Phi}{\partial u_1}\mathbf{e}_1 + \frac{1}{h_2}\frac{\partial\Phi}{\partial u_2}\mathbf{e}_2 + \frac{1}{h_3}\frac{\partial\Phi}{\partial u_3}\mathbf{e}_3$$

*2.* $$\nabla\cdot\mathbf{A} = \text{div } \mathbf{A} = \frac{1}{h_1 h_2 h_3}\left[\frac{\partial}{\partial u_1}(h_2 h_3 A_1) + \frac{\partial}{\partial u_2}(h_3 h_1 A_2) + \frac{\partial}{\partial u_3}(h_1 h_2 A_3)\right]$$

*3.* $$\nabla\times\mathbf{A} = \text{curl } \mathbf{A} = \frac{1}{h_1 h_2 h_3}\begin{vmatrix} h_1\mathbf{e}_1 & h_2\mathbf{e}_2 & h_3\mathbf{e}_3 \\ \dfrac{\partial}{\partial u_1} & \dfrac{\partial}{\partial u_2} & \dfrac{\partial}{\partial u_3} \\ h_1 A_1 & h_2 A_2 & h_3 A_3 \end{vmatrix}$$

*4.* $$\nabla^2\Phi = \text{Laplacian of } \Phi = \frac{1}{h_1 h_2 h_3}\left[\frac{\partial}{\partial u_1}\left(\frac{h_2 h_3}{h_1}\frac{\partial\Phi}{\partial u_1}\right) + \frac{\partial}{\partial u_2}\left(\frac{h_3 h_1}{h_2}\frac{\partial\Phi}{\partial u_2}\right) + \frac{\partial}{\partial u_3}\left(\frac{h_1 h_2}{h_3}\frac{\partial\Phi}{\partial u_3}\right)\right]$$

If $h_1 = h_2 = h_3 = 1$ and $\mathbf{e}_1, \mathbf{e}_2, \mathbf{e}_3$ are replaced by $\mathbf{i}, \mathbf{j}, \mathbf{k}$, these reduce to the usual expressions in rectangular coordinates where $(u_1, u_2, u_3)$ is replaced by $(x, y, z)$.

Extensions of the above results are achieved by a more general theory of curvilinear systems using the methods of *tensor analysis* which is considered in Chapter 8.

**SPECIAL ORTHOGONAL COORDINATE SYSTEMS.**

**1. Cylindrical Coordinates** $(\rho, \phi, z)$. See Fig.4 below.

$$x = \rho\cos\phi, \quad y = \rho\sin\phi, \quad z = z$$

$$\text{where } \rho \geqq 0, \quad 0 \leqq \phi < 2\pi, \quad -\infty < z < \infty$$

$$h_\rho = 1, \quad h_\phi = \rho, \quad h_z = 1$$

**2. Spherical Coordinates** $(r, \theta, \phi)$. See Fig.5 below.

$$x = r\sin\theta\cos\phi, \quad y = r\sin\theta\sin\phi, \quad z = r\cos\theta$$

$$\text{where } r \geqq 0, \quad 0 \leqq \phi < 2\pi, \quad 0 \leqq \theta \leqq \pi$$

$$h_r = 1, \quad h_\theta = r, \quad h_\phi = r\sin\theta$$

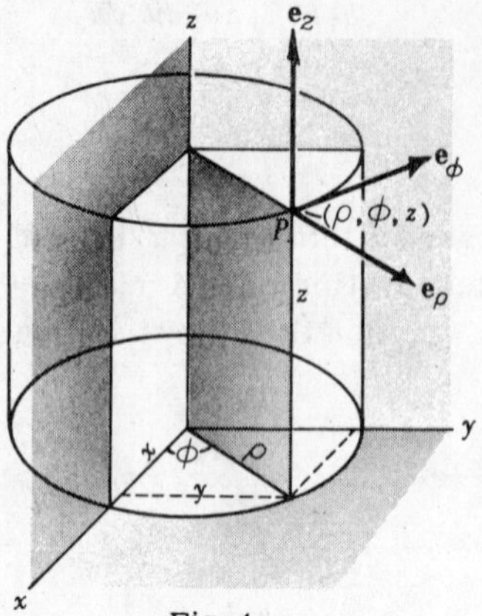


Fig. 4

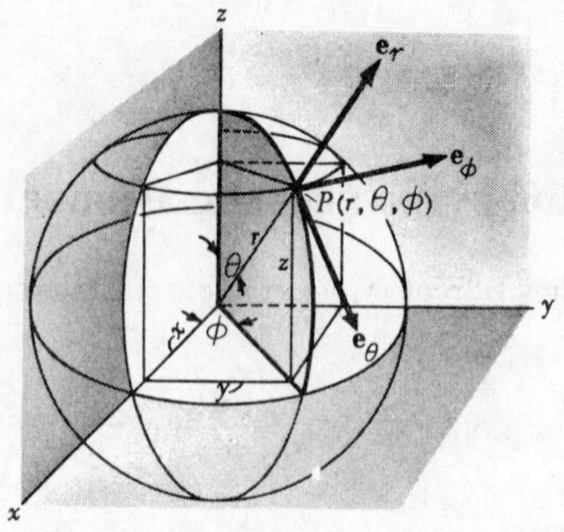


Fig. 5

3. **Parabolic Cylindrical Coordinates** $(u, v, z)$. See Fig.6 below.

$$x = \tfrac{1}{2}(u^2 - v^2), \qquad y = uv, \qquad z = z$$

$$\text{where} \quad -\infty < u < \infty, \quad v \geqq 0, \quad -\infty < z < \infty$$

$$h_u = h_v = \sqrt{u^2 + v^2}, \quad h_z = 1$$

In cylindrical coordinates, $u = \sqrt{2\rho}\,\cos\dfrac{\phi}{2}, \quad v = \sqrt{2\rho}\,\sin\dfrac{\phi}{2}, \quad z = z$

The traces of the coordinate surfaces on the $xy$ plane are shown in Fig.6 below. They are confocal parabolas with a common axis.

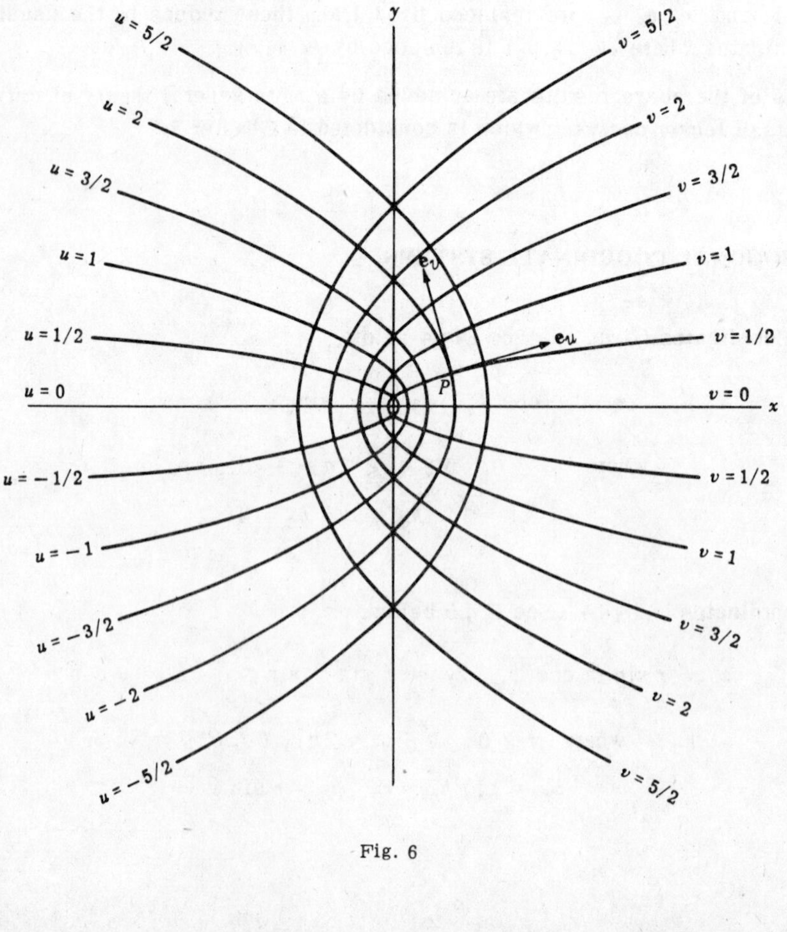


Fig. 6

**4. Paraboloidal Coordinates** $(u, v, \phi)$.

$$x = uv\cos\phi, \quad y = uv\sin\phi, \quad z = \tfrac{1}{2}(u^2 - v^2)$$

$$\text{where} \quad u \geqq 0, \quad v \geqq 0, \quad 0 \leqq \phi < 2\pi$$

$$h_u = h_v = \sqrt{u^2 + v^2}, \quad h_\phi = uv$$

Two sets of coordinate surfaces are obtained by revolving the parabolas of Fig.6 above about the $x$ axis which is relabelled the $z$ axis. The third set of coordinate surfaces are planes passing through this axis.

**5. Elliptic Cylindrical Coordinates** $(u, v, z)$. See Fig.7 below.

$$x = a\cosh u\cos v, \quad y = a\sinh u\sin v, \quad z = z$$

$$\text{where} \quad u \geqq 0, \quad 0 \leqq v < 2\pi, \quad -\infty < z < \infty$$

$$h_u = h_v = a\sqrt{\sinh^2 u + \sin^2 v}, \quad h_z = 1$$

The traces of the coordinate surfaces on the $xy$ plane are shown in Fig.7 below. They are confocal ellipses and hyperbolas.

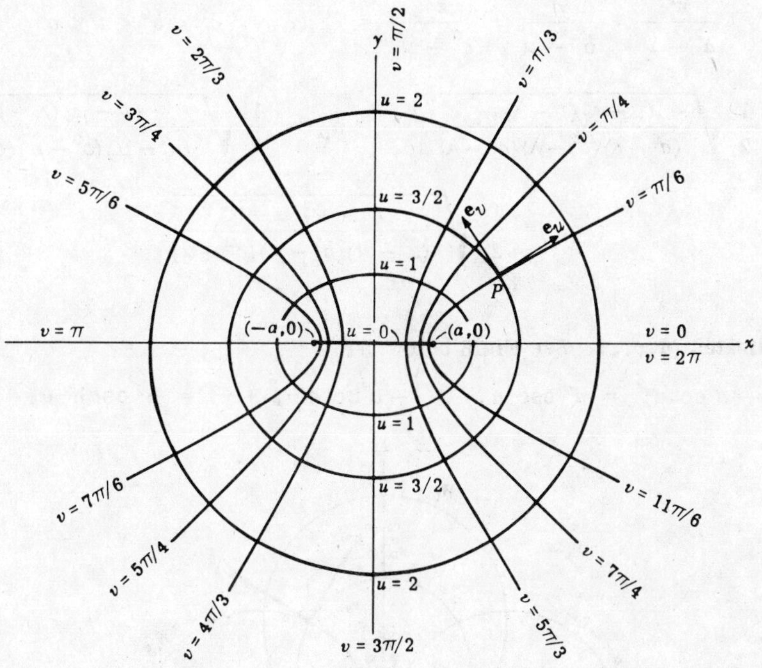


Fig. 7

**6. Prolate Spheroidal Coordinates** $(\xi, \eta, \phi)$.

$$x = a\sinh\xi\,\sin\eta\,\cos\phi, \quad y = a\sinh\xi\,\sin\eta\,\sin\phi, \quad z = a\cosh\xi\,\cos\eta$$

$$\text{where} \quad \xi \geqq 0, \quad 0 \leqq \eta \leqq \pi, \quad 0 \leqq \phi < 2\pi$$

$$h_\xi = h_\eta = a\sqrt{\sinh^2\xi + \sin^2\eta}, \quad h_\phi = a\sinh\xi\,\sin\eta$$

Two sets of coordinate surfaces are obtained by revolving the curves of Fig.7 above about the $x$ axis which is relabelled the $z$ axis. The third set of coordinate surfaces are planes passing through this axis.

**7. Oblate Spheroidal Coordinates** $(\xi, \eta, \phi)$.

$$x = a\cosh\xi\,\cos\eta\,\cos\phi, \qquad y = a\cosh\xi\,\cos\eta\,\sin\phi, \qquad z = a\sinh\xi\,\sin\eta$$

$$\text{where} \quad \xi \geqq 0, \quad -\frac{\pi}{2} \leqq \eta \leqq \frac{\pi}{2}, \quad 0 \leqq \phi < 2\pi$$

$$h_\xi = h_\eta = a\sqrt{\sinh^2\xi + \sin^2\eta}, \quad h_\phi = a\cosh\xi\,\cos\eta$$

Two sets of coordinate surfaces are obtained by revolving the curves of Fig.7 above about the $y$ axis which is relabelled the $z$ axis. The third set of coordinate surfaces are planes passing through this axis.

**8. Ellipsoidal Coordinates** $(\lambda, \mu, \nu)$.

$$\frac{x^2}{a^2 - \lambda} + \frac{y^2}{b^2 - \lambda} + \frac{z^2}{c^2 - \lambda} = 1, \qquad \lambda < c^2 < b^2 < a^2$$

$$\frac{x^2}{a^2 - \mu} + \frac{y^2}{b^2 - \mu} + \frac{z^2}{c^2 - \mu} = 1, \qquad c^2 < \mu < b^2 < a^2$$

$$\frac{x^2}{a^2 - \nu} + \frac{y^2}{b^2 - \nu} + \frac{z^2}{c^2 - \nu} = 1, \qquad c^2 < b^2 < \nu < a^2$$

$$h_\lambda = \frac{1}{2}\sqrt{\frac{(\mu-\lambda)(\nu-\lambda)}{(a^2-\lambda)(b^2-\lambda)(c^2-\lambda)}}, \qquad h_\mu = \frac{1}{2}\sqrt{\frac{(\nu-\mu)(\lambda-\mu)}{(a^2-\mu)(b^2-\mu)(c^2-\mu)}}$$

$$h_\nu = \frac{1}{2}\sqrt{\frac{(\lambda-\nu)(\mu-\nu)}{(a^2-\nu)(b^2-\nu)(c^2-\nu)}}$$

**9. Bipolar Coordinates** $(u, v, z)$. See Fig.8 below.

$$x^2 + (y - a\cot u)^2 = a^2\csc^2 u, \qquad (x - a\coth v)^2 + y^2 = a^2\operatorname{csch}^2 v, \qquad z = z$$

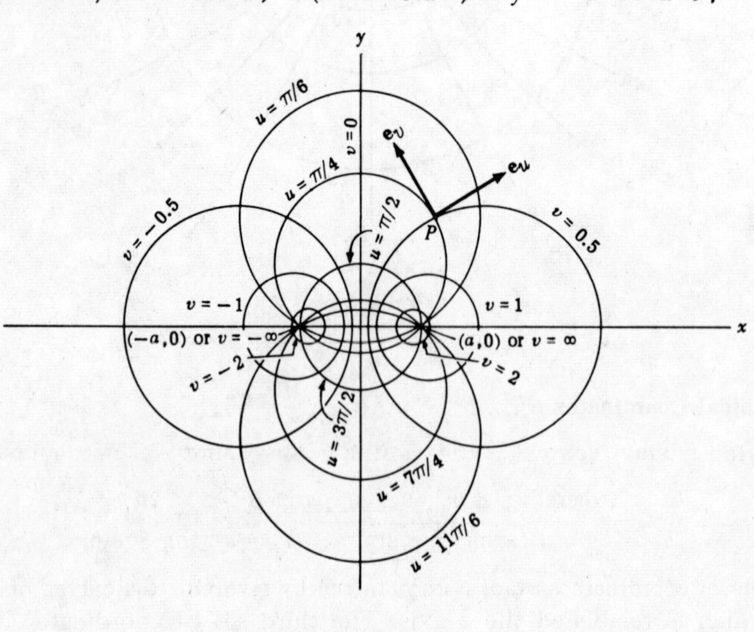


Fig. 8

or $$x = \frac{a \sinh v}{\cosh v - \cos u}, \quad y = \frac{a \sin u}{\cosh v - \cos u}, \quad z = z$$

$$\text{where} \quad 0 \leq u < 2\pi, \quad -\infty < v < \infty, \quad -\infty < z < \infty$$

$$h_u = h_v = \frac{a}{\cosh v - \cos u}, \quad h_z = 1$$

The traces of the coordinate surfaces on the $xy$ plane are shown in Fig.8 above. By revolving the curves of Fig.8 about the $y$ axis and relabelling this the $z$ axis a *toroidal coordinate system* is obtained.

## SOLVED PROBLEMS

1. Describe the coordinate surfaces and coordinate curves for (*a*) cylindrical and (*b*) spherical coordinates.

(*a*) The coordinate surfaces (or level surfaces) are:

$\rho = c_1$ cylinders coaxial with the $z$ axis (or $z$ axis if $c_1 = 0$).
$\phi = c_2$ planes through the $z$ axis.
$z = c_3$ planes perpendicular to the $z$ axis.

The coordinate curves are:

Intersection of $\rho = c_1$ and $\phi = c_2$ ($z$ curve) is a straight line.
Intersection of $\rho = c_1$ and $z = c_3$ ($\phi$ curve) is a circle (or point).
Intersection of $\phi = c_2$ and $z = c_3$ ($\rho$ curve) is a straight line.

(*b*) The coordinate surfaces are:

$r = c_1$ spheres having centre at the origin (or origin if $c_1 = 0$).
$\theta = c_2$ cones having vertex at the origin (lines if $c_2 = 0$ or $\pi$, and the $xy$ plane if $c_2 = \pi/2$).
$\phi = c_3$ planes through the $z$ axis.

The coordinate curves are:

Intersection of $r = c_1$ and $\theta = c_2$ ($\phi$ curve) is a circle (or point).
Intersection of $r = c_1$ and $\phi = c_3$ ($\theta$ curve) is a semi-circle ($c_1 \neq 0$).
Intersection of $\theta = c_2$ and $\phi = c_3$ ($r$ curve) is a line.

2. Determine the transformation from cylindrical to rectangular coordinates.

The equations defining the transformation from rectangular to cylindrical coordinates are

$$(1)\ x = \rho\cos\phi, \qquad (2)\ y = \rho\sin\phi, \qquad (3)\ z = z$$

Squaring (*1*) and (*2*) and adding, $\rho^2(\cos^2\phi + \sin^2\phi) = x^2 + y^2$ or
$\rho = \sqrt{x^2+y^2}$, since $\cos^2\phi + \sin^2\phi = 1$ and $\rho$ is positive.

Dividing equation (*2*) by (*1*), $\dfrac{y}{x} = \dfrac{\rho\sin\phi}{\rho\cos\phi} = \tan\phi$ or $\phi = \arctan\dfrac{y}{x}$.

Then the required transformation is (*4*) $\rho = \sqrt{x^2+y^2}$, (*5*) $\phi = \arctan\dfrac{y}{x}$, (*6*) $z = z$.

For points on the $z$ axis ($x = 0, y = 0$), note that $\phi$ is indeterminate. Such points are called *singular points* of the transformation.

3. Prove that a cylindrical coordinate system is orthogonal.

The position vector of any point in cylindrical coordinates is

$$\mathbf{r} = x\mathbf{i} + y\mathbf{j} + z\mathbf{k} = \rho\cos\phi\,\mathbf{i} + \rho\sin\phi\,\mathbf{j} + z\mathbf{k}$$

The tangent vectors to the $\rho$, $\phi$ and $z$ curves are given respectively by $\frac{\partial\mathbf{r}}{\partial\rho}$, $\frac{\partial\mathbf{r}}{\partial\phi}$ and $\frac{\partial\mathbf{r}}{\partial z}$ where

$$\frac{\partial\mathbf{r}}{\partial\rho} = \cos\phi\,\mathbf{i} + \sin\phi\,\mathbf{j}, \qquad \frac{\partial\mathbf{r}}{\partial\phi} = -\rho\sin\phi\,\mathbf{i} + \rho\cos\phi\,\mathbf{j}, \qquad \frac{\partial\mathbf{r}}{\partial z} = \mathbf{k}$$

The unit vectors in these directions are

$$\mathbf{e}_1 = \mathbf{e}_\rho = \frac{\partial\mathbf{r}/\partial\rho}{|\partial\mathbf{r}/\partial\rho|} = \frac{\cos\phi\,\mathbf{i} + \sin\phi\,\mathbf{j}}{\sqrt{\cos^2\phi + \sin^2\phi}} = \cos\phi\,\mathbf{i} + \sin\phi\,\mathbf{j}$$

$$\mathbf{e}_2 = \mathbf{e}_\phi = \frac{\partial\mathbf{r}/\partial\phi}{|\partial\mathbf{r}/\partial\phi|} = \frac{-\rho\sin\phi\,\mathbf{i} + \rho\cos\phi\,\mathbf{j}}{\sqrt{\rho^2\sin^2\phi + \rho^2\cos^2\phi}} = -\sin\phi\,\mathbf{i} + \cos\phi\,\mathbf{j}$$

$$\mathbf{e}_3 = \mathbf{e}_z = \frac{\partial\mathbf{r}/\partial z}{|\partial\mathbf{r}/\partial z|} = \mathbf{k}$$

Then

$$\mathbf{e}_1\cdot\mathbf{e}_2 = (\cos\phi\,\mathbf{i} + \sin\phi\,\mathbf{j})\cdot(-\sin\phi\,\mathbf{i} + \cos\phi\,\mathbf{j}) = 0$$

$$\mathbf{e}_1\cdot\mathbf{e}_3 = (\cos\phi\,\mathbf{i} + \sin\phi\,\mathbf{j})\cdot(\mathbf{k}) = 0$$

$$\mathbf{e}_2\cdot\mathbf{e}_3 = (-\sin\phi\,\mathbf{i} + \cos\phi\,\mathbf{j})\cdot(\mathbf{k}) = 0$$

and so $\mathbf{e}_1$, $\mathbf{e}_2$ and $\mathbf{e}_3$ are mutually perpendicular and the coordinate system is orthogonal.

4. Represent the vector $\mathbf{A} = z\mathbf{i} - 2x\mathbf{j} + y\mathbf{k}$ in cylindrical coordinates. Thus determine $A_\rho$, $A_\phi$ and $A_z$.

From Problem 3,

$$(1)\ \mathbf{e}_\rho = \cos\phi\,\mathbf{i} + \sin\phi\,\mathbf{j} \qquad (2)\ \mathbf{e}_\phi = -\sin\phi\,\mathbf{i} + \cos\phi\,\mathbf{j} \qquad (3)\ \mathbf{e}_z = \mathbf{k}$$

Solving (1) and (2) simultaneously,

$$\mathbf{i} = \cos\phi\,\mathbf{e}_\rho - \sin\phi\,\mathbf{e}_\phi, \qquad \mathbf{j} = \sin\phi\,\mathbf{e}_\rho + \cos\phi\,\mathbf{e}_\phi$$

Then

$$\begin{aligned}\mathbf{A} &= z\mathbf{i} - 2x\mathbf{j} + y\mathbf{k} \\ &= z(\cos\phi\,\mathbf{e}_\rho - \sin\phi\,\mathbf{e}_\phi) - 2\rho\cos\phi(\sin\phi\,\mathbf{e}_\rho + \cos\phi\,\mathbf{e}_\phi) + \rho\sin\phi\,\mathbf{e}_z \\ &= (z\cos\phi - 2\rho\cos\phi\sin\phi)\mathbf{e}_\rho - (z\sin\phi + 2\rho\cos^2\phi)\mathbf{e}_\phi + \rho\sin\phi\,\mathbf{e}_z\end{aligned}$$

and $A_\rho = z\cos\phi - 2\rho\cos\phi\sin\phi$, $A_\phi = -z\sin\phi - 2\rho\cos^2\phi$, $A_z = \rho\sin\phi$.

5. Prove $\frac{d}{dt}\mathbf{e}_\rho = \dot\phi\,\mathbf{e}_\phi$, $\frac{d}{dt}\mathbf{e}_\phi = -\dot\phi\,\mathbf{e}_\rho$ where dots denote differentiation with respect to time $t$.

From Problem 3,

$$\mathbf{e}_\rho = \cos\phi\,\mathbf{i} + \sin\phi\,\mathbf{j}, \qquad \mathbf{e}_\phi = -\sin\phi\,\mathbf{i} + \cos\phi\,\mathbf{j}$$

Then

$$\frac{d}{dt}\mathbf{e}_\rho = -(\sin\phi)\dot\phi\,\mathbf{i} + (\cos\phi)\dot\phi\,\mathbf{j} = (-\sin\phi\,\mathbf{i} + \cos\phi\,\mathbf{j})\dot\phi = \dot\phi\,\mathbf{e}_\phi$$

$$\frac{d}{dt}\mathbf{e}_\phi = -(\cos\phi)\dot\phi\,\mathbf{i} - (\sin\phi)\dot\phi\,\mathbf{j} = -(\cos\phi\,\mathbf{i} + \sin\phi\,\mathbf{j})\dot\phi = -\dot\phi\,\mathbf{e}_\rho$$

6. Express the velocity $\mathbf{v}$ and acceleration $\mathbf{a}$ of a particle in cylindrical coordinates.

In rectangular coordinates the position vector is $\mathbf{r} = x\mathbf{i} + y\mathbf{j} + z\mathbf{k}$ and the velocity and acceleration vectors are

$$\mathbf{v} = \frac{d\mathbf{r}}{dt} = \dot{x}\mathbf{i} + \dot{y}\mathbf{j} + \dot{z}\mathbf{k} \quad \text{and} \quad \mathbf{a} = \frac{d^2\mathbf{r}}{dt^2} = \ddot{x}\mathbf{i} + \ddot{y}\mathbf{j} + \ddot{z}\mathbf{k}$$

In cylindrical coordinates, using Problem 4,

$$\begin{aligned} \mathbf{r} = x\mathbf{i} + y\mathbf{j} + z\mathbf{k} &= (\rho\cos\phi)(\cos\phi\,\mathbf{e}_\rho - \sin\phi\,\mathbf{e}_\phi) \\ &\quad + (\rho\sin\phi)(\sin\phi\,\mathbf{e}_\rho + \cos\phi\,\mathbf{e}_\phi) + z\,\mathbf{e}_z \\ &= \rho\,\mathbf{e}_\rho + z\,\mathbf{e}_z \end{aligned}$$

Then $$\mathbf{v} = \frac{d\mathbf{r}}{dt} = \frac{d\rho}{dt}\mathbf{e}_\rho + \rho\frac{d\mathbf{e}_\rho}{dt} + \frac{dz}{dt}\mathbf{e}_z = \dot{\rho}\,\mathbf{e}_\rho + \rho\dot{\phi}\,\mathbf{e}_\phi + \dot{z}\,\mathbf{e}_z$$

using Problem 5. Differentiating again,

$$\begin{aligned} \mathbf{a} = \frac{d^2\mathbf{r}}{dt^2} &= \frac{d}{dt}(\dot{\rho}\,\mathbf{e}_\rho + \rho\dot{\phi}\,\mathbf{e}_\phi + \dot{z}\,\mathbf{e}_z) \\ &= \dot{\rho}\frac{d\mathbf{e}_\rho}{dt} + \ddot{\rho}\,\mathbf{e}_\rho + \rho\dot{\phi}\frac{d\mathbf{e}_\phi}{dt} + \rho\ddot{\phi}\,\mathbf{e}_\phi + \dot{\rho}\dot{\phi}\,\mathbf{e}_\phi + \ddot{z}\,\mathbf{e}_z \\ &= \dot{\rho}\dot{\phi}\,\mathbf{e}_\phi + \ddot{\rho}\,\mathbf{e}_\rho + \rho\dot{\phi}(-\dot{\phi}\,\mathbf{e}_\rho) + \rho\ddot{\phi}\,\mathbf{e}_\phi + \dot{\rho}\dot{\phi}\,\mathbf{e}_\phi + \ddot{z}\,\mathbf{e}_z \\ &= (\ddot{\rho} - \rho\dot{\phi}^2)\,\mathbf{e}_\rho + (\rho\ddot{\phi} + 2\dot{\rho}\dot{\phi})\,\mathbf{e}_\phi + \ddot{z}\,\mathbf{e}_z \end{aligned}$$

using Problem 5.

7. Find the square of the element of arc length in cylindrical coordinates and determine the corresponding scale factors.

*First Method.*

$$x = \rho\cos\phi, \quad y = \rho\sin\phi, \quad z = z$$

$$dx = -\rho\sin\phi\,d\phi + \cos\phi\,d\rho, \quad dy = \rho\cos\phi\,d\phi + \sin\phi\,d\rho, \quad dz = dz$$

Then $$\begin{aligned} ds^2 = dx^2 + dy^2 + dz^2 &= (-\rho\sin\phi\,d\phi + \cos\phi\,d\rho)^2 + (\rho\cos\phi\,d\phi + \sin\phi\,d\rho)^2 + (dz)^2 \\ &= (d\rho)^2 + \rho^2(d\phi)^2 + (dz)^2 = h_1^2(d\rho)^2 + h_2^2(d\phi)^2 + h_3^2(dz)^2 \end{aligned}$$

and $h_1 = h_\rho = 1$, $h_2 = h_\phi = \rho$, $h_3 = h_z = 1$ are the scale factors.

*Second Method.* The position vector is $\mathbf{r} = \rho\cos\phi\,\mathbf{i} + \rho\sin\phi\,\mathbf{j} + z\mathbf{k}$. Then

$$\begin{aligned} d\mathbf{r} &= \frac{\partial\mathbf{r}}{\partial\rho}d\rho + \frac{\partial\mathbf{r}}{\partial\phi}d\phi + \frac{\partial\mathbf{r}}{\partial z}dz \\ &= (\cos\phi\,\mathbf{i} + \sin\phi\,\mathbf{j})\,d\rho + (-\rho\sin\phi\,\mathbf{i} + \rho\cos\phi\,\mathbf{j})\,d\phi + \mathbf{k}\,dz \\ &= (\cos\phi\,d\rho - \rho\sin\phi\,d\phi)\,\mathbf{i} + (\sin\phi\,d\rho + \rho\cos\phi\,d\phi)\,\mathbf{j} + \mathbf{k}\,dz \end{aligned}$$

Thus $$\begin{aligned} ds^2 = d\mathbf{r}\cdot d\mathbf{r} &= (\cos\phi\,d\rho - \rho\sin\phi\,d\phi)^2 + (\sin\phi\,d\rho + \rho\cos\phi\,d\phi)^2 + (dz)^2 \\ &= (d\rho)^2 + \rho^2(d\phi)^2 + (dz)^2 \end{aligned}$$

**8.** Work Problem 7 for (*a*) spherical and (*b*) parabolic cylindrical coordinates.

(*a*) $$x = r\sin\theta\cos\phi, \quad y = r\sin\theta\sin\phi, \quad z = r\cos\theta$$

Then
$$dx = -r\sin\theta\sin\phi\,d\phi + r\cos\theta\cos\phi\,d\theta + \sin\theta\cos\phi\,dr$$
$$dy = r\sin\theta\cos\phi\,d\phi + r\cos\theta\sin\phi\,d\theta + \sin\theta\sin\phi\,dr$$
$$dz = -r\sin\theta\,d\theta + \cos\theta\,dr$$

and $$(ds)^2 = (dx)^2 + (dy)^2 + (dz)^2 = (dr)^2 + r^2(d\theta)^2 + r^2\sin^2\theta\,(d\phi)^2$$

The scale factors are $h_1 = h_r = 1$, $h_2 = h_\theta = r$, $h_3 = h_\phi = r\sin\theta$.

(*b*) $$x = \tfrac{1}{2}(u^2 - v^2), \quad y = uv, \quad z = z$$

Then $$dx = u\,du - v\,dv, \quad dy = u\,dv + v\,du, \quad dz = dz$$

and $$(ds)^2 = (dx)^2 + (dy)^2 + (dz)^2 = (u^2+v^2)(du)^2 + (u^2+v^2)(dv)^2 + (dz)^2$$

The scale factors are $h_1 = h_u = \sqrt{u^2+v^2}$, $h_2 = h_v = \sqrt{u^2+v^2}$, $h_3 = h_z = 1$.

**9.** Sketch a volume element in (*a*) cylindrical and (*b*) spherical coordinates giving the magnitudes of its edges.

(*a*) The edges of the volume element in cylindrical coordinates (Fig.(*a*) below) have magnitudes $\rho\,d\phi$, $d\rho$ and $dz$. This could also be seen from the fact that the edges are given by

$$ds_1 = h_1 du_1 = (1)(d\rho) = d\rho, \quad ds_2 = h_2 du_2 = \rho\,d\phi, \quad ds_3 = (1)(dz) = dz$$

using the scale factors obtained from Problem 7.

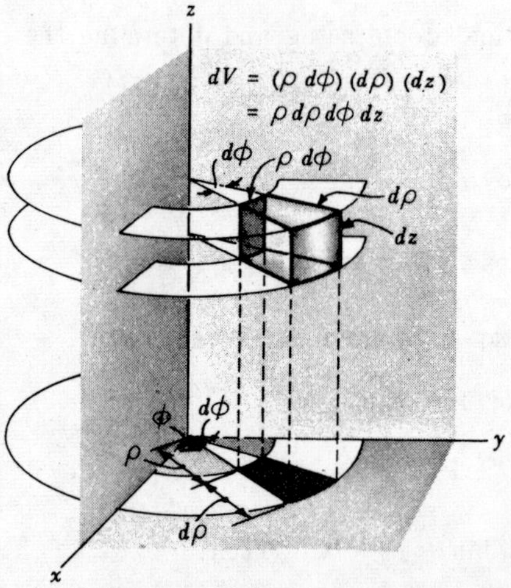


Fig. (*a*) **Volume element in cylindrical coordinates.**

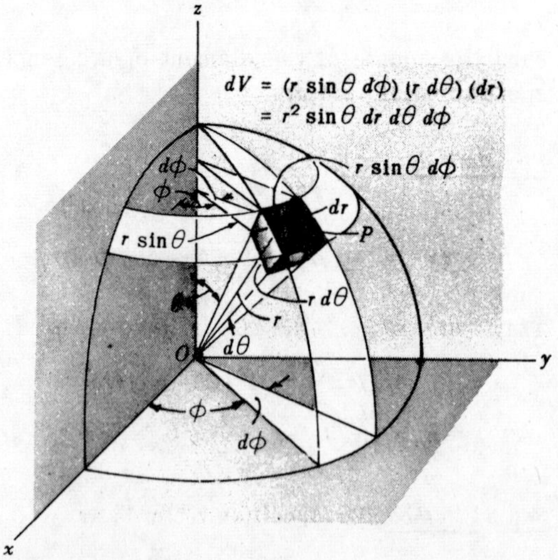


Fig. (*b*) **Volume element in spherical coordinates.**

(*b*) The edges of the volume element in spherical coordinates (Fig.(*b*) above) have magnitudes $dr$, $r\,d\theta$ and $r\sin\theta\,d\phi$. This could also be seen from the fact that the edges are given by

$$ds_1 = h_1 du_1 = (1)(dr) = dr, \quad ds_2 = h_2 du_2 = r\,d\theta, \quad ds_3 = h_3 du_3 = r\sin\theta\,d\phi$$

using the scale factors obtained from Problem 8(*a*).

**10.** Find the volume element $dV$ in (*a*) cylindrical, (*b*) spherical and (*c*) parabolic cylindrical coordinates.

The volume element in orthogonal curvilinear coordinates $u_1, u_2, u_3$ is

$$dV = h_1 h_2 h_3 \, du_1 du_2 du_3$$

(*a*) In cylindrical coordinates $u_1 = \rho$, $u_2 = \phi$, $u_3 = z$, $h_1 = 1$, $h_2 = \rho$, $h_3 = 1$ (see Problem 7). Then

$$dV = (1)(\rho)(1)\, d\rho\, d\phi\, dz = \rho\, d\rho\, d\phi\, dz$$

This can also be observed directly from Fig. (*a*) of Problem 9.

(*b*) In spherical coordinates $u_1 = r$, $u_2 = \theta$, $u_3 = \phi$, $h_1 = 1$, $h_2 = r$, $h_3 = r \sin\theta$ (see Problem 8(*a*)). Then

$$dV = (1)(r)(r\sin\theta)\, dr\, d\theta\, d\phi = r^2 \sin\theta\, dr\, d\theta\, d\phi$$

This can also be observed directly from Fig. (*b*) of Problem 9.

(*c*) In parabolic cylindrical coordinates $u_1 = u$, $u_2 = v$, $u_3 = z$, $h_1 = \sqrt{u^2+v^2}$, $h_2 = \sqrt{u^2+v^2}$, $h_3 = 1$ (see Problem 8(*b*)). Then

$$dV = (\sqrt{u^2+v^2})(\sqrt{u^2+v^2})(1)\, du\, dv\, dz = (u^2+v^2)\, du\, dv\, dz$$

**11.** Find (*a*) the scale factors and (*b*) the volume element $dV$ in oblate spheroidal coordinates.

(*a*) $x = a\cosh\xi\cos\eta\cos\phi, \quad y = a\cosh\xi\cos\eta\sin\phi, \quad z = a\sinh\xi\sin\eta$

$$dx = -a\cosh\xi\cos\eta\sin\phi\, d\phi - a\cosh\xi\sin\eta\cos\phi\, d\eta + a\sinh\xi\cos\eta\cos\phi\, d\xi$$

$$dy = a\cosh\xi\cos\eta\cos\phi\, d\phi - a\cosh\xi\sin\eta\sin\phi\, d\eta + a\sinh\xi\cos\eta\sin\phi\, d\xi$$

$$dz = a\sinh\xi\cos\eta\, d\eta + a\cosh\xi\sin\eta\, d\xi$$

Then

$$\begin{aligned}(ds)^2 = (dx)^2 + (dy)^2 + (dz)^2 &= a^2(\sinh^2\xi + \sin^2\eta)(d\xi)^2 \\ &\quad + a^2(\sinh^2\xi + \sin^2\eta)(d\eta)^2 \\ &\quad + a^2\cosh^2\xi\cos^2\eta\,(d\phi)^2\end{aligned}$$

and $h_1 = h_\xi = a\sqrt{\sinh^2\xi + \sin^2\eta}$, $h_2 = h_\eta = a\sqrt{\sinh^2\xi + \sin^2\eta}$, $h_3 = h_\phi = a\cosh\xi\cos\eta$.

(*b*)

$$\begin{aligned}dV &= (a\sqrt{\sinh^2\xi + \sin^2\eta})(a\sqrt{\sinh^2\xi + \sin^2\eta})(a\cosh\xi\cos\eta)\, d\xi\, d\eta\, d\phi \\ &= a^3(\sinh^2\xi + \sin^2\eta)\cosh\xi\cos\eta\, d\xi\, d\eta\, d\phi\end{aligned}$$

**12.** Find expressions for the elements of area in orthogonal curvilinear coordinates.

Referring to Figure 3, p.136, the area elements are given by

$$dA_1 = |(h_2\, du_2\, \mathbf{e}_2) \times (h_3\, du_3\, \mathbf{e}_3)| = h_2 h_3 |\mathbf{e}_2 \times \mathbf{e}_3|\, du_2\, du_3 = h_2 h_3\, du_2\, du_3$$

since $|\mathbf{e}_2 \times \mathbf{e}_3| = |\mathbf{e}_1| = 1$. Similarly

$$dA_2 = |(h_1\, du_1\, \mathbf{e}_1) \times (h_3\, du_3\, \mathbf{e}_3)| = h_1 h_3\, du_1\, du_3$$

$$dA_3 = |(h_1\, du_1\, \mathbf{e}_1) \times (h_2\, du_2\, \mathbf{e}_2)| = h_1 h_2\, du_1\, du_2$$

**13.** If $u_1, u_2, u_3$ are orthogonal curvilinear coordinates, show that the Jacobian of $x, y, z$ with respect to $u_1, u_2, u_3$ is

$$J\left(\frac{x, y, z}{u_1, u_2, u_3}\right) = \frac{\partial(x,y,z)}{\partial(u_1,u_2,u_3)} = \begin{vmatrix} \frac{\partial x}{\partial u_1} & \frac{\partial y}{\partial u_1} & \frac{\partial z}{\partial u_1} \\ \frac{\partial x}{\partial u_2} & \frac{\partial y}{\partial u_2} & \frac{\partial z}{\partial u_2} \\ \frac{\partial x}{\partial u_3} & \frac{\partial y}{\partial u_3} & \frac{\partial z}{\partial u_3} \end{vmatrix} = h_1 h_2 h_3$$

By Problem 38 of Chapter 2, the given determinant equals

$$\left(\frac{\partial x}{\partial u_1}\mathbf{i} + \frac{\partial y}{\partial u_1}\mathbf{j} + \frac{\partial z}{\partial u_1}\mathbf{k}\right)\cdot\left(\frac{\partial x}{\partial u_2}\mathbf{i} + \frac{\partial y}{\partial u_2}\mathbf{j} + \frac{\partial z}{\partial u_2}\mathbf{k}\right)\times\left(\frac{\partial x}{\partial u_3}\mathbf{i} + \frac{\partial y}{\partial u_3}\mathbf{j} + \frac{\partial z}{\partial u_3}\mathbf{k}\right)$$

$$= \frac{\partial \mathbf{r}}{\partial u_1}\cdot\frac{\partial \mathbf{r}}{\partial u_2}\times\frac{\partial \mathbf{r}}{\partial u_3} = h_1\mathbf{e}_1 \cdot h_2\mathbf{e}_2 \times h_3\mathbf{e}_3$$

$$= h_1 h_2 h_3\, \mathbf{e}_1 \cdot \mathbf{e}_2 \times \mathbf{e}_3 = h_1 h_2 h_3$$

If the Jacobian equals zero identically then $\frac{\partial \mathbf{r}}{\partial u_1}, \frac{\partial \mathbf{r}}{\partial u_2}, \frac{\partial \mathbf{r}}{\partial u_3}$ are coplanar vectors and the curvilinear coordinate transformation breaks down, i.e. there is a relation between $x, y, z$ having the form $F(x,y,z) = 0$. We shall therefore require the Jacobian to be different from zero.

**14.** Evaluate $\iiint_V (x^2 + y^2 + z^2)\, dx\, dy\, dz$ where $V$ is a sphere having centre at the origin and radius equal to $a$.

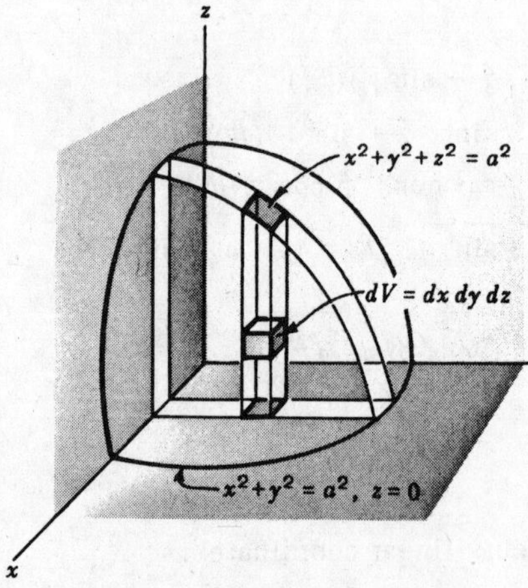


Fig. (a)

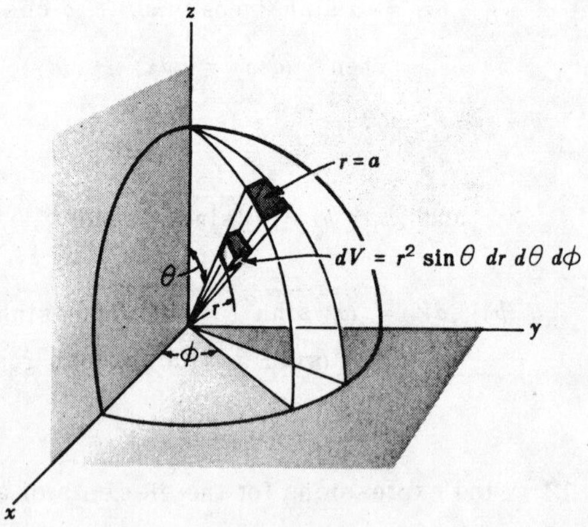


Fig. (b)

The required integral is equal to eight times the integral evaluated over that part of the sphere contained in the first octant (see Fig. (a) above).

Then in rectangular coordinates the integral equals

$$8\int_{x=0}^{a}\int_{y=0}^{\sqrt{a^2-x^2}}\int_{z=0}^{\sqrt{a^2-x^2-y^2}} (x^2 + y^2 + z^2)\, dz\, dy\, dx$$

but the evaluation, although possible, is tedious. It is easier to use spherical coordinates for the evaluation. In changing to spherical coordinates, the integrand $x^2+y^2+z^2$ is replaced by its equivalent $r^2$ while the volume element $dx\,dy\,dz$ is replaced by the volume element $r^2 \sin\theta\, dr\,d\theta\,d\phi$ (see Problem 10(*b*)). To cover the required region in the first octant, fix $\theta$ and $\phi$ (see Fig. (*b*) above) and integrate from $r=0$ to $r=a$; then keep $\phi$ constant and integrate from $\theta=0$ to $\pi/2$; finally integrate with respect to $\phi$ from $\phi=0$ to $\phi=\pi/2$. Here we have performed the integration in the order $r,\theta,\phi$ although any order can be used. The result is

$$8\int_{\phi=0}^{\pi/2}\int_{\theta=0}^{\pi/2}\int_{r=0}^{a}(r^2)(r^2\sin\theta\,dr\,d\theta\,d\phi) \;=\; 8\int_{\phi=0}^{\pi/2}\int_{\theta=0}^{\pi/2}\int_{r=0}^{a} r^4\sin\theta\,dr\,d\theta\,d\phi$$

$$= \;8\int_{\phi=0}^{\pi/2}\int_{\theta=0}^{\pi/2}\frac{r^5}{5}\sin\theta\,\Big|_{r=0}^{a}d\theta\,d\phi \;=\; \frac{8a^5}{5}\int_{\phi=0}^{\pi/2}\int_{\theta=0}^{\pi/2}\sin\theta\,d\theta\,d\phi$$

$$= \;\frac{8a^5}{5}\int_{\phi=0}^{\pi/2} -\cos\theta\,\Big|_{\theta=0}^{\pi/2}d\phi \;=\; \frac{8a^5}{5}\int_{\phi=0}^{\pi/2}d\phi \;=\; \frac{4\pi a^5}{5}$$

Physically the integral represents the moment of inertia of the sphere with respect to the origin, i.e. the polar moment of inertia, if the sphere has unit density.

In general, when transforming multiple integrals from rectangular to orthogonal curvilinear coordinates the volume element $dx\,dy\,dz$ is replaced by $h_1h_2h_3\,du_1du_2du_3$ or the equivalent $J\left(\frac{x,y,z}{u_1,u_2,u_3}\right)du_1du_2du_3$ where $J$ is the Jacobian of the transformation from $x,y,z$ to $u_1,u_2,u_3$ (see Problem 13).

**15.** If $u_1,u_2,u_3$ are general coordinates, show that $\frac{\partial\mathbf{r}}{\partial u_1}, \frac{\partial\mathbf{r}}{\partial u_2}, \frac{\partial\mathbf{r}}{\partial u_3}$ and $\nabla u_1, \nabla u_2, \nabla u_3$ are reciprocal systems of vectors.

We must show that $\frac{\partial\mathbf{r}}{\partial u_p}\cdot\nabla u_q = \begin{cases}1 & \text{if } p=q\\ 0 & \text{if } p\neq q\end{cases}$ where $p$ and $q$ can have any of the values 1,2,3. We have

$$d\mathbf{r} \;=\; \frac{\partial\mathbf{r}}{\partial u_1}du_1 + \frac{\partial\mathbf{r}}{\partial u_2}du_2 + \frac{\partial\mathbf{r}}{\partial u_3}du_3$$

Multiply by $\nabla u_1\,\cdot$ . Then

$$\nabla u_1\cdot d\mathbf{r} \;=\; du_1 \;=\; \left(\nabla u_1\cdot\frac{\partial\mathbf{r}}{\partial u_1}\right)du_1 + \left(\nabla u_1\cdot\frac{\partial\mathbf{r}}{\partial u_2}\right)du_2 + \left(\nabla u_1\cdot\frac{\partial\mathbf{r}}{\partial u_3}\right)du_3$$

or

$$\nabla u_1\cdot\frac{\partial\mathbf{r}}{\partial u_1} = 1,\qquad \nabla u_1\cdot\frac{\partial\mathbf{r}}{\partial u_2} = 0,\qquad \nabla u_1\cdot\frac{\partial\mathbf{r}}{\partial u_3} = 0$$

Similarly, upon multiplying by $\nabla u_2\,\cdot$ and $\nabla u_3\,\cdot$ the remaining relations are proved.

**16.** Prove $\left\{\frac{\partial\mathbf{r}}{\partial u_1}\cdot\frac{\partial\mathbf{r}}{\partial u_2}\times\frac{\partial\mathbf{r}}{\partial u_3}\right\}\left\{\nabla u_1\cdot\nabla u_2\times\nabla u_3\right\} = 1.$

From Problem 15, $\frac{\partial\mathbf{r}}{\partial u_1}, \frac{\partial\mathbf{r}}{\partial u_2}, \frac{\partial\mathbf{r}}{\partial u_3}$ and $\nabla u_1, \nabla u_2, \nabla u_3$ are reciprocal systems of vectors. Then the required result follows from Problem 53(*c*) of Chapter 2.

The result is equivalent to a theorem on Jacobians for

$$\nabla u_1 \cdot \nabla u_2 \times \nabla u_3 = \begin{vmatrix} \frac{\partial u_1}{\partial x} & \frac{\partial u_1}{\partial y} & \frac{\partial u_1}{\partial z} \\ \frac{\partial u_2}{\partial x} & \frac{\partial u_2}{\partial y} & \frac{\partial u_2}{\partial z} \\ \frac{\partial u_3}{\partial x} & \frac{\partial u_3}{\partial y} & \frac{\partial u_3}{\partial z} \end{vmatrix} = J\left(\frac{u_1, u_2, u_3}{x, y, z}\right)$$

and so $J\left(\frac{x,y,z}{u_1,u_2,u_3}\right) J\left(\frac{u_1,u_2,u_3}{x,y,z}\right) = 1$ using Problem 13.

**17.** Show that the square of the element of arc length in general curvilinear coordinates can be expressed by

$$ds^2 = \sum_{p=1}^{3} \sum_{q=1}^{3} g_{pq}\, du_p\, du_q$$

We have

$$d\mathbf{r} = \frac{\partial \mathbf{r}}{\partial u_1} du_1 + \frac{\partial \mathbf{r}}{\partial u_2} du_2 + \frac{\partial \mathbf{r}}{\partial u_3} du_3 = \boldsymbol{\alpha}_1 du_1 + \boldsymbol{\alpha}_2 du_2 + \boldsymbol{\alpha}_3 du_3$$

Then

$$\begin{aligned} ds^2 = d\mathbf{r}\cdot d\mathbf{r} &= \boldsymbol{\alpha}_1\cdot\boldsymbol{\alpha}_1\, du_1^2 + \boldsymbol{\alpha}_1\cdot\boldsymbol{\alpha}_2\, du_1\, du_2 + \boldsymbol{\alpha}_1\cdot\boldsymbol{\alpha}_3\, du_1\, du_3 \\ &\quad + \boldsymbol{\alpha}_2\cdot\boldsymbol{\alpha}_1\, du_2\, du_1 + \boldsymbol{\alpha}_2\cdot\boldsymbol{\alpha}_2\, du_2^2 + \boldsymbol{\alpha}_2\cdot\boldsymbol{\alpha}_3\, du_2\, du_3 \\ &\quad + \boldsymbol{\alpha}_3\cdot\boldsymbol{\alpha}_1\, du_3\, du_1 + \boldsymbol{\alpha}_3\cdot\boldsymbol{\alpha}_2\, du_3\, du_2 + \boldsymbol{\alpha}_3\cdot\boldsymbol{\alpha}_3\, du_3^2 \\ &= \sum_{p=1}^{3}\sum_{q=1}^{3} g_{pq}\, du_p\, du_q \quad \text{where } g_{pq} = \boldsymbol{\alpha}_p\cdot\boldsymbol{\alpha}_q \end{aligned}$$

This is called the *fundamental quadratic form* or *metric form*. The quantities $g_{pq}$ are called *metric coefficients* and are symmetric, i.e. $g_{pq} = g_{qp}$. If $g_{pq} = 0$, $p \neq q$, then the coordinate system is orthogonal. In this case $g_{11} = h_1^2$, $g_{22} = h_2^2$, $g_{33} = h_3^2$. The metric form extended to higher dimensional space is of fundamental importance in the theory of relativity (see Chapter 8).

## GRADIENT, DIVERGENCE AND CURL IN ORTHOGONAL COORDINATES.

**18.** Derive an expresssion for $\nabla\Phi$ in orthogonal curvilinear coordinates.

Let $\nabla\Phi = f_1\mathbf{e}_1 + f_2\mathbf{e}_2 + f_3\mathbf{e}_3$ where $f_1, f_2, f_3$ are to be determined.

Since

$$\begin{aligned} d\mathbf{r} &= \frac{\partial \mathbf{r}}{\partial u_1} du_1 + \frac{\partial \mathbf{r}}{\partial u_2} du_2 + \frac{\partial \mathbf{r}}{\partial u_3} du_3 \\ &= h_1\mathbf{e}_1\, du_1 + h_2\mathbf{e}_2\, du_2 + h_3\mathbf{e}_3\, du_3 \end{aligned}$$

we have

$$(1) \quad d\Phi = \nabla\Phi\cdot d\mathbf{r} = h_1 f_1\, du_1 + h_2 f_2\, du_2 + h_3 f_3\, du_3$$

But

$$(2) \quad d\Phi = \frac{\partial\Phi}{\partial u_1} du_1 + \frac{\partial\Phi}{\partial u_2} du_2 + \frac{\partial\Phi}{\partial u_3} du_3$$

Equating (1) and (2), $f_1 = \frac{1}{h_1}\frac{\partial\Phi}{\partial u_1}, \quad f_2 = \frac{1}{h_2}\frac{\partial\Phi}{\partial u_2}, \quad f_3 = \frac{1}{h_3}\frac{\partial\Phi}{\partial u_3}.$

Then
$$\nabla\Phi = \frac{\mathbf{e}_1}{h_1}\frac{\partial\Phi}{\partial u_1} + \frac{\mathbf{e}_2}{h_2}\frac{\partial\Phi}{\partial u_2} + \frac{\mathbf{e}_3}{h_3}\frac{\partial\Phi}{\partial u_3}$$

This indicates the operator equivalence
$$\nabla \equiv \frac{\mathbf{e}_1}{h_1}\frac{\partial}{\partial u_1} + \frac{\mathbf{e}_2}{h_2}\frac{\partial}{\partial u_2} + \frac{\mathbf{e}_3}{h_3}\frac{\partial}{\partial u_3}$$

which reduces to the usual expression for the operator $\nabla$ in rectangular coordinates.

**19.** Let $u_1, u_2, u_3$ be orthogonal coordinates. (*a*) Prove that $|\nabla u_p| = h_p^{-1}$, $p = 1,2,3$.
(*b*) Show that $\mathbf{e}_p = \mathbf{E}_p$.

(*a*) Let $\Phi = u_1$ in Problem 18. Then $\nabla u_1 = \frac{\mathbf{e}_1}{h_1}$ and so $|\nabla u_1| = |\mathbf{e}_1|/h_1 = h_1^{-1}$, since $|\mathbf{e}_1| = 1$. Similarly by letting $\Phi = u_2$ and $u_3$, $|\nabla u_2| = h_2^{-1}$ and $|\nabla u_3| = h_3^{-1}$.

(*b*) By definition $\mathbf{E}_p = \frac{\nabla u_p}{|\nabla u_p|}.$ From part (*a*), this can be written $\mathbf{E}_p = h_p \nabla u_p = \mathbf{e}_p$ and the result is proved.

**20.** Prove $\mathbf{e}_1 = h_2 h_3 \nabla u_2 \times \nabla u_3$ with similar equations for $\mathbf{e}_2$ and $\mathbf{e}_3$, where $u_1, u_2, u_3$ are orthogonal coordinates.

From Problem 19, $\nabla u_1 = \frac{\mathbf{e}_1}{h_1}, \quad \nabla u_2 = \frac{\mathbf{e}_2}{h_2}, \quad \nabla u_3 = \frac{\mathbf{e}_3}{h_3}.$

Then $\nabla u_2 \times \nabla u_3 = \frac{\mathbf{e}_2 \times \mathbf{e}_3}{h_2 h_3} = \frac{\mathbf{e}_1}{h_2 h_3}$ and $\mathbf{e}_1 = h_2 h_3 \nabla u_2 \times \nabla u_3$.

Similarly $\mathbf{e}_2 = h_3 h_1 \nabla u_3 \times \nabla u_1$ and $\mathbf{e}_3 = h_1 h_2 \nabla u_1 \times \nabla u_2$.

**21.** Show that in orthogonal coordinates

(*a*) $\nabla \cdot (A_1 \mathbf{e}_1) = \frac{1}{h_1 h_2 h_3}\frac{\partial}{\partial u_1}(A_1 h_2 h_3)$

(*b*) $\nabla \times (A_1 \mathbf{e}_1) = \frac{\mathbf{e}_2}{h_3 h_1}\frac{\partial}{\partial u_3}(A_1 h_1) - \frac{\mathbf{e}_3}{h_1 h_2}\frac{\partial}{\partial u_2}(A_1 h_1)$

with similar results for vectors $A_2 \mathbf{e}_2$ and $A_3 \mathbf{e}_3$.

(*a*) From Problem 20,

$$\begin{aligned}
\nabla \cdot (A_1 \mathbf{e}_1) &= \nabla \cdot (A_1 h_2 h_3 \nabla u_2 \times \nabla u_3) \\
&= \nabla(A_1 h_2 h_3) \cdot \nabla u_2 \times \nabla u_3 + A_1 h_2 h_3 \nabla \cdot (\nabla u_2 \times \nabla u_3) \\
&= \nabla(A_1 h_2 h_3) \cdot \frac{\mathbf{e}_2}{h_2} \times \frac{\mathbf{e}_3}{h_3} + 0 = \nabla(A_1 h_2 h_3) \cdot \frac{\mathbf{e}_1}{h_2 h_3} \\
&= \left[\frac{\mathbf{e}_1}{h_1}\frac{\partial}{\partial u_1}(A_1 h_2 h_3) + \frac{\mathbf{e}_2}{h_2}\frac{\partial}{\partial u_2}(A_1 h_2 h_3) + \frac{\mathbf{e}_3}{h_3}\frac{\partial}{\partial u_3}(A_1 h_2 h_3)\right] \cdot \frac{\mathbf{e}_1}{h_2 h_3} \\
&= \frac{1}{h_1 h_2 h_3}\frac{\partial}{\partial u_1}(A_1 h_2 h_3)
\end{aligned}$$

(*b*) $$\begin{aligned}\nabla\times(A_1\mathbf{e}_1) &= \nabla\times(A_1h_1\nabla u_1)\\ &= \nabla(A_1h_1)\times\nabla u_1 + A_1h_1\nabla\times\nabla u_1\\ &= \nabla(A_1h_1)\times\frac{\mathbf{e}_1}{h_1} + \mathbf{0}\\ &= \left[\frac{\mathbf{e}_1}{h_1}\frac{\partial}{\partial u_1}(A_1h_1) + \frac{\mathbf{e}_2}{h_2}\frac{\partial}{\partial u_2}(A_1h_1) + \frac{\mathbf{e}_3}{h_3}\frac{\partial}{\partial u_3}(A_1h_1)\right]\times\frac{\mathbf{e}_1}{h_1}\\ &= \frac{\mathbf{e}_2}{h_3h_1}\frac{\partial}{\partial u_3}(A_1h_1) - \frac{\mathbf{e}_3}{h_1h_2}\frac{\partial}{\partial u_2}(A_1h_1)\end{aligned}$$

**22.** Express div $\mathbf{A} = \nabla\cdot\mathbf{A}$ in orthogonal coordinates.

$$\begin{aligned}\nabla\cdot\mathbf{A} &= \nabla\cdot(A_1\mathbf{e}_1 + A_2\mathbf{e}_2 + A_3\mathbf{e}_3) = \nabla\cdot(A_1\mathbf{e}_1) + \nabla\cdot(A_2\mathbf{e}_2) + \nabla\cdot(A_3\mathbf{e}_3)\\ &= \frac{1}{h_1h_2h_3}\left[\frac{\partial}{\partial u_1}(A_1h_2h_3) + \frac{\partial}{\partial u_2}(A_2h_3h_1) + \frac{\partial}{\partial u_3}(A_3h_1h_2)\right]\end{aligned}$$

using Problem 21(*a*).

**23.** Express curl $\mathbf{A} = \nabla\times\mathbf{A}$ in orthogonal coordinates.

$$\begin{aligned}\nabla\times\mathbf{A} &= \nabla\times(A_1\mathbf{e}_1 + A_2\mathbf{e}_2 + A_3\mathbf{e}_3) = \nabla\times(A_1\mathbf{e}_1) + \nabla\times(A_2\mathbf{e}_2) + \nabla\times(A_3\mathbf{e}_3)\\ &= \frac{\mathbf{e}_2}{h_3h_1}\frac{\partial}{\partial u_3}(A_1h_1) - \frac{\mathbf{e}_3}{h_1h_2}\frac{\partial}{\partial u_2}(A_1h_1)\\ &\quad + \frac{\mathbf{e}_3}{h_1h_2}\frac{\partial}{\partial u_1}(A_2h_2) - \frac{\mathbf{e}_1}{h_2h_3}\frac{\partial}{\partial u_3}(A_2h_2)\\ &\quad + \frac{\mathbf{e}_1}{h_2h_3}\frac{\partial}{\partial u_2}(A_3h_3) - \frac{\mathbf{e}_2}{h_3h_1}\frac{\partial}{\partial u_1}(A_3h_3)\\ &= \frac{\mathbf{e}_1}{h_2h_3}\left[\frac{\partial}{\partial u_2}(A_3h_3) - \frac{\partial}{\partial u_3}(A_2h_2)\right] + \frac{\mathbf{e}_2}{h_3h_1}\left[\frac{\partial}{\partial u_3}(A_1h_1) - \frac{\partial}{\partial u_1}(A_3h_3)\right]\\ &\quad + \frac{\mathbf{e}_3}{h_1h_2}\left[\frac{\partial}{\partial u_1}(A_2h_2) - \frac{\partial}{\partial u_2}(A_1h_1)\right]\end{aligned}$$

using Problem 21(*b*). This can be written

$$\nabla\times\mathbf{A} = \frac{1}{h_1h_2h_3}\begin{vmatrix} h_1\mathbf{e}_1 & h_2\mathbf{e}_2 & h_3\mathbf{e}_3\\ \dfrac{\partial}{\partial u_1} & \dfrac{\partial}{\partial u_2} & \dfrac{\partial}{\partial u_3}\\ A_1h_1 & A_2h_2 & A_3h_3\end{vmatrix}$$

**24.** Express $\nabla^2\psi$ in orthogonal curvilinear coordinates.

From Problem 18, $$\nabla\psi = \frac{\mathbf{e}_1}{h_1}\frac{\partial\psi}{\partial u_1} + \frac{\mathbf{e}_2}{h_2}\frac{\partial\psi}{\partial u_2} + \frac{\mathbf{e}_3}{h_3}\frac{\partial\psi}{\partial u_3}.$$

If $\mathbf{A} = \nabla\psi$, then $A_1 = \frac{1}{h_1}\frac{\partial\psi}{\partial u_1}$, $A_2 = \frac{1}{h_2}\frac{\partial\psi}{\partial u_2}$, $A_3 = \frac{1}{h_3}\frac{\partial\psi}{\partial u_3}$ and by Problem 22,

$$\begin{aligned}\nabla\cdot\mathbf{A} &= \nabla\cdot\nabla\psi = \nabla^2\psi \\ &= \frac{1}{h_1h_2h_3}\left[\frac{\partial}{\partial u_1}\left(\frac{h_2h_3}{h_1}\frac{\partial\psi}{\partial u_1}\right) + \frac{\partial}{\partial u_2}\left(\frac{h_3h_1}{h_2}\frac{\partial\psi}{\partial u_2}\right) + \frac{\partial}{\partial u_3}\left(\frac{h_1h_2}{h_3}\frac{\partial\psi}{\partial u_3}\right)\right]\end{aligned}$$

25. Use the integral definition

$$\operatorname{div}\mathbf{A} = \nabla\cdot\mathbf{A} = \lim_{\Delta V\to 0}\frac{\iint_{\Delta S}\mathbf{A}\cdot\mathbf{n}\,dS}{\Delta V}$$

(see Problem 19, Chapter 6) to express $\nabla\cdot\mathbf{A}$ in orthogonal curvilinear coordinates.

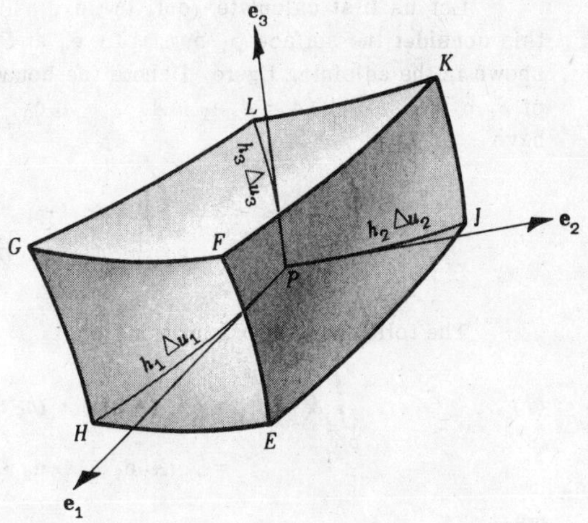


Consider the volume element $\Delta V$ (see adjacent figure) having edges $h_1\Delta u_1$, $h_2\Delta u_2$, $h_3\Delta u_3$.

Let $\mathbf{A} = A_1\mathbf{e}_1 + A_2\mathbf{e}_2 + A_3\mathbf{e}_3$ and let $\mathbf{n}$ be the outward drawn unit normal to the surface $\Delta S$ of $\Delta V$. On face $JKLP$, $\mathbf{n} = -\mathbf{e}_1$. Then we have approximately,

$$\begin{aligned}\iint_{JKLP}\mathbf{A}\cdot\mathbf{n}\,dS &= (\mathbf{A}\cdot\mathbf{n} \text{ at point } P)(\text{Area of } JKLP) \\ &= [(A_1\mathbf{e}_1 + A_2\mathbf{e}_2 + A_3\mathbf{e}_3)\cdot(-\mathbf{e}_1)]\,(h_2h_3\,\Delta u_2\Delta u_3) \\ &= -A_1\,h_2h_3\,\Delta u_2\Delta u_3\end{aligned}$$

On face $EFGH$, the surface integral is

$$A_1\,h_2h_3\,\Delta u_2\Delta u_3 + \frac{\partial}{\partial u_1}(A_1\,h_2h_3\,\Delta u_2\Delta u_3)\,\Delta u_1$$

apart from infinitesimals of order higher than $\Delta u_1\,\Delta u_2\,\Delta u_3$. Then the net contribution to the surface integral from these two faces is

$$\frac{\partial}{\partial u_1}(A_1\,h_2h_3\,\Delta u_2\Delta u_3)\,\Delta u_1 = \frac{\partial}{\partial u_1}(A_1\,h_2h_3)\,\Delta u_1\,\Delta u_2\,\Delta u_3$$

The contribution from all six faces of $\Delta V$ is

$$\left[\frac{\partial}{\partial u_1}(A_1\,h_2h_3) + \frac{\partial}{\partial u_2}(A_2\,h_1h_3) + \frac{\partial}{\partial u_3}(A_3\,h_1h_2)\right]\Delta u_1\,\Delta u_2\,\Delta u_3$$

Dividing this by the volume $h_1h_2h_3\,\Delta u_1\,\Delta u_2\,\Delta u_3$ and taking the limit as $\Delta u_1, \Delta u_2, \Delta u_3$ approach zero, we find

$$\operatorname{div}\mathbf{A} = \nabla\cdot\mathbf{A} = \frac{1}{h_1h_2h_3}\left[\frac{\partial}{\partial u_1}(A_1\,h_2h_3) + \frac{\partial}{\partial u_2}(A_2\,h_1h_3) + \frac{\partial}{\partial u_3}(A_3\,h_1h_2)\right]$$

Note that the same result would be obtained had we chosen the volume element $\Delta V$ such that $P$ is at its centre. In this case the calculation would proceed in a manner analogous to that of Problem 21, Chapter 4.

**26.** Use the integral definition

$$(\text{curl}\,\mathbf{A})\cdot\mathbf{n} = (\nabla\times\mathbf{A})\cdot\mathbf{n} = \lim_{\Delta S\to 0}\frac{\oint_C \mathbf{A}\cdot d\mathbf{r}}{\Delta S}$$

(see Problem 35, Chapter 6) to express $\nabla\times\mathbf{A}$ in orthogonal curvilinear coordinates.

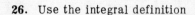

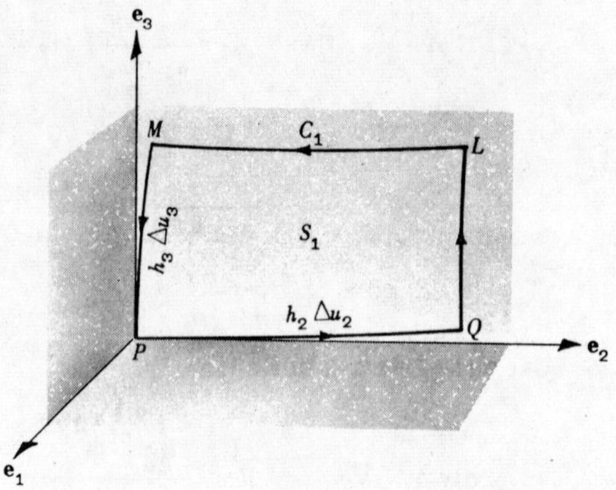


Let us first calculate $(\text{curl}\,\mathbf{A})\cdot\mathbf{e}_1$. To do this consider the surface $S_1$ normal to $\mathbf{e}_1$ at $P$, as shown in the adjoining figure. Denote the boundary of $S_1$ by $C_1$. Let $\mathbf{A} = A_1\mathbf{e}_1 + A_2\mathbf{e}_2 + A_3\mathbf{e}_3$. We have

$$\oint_{C_1}\mathbf{A}\cdot d\mathbf{r} = \int_{PQ}\mathbf{A}\cdot d\mathbf{r} + \int_{QL}\mathbf{A}\cdot d\mathbf{r} + \int_{LM}\mathbf{A}\cdot d\mathbf{r} + \int_{MP}\mathbf{A}\cdot d\mathbf{r}$$

The following approximations hold

(1) $$\int_{PQ}\mathbf{A}\cdot d\mathbf{r} = (\mathbf{A}\text{ at }P)\cdot(h_2\,\Delta u_2\,\mathbf{e}_2)$$
$$= (A_1\mathbf{e}_1 + A_2\mathbf{e}_2 + A_3\mathbf{e}_3)\cdot(h_2\,\Delta u_2\,\mathbf{e}_2) = A_2\,h_2\,\Delta u_2$$

Then

$$\int_{ML}\mathbf{A}\cdot d\mathbf{r} = A_2\,h_2\,\Delta u_2 + \frac{\partial}{\partial u_3}(A_2\,h_2\,\Delta u_2)\,\Delta u_3$$

or

(2) $$\int_{LM}\mathbf{A}\cdot d\mathbf{r} = -A_2\,h_2\,\Delta u_2 - \frac{\partial}{\partial u_3}(A_2\,h_2\,\Delta u_2)\,\Delta u_3$$

Similarly,

$$\int_{PM}\mathbf{A}\cdot d\mathbf{r} = (\mathbf{A}\text{ at }P)\cdot(h_3\,\Delta u_3\,\mathbf{e}_3) = A_3\,h_3\,\Delta u_3$$

or

(3) $$\int_{MP}\mathbf{A}\cdot d\mathbf{r} = -A_3\,h_3\,\Delta u_3$$

and

(4) $$\int_{QL}\mathbf{A}\cdot d\mathbf{r} = A_3\,h_3\,\Delta u_3 + \frac{\partial}{\partial u_2}(A_3\,h_3\,\Delta u_3)\,\Delta u_2$$

Adding (1), (2), (3), (4) we have

$$\oint_{C_1}\mathbf{A}\cdot d\mathbf{r} = \frac{\partial}{\partial u_2}(A_3\,h_3\,\Delta u_3)\,\Delta u_2 - \frac{\partial}{\partial u_3}(A_2\,h_2\,\Delta u_2)\,\Delta u_3$$

$$= \left[\frac{\partial}{\partial u_2}(A_3h_3) - \frac{\partial}{\partial u_3}(A_2h_2)\right]\Delta u_2\,\Delta u_3$$

apart from infinitesimals of order higher than $\Delta u_2\,\Delta u_3$.

Dividing by the area of $S_1$ equal to $h_2h_3\Delta u_2\Delta u_3$ and taking the limit as $\Delta u_2$ and $\Delta u_3$ approach zero,

$$(\text{curl}\,\mathbf{A})\cdot\mathbf{e}_1 = \frac{1}{h_2h_3}\left[\frac{\partial}{\partial u_2}(A_3h_3) - \frac{\partial}{\partial u_3}(A_2h_2)\right]$$

Similarly, by choosing areas $S_2$ and $S_3$ perpendicular to $\mathbf{e}_2$ and $\mathbf{e}_3$ at $P$ respectively, we find $(\text{curl}\,\mathbf{A})\cdot\mathbf{e}_2$ and $(\text{curl}\,\mathbf{A})\cdot\mathbf{e}_3$. This leads to the required result

$$\begin{aligned}\text{curl}\,\mathbf{A} &= \frac{\mathbf{e}_1}{h_2h_3}\left[\frac{\partial}{\partial u_2}(A_3h_3) - \frac{\partial}{\partial u_3}(A_2h_2)\right]\\ &\quad + \frac{\mathbf{e}_2}{h_3h_1}\left[\frac{\partial}{\partial u_3}(A_1h_1) - \frac{\partial}{\partial u_1}(A_3h_3)\right]\\ &\quad + \frac{\mathbf{e}_3}{h_1h_2}\left[\frac{\partial}{\partial u_1}(A_2h_2) - \frac{\partial}{\partial u_2}(A_1h_1)\right] = \frac{1}{h_1h_2h_3}\begin{vmatrix} h_1\mathbf{e}_1 & h_2\mathbf{e}_2 & h_3\mathbf{e}_3\\ \frac{\partial}{\partial u_1} & \frac{\partial}{\partial u_2} & \frac{\partial}{\partial u_3}\\ h_1A_1 & h_2A_2 & h_3A_3\end{vmatrix}\end{aligned}$$

The result could also have been derived by choosing $P$ as the centre of area $S_1$; the calculation would then proceed as in Problem 36, Chapter 6.

**27.** Express in cylindrical coordinates the quantities (*a*) $\nabla\Phi$, (*b*) $\nabla\cdot\mathbf{A}$, (*c*) $\nabla\times\mathbf{A}$, (*d*) $\nabla^2\Phi$.

For cylindrical coordinates $(\rho,\phi,z)$,

$$u_1=\rho,\ u_2=\phi,\ u_3=z;\quad \mathbf{e}_1=\mathbf{e}_\rho,\ \mathbf{e}_2=\mathbf{e}_\phi,\ \mathbf{e}_3=\mathbf{e}_z;$$

and

$$h_1=h_\rho=1,\quad h_2=h_\phi=\rho,\quad h_3=h_z=1$$

(*a*)
$$\begin{aligned}\nabla\Phi &= \frac{1}{h_1}\frac{\partial\Phi}{\partial u_1}\mathbf{e}_1 + \frac{1}{h_2}\frac{\partial\Phi}{\partial u_2}\mathbf{e}_2 + \frac{1}{h_3}\frac{\partial\Phi}{\partial u_3}\mathbf{e}_3\\ &= \frac{1}{1}\frac{\partial\Phi}{\partial\rho}\mathbf{e}_\rho + \frac{1}{\rho}\frac{\partial\Phi}{\partial\phi}\mathbf{e}_\phi + \frac{1}{1}\frac{\partial\Phi}{\partial z}\mathbf{e}_z\\ &= \frac{\partial\Phi}{\partial\rho}\mathbf{e}_\rho + \frac{1}{\rho}\frac{\partial\Phi}{\partial\phi}\mathbf{e}_\phi + \frac{\partial\Phi}{\partial z}\mathbf{e}_z\end{aligned}$$

(*b*)
$$\begin{aligned}\nabla\cdot\mathbf{A} &= \frac{1}{h_1h_2h_3}\left[\frac{\partial}{\partial u_1}(h_2h_3A_1) + \frac{\partial}{\partial u_2}(h_3h_1A_2) + \frac{\partial}{\partial u_3}(h_1h_2A_3)\right]\\ &= \frac{1}{(1)(\rho)(1)}\left[\frac{\partial}{\partial\rho}\Big((\rho)(1)A_\rho\Big) + \frac{\partial}{\partial\phi}\Big((1)(1)A_\phi\Big) + \frac{\partial}{\partial z}\Big((1)(\rho)A_z\Big)\right]\\ &= \frac{1}{\rho}\left[\frac{\partial}{\partial\rho}(\rho A_\rho) + \frac{\partial A_\phi}{\partial\phi} + \frac{\partial}{\partial z}(\rho A_z)\right]\end{aligned}$$

where $\mathbf{A} = A_\rho\mathbf{e}_1 + A_\phi\mathbf{e}_2 + A_z\mathbf{e}_3$, i.e. $A_1=A_\rho$, $A_2=A_\phi$, $A_3=A_z$.

(*c*)
$$\nabla\times\mathbf{A} = \frac{1}{h_1h_2h_3}\begin{vmatrix} h_1\mathbf{e}_1 & h_2\mathbf{e}_2 & h_3\mathbf{e}_3\\ \frac{\partial}{\partial u_1} & \frac{\partial}{\partial u_2} & \frac{\partial}{\partial u_3}\\ h_1A_1 & h_2A_2 & h_3A_3\end{vmatrix} = \frac{1}{\rho}\begin{vmatrix} \mathbf{e}_\rho & \rho\mathbf{e}_\phi & \mathbf{e}_z\\ \frac{\partial}{\partial\rho} & \frac{\partial}{\partial\phi} & \frac{\partial}{\partial z}\\ A_\rho & \rho A_\phi & A_z\end{vmatrix}$$

$$= \frac{1}{\rho}\left[\left(\frac{\partial A_z}{\partial\phi} - \frac{\partial}{\partial z}(\rho A_\phi)\right)\mathbf{e}_\rho + \left(\rho\frac{\partial A_\rho}{\partial z} - \rho\frac{\partial A_z}{\partial\rho}\right)\mathbf{e}_\phi + \left(\frac{\partial}{\partial\rho}(\rho A_\phi) - \frac{\partial A_\rho}{\partial\phi}\right)\mathbf{e}_z\right]$$

(d) $$\nabla^2\Phi = \frac{1}{h_1h_2h_3}\left[\frac{\partial}{\partial u_1}\left(\frac{h_2h_3}{h_1}\frac{\partial\Phi}{\partial u_1}\right) + \frac{\partial}{\partial u_2}\left(\frac{h_3h_1}{h_2}\frac{\partial\Phi}{\partial u_2}\right) + \frac{\partial}{\partial u_3}\left(\frac{h_1h_2}{h_3}\frac{\partial\Phi}{\partial u_3}\right)\right]$$

$$= \frac{1}{(1)(\rho)(1)}\left[\frac{\partial}{\partial\rho}\left(\frac{(\rho)(1)}{(1)}\frac{\partial\Phi}{\partial\rho}\right) + \frac{\partial}{\partial\phi}\left(\frac{(1)(1)}{\rho}\frac{\partial\Phi}{\partial\phi}\right) + \frac{\partial}{\partial z}\left(\frac{(1)(\rho)}{(1)}\frac{\partial\Phi}{\partial z}\right)\right]$$

$$= \frac{1}{\rho}\frac{\partial}{\partial\rho}\left(\rho\frac{\partial\Phi}{\partial\rho}\right) + \frac{1}{\rho^2}\frac{\partial^2\Phi}{\partial\phi^2} + \frac{\partial^2\Phi}{\partial z^2}$$

**28.** Express (*a*) $\nabla\times\mathbf{A}$ and (*b*) $\nabla^2\psi$ in spherical coordinates.

Here $u_1 = r,\ u_2 = \theta,\ u_3 = \phi;\ \mathbf{e}_1 = \mathbf{e}_r,\ \mathbf{e}_2 = \mathbf{e}_\theta,\ \mathbf{e}_3 = \mathbf{e}_\phi;\ h_1 = h_r = 1,\ h_2 = h_\theta = r,\ h_3 = h_\phi = r\sin\theta.$

(a) $$\nabla\times\mathbf{A} = \frac{1}{h_1h_2h_3}\begin{vmatrix} h_1\mathbf{e}_1 & h_2\mathbf{e}_2 & h_3\mathbf{e}_3 \\ \frac{\partial}{\partial u_1} & \frac{\partial}{\partial u_2} & \frac{\partial}{\partial u_3} \\ h_1A_1 & h_2A_2 & h_3A_3 \end{vmatrix} = \frac{1}{(1)(r)(r\sin\theta)}\begin{vmatrix} \mathbf{e}_r & r\mathbf{e}_\theta & r\sin\theta\,\mathbf{e}_\phi \\ \frac{\partial}{\partial r} & \frac{\partial}{\partial\theta} & \frac{\partial}{\partial\phi} \\ A_r & rA_\theta & r\sin\theta\,A_\phi \end{vmatrix}$$

$$= \frac{1}{r^2\sin\theta}\left[\left\{\frac{\partial}{\partial\theta}(r\sin\theta\,A_\phi) - \frac{\partial}{\partial\phi}(rA_\theta)\right\}\mathbf{e}_r\right.$$

$$\left. + \left\{\frac{\partial A_r}{\partial\phi} - \frac{\partial}{\partial r}(r\sin\theta\,A_\phi)\right\}r\mathbf{e}_\theta + \left\{\frac{\partial}{\partial r}(rA_\theta) - \frac{\partial A_r}{\partial\theta}\right\}r\sin\theta\,\mathbf{e}_\phi\right]$$

(b) $$\nabla^2\psi = \frac{1}{h_1h_2h_3}\left[\frac{\partial}{\partial u_1}\left(\frac{h_2h_3}{h_1}\frac{\partial\psi}{\partial u_1}\right) + \frac{\partial}{\partial u_2}\left(\frac{h_3h_1}{h_2}\frac{\partial\psi}{\partial u_2}\right) + \frac{\partial}{\partial u_3}\left(\frac{h_1h_2}{h_3}\frac{\partial\psi}{\partial u_3}\right)\right]$$

$$= \frac{1}{(1)(r)(r\sin\theta)}\left[\frac{\partial}{\partial r}\left(\frac{(r)(r\sin\theta)}{(1)}\frac{\partial\psi}{\partial r}\right) + \frac{\partial}{\partial\theta}\left(\frac{(r\sin\theta)(1)}{r}\frac{\partial\psi}{\partial\theta}\right)\right.$$

$$\left. + \frac{\partial}{\partial\phi}\left(\frac{(1)(r)}{r\sin\theta}\frac{\partial\psi}{\partial\phi}\right)\right]$$

$$= \frac{1}{r^2\sin\theta}\left[\sin\theta\frac{\partial}{\partial r}\left(r^2\frac{\partial\psi}{\partial r}\right) + \frac{\partial}{\partial\theta}\left(\sin\theta\frac{\partial\psi}{\partial\theta}\right) + \frac{1}{\sin\theta}\frac{\partial^2\psi}{\partial\phi^2}\right]$$

$$= \frac{1}{r^2}\frac{\partial}{\partial r}\left(r^2\frac{\partial\psi}{\partial r}\right) + \frac{1}{r^2\sin\theta}\frac{\partial}{\partial\theta}\left(\sin\theta\frac{\partial\psi}{\partial\theta}\right) + \frac{1}{r^2\sin^2\theta}\frac{\partial^2\psi}{\partial\phi^2}$$

**29.** Write Laplace's equation in parabolic cylindrical coordinates.

From Problem 8(*b*),

$$u_1 = u,\ u_2 = v,\ u_3 = z;\quad h_1 = \sqrt{u^2+v^2},\ h_2 = \sqrt{u^2+v^2},\ h_3 = 1$$

Then $$\nabla^2\psi = \frac{1}{u^2+v^2}\left[\frac{\partial}{\partial u}\left(\frac{\partial\psi}{\partial u}\right) + \frac{\partial}{\partial v}\left(\frac{\partial\psi}{\partial v}\right) + \frac{\partial}{\partial z}\left((u^2+v^2)\frac{\partial\psi}{\partial z}\right)\right]$$
$$= \frac{1}{u^2+v^2}\left(\frac{\partial^2\psi}{\partial u^2} + \frac{\partial^2\psi}{\partial v^2}\right) + \frac{\partial^2\psi}{\partial z^2}$$

and Laplace's equation is $\nabla^2\psi = 0$ or
$$\frac{\partial^2\psi}{\partial u^2} + \frac{\partial^2\psi}{\partial v^2} + (u^2+v^2)\frac{\partial^2\psi}{\partial z^2} = 0$$

**30.** Express the heat conduction equation $\dfrac{\partial U}{\partial t} = \kappa\nabla^2 U$ in elliptic cylindrical coordinates.

Here $u_1 = u,\ u_2 = v,\ u_3 = z;\quad h_1 = h_2 = a\sqrt{\sinh^2 u + \sin^2 v},\ h_3 = 1$. Then

$$\nabla^2 U = \frac{1}{a^2(\sinh^2 u + \sin^2 v)}\left[\frac{\partial}{\partial u}\left(\frac{\partial U}{\partial u}\right) + \frac{\partial}{\partial v}\left(\frac{\partial U}{\partial v}\right) + \frac{\partial}{\partial z}\left(a^2(\sinh^2 u + \sin^2 v)\frac{\partial U}{\partial z}\right)\right]$$
$$= \frac{1}{a^2(\sinh^2 u + \sin^2 v)}\left[\frac{\partial^2 U}{\partial u^2} + \frac{\partial^2 U}{\partial v^2}\right] + \frac{\partial^2 U}{\partial z^2}$$

and the heat conduction equation is
$$\frac{\partial U}{\partial t} = \kappa\left\{\frac{1}{a^2(\sinh^2 u + \sin^2 v)}\left[\frac{\partial^2 U}{\partial u^2} + \frac{\partial^2 U}{\partial v^2}\right] + \frac{\partial^2 U}{\partial z^2}\right\}$$

## SURFACE CURVILINEAR COORDINATES

**31.** Show that the square of the element of arc length on the surface $\mathbf{r} = \mathbf{r}(u,v)$ can be written
$$ds^2 = E\,du^2 + 2F\,du\,dv + G\,dv^2$$

We have $$d\mathbf{r} = \frac{\partial\mathbf{r}}{\partial u}du + \frac{\partial\mathbf{r}}{\partial v}dv$$

Then $$ds^2 = d\mathbf{r}\cdot d\mathbf{r}$$
$$= \frac{\partial\mathbf{r}}{\partial u}\cdot\frac{\partial\mathbf{r}}{\partial u}du^2 + 2\frac{\partial\mathbf{r}}{\partial u}\cdot\frac{\partial\mathbf{r}}{\partial v}du\,dv + \frac{\partial\mathbf{r}}{\partial v}\cdot\frac{\partial\mathbf{r}}{\partial v}dv^2$$
$$= E\,du^2 + 2F\,du\,dv + G\,dv^2$$

**32.** Show that the element of surface area of the surface $\mathbf{r} = \mathbf{r}(u,v)$ is given by
$$dS = \sqrt{EG - F^2}\,du\,dv$$

The element of area is given by

$$dS = \left|\left(\frac{\partial\mathbf{r}}{\partial u}du\right)\times\left(\frac{\partial\mathbf{r}}{\partial v}dv\right)\right| = \left|\frac{\partial\mathbf{r}}{\partial u}\times\frac{\partial\mathbf{r}}{\partial v}\right|du\,dv = \sqrt{\left(\frac{\partial\mathbf{r}}{\partial u}\times\frac{\partial\mathbf{r}}{\partial v}\right)\cdot\left(\frac{\partial\mathbf{r}}{\partial u}\times\frac{\partial\mathbf{r}}{\partial v}\right)}\,du\,dv$$

The quantity under the square root sign is equal to (see Problem 48, Chapter 2)
$$\left(\frac{\partial\mathbf{r}}{\partial u}\cdot\frac{\partial\mathbf{r}}{\partial u}\right)\left(\frac{\partial\mathbf{r}}{\partial v}\cdot\frac{\partial\mathbf{r}}{\partial v}\right) - \left(\frac{\partial\mathbf{r}}{\partial u}\cdot\frac{\partial\mathbf{r}}{\partial v}\right)\left(\frac{\partial\mathbf{r}}{\partial v}\cdot\frac{\partial\mathbf{r}}{\partial u}\right) = EG - F^2$$ and the result follows.

## MISCELLANEOUS PROBLEMS ON GENERAL COORDINATES.

**33.** Let **A** be a given vector defined with respect to two general curvilinear coordinate systems $(u_1, u_2, u_3)$ and $(\bar{u}_1, \bar{u}_2, \bar{u}_3)$. Find the relation between the contravariant components of the vector in the two coordinate systems.

Suppose the transformation equations from a rectangular $(x,y,z)$ system to the $(u_1,u_2,u_3)$ and $(\bar{u}_1,\bar{u}_2,\bar{u}_3)$ systems are given by

$$(1)\qquad \begin{cases} x = x_1(u_1,u_2,u_3), & y = y_1(u_1,u_2,u_3), & z = z_1(u_1,u_2,u_3) \\ x = x_2(\bar{u}_1,\bar{u}_2,\bar{u}_3), & y = y_2(\bar{u}_1,\bar{u}_2,\bar{u}_3), & z = z_2(\bar{u}_1,\bar{u}_2,\bar{u}_3) \end{cases}$$

Then there exists a transformation directly from the $(u_1,u_2,u_3)$ system to the $(\bar{u}_1,\bar{u}_2,\bar{u}_3)$ system defined by

$$(2)\qquad u_1 = u_1(\bar{u}_1,\bar{u}_2,\bar{u}_3), \quad u_2 = u_2(\bar{u}_1,\bar{u}_2,\bar{u}_3), \quad u_3 = u_3(\bar{u}_1,\bar{u}_2,\bar{u}_3)$$

and conversely. From (*1*),

$$d\mathbf{r} = \frac{\partial \mathbf{r}}{\partial u_1}du_1 + \frac{\partial \mathbf{r}}{\partial u_2}du_2 + \frac{\partial \mathbf{r}}{\partial u_3}du_3 = \boldsymbol{\alpha}_1\,du_1 + \boldsymbol{\alpha}_2\,du_2 + \boldsymbol{\alpha}_3\,du_3$$

$$d\mathbf{r} = \frac{\partial \mathbf{r}}{\partial \bar{u}_1}d\bar{u}_1 + \frac{\partial \mathbf{r}}{\partial \bar{u}_2}d\bar{u}_2 + \frac{\partial \mathbf{r}}{\partial \bar{u}_3}d\bar{u}_3 = \bar{\boldsymbol{\alpha}}_1\,d\bar{u}_1 + \bar{\boldsymbol{\alpha}}_2\,d\bar{u}_2 + \bar{\boldsymbol{\alpha}}_3\,d\bar{u}_3$$

Then

$$(3)\qquad \boldsymbol{\alpha}_1\,du_1 + \boldsymbol{\alpha}_2\,du_2 + \boldsymbol{\alpha}_3\,du_3 = \bar{\boldsymbol{\alpha}}_1\,d\bar{u}_1 + \bar{\boldsymbol{\alpha}}_2\,d\bar{u}_2 + \bar{\boldsymbol{\alpha}}_3\,d\bar{u}_3$$

From (*2*),
$$du_1 = \frac{\partial u_1}{\partial \bar{u}_1}d\bar{u}_1 + \frac{\partial u_1}{\partial \bar{u}_2}d\bar{u}_2 + \frac{\partial u_1}{\partial \bar{u}_3}d\bar{u}_3$$

$$du_2 = \frac{\partial u_2}{\partial \bar{u}_1}d\bar{u}_1 + \frac{\partial u_2}{\partial \bar{u}_2}d\bar{u}_2 + \frac{\partial u_2}{\partial \bar{u}_3}d\bar{u}_3$$

$$du_3 = \frac{\partial u_3}{\partial \bar{u}_1}d\bar{u}_1 + \frac{\partial u_3}{\partial \bar{u}_2}d\bar{u}_2 + \frac{\partial u_3}{\partial \bar{u}_3}d\bar{u}_3$$

Substituting into (*3*) and equating coefficients of $d\bar{u}_1$, $d\bar{u}_2$, $d\bar{u}_3$ on both sides, we find

$$(4)\qquad \begin{cases} \bar{\boldsymbol{\alpha}}_1 = \boldsymbol{\alpha}_1 \dfrac{\partial u_1}{\partial \bar{u}_1} + \boldsymbol{\alpha}_2 \dfrac{\partial u_2}{\partial \bar{u}_1} + \boldsymbol{\alpha}_3 \dfrac{\partial u_3}{\partial \bar{u}_1} \\ \bar{\boldsymbol{\alpha}}_2 = \boldsymbol{\alpha}_1 \dfrac{\partial u_1}{\partial \bar{u}_2} + \boldsymbol{\alpha}_2 \dfrac{\partial u_2}{\partial \bar{u}_2} + \boldsymbol{\alpha}_3 \dfrac{\partial u_3}{\partial \bar{u}_2} \\ \bar{\boldsymbol{\alpha}}_3 = \boldsymbol{\alpha}_1 \dfrac{\partial u_1}{\partial \bar{u}_3} + \boldsymbol{\alpha}_2 \dfrac{\partial u_2}{\partial \bar{u}_3} + \boldsymbol{\alpha}_3 \dfrac{\partial u_3}{\partial \bar{u}_3} \end{cases}$$

Now **A** can be expressed in the two coordinate systems as

$$(5)\qquad \mathbf{A} = C_1\boldsymbol{\alpha}_1 + C_2\boldsymbol{\alpha}_2 + C_3\boldsymbol{\alpha}_3 \quad \text{and} \quad \mathbf{A} = \bar{C}_1\bar{\boldsymbol{\alpha}}_1 + \bar{C}_2\bar{\boldsymbol{\alpha}}_2 + \bar{C}_3\bar{\boldsymbol{\alpha}}_3$$

where $C_1, C_2, C_3$ and $\bar{C}_1, \bar{C}_2, \bar{C}_3$ are the contravariant components of **A** in the two systems. Substituting (*4*) into (*5*),

$$C_1\boldsymbol{\alpha}_1 + C_2\boldsymbol{\alpha}_2 + C_3\boldsymbol{\alpha}_3 = \bar{C}_1\bar{\boldsymbol{\alpha}}_1 + \bar{C}_2\bar{\boldsymbol{\alpha}}_2 + \bar{C}_3\bar{\boldsymbol{\alpha}}_3$$

$$= (\bar{C}_1\frac{\partial u_1}{\partial \bar{u}_1} + \bar{C}_2\frac{\partial u_1}{\partial \bar{u}_2} + \bar{C}_3\frac{\partial u_1}{\partial \bar{u}_3})\boldsymbol{\alpha}_1 + (\bar{C}_1\frac{\partial u_2}{\partial \bar{u}_1} + \bar{C}_2\frac{\partial u_2}{\partial \bar{u}_2} + \bar{C}_3\frac{\partial u_2}{\partial \bar{u}_3})\boldsymbol{\alpha}_2 + (\bar{C}_1\frac{\partial u_3}{\partial \bar{u}_1} + \bar{C}_2\frac{\partial u_3}{\partial \bar{u}_2} + \bar{C}_3\frac{\partial u_3}{\partial \bar{u}_3})\boldsymbol{\alpha}_3$$

Then

$$(6) \quad \begin{cases} C_1 = \bar{C}_1 \dfrac{\partial u_1}{\partial \bar{u}_1} + \bar{C}_2 \dfrac{\partial u_1}{\partial \bar{u}_2} + \bar{C}_3 \dfrac{\partial u_1}{\partial \bar{u}_3} \\ C_2 = \bar{C}_1 \dfrac{\partial u_2}{\partial \bar{u}_1} + \bar{C}_2 \dfrac{\partial u_2}{\partial \bar{u}_2} + \bar{C}_3 \dfrac{\partial u_2}{\partial \bar{u}_3} \\ C_3 = \bar{C}_1 \dfrac{\partial u_3}{\partial \bar{u}_1} + \bar{C}_2 \dfrac{\partial u_3}{\partial \bar{u}_2} + \bar{C}_3 \dfrac{\partial u_3}{\partial \bar{u}_3} \end{cases}$$

or in shorter notation

$$(7) \quad C_p = \bar{C}_1 \frac{\partial u_p}{\partial \bar{u}_1} + \bar{C}_2 \frac{\partial u_p}{\partial \bar{u}_2} + \bar{C}_3 \frac{\partial u_p}{\partial \bar{u}_3} \qquad p = 1,2,3$$

and in even shorter notation

$$(8) \quad C_p = \sum_{q=1}^{3} \bar{C}_q \frac{\partial u_p}{\partial \bar{u}_q} \qquad p = 1,2,3$$

Similarly, by interchanging the coordinates we see that

$$(9) \quad \bar{C}_p = \sum_{q=1}^{3} C_q \frac{\partial \bar{u}_p}{\partial u_q} \qquad p = 1,2,3$$

The above results lead us to adopt the following definition. If three quantities $C_1, C_2, C_3$ of a coordinate system $(u_1, u_2, u_3)$ are related to three other quantities $\bar{C}_1, \bar{C}_2, \bar{C}_3$ of another coordinate system $(\bar{u}_1, \bar{u}_2, \bar{u}_3)$ by the transformation equations (6), (7), (8) or (9), then the quantities are called *components of a contravariant vector* or a *contravariant tensor of the first rank.*

**34.** Work Problem 33 for the covariant components of **A**.

Write the covariant components of **A** in the systems $(u_1, u_2, u_3)$ and $(\bar{u}_1, \bar{u}_2, \bar{u}_3)$ as $c_1, c_2, c_3$ and $\bar{c}_1, \bar{c}_2, \bar{c}_3$ respectively. Then

$$(1) \quad \mathbf{A} = c_1 \nabla u_1 + c_2 \nabla u_2 + c_3 \nabla u_3 = \bar{c}_1 \nabla \bar{u}_1 + \bar{c}_2 \nabla \bar{u}_2 + \bar{c}_3 \nabla \bar{u}_3$$

Now since $\bar{u}_p = \bar{u}_p(u_1, u_2, u_3)$ with $p = 1,2,3$,

$$(2) \quad \begin{cases} \dfrac{\partial \bar{u}_p}{\partial x} = \dfrac{\partial \bar{u}_p}{\partial u_1}\dfrac{\partial u_1}{\partial x} + \dfrac{\partial \bar{u}_p}{\partial u_2}\dfrac{\partial u_2}{\partial x} + \dfrac{\partial \bar{u}_p}{\partial u_3}\dfrac{\partial u_3}{\partial x} \\ \dfrac{\partial \bar{u}_p}{\partial y} = \dfrac{\partial \bar{u}_p}{\partial u_1}\dfrac{\partial u_1}{\partial y} + \dfrac{\partial \bar{u}_p}{\partial u_2}\dfrac{\partial u_2}{\partial y} + \dfrac{\partial \bar{u}_p}{\partial u_3}\dfrac{\partial u_3}{\partial y} \\ \dfrac{\partial \bar{u}_p}{\partial z} = \dfrac{\partial \bar{u}_p}{\partial u_1}\dfrac{\partial u_1}{\partial z} + \dfrac{\partial \bar{u}_p}{\partial u_2}\dfrac{\partial u_2}{\partial z} + \dfrac{\partial \bar{u}_p}{\partial u_3}\dfrac{\partial u_3}{\partial z} \end{cases} \qquad p = 1,2,3$$

Also,

$$(3) \quad c_1 \nabla u_1 + c_2 \nabla u_2 + c_3 \nabla u_3 = (c_1 \frac{\partial u_1}{\partial x} + c_2 \frac{\partial u_2}{\partial x} + c_3 \frac{\partial u_3}{\partial x})\mathbf{i}$$

$$+ (c_1 \frac{\partial u_1}{\partial y} + c_2 \frac{\partial u_2}{\partial y} + c_3 \frac{\partial u_3}{\partial y})\mathbf{j} + (c_1 \frac{\partial u_1}{\partial z} + c_2 \frac{\partial u_2}{\partial z} + c_3 \frac{\partial u_3}{\partial z})\mathbf{k}$$

and

$$(4) \quad \bar{c}_1 \nabla\bar{u}_1 + \bar{c}_2 \nabla\bar{u}_2 + \bar{c}_3 \nabla\bar{u}_3 = (\bar{c}_1 \frac{\partial\bar{u}_1}{\partial x} + \bar{c}_2 \frac{\partial\bar{u}_2}{\partial x} + \bar{c}_3 \frac{\partial\bar{u}_3}{\partial x})\,\mathbf{i}$$
$$+ (\bar{c}_1 \frac{\partial\bar{u}_1}{\partial y} + \bar{c}_2 \frac{\partial\bar{u}_2}{\partial y} + \bar{c}_3 \frac{\partial\bar{u}_3}{\partial y})\,\mathbf{j} + (\bar{c}_1 \frac{\partial\bar{u}_1}{\partial z} + \bar{c}_2 \frac{\partial\bar{u}_2}{\partial z} + \bar{c}_3 \frac{\partial\bar{u}_3}{\partial z})\,\mathbf{k}$$

Equating coefficients of $\mathbf{i}, \mathbf{j}, \mathbf{k}$ in (3) and (4),

$$(5) \quad \begin{cases} c_1 \dfrac{\partial u_1}{\partial x} + c_2 \dfrac{\partial u_2}{\partial x} + c_3 \dfrac{\partial u_3}{\partial x} = \bar{c}_1 \dfrac{\partial\bar{u}_1}{\partial x} + \bar{c}_2 \dfrac{\partial\bar{u}_2}{\partial x} + \bar{c}_3 \dfrac{\partial\bar{u}_3}{\partial x} \\ c_1 \dfrac{\partial u_1}{\partial y} + c_2 \dfrac{\partial u_2}{\partial y} + c_3 \dfrac{\partial u_3}{\partial y} = \bar{c}_1 \dfrac{\partial\bar{u}_1}{\partial y} + \bar{c}_2 \dfrac{\partial\bar{u}_2}{\partial y} + \bar{c}_3 \dfrac{\partial\bar{u}_3}{\partial y} \\ c_1 \dfrac{\partial u_1}{\partial z} + c_2 \dfrac{\partial u_2}{\partial z} + c_3 \dfrac{\partial u_3}{\partial z} = \bar{c}_1 \dfrac{\partial\bar{u}_1}{\partial z} + \bar{c}_2 \dfrac{\partial\bar{u}_2}{\partial z} + \bar{c}_3 \dfrac{\partial\bar{u}_3}{\partial z} \end{cases}$$

Substituting equations (2) with $p = 1,2,3$ in any of the equations (5) and equating coefficients of $\frac{\partial u_1}{\partial x}$, $\frac{\partial u_2}{\partial x}$, $\frac{\partial u_3}{\partial x}$, $\frac{\partial u_1}{\partial y}$, $\frac{\partial u_2}{\partial y}$, $\frac{\partial u_3}{\partial y}$, $\frac{\partial u_1}{\partial z}$, $\frac{\partial u_2}{\partial z}$, $\frac{\partial u_3}{\partial z}$ on each side, we find

$$(6) \quad \begin{cases} c_1 = \bar{c}_1 \dfrac{\partial\bar{u}_1}{\partial u_1} + \bar{c}_2 \dfrac{\partial\bar{u}_2}{\partial u_1} + \bar{c}_3 \dfrac{\partial\bar{u}_3}{\partial u_1} \\ c_2 = \bar{c}_1 \dfrac{\partial\bar{u}_1}{\partial u_2} + \bar{c}_2 \dfrac{\partial\bar{u}_2}{\partial u_2} + \bar{c}_3 \dfrac{\partial\bar{u}_3}{\partial u_2} \\ c_3 = \bar{c}_1 \dfrac{\partial\bar{u}_1}{\partial u_3} + \bar{c}_2 \dfrac{\partial\bar{u}_2}{\partial u_3} + \bar{c}_3 \dfrac{\partial\bar{u}_3}{\partial u_3} \end{cases}$$

which can be written

$$(7) \quad c_p = \bar{c}_1 \frac{\partial\bar{u}_1}{\partial u_p} + \bar{c}_2 \frac{\partial\bar{u}_2}{\partial u_p} + \bar{c}_3 \frac{\partial\bar{u}_3}{\partial u_p} \qquad p = 1,2,3$$

or

$$(8) \quad c_p = \sum_{q=1}^{3} \bar{c}_q \frac{\partial\bar{u}_q}{\partial u_p} \qquad p = 1,2,3$$

Similarly, we can show that

$$(9) \quad \bar{c}_p = \sum_{q=1}^{3} c_q \frac{\partial u_q}{\partial\bar{u}_p} \qquad p = 1,2,3$$

The above results lead us to adopt the following definition. If three quantities $c_1, c_2, c_3$ of a coordinate system $(u_1, u_2, u_3)$ are related to three other quantities $\bar{c}_1, \bar{c}_2, \bar{c}_3$ of another coordinate system $(\bar{u}_1, \bar{u}_2, \bar{u}_3)$ by the transformation equations (6), (7), (8) or (9), then the quantities are called *components of a covariant vector* or a *covariant tensor of the first rank*.

In generalizing the concepts in this Problem and in Problem 33 to higher dimensional spaces, and in generalizing the concept of vector, we are led to *tensor analysis* which we treat in Chapter 8. In the process of generalization it is convenient to use a concise notation in order to express fundamental ideas in compact form. It should be remembered, however, that despite the notation used, the basic ideas treated in Chapter 8 are intimately connected with those treated in this chapter.

**35.** (*a*) Prove that in general coordinates $(u_1, u_2, u_3)$,

$$g = \begin{vmatrix} g_{11} & g_{12} & g_{13} \\ g_{21} & g_{22} & g_{23} \\ g_{31} & g_{32} & g_{33} \end{vmatrix} = \left(\frac{\partial \mathbf{r}}{\partial u_1} \cdot \frac{\partial \mathbf{r}}{\partial u_2} \times \frac{\partial \mathbf{r}}{\partial u_3}\right)^2$$

where $g_{pq}$ are the coefficients of $du_p\, du_q$ in $ds^2$ (Problem 17).

(*b*) Show that the volume element in general coordinates is $\sqrt{g}\, du_1\, du_2\, du_3$.

(*a*) From Problem 17,

$$(1) \qquad g_{pq} = \boldsymbol{\alpha}_p \cdot \boldsymbol{\alpha}_q = \frac{\partial \mathbf{r}}{\partial u_p} \cdot \frac{\partial \mathbf{r}}{\partial u_q} = \frac{\partial x}{\partial u_p}\frac{\partial x}{\partial u_q} + \frac{\partial y}{\partial u_p}\frac{\partial y}{\partial u_q} + \frac{\partial z}{\partial u_p}\frac{\partial z}{\partial u_q} \qquad p, q = 1,2,3$$

Then, using the following theorem on multiplication of determinants,

$$\begin{vmatrix} a_1 & a_2 & a_3 \\ b_1 & b_2 & b_3 \\ c_1 & c_2 & c_3 \end{vmatrix} \begin{vmatrix} A_1 & B_1 & C_1 \\ A_2 & B_2 & C_2 \\ A_3 & B_3 & C_3 \end{vmatrix} = \begin{vmatrix} a_1A_1 + a_2A_2 + a_3A_3 & a_1B_1 + a_2B_2 + a_3B_3 & a_1C_1 + a_2C_2 + a_3C_3 \\ b_1A_1 + b_2A_2 + b_3A_3 & b_1B_1 + b_2B_2 + b_3B_3 & b_1C_1 + b_2C_2 + b_3C_3 \\ c_1A_1 + c_2A_2 + c_3A_3 & c_1B_1 + c_2B_2 + c_3B_3 & c_1C_1 + c_2C_2 + c_3C_3 \end{vmatrix}$$

we have

$$\left(\frac{\partial \mathbf{r}}{\partial u_1} \cdot \frac{\partial \mathbf{r}}{\partial u_2} \times \frac{\partial \mathbf{r}}{\partial u_3}\right)^2 = \begin{vmatrix} \frac{\partial x}{\partial u_1} & \frac{\partial y}{\partial u_1} & \frac{\partial z}{\partial u_1} \\ \frac{\partial x}{\partial u_2} & \frac{\partial y}{\partial u_2} & \frac{\partial z}{\partial u_2} \\ \frac{\partial x}{\partial u_3} & \frac{\partial y}{\partial u_3} & \frac{\partial z}{\partial u_3} \end{vmatrix}^2$$

$$= \begin{vmatrix} \frac{\partial x}{\partial u_1} & \frac{\partial y}{\partial u_1} & \frac{\partial z}{\partial u_1} \\ \frac{\partial x}{\partial u_2} & \frac{\partial y}{\partial u_2} & \frac{\partial z}{\partial u_2} \\ \frac{\partial x}{\partial u_3} & \frac{\partial y}{\partial u_3} & \frac{\partial z}{\partial u_3} \end{vmatrix} \begin{vmatrix} \frac{\partial x}{\partial u_1} & \frac{\partial x}{\partial u_2} & \frac{\partial x}{\partial u_3} \\ \frac{\partial y}{\partial u_1} & \frac{\partial y}{\partial u_2} & \frac{\partial y}{\partial u_3} \\ \frac{\partial z}{\partial u_1} & \frac{\partial z}{\partial u_2} & \frac{\partial z}{\partial u_3} \end{vmatrix} = \begin{vmatrix} g_{11} & g_{12} & g_{13} \\ g_{21} & g_{22} & g_{23} \\ g_{31} & g_{32} & g_{33} \end{vmatrix}$$

(*b*) The volume element is given by

$$dV = \left|\left(\frac{\partial \mathbf{r}}{\partial u_1} du_1\right) \cdot \left(\frac{\partial \mathbf{r}}{\partial u_2} du_2\right) \times \left(\frac{\partial \mathbf{r}}{\partial u_3} du_3\right)\right| = \left|\frac{\partial \mathbf{r}}{\partial u_1} \cdot \frac{\partial \mathbf{r}}{\partial u_2} \times \frac{\partial \mathbf{r}}{\partial u_3}\right| du_1\, du_2\, du_3$$

$$= \sqrt{g}\, du_1\, du_2\, du_3 \quad \text{by part } (a).$$

Note that $\sqrt{g}$ is the absolute value of the Jacobian of $x, y, z$ with respect to $u_1, u_2, u_3$ (see Prob.13).

## SUPPLEMENTARY PROBLEMS

Answers to the Supplementary Problems are given at the end of this Chapter.

**36.** Describe and sketch the coordinate surfaces and coordinate curves for (*a*) elliptic cylindrical, (*b*) bipolar, and (*c*) parabolic cylindrical coordinates.

**37.** Determine the transformation from (*a*) spherical to rectangular coordinates, (*b*) spherical to cylindrical coordinates.

**38.** Express each of the following loci in spherical coordinates:
(*a*) the sphere $x^2+y^2+z^2 = 9$
(*b*) the cone $z^2 = 3(x^2+y^2)$
(*c*) the paraboloid $z = x^2+y^2$
(*d*) the plane $z = 0$
(*e*) the plane $y = x$.

**39.** If $\rho, \phi, z$ are cylindrical coordinates, describe each of the following loci and write the equation of each locus in rectangular coordinates: (*a*) $\rho = 4,\ z = 0$; (*b*) $\rho = 4$; (*c*) $\phi = \pi/2$; (*d*) $\phi = \pi/3,\ z = 1$.

**40.** If $u, v, z$ are elliptic cylindrical coordinates where $a = 4$, describe each of the following loci and write the equation of each locus in rectangular coordinates:
(*a*) $v = \pi/4$; (*b*) $u = 0,\ z = 0$; (*c*) $u = \ln 2,\ z = 2$; (*d*) $v = 0,\ z = 0$.

**41.** If $u, v, z$ are parabolic cylindrical coordinates, graph the curves or regions described by each of the following: (*a*) $u = 2,\ z = 0$; (*b*) $v = 1,\ z = 2$; (*c*) $1 \leqq u \leqq 2,\ 2 \leqq v \leqq 3,\ z = 0$; (*d*) $1 < u < 2,\ 2 < v < 3,\ z = 0$.

**42.** (*a*) Find the unit vectors $\mathbf{e}_r$, $\mathbf{e}_\theta$ and $\mathbf{e}_\phi$ of a spherical coordinate system in terms of $\mathbf{i}, \mathbf{j}$ and $\mathbf{k}$.
(*b*) Solve for $\mathbf{i}, \mathbf{j}$ and $\mathbf{k}$ in terms of $\mathbf{e}_r$, $\mathbf{e}_\theta$ and $\mathbf{e}_\phi$.

**43.** Represent the vector $\mathbf{A} = 2y\,\mathbf{i} - z\,\mathbf{j} + 3x\,\mathbf{k}$ in spherical coordinates and determine $A_r$, $A_\theta$ and $A_\phi$.

**44.** Prove that a spherical coordinate system is orthogonal.

**45.** Prove that (*a*) parabolic cylindrical, (*b*) elliptic cylindrical, and (*c*) oblate spheroidal coordinate systems are orthogonal.

**46.** Prove $\dot{\mathbf{e}}_r = \dot{\theta}\,\mathbf{e}_\theta + \sin\theta\,\dot{\phi}\,\mathbf{e}_\phi$, $\dot{\mathbf{e}}_\theta = -\dot{\theta}\,\mathbf{e}_r + \cos\theta\,\dot{\phi}\,\mathbf{e}_\phi$, $\dot{\mathbf{e}}_\phi = -\sin\theta\,\dot{\phi}\,\mathbf{e}_r - \cos\theta\,\dot{\phi}\,\mathbf{e}_\theta$.

**47.** Express the velocity $\mathbf{v}$ and acceleration $\mathbf{a}$ of a particle in spherical coordinates.

**48.** Find the square of the element of arc length and the corresponding scale factors in (*a*) paraboloidal, (*b*) elliptic cylindrical, and (*c*) oblate spheroidal coordinates.

**49.** Find the volume element $dV$ in (*a*) paraboloidal, (*b*) elliptic cylindrical, and (*c*) bipolar coordinates.

**50.** Find (*a*) the scale factors and (*b*) the volume element $dV$ for prolate spheroidal coordinates.

**51.** Derive expressions for the scale factors in (*a*) ellipsoidal and (*b*) bipolar coordinates.

**52.** Find the elements of area of a volume element in (*a*) cylindrical, (*b*) spherical, and (*c*) paraboloidal coordinates.

**53.** Prove that a necessary and sufficient condition that a curvilinear coordinate system be orthogonal is that $g_{pq} = 0$ for $p \neq q$.

**54.** Find the Jacobian $J(\frac{x,y,z}{u_1,u_2,u_3})$ for (*a*) cylindrical, (*b*) spherical, (*c*) parabolic cylindrical, (*d*) elliptic cylindrical, and (*e*) prolate spheroidal coordinates.

**55.** Evaluate $\iiint\limits_V \sqrt{x^2+y^2}\,dx\,dy\,dz$, where $V$ is the region bounded by $z = x^2+y^2$ and $z = 8-(x^2+y^2)$. Hint: Use cylindrical coordinates.

**56.** Find the volume of the smaller of the two regions bounded by the sphere $x^2+y^2+z^2 = 16$ and the cone $z^2 = x^2+y^2$.

**57.** Use spherical coordinates to find the volume of the smaller of the two regions bounded by a sphere of radius $a$ and a plane intersecting the sphere at a distance $h$ from its centre.

**58.** (*a*) Describe the coordinate surfaces and coordinate curves for the system

$$x^2-y^2 = 2u_1\cos u_2, \quad xy = u_1\sin u_2, \quad z = u_3$$

(*b*) Show that the system is orthogonal. (*c*) Determine $J(\frac{x,y,z}{u_1,u_2,u_3})$ for the system. (*d*) Show that $u_1$ and $u_2$ are related to the cylindrical coordinates $\rho$ and $\phi$ and determine the relationship.

**59.** Find the moment of inertia of the region bounded by $x^2-y^2 = 2$, $x^2-y^2 = 4$, $xy=1$, $xy=2$, $z=1$ and $z=3$ with respect to the $z$ axis if the density is constant and equal to $\kappa$. Hint: Let $x^2-y^2 = 2u$, $xy = v$.

**60.** Find $\frac{\partial \mathbf{r}}{\partial u_1}, \frac{\partial \mathbf{r}}{\partial u_2}, \frac{\partial \mathbf{r}}{\partial u_3}, \nabla u_1, \nabla u_2, \nabla u_3$ in (*a*) cylindrical, (*b*) spherical, and (*c*) parabolic cylindrical coordinates. Show that $\mathbf{e}_1 = \mathbf{E}_1$, $\mathbf{e}_2 = \mathbf{E}_2$, $\mathbf{e}_3 = \mathbf{E}_3$ for these systems.

**61.** Given the coordinate transformation $u_1 = xy$, $2u_2 = x^2+y^2$, $u_3 = z$. (*a*) Show that the coordinate system is not orthogonal. (*b*) Find $J(\frac{x,y,z}{u_1,u_2,u_3})$. (*c*) Find $ds^2$.

**62.** Find $\nabla\Phi$, div **A** and curl **A** in parabolic cylindrical coordinates.

**63.** Express (*a*) $\nabla\psi$ and (*b*) $\nabla\cdot\mathbf{A}$ in spherical coordinates.

**64.** Find $\nabla^2\psi$ in oblate spheroidal coordinates.

**65.** Write the equation $\frac{\partial^2\Phi}{\partial x^2} + \frac{\partial^2\Phi}{\partial y^2} = \Phi$ in elliptic cylindrical coordinates.

**66.** Express Maxwell's equation $\nabla\times\mathbf{E} = -\frac{1}{c}\frac{\partial\mathbf{H}}{\partial t}$ in prolate spheroidal coordinates.

**67.** Express Schroedinger's equation of quantum mechanics $\nabla^2\psi + \frac{8\pi^2 m}{h^2}(E - V(x,y,z))\psi = 0$ in parabolic cylindrical coordinates where $m, h$ and $E$ are constants.

**68.** Write Laplace's equation in paraboloidal coordinates.

**69.** Express the heat equation $\frac{\partial U}{\partial t} = \kappa\nabla^2 U$ in spherical coordinates if $U$ is independent of (*a*) $\phi$, (*b*) $\phi$ and $\theta$, (*c*) $r$ and $t$, (*d*) $\phi$, $\theta$ and $t$.

**70.** Find the element of arc length on a sphere of radius $a$.

**71.** Prove that in any orthogonal curvilinear coordinate system, div curl $\mathbf{A} = 0$ and curl grad $\Phi = \mathbf{0}$.

72. Prove that the surface area of a given region $R$ of the surface $\mathbf{r} = \mathbf{r}(u,v)$ is $\iint_R \sqrt{EG-F^2}\, du\, dv$. Use this to determine the surface area of a sphere.

73. Prove that a vector of length $p$ which is everywhere normal to the surface $\mathbf{r} = \mathbf{r}(u,v)$ is given by

$$\mathbf{A} = \pm\, p\left(\frac{\partial \mathbf{r}}{\partial u} \times \frac{\partial \mathbf{r}}{\partial v}\right)\Big/\sqrt{EG-F^2}$$

74. (a) Describe the plane transformation $x = x(u,v)$, $y = y(u,v)$.
(b) Under what conditions will the $u, v$ coordinate lines be orthogonal?

75. Let $(x,y)$ be coordinates of a point $P$ in a rectangular $xy$ plane and $(u,v)$ the coordinates of a point $Q$ in a rectangular $uv$ plane. If $x = x(u,v)$ and $y = y(u,v)$ we say that there is a *correspondence* or *mapping* between points $P$ and $Q$.
(a) If $x = 2u + v$ and $y = u - 2v$, show that the lines in the $xy$ plane correspond to lines in the $uv$ plane.
(b) What does the square bounded by $x = 0$, $x = 5$, $y = 0$ and $y = 5$ correspond to in the $uv$ plane?
(c) Compute the Jacobian $J(\frac{x,y}{u,v})$ and show that this is related to the ratios of the areas of the square and its image in the $uv$ plane.

76. If $x = \frac{1}{2}(u^2 - v^2)$, $y = uv$ determine the image (or images) in the $uv$ plane of a square bounded by $x = 0$, $x = 1$, $y = 0$, $y = 1$ in the $xy$ plane.

77. Show that under suitable conditions on $F$ and $G$,

$$\int_0^\infty \int_0^\infty e^{-s(x+y)}\, F(x)\, G(y)\, dx\, dy \quad = \quad \int_0^\infty e^{-st} \left\{ \int_0^t F(u)\, G(t-u)\, du \right\} dt$$

Hint: Use the transformation $x + y = t$, $x = v$ from the $xy$ plane to the $vt$ plane. The result is important in the theory of Laplace transforms.

78. (a) If $x = 3u_1 + u_2 - u_3$, $y = u_1 + 2u_2 + 2u_3$, $z = 2u_1 - u_2 - u_3$, find the volumes of the cube bounded by $x = 0$, $x = 15$, $y = 0$, $y = 10$, $z = 0$ and $z = 5$, and the image of this cube in the $u_1u_2u_3$ rectangular coordinate system.
(b) Relate the ratio of these volumes to the Jacobian of the transformation.

79. Let $(x,y,z)$ and $(u_1,u_2,u_3)$ be respectively the rectangular and curvilinear coordinates of a point.
(a) If $x = 3u_1 + u_2 - u_3$, $y = u_1 + 2u_2 + 2u_3$, $z = 2u_1 - u_2 - u_3$, is the system $u_1u_2u_3$ orthogonal?
(b) Find $ds^2$ and $g$ for the system.
(c) What is the relation between this and the preceding problem?

80. If $x = u_1^2 + 2$, $y = u_1 + u_2$, $z = u_3^2 - u_1$ find (a) $g$ and (b) the Jacobian $J = \dfrac{\partial(x,y,z)}{\partial(u_1,u_2,u_3)}$. Verify that $J^2 = g$.

## ANSWERS TO SUPPLEMENTARY PROBLEMS.

36. (a) $u = c_1$ and $v = c_2$ are elliptic and hyperbolic cylinders respectively, having $z$ axis as common axis. $z = c_3$ are planes. See Fig. 7, page 139.
(b) $u = c_1$ and $v = c_2$ are circular cylinders whose intersections with the $xy$ plane are circles with centres on the $y$ and $x$ axes respectively and intersecting at right angles. The cylinders $u = c_1$ all pass through the points $(-a,0,0)$ and $(a,0,0)$. $z = c_3$ are planes. See Fig. 8, page 140.
(c) $u = c_1$ and $v = c_2$ are parabolic cylinders whose traces on the $xy$ plane are intersecting mutually perpendicular coaxial parabolas with vertices on the $x$ axis but on opposite sides of the origin. $z = c_3$ are planes. See Fig. 6, page 138.

The coordinate curves are the intersections of the coordinate surfaces.

**37.** (a) $r = \sqrt{x^2+y^2+z^2}$, $\theta = \arc\tan \dfrac{\sqrt{x^2+y^2}}{z}$, $\phi = \arc\tan \dfrac{y}{x}$

(b) $r = \sqrt{\rho^2+z^2}$, $\theta = \arc\tan \dfrac{\rho}{z}$, $\phi = \phi$

**38.** (a) $r = 3$, (b) $\theta = \pi/6$, (c) $r \sin^2\theta = \cos\theta$, (d) $\theta = \pi/2$,
(e) the plane $y = x$ is made up of the two half planes $\phi = \pi/4$ and $\phi = 5\pi/4$.

**39.** (a) Circle in the $xy$ plane $x^2+y^2 = 16$, $z = 0$. (b) Cylinder $x^2+y^2 = 16$ whose axis coincides with $z$ axis.
(c) The $yz$ plane where $y \geqq 0$. (d) The straight line $y = \sqrt{3}\,x$, $z = 1$ where $x \geqq 0$, $y \geqq 0$.

**40.** (a) Hyperbolic cylinder $x^2-y^2 = 8$. (b) The line joining points $(-4,0,0)$ and $(4,0,0)$, i.e. $x = t$, $y = 0$, $z = 0$ where $-4 \leqq t \leqq 4$. (c) Ellipse $\dfrac{x^2}{25} + \dfrac{y^2}{9} = 1$, $z = 2$. (d) The portion of the $x$ axis defined by $x \geqq 4$, $y = 0$, $z = 0$.

**41.** (a) Parabola $y^2 = -8(x-2)$, $z = 0$. (b) Parabola $y^2 = 2x+1$, $z = 2$. (c) Region in $xy$ plane bounded by parabolas $y^2 = -2(x-1/2)$, $y^2 = -8(x-2)$, $y^2 = 8(x+2)$ and $y^2 = 18(x+9/2)$ *including* the boundary.
(d) Same as (c) but excluding the boundary.

**42.** (a)
$$\mathbf{e}_r = \sin\theta\cos\phi\,\mathbf{i} + \sin\theta\sin\phi\,\mathbf{j} + \cos\theta\,\mathbf{k}$$
$$\mathbf{e}_\theta = \cos\theta\cos\phi\,\mathbf{i} + \cos\theta\sin\phi\,\mathbf{j} - \sin\theta\,\mathbf{k}$$
$$\mathbf{e}_\phi = -\sin\phi\,\mathbf{i} + \cos\phi\,\mathbf{j}$$

(b)
$$\mathbf{i} = \sin\theta\cos\phi\,\mathbf{e}_r + \cos\theta\cos\phi\,\mathbf{e}_\theta - \sin\phi\,\mathbf{e}_\phi$$
$$\mathbf{j} = \sin\theta\sin\phi\,\mathbf{e}_r + \cos\theta\sin\phi\,\mathbf{e}_\theta + \cos\phi\,\mathbf{e}_\phi$$
$$\mathbf{k} = \cos\theta\,\mathbf{e}_r - \sin\theta\,\mathbf{e}_\theta$$

**43.** $\mathbf{A} = A_r\mathbf{e}_r + A_\theta\mathbf{e}_\theta + A_\phi\mathbf{e}_\phi$ where
$$A_r = 2r\sin^2\theta\sin\phi\cos\phi - r\sin\theta\cos\theta\sin\phi + 3r\sin\theta\cos\theta\cos\phi$$
$$A_\theta = 2r\sin\theta\cos\theta\sin\phi\cos\phi - r\cos^2\theta\sin\phi - 3r\sin^2\theta\cos\phi$$
$$A_\phi = -2r\sin\theta\sin^2\phi - r\cos\theta\cos\phi$$

**47.** $\mathbf{v} = v_r\mathbf{e}_r + v_\theta\mathbf{e}_\theta + v_\phi\mathbf{e}_\phi$ where $v_r = \dot{r}$, $v_\theta = r\dot{\theta}$, $v_\phi = r\sin\theta\,\dot{\phi}$

$\mathbf{a} = a_r\mathbf{e}_r + a_\theta\mathbf{e}_\theta + a_\phi\mathbf{e}_\phi$ where $a_r = \ddot{r} - r\dot{\theta}^2 - r\sin^2\theta\,\dot{\phi}^2$,
$$a_\theta = \frac{1}{r}\frac{d}{dt}(r^2\dot{\theta}) - r\sin\theta\cos\theta\,\dot{\phi}^2,$$
$$a_\phi = \frac{1}{r\sin\theta}\frac{d}{dt}(r^2\sin^2\theta\,\dot{\phi})$$

**48.** (a) $ds^2 = (u^2+v^2)(du^2+dv^2) + u^2v^2\,d\phi^2$, $h_u = h_v = \sqrt{u^2+v^2}$, $h_\phi = uv$

(b) $ds^2 = a^2(\sinh^2 u + \sin^2 v)(du^2+dv^2) + dz^2$, $h_u = h_v = a\sqrt{\sinh^2 u + \sin^2 v}$, $h_z = 1$

(c) $ds^2 = a^2(\sinh^2\xi + \sin^2\eta)(d\xi^2+d\eta^2) + a^2\cosh^2\xi\cos^2\eta\,d\phi^2$,
$h_\xi = h_\eta = a\sqrt{\sinh^2\xi + \sin^2\eta}$, $h_\phi = a\cosh\xi\cos\eta$

**49.** (a) $uv(u^2+v^2)\,du\,dv\,d\phi$, (b) $a^2(\sinh^2 u + \sin^2 v)\,du\,dv\,dz$, (c) $\dfrac{a^2\,du\,dv\,dz}{(\cosh v - \cos u)^2}$

**50.** (a) $h_\xi = h_\eta = a\sqrt{\sinh^2\xi + \sin^2\eta}$, $h_\phi = a\sinh\xi\sin\eta$

(b) $a^3(\sinh^2\xi + \sin^2\eta)\sinh\xi\sin\eta\,d\xi\,d\eta\,d\phi$

**52.** (a) $\rho\, d\rho\, d\phi$, $\rho\, d\phi\, dz$, $d\rho\, dz$
(b) $r \sin\theta\, dr\, d\phi$, $r^2 \sin\theta\, d\theta\, d\phi$, $r\, dr\, d\theta$
(c) $(u^2+v^2)\, du\, dv$, $uv\sqrt{u^2+v^2}\, du\, d\phi$, $uv\sqrt{u^2+v^2}\, dv\, d\phi$

**54.** (a) $\rho$, (b) $r^2 \sin\theta$, (c) $u^2+v^2$, (d) $a^2(\sinh^2 u + \sin^2 v)$, (e) $a^3(\sinh^2\xi + \sin^2\eta) \sinh\xi \sin\eta$

**55.** $\dfrac{256\pi}{15}$ **56.** $\dfrac{64\pi(2-\sqrt{2})}{3}$ **57.** $\dfrac{\pi}{3}(2a^3 - 3a^2h + h^3)$ **58.** (c) $\frac{1}{2}$; (d) $u_1 = \frac{1}{2}\rho^2$, $u_2 = 2\phi$

**59.** $2\kappa$

**60.** (a)
$$\frac{\partial \mathbf{r}}{\partial \rho} = \cos\phi\,\mathbf{i} + \sin\phi\,\mathbf{j}, \qquad \nabla\rho = \frac{x\,\mathbf{i} + y\,\mathbf{j}}{\sqrt{x^2+y^2}} = \cos\phi\,\mathbf{i} + \sin\phi\,\mathbf{j}$$
$$\frac{\partial \mathbf{r}}{\partial \phi} = -\rho\sin\phi\,\mathbf{i} + \rho\cos\phi\,\mathbf{j}, \qquad \nabla\phi = \frac{-\sin\phi\,\mathbf{i} + \cos\phi\,\mathbf{j}}{\rho}$$
$$\frac{\partial \mathbf{r}}{\partial z} = \mathbf{k}, \qquad \nabla z = \mathbf{k}$$

(b)
$$\frac{\partial \mathbf{r}}{\partial r} = \sin\theta\cos\phi\,\mathbf{i} + \sin\theta\sin\phi\,\mathbf{j} + \cos\theta\,\mathbf{k}$$
$$\frac{\partial \mathbf{r}}{\partial \theta} = r\cos\theta\cos\phi\,\mathbf{i} + r\cos\theta\sin\phi\,\mathbf{j} - r\sin\theta\,\mathbf{k}$$
$$\frac{\partial \mathbf{r}}{\partial \phi} = -r\sin\theta\sin\phi\,\mathbf{i} + r\sin\theta\cos\phi\,\mathbf{j}$$
$$\nabla r = \frac{x\,\mathbf{i} + y\,\mathbf{j} + z\,\mathbf{k}}{\sqrt{x^2+y^2+z^2}} = \sin\theta\cos\phi\,\mathbf{i} + \sin\theta\sin\phi\,\mathbf{j} + \cos\theta\,\mathbf{k}$$
$$\nabla\theta = \frac{xz\,\mathbf{i} + yz\,\mathbf{j} - (x^2+y^2)\mathbf{k}}{(x^2+y^2+z^2)\sqrt{x^2+y^2}} = \frac{\cos\theta\cos\phi\,\mathbf{i} + \cos\theta\sin\phi\,\mathbf{j} - \sin\theta\,\mathbf{k}}{r}$$
$$\nabla\phi = \frac{-y\,\mathbf{i} + x\,\mathbf{j}}{x^2+y^2} = \frac{-\sin\phi\,\mathbf{i} + \cos\phi\,\mathbf{j}}{r\sin\theta}$$

(c)
$$\frac{\partial \mathbf{r}}{\partial u} = u\,\mathbf{i} + v\,\mathbf{j}, \qquad \frac{\partial \mathbf{r}}{\partial v} = -v\,\mathbf{i} + u\,\mathbf{j}, \qquad \frac{\partial \mathbf{r}}{\partial z} = \mathbf{k}$$
$$\nabla u = \frac{u\,\mathbf{i} + v\,\mathbf{j}}{u^2+v^2}, \qquad \nabla v = \frac{-v\,\mathbf{i} + u\,\mathbf{j}}{u^2+v^2}, \qquad \nabla z = \mathbf{k}$$

**61.** (b) $\dfrac{1}{y^2-x^2}$, (c) $ds^2 = \dfrac{(x^2+y^2)(du_1^2+du_2^2) - 4xy\,du_1 du_2}{(x^2-y^2)^2} + du_3^2 = \dfrac{u_2(du_1^2+du_2^2) - 2u_1\,du_1 du_2}{2(u_2^2-u_1^2)} + du_3^2$

**62.**
$$\nabla\Phi = \frac{1}{\sqrt{u^2+v^2}}\frac{\partial\Phi}{\partial u}\mathbf{e}_u + \frac{1}{\sqrt{u^2+v^2}}\frac{\partial\Phi}{\partial v}\mathbf{e}_v + \frac{\partial\Phi}{\partial z}\mathbf{e}_z$$
$$\text{div}\,\mathbf{A} = \frac{1}{u^2+v^2}\left[\frac{\partial}{\partial u}\left(\sqrt{u^2+v^2}\,A_u\right) + \frac{\partial}{\partial v}\left(\sqrt{u^2+v^2}\,A_v\right)\right] + \frac{\partial A_z}{\partial z}$$
$$\text{curl}\,\mathbf{A} = \frac{1}{u^2+v^2}\left[\left\{\frac{\partial A_z}{\partial v} - \frac{\partial}{\partial z}\left(\sqrt{u^2+v^2}\,A_v\right)\right\}\sqrt{u^2+v^2}\,\mathbf{e}_u\right.$$
$$+ \left\{\frac{\partial}{\partial z}\left(\sqrt{u^2+v^2}\,A_u\right) - \frac{\partial A_z}{\partial u}\right\}\sqrt{u^2+v^2}\,\mathbf{e}_v$$
$$\left. + \left\{\frac{\partial}{\partial u}\left(\sqrt{u^2+v^2}\,A_v\right) - \frac{\partial}{\partial v}\left(\sqrt{u^2+v^2}\,A_u\right)\right\}\mathbf{e}_z\right]$$

**63.** (a) $\nabla\psi = \frac{\partial\psi}{\partial r}\mathbf{e}_r + \frac{1}{r}\frac{\partial\psi}{\partial\theta}\mathbf{e}_\theta + \frac{1}{r\sin\theta}\frac{\partial\psi}{\partial\phi}\mathbf{e}_\phi$

(b) $\nabla\cdot\mathbf{A} = \frac{1}{r^2}\frac{\partial}{\partial r}(r^2A_r) + \frac{1}{r\sin\theta}\frac{\partial}{\partial\theta}(\sin\theta\, A_\theta) + \frac{1}{r\sin\theta}\frac{\partial A_\phi}{\partial\phi}$

**64.** $$\nabla^2\psi = \frac{1}{a^2\cosh\xi\,(\sinh^2\xi + \sin^2\eta)}\frac{\partial}{\partial\xi}(\cosh\xi\,\frac{\partial\psi}{\partial\xi}) + \frac{1}{a^2\cos\eta\,(\sinh^2\xi+\sin^2\eta)}\frac{\partial}{\partial\eta}(\cos\eta\,\frac{\partial\psi}{\partial\eta}) + \frac{1}{a^2\cosh^2\xi\,\cos^2\eta}\frac{\partial^2\psi}{\partial\phi^2}$$

**65.** $\frac{\partial^2\Phi}{\partial u^2} + \frac{\partial^2\Phi}{\partial v^2} = a^2(\sinh^2 u + \sin^2 v)\Phi$

**66.** $$\frac{1}{aRS^2}\left[\left\{\frac{\partial}{\partial\eta}(RE_\phi) - \frac{\partial}{\partial\phi}(SE_\eta)\right\} S\,\mathbf{e}_\xi + \left\{\frac{\partial}{\partial\phi}(SE_\xi) - \frac{\partial}{\partial\xi}(RE_\phi)\right\} S\,\mathbf{e}_\eta + \left\{\frac{\partial}{\partial\xi}(SE_\eta) - \frac{\partial}{\partial\eta}(SE_\xi)\right\} R\,\mathbf{e}_\phi\right]$$
$$= -\frac{1}{c}\frac{\partial H_\xi}{\partial t}\mathbf{e}_\xi - \frac{1}{c}\frac{\partial H_\eta}{\partial t}\mathbf{e}_\eta - \frac{1}{c}\frac{\partial H_\phi}{\partial t}\mathbf{e}_\phi$$

where $R \equiv \sinh\xi\,\sin\eta$ and $S \equiv \sqrt{\sinh^2\xi + \sin^2\eta}$ .

**67.** $\frac{1}{u^2+v^2}\left[\frac{\partial^2\psi}{\partial u^2} + \frac{\partial^2\psi}{\partial v^2}\right] + \frac{\partial^2\psi}{\partial z^2} + \frac{8\pi^2 m}{h^2}\Big(E - W(u,v,z)\Big)\psi = 0$, where $W(u,v,z) = V(x,y,z)$.

**68.** $uv^2\frac{\partial}{\partial u}(u\frac{\partial\psi}{\partial u}) + u^2v\frac{\partial}{\partial v}(v\frac{\partial\psi}{\partial v}) + (u^2+v^2)\frac{\partial^2\psi}{\partial\phi^2} = 0$

**69.** (a) $\frac{\partial U}{\partial t} = \kappa\left[\frac{1}{r^2}\frac{\partial}{\partial r}(r^2\frac{\partial U}{\partial r}) + \frac{1}{r^2\sin\theta}\frac{\partial}{\partial\theta}(\sin\theta\frac{\partial U}{\partial\theta})\right]$

(b) $\frac{\partial U}{\partial t} = \kappa\left[\frac{1}{r^2}\frac{\partial}{\partial r}(r^2\frac{\partial U}{\partial r})\right]$ (c) $\sin\theta\frac{\partial}{\partial\theta}(\sin\theta\frac{\partial U}{\partial\theta}) + \frac{\partial^2 U}{\partial\phi^2} = 0$ (d) $\frac{d}{dr}(r^2\frac{dU}{dr}) = 0$

**70.** $ds^2 = a^2\,[\,d\theta^2 + \sin^2\theta\,d\phi^2]$

**74.** (b) $\frac{\partial x}{\partial u}\frac{\partial x}{\partial v} + \frac{\partial y}{\partial u}\frac{\partial y}{\partial v} = 0$

**78.** (a) 750, 75; (b) Jacobian = 10

**79.** (a) No. (b) $ds^2 = 14\,du_1^2 + 6\,du_2^2 + 6\,du_3^2 + 6\,du_1du_2 - 6\,du_1du_3 + 8\,du_2du_3$, $g = 100$

**80.** (a) $g = 16u_1^2u_3^2$, (b) $J = 4u_1u_3$

# Chapter 8

# TENSOR ANALYSIS

**PHYSICAL LAWS** must be independent of any particular coordinate systems used in describing them mathematically, if they are to be valid. A study of the consequences of this requirement leads to *tensor analysis*, of great use in general relativity theory, differential geometry, mechanics, elasticity, hydrodynamics, electromagnetic theory and numerous other fields of science and engineering.

**SPACES OF $N$ DIMENSIONS.** In three dimensional space a point is a set of three numbers, called *coordinates*, determined by specifying a particular coordinate system or frame of reference. For example $(x,y,z)$, $(\rho,\phi,z)$, $(r,\theta,\phi)$ are coordinates of a point in rectangular, cylindrical and spherical coordinate systems respectively. A point in $N$ dimensional space is, by analogy, a set of $N$ numbers denoted by $(x^1, x^2, \ldots, x^N)$ where $1, 2, \ldots, N$ are taken not as exponents but as *superscripts*, a policy which will prove useful.

The fact that we cannot visualize points in spaces of dimension higher than three has of course nothing whatsoever to do with their existence.

**COORDINATE TRANSFORMATIONS.** Let $(x^1, x^2, \ldots, x^N)$ and $(\bar{x}^1, \bar{x}^2, \ldots, \bar{x}^N)$ be coordinates of a point in two different frames of reference. Suppose there exists $N$ independent relations between the coordinates of the two systems having the form

$$\begin{array}{lcl} \bar{x}^1 & = & \bar{x}^1(x^1, x^2, \ldots, x^N) \\ \bar{x}^2 & = & \bar{x}^2(x^1, x^2, \ldots, x^N) \\ \vdots & \vdots & \quad \vdots \\ \bar{x}^N & = & \bar{x}^N(x^1, x^2, \ldots, x^N) \end{array} \tag{1}$$

which we can indicate briefly by

$$\bar{x}^k = \bar{x}^k(x^1, x^2, \ldots, x^N) \qquad k = 1, 2, \ldots, N \tag{2}$$

where it is supposed that the functions involved are single-valued, continuous, and have continuous derivatives. Then conversely to each set of coordinates $(\bar{x}^1, \bar{x}^2, \ldots, \bar{x}^N)$ there will correspond a unique set $(x^1, x^2, \ldots, x^N)$ given by

$$x^k = x^k(\bar{x}^1, \bar{x}^2, \ldots, \bar{x}^N) \qquad k = 1, 2, \ldots, N \tag{3}$$

The relations (*2*) or (*3*) define a *transformation of coordinates* from one frame of reference to another.

**THE SUMMATION CONVENTION.** In writing an expression such as $a_1x^1 + a_2x^2 + \ldots + a_Nx^N$ we can use the short notation $\sum_{j=1}^{N} a_j x^j$. An even shorter notation is simply to write it as $a_jx^j$, where we adopt the convention that whenever an index (subscript or superscript) is repeated in a given term we are to sum over that index from 1 to $N$ unless otherwise specified. This is called the *summation convention.* Clearly, instead of using the index $j$ we could have used another letter, say $p$, and the sum could be written $a_px^p$. Any index which is repeated in a given term, so that the summation convention applies, is called a *dummy index* or *umbral index.*

An index occurring only once in a given term is called a *free index* and can stand for any of the numbers $1, 2, \ldots, N$ such as $k$ in equation (*2*) or (*3*), each of which represents $N$ equations.

**CONTRAVARIANT AND COVARIANT VECTORS.** If $N$ quantities $A^1, A^2, \ldots, A^N$ in a coordinate system $(x^1, x^2, \ldots, x^N)$ are related to $N$ other quantities $\bar{A}^1, \bar{A}^2, \ldots, \bar{A}^N$ in another coordinate system $(\bar{x}^1, \bar{x}^2, \ldots, \bar{x}^N)$ by the transformation equations

$$\bar{A}^p = \sum_{q=1}^{N} \frac{\partial \bar{x}^p}{\partial x^q} A^q \qquad p = 1, 2, \ldots, N$$

which by the conventions adopted can simply be written as

$$\bar{A}^p = \frac{\partial \bar{x}^p}{\partial x^q} A^q$$

they are called components of a *contravariant vector* or *contravariant tensor of the first rank or first order.* To provide motivation for this and later transformations, see Problems 33 and 34 of Chapter 7.

If $N$ quantities $A_1, A_2, \ldots, A_N$ in a coordinate system $(x^1, x^2, \ldots, x^N)$ are related to $N$ other quantities $\bar{A}_1, \bar{A}_2, \ldots, \bar{A}_N$ in another coordinate system $(\bar{x}^1, \bar{x}^2, \ldots, \bar{x}^N)$ by the transformation equations

$$\bar{A}_p = \sum_{q=1}^{N} \frac{\partial x^q}{\partial \bar{x}^p} A_q \qquad p = 1, 2, \ldots, N$$

or

$$\bar{A}_p = \frac{\partial x^q}{\partial \bar{x}^p} A_q$$

they are called components of a *covariant vector* or *covariant tensor of the first rank or first order.*

Note that a superscript is used to indicate contravariant components whereas a subscript is used to indicate covariant components; an exception occurs in the notation for coordinates.

Instead of speaking of a tensor whose components are $A^p$ or $A_p$ we shall often refer simply to the tensor $A^p$ or $A_p$. No confusion should arise from this.

**CONTRAVARIANT, COVARIANT AND MIXED TENSORS.** If $N^2$ quantities $A^{qs}$ in a coordinate system $(x^1, x^2, \ldots, x^N)$ are related to $N^2$ other quantities $\bar{A}^{pr}$ in another coordinate system $(\bar{x}^1, \bar{x}^2, \ldots, \bar{x}^N)$ by the transformation equations

$$\bar{A}^{pr} = \sum_{s=1}^{N} \sum_{q=1}^{N} \frac{\partial \bar{x}^p}{\partial x^q} \frac{\partial \bar{x}^r}{\partial x^s} A^{qs} \qquad p,r = 1, 2, \ldots, N$$

or

$$\bar{A}^{pr} = \frac{\partial \bar{x}^p}{\partial x^q}\frac{\partial \bar{x}^r}{\partial x^s} A^{qs}$$

by the adopted conventions, they are called *contravariant components of a tensor of the second rank* or rank two.

The $N^2$ quantities $A_{qs}$ are called *covariant components of a tensor of the second rank* if

$$\bar{A}_{pr} = \frac{\partial x^q}{\partial \bar{x}^p}\frac{\partial x^s}{\partial \bar{x}^r} A_{qs}$$

Similarly the $N^2$ quantities $A^q_s$ are called *components of a mixed tensor of the second rank* if

$$\bar{A}^p_r = \frac{\partial \bar{x}^p}{\partial x^q}\frac{\partial x^s}{\partial \bar{x}^r} A^q_s$$

**THE KRONECKER DELTA**, written $\delta^j_k$, is defined by

$$\delta^j_k = \begin{cases} 0 & \text{if } j \neq k \\ 1 & \text{if } j = k \end{cases}$$

As its notation indicates, it is a mixed tensor of the second rank.

**TENSORS OF RANK GREATER THAN TWO** are easily defined. For example, $A^{qst}_{kl}$ are the components of a mixed tensor of rank 5, contravariant of order 3 and covariant of order 2, if they transform according to the relations

$$\bar{A}^{prm}_{ij} = \frac{\partial \bar{x}^p}{\partial x^q}\frac{\partial \bar{x}^r}{\partial x^s}\frac{\partial \bar{x}^m}{\partial x^t}\frac{\partial x^k}{\partial \bar{x}^i}\frac{\partial x^l}{\partial \bar{x}^j} A^{qst}_{kl}$$

**SCALARS OR INVARIANTS.** Suppose $\phi$ is a function of the coordinates $x^k$, and let $\bar{\phi}$ denote the functional value under a transformation to a new set of coordinates $\bar{x}^k$. Then $\phi$ is called a *scalar* or *invariant* with respect to the coordinate transformation if $\phi = \bar{\phi}$. A scalar or invariant is also called a *tensor of rank zero.*

**TENSOR FIELDS.** If to each point of a region in $N$ dimensional space there corresponds a definite tensor, we say that a *tensor field* has been defined. This is a *vector field* or a *scalar field* according as the tensor is of rank one or zero. It should be noted that a tensor or tensor field is not just the set of its components in one special coordinate system but *all the possible sets* under *any* transformation of coordinates.

**SYMMETRIC AND SKEW-SYMMETRIC TENSORS.** A tensor is called *symmetric with respect to two contravariant or two covariant indices* if its components remain unaltered upon interchange of the indices. Thus if $A^{mpr}_{qs} = A^{pmr}_{qs}$ the tensor is symmetric in $m$ and $p$. If a tensor is symmetric with respect to *any* two contravariant *and any* two covariant indices, it is called *symmetric.*

A tensor is called *skew-symmetric with respect to two contravariant or two covariant indices* if its components change sign upon interchange of the indices. Thus if $A^{mpr}_{qs} = -A^{pmr}_{qs}$ the tensor is skew-symmetric in $m$ and $p$. If a tensor is skew-symmetric with respect to *any* two contravariant *and any* two covariant indices it is called *skew-symmetric.*

## FUNDAMENTAL OPERATIONS WITH TENSORS.

1. **Addition.** The *sum* of two or more tensors of the same rank and type (i.e. same number of contravariant indices and same number of covariant indices) is also a tensor of the same rank and type. Thus if $A_q^{mp}$ and $B_q^{mp}$ are tensors, then $C_q^{mp} = A_q^{mp} + B_q^{mp}$ is also a tensor. Addition of tensors is commutative and associative.

2. **Subtraction.** The *difference* of two tensors of the same rank and type is also a tensor of the same rank and type. Thus if $A_q^{mp}$ and $B_q^{mp}$ are tensors, then $D_q^{mp} = A_q^{mp} - B_q^{mp}$ is also a tensor.

3. **Outer Multiplication.** The *product* of two tensors is a tensor whose rank is the sum of the ranks of the given tensors. This product which involves ordinary multiplication of the components of the tensor is called the *outer product.* For example, $A_q^{pr}\, B_s^m = C_{qs}^{prm}$ is the outer product of $A_q^{pr}$ and $B_s^m$. However, note that not every tensor can be written as a product of two tensors of lower rank. For this reason division of tensors is not always possible.

4. **Contraction.** If one contravariant and one covariant index of a tensor are set equal, the result indicates that a summation over the equal indices is to be taken according to the summation convention. This resulting sum is a tensor of rank two less than that of the original tensor. The process is called *contraction.* For example, in the tensor of rank 5, $A_{qs}^{mpr}$, set $r=s$ to obtain $A_{qr}^{mpr} = B_q^{mp}$ a tensor of rank 3. Further, by setting $p=q$ we obtain $B_p^{mp} = C^m$ a tensor of rank 1.

5. **Inner Multiplication.** By the process of outer multiplication of two tensors followed by a contraction, we obtain a new tensor called an *inner product* of the given tensors. The process is called *inner multiplication.* For example, given the tensors $A_q^{mp}$ and $B_{st}^r$, the outer product is $A_q^{mp} B_{st}^r$. Letting $q=r$, we obtain the inner product $A_r^{mp} B_{st}^r$. Letting $q=r$ and $p=s$, another inner product $A_r^{mp} B_{pt}^r$ is obtained. Inner and outer multiplication of tensors is commutative and associative.

6. **Quotient Law.** Suppose it is not known whether a quantity $X$ is a tensor or not. If an inner product of $X$ with an arbitrary tensor is itself a tensor, then $X$ is also a tensor. This is called the *quotient law.*

**MATRICES.** A matrix of order $m$ by $n$ is an array of quantities $a_{pq}$, called *elements*, arranged in $m$ rows and $n$ columns and generally denoted by

$$\begin{pmatrix} a_{11} & a_{12} & \cdots & a_{1n} \\ a_{21} & a_{22} & \cdots & a_{2n} \\ \vdots & \vdots & & \vdots \\ a_{m1} & a_{m2} & \cdots & a_{mn} \end{pmatrix} \quad \text{or} \quad \begin{bmatrix} a_{11} & a_{12} & \cdots & a_{1n} \\ a_{21} & a_{22} & \cdots & a_{2n} \\ \vdots & \vdots & & \vdots \\ a_{m1} & a_{m2} & \cdots & a_{mn} \end{bmatrix}$$

or in abbreviated form by $(a_{pq})$ or $[a_{pq}]$, $p=1,\ldots,m$; $q=1,\ldots,n$. If $m=n$ the matrix is a *square matrix* of order $m$ by $m$ or simply $m$; if $m=1$ it is a *row matrix* or *row vector*; if $n=1$ it is a *column matrix* or *column vector.*

The diagonal of a square matrix containing the elements $a_{11}, a_{22}, \ldots, a_{mm}$ is called the *principal* or *main diagonal.* A square matrix whose elements are equal to one in the principal diagonal and zero elsewhere is called a *unit matrix* and is denoted by $I$. A *null matrix*, denoted by $O$, is a matrix all of whose elements are zero.

**MATRIX ALGEBRA.** If $A = (a_{pq})$ and $B = (b_{pq})$ are matrices having the same order ($m$ by $n$) then

1. $A = B$ if and only if $a_{pq} = b_{pq}$.

2. The *sum* $S$ and *difference* $D$ are the matrices defined by
$$S = A+B = (a_{pq} + b_{pq}), \qquad D = A - B = (a_{pq} - b_{pq})$$

3. The *product* $P = AB$ is defined only when the number $n$ of columns in $A$ equals the number of rows in $B$ and is then given by
$$P = AB = (a_{pq})(b_{pq}) = (a_{pr}\, b_{rq})$$
where $a_{pr}\, b_{rq} = \sum_{r=1}^{n} a_{pr} b_{rq}$ by the summation convention. Matrices whose product is defined are called *conformable.*

   In general, multiplication of matrices is not commutative, i.e. $AB \neq BA$. However the associative law for multiplication of matrices holds, i.e. $A(BC) = (AB)C$ provided the matrices are conformable. Also the distributive laws hold, i.e. $A(B+C) = AB + AC$, $(A+B)C = AC + BC$.

4. The *determinant* of a square matrix $A = (a_{pq})$ is denoted by $|A|$, det $A$, $|a_{pq}|$ or $\det(a_{pq})$.
   If $P = AB$ then $|P| = |A|\,|B|$.

5. The *inverse* of a square matrix $A$ is a matrix $A^{-1}$ such that $AA^{-1} = I$, where $I$ is the unit matrix. A necessary and sufficient condition that $A^{-1}$ exist is that $\det A \neq 0$. If $\det A = 0$, $A$ is called *singular.*

6. The product of a scalar $\lambda$ by a matrix $A = (a_{pq})$, denoted by $\lambda A$, is the matrix $(\lambda a_{pq})$ where each element of $A$ is multiplied by $\lambda$.

7. The *transpose* of a matrix $A$ is a matrix $A^T$ which is formed from $A$ by interchanging its rows and columns. Thus if $A = (a_{pq})$, then $A^T = (a_{qp})$. The transpose of $A$ is also denoted by $\tilde{A}$.

**THE LINE ELEMENT AND METRIC TENSOR.** In rectangular coordinates $(x, y, z)$ the differential of arc length $ds$ is obtained from $ds^2 = dx^2 + dy^2 + dz^2$. By transforming to general curvilinear coordinates (see Problem 17, Chapter 7) this becomes $ds^2 = \sum_{p=1}^{3} \sum_{q=1}^{3} g_{pq}\, du_p\, du_q$. Such spaces are called *three dimensional Euclidean spaces.*

A generalization to $N$ dimensional space with coordinates $(x^1, x^2, \ldots, x^N)$ is immediate. We define the *line element ds* in this space to be given by the quadratic form, called the *metric form* or *metric*,

$$ds^2 = \sum_{p=1}^{N} \sum_{q=1}^{N} g_{pq}\, dx^p\, dx^q$$

or, using the summation convention,

$$ds^2 = g_{pq}\, dx^p\, dx^q$$

In the special case where there exists a transformation of coordinates from $x^j$ to $\bar{x}^k$ such that the metric form is transformed into $(d\bar{x}^1)^2 + (d\bar{x}^2)^2 + \ldots + (d\bar{x}^N)^2$ or $d\bar{x}^k d\bar{x}^k$, then the space is called *N dimensional Euclidean space*. In the general case, however, the space is called *Riemannian*.

The quantities $g_{pq}$ are the components of a covariant tensor of rank two called the *metric tensor* or *fundamental tensor*. We can and always will choose this tensor to be symmetric (see Problem 29).

**CONJUGATE OR RECIPROCAL TENSORS.** Let $g = |g_{pq}|$ denote the determinant with elements $g_{pq}$ and suppose $g \neq 0$. Define $g^{pq}$ by

$$g^{pq} = \frac{\text{cofactor of } g_{pq}}{g}$$

Then $g^{pq}$ is a symmetric contravariant tensor of rank two called the *conjugate* or *reciprocal tensor* of $g_{pq}$ (see Problem 34). It can be shown (Problem 33) that

$$g^{pq} g_{rq} = \delta^p_r$$

**ASSOCIATED TENSORS.** Given a tensor, we can derive other tensors by raising or lowering indices. For example, given the tensor $A_{pq}$ we obtain by raising the index $p$, the tensor $A^p_{\cdot q}$, the dot indicating the original position of the moved index. By raising the index $q$ also we obtain $A^{pq}_{\cdot\cdot}$. Where no confusion can arise we shall often omit the dots; thus $A^{pq}_{\cdot\cdot}$ can be written $A^{pq}$. These derived tensors can be obtained by forming inner products of the given tensor with the metric tensor $g_{pq}$ or its conjugate $g^{pq}$. Thus, for example

$$A^p_{\cdot q} = g^{rp} A_{rq}, \qquad A^{pq} = g^{rp} g^{sq} A_{rs}, \qquad A^p_{\cdot rs} = g_{rq} A^{pq}_{\cdot\cdot s}$$

$$A^{qm\cdot tk}_{\cdot\cdot n} = g^{pk} g_{sn} g^{rm} A^{q\cdot st}_{\cdot r\cdot\cdot p}$$

These become clear if we interpret multiplication by $g^{rp}$ as meaning: let $r = p$ (or $p = r$) in whatever follows and *raise* this index. Similarly we interpret multiplication by $g_{rq}$ as meaning: let $r = q$ (or $q = r$) in whatever follows and lower this index.

All tensors obtained from a given tensor by forming inner products with the metric tensor and its conjugate are called *associated tensors* of the given tensor. For example $A^m$ and $A_m$ are associated tensors, the first are contravariant and the second covariant components. The relation between them is given by

$$A_p = g_{pq} A^q \qquad \text{or} \qquad A^p = g^{pq} A_q$$

For rectangular coordinates $g_{pq} = 1$ if $p = q$, and 0 if $p \neq q$, so that $A_p = A^p$, which explains why no distinction was made between contravariant and covariant components of a vector in earlier chapters.

**LENGTH OF A VECTOR, ANGLE BETWEEN VECTORS.** The quantity $A^p B_p$, which is the inner product of $A^p$ and $B_q$, is a scalar analogous to the scalar product in rectangular coordinates. We define the length $L$ of the vector $A^p$ or $A_p$ as given by

$$L^2 = A^p A_p = g^{pq} A_p A_q = g_{pq} A^p A^q$$

We can define the angle $\theta$ between $A^p$ and $B_p$ as given by

$$\cos\theta = \frac{A^p B_p}{\sqrt{(A^p A_p)(B^p B_p)}}$$

**THE PHYSICAL COMPONENTS** of a vector $A^p$ or $A_p$, denoted by $A_u, A_v$, and $A_w$ are the projections of the vector on the tangents to the coordinate curves and are given in the case of orthogonal coordinates by

$$A_u = \sqrt{g_{11}}\, A^1 = \frac{A_1}{\sqrt{g_{11}}}, \qquad A_v = \sqrt{g_{22}}\, A^2 = \frac{A_2}{\sqrt{g_{22}}}, \qquad A_w = \sqrt{g_{33}}\, A^3 = \frac{A_3}{\sqrt{g_{33}}}$$

Similarly the physical components of a tensor $A^{pq}$ or $A_{pq}$ are given by

$$A_{uu} = g_{11} A^{11} = \frac{A_{11}}{g_{11}}, \qquad A_{uv} = \sqrt{g_{11} g_{22}}\, A^{12} = \frac{A_{12}}{\sqrt{g_{11} g_{22}}}, \qquad A_{uw} = \sqrt{g_{11} g_{33}}\, A^{13} = \frac{A_{13}}{\sqrt{g_{11} g_{33}}}, \qquad \text{etc.}$$

**CHRISTOFFEL'S SYMBOLS.** The symbols

$$[pq,r] = \frac{1}{2}\left(\frac{\partial g_{pr}}{\partial x^q} + \frac{\partial g_{qr}}{\partial x^p} - \frac{\partial g_{pq}}{\partial x^r}\right)$$

$$\left\{ {s \atop pq} \right\} = g^{sr}[pq,r]$$

are called the *Christoffel symbols of the first and second kind* respectively. Other symbols used instead of $\left\{ {s \atop pq} \right\}$ are $\{pq,s\}$ and $\Gamma^s_{pq}$. The latter symbol suggests however a tensor character, which is not true in general.

**TRANSFORMATION LAWS OF CHRISTOFFEL'S SYMBOLS.** If we denote by a bar a symbol in a coordinate system $\bar{x}^k$, then

$$\overline{[jk,m]} = [pq,r]\frac{\partial x^p}{\partial \bar{x}^j}\frac{\partial x^q}{\partial \bar{x}^k}\frac{\partial x^r}{\partial \bar{x}^m} + g_{pq}\frac{\partial x^p}{\partial \bar{x}^m}\frac{\partial^2 x^q}{\partial \bar{x}^j \partial \bar{x}^k}$$

$$\overline{\left\{ {n \atop jk} \right\}} = \left\{ {s \atop pq} \right\}\frac{\partial \bar{x}^n}{\partial x^s}\frac{\partial x^p}{\partial \bar{x}^j}\frac{\partial x^q}{\partial \bar{x}^k} + \frac{\partial \bar{x}^n}{\partial x^q}\frac{\partial^2 x^q}{\partial \bar{x}^j \partial \bar{x}^k}$$

are the laws of transformation of the Christoffel symbols showing that they are not tensors unless the second terms on the right are zero.

**GEODESICS.** The distance $s$ between two points $t_1$ and $t_2$ on a curve $x^r = x^r(t)$ in a Riemannian space is given by

$$s = \int_{t_1}^{t_2} \sqrt{g_{pq}\frac{dx^p}{dt}\frac{dx^q}{dt}}\, dt$$

That curve in the space which makes the distance a minimum is called a *geodesic* of the space. By use of the *calculus of variations* (see Problems 50 and 51) the geodesics are found from the differential equation

$$\frac{d^2 x^r}{ds^2} + \left\{ {r \atop pq} \right\} \frac{dx^p}{ds} \frac{dx^q}{ds} = 0$$

where $s$ is the arc length parameter. As examples, the geodesics on a plane are straight lines whereas the geodesics on a sphere are arcs of great circles.

**THE COVARIANT DERIVATIVE** of a tensor $A_p$ with repect to $x^q$ is denoted by $A_{p,q}$ and is defined by

$$A_{p,q} \equiv \frac{\partial A_p}{\partial x^q} - \left\{ {s \atop pq} \right\} A_s$$

a covariant tensor of rank two.

The covariant derivative of a tensor $A^p$ with respect to $x^q$ is denoted by $A^p_{\ ,q}$ and is defined by

$$A^p_{\ ,q} \equiv \frac{\partial A^p}{\partial x^q} + \left\{ {p \atop qs} \right\} A^s$$

a mixed tensor of rank two.

For rectangular systems, the Christoffel symbols are zero and the covariant derivatives are the usual partial derivatives. Covariant derivatives of tensors are also tensors (see Problem 52).

The above results can be extended to covariant derivatives of higher rank tensors. Thus

$$\begin{aligned} A^{p_1 \cdots p_m}_{r_1 \cdots r_n, q} \equiv & \frac{\partial A^{p_1 \cdots p_m}_{r_1 \cdots r_n}}{\partial x^q} \\ & - \left\{ {s \atop r_1 q} \right\} A^{p_1 \cdots p_m}_{s r_2 \cdots r_n} - \left\{ {s \atop r_2 q} \right\} A^{p_1 \cdots p_m}_{r_1 s r_3 \cdots r_n} - \cdots - \left\{ {s \atop r_n q} \right\} A^{p_1 \cdots p_m}_{r_1 \cdots r_{n-1} s} \\ & + \left\{ {p_1 \atop qs} \right\} A^{s p_2 \cdots p_m}_{r_1 \cdots r_n} + \left\{ {p_2 \atop qs} \right\} A^{p_1 s p_3 \cdots p_m}_{r_1 \cdots r_n} + \cdots + \left\{ {p_m \atop qs} \right\} A^{p_1 \cdots p_{m-1} s}_{r_1 \cdots r_n} \end{aligned}$$

is the covariant derivative of $A^{p_1 \cdots p_m}_{r_1 \cdots r_n}$ with respect to $x^q$.

The rules of covariant differentiation for sums and products of tensors are the same as those for ordinary differentiation. In performing the differentiations, the tensors $g_{pq}, g^{pq}$ and $\delta^p_q$ may be treated as constants since their covariant derivatives are zero (see Problem 54). Since covariant derivatives express rates of change of physical quantities independent of any frames of reference, they are of great importance in expressing physical laws.

**PERMUTATION SYMBOLS AND TENSORS.** Define $e_{pqr}$ by the relations

$e_{123} = e_{231} = e_{312} = +1$, $\quad e_{213} = e_{132} = e_{321} = -1$, $\quad e_{pqr} = 0$ if two or more indices are equal

and define $e^{pqr}$ in the same manner. The symbols $e_{pqr}$ and $e^{pqr}$ are called *permutation symbols* in three dimensional space.

Further, let us define

$$\epsilon_{pqr} = \frac{1}{\sqrt{g}} e_{pqr}, \qquad \epsilon^{pqr} = \sqrt{g}\, e^{pqr}$$

It can be shown that $\epsilon_{pqr}$ and $\epsilon^{pqr}$ are covariant and contravariant tensors respectively, called *permutation tensors* in three dimensional space. Generalizations to higher dimensions are possible.

**TENSOR FORM OF GRADIENT, DIVERGENCE AND CURL.**

1. **Gradient.** If $\Phi$ is a scalar or invariant the gradient of $\Phi$ is defined by

$$\nabla\Phi = \text{grad}\,\Phi = \Phi_{,p} = \frac{\partial\Phi}{\partial x^p}$$

where $\Phi_{,p}$ is the covariant derivative of $\Phi$ with respect to $x^p$.

2. **Divergence.** The divergence of $A^p$ is the contraction of its covariant derivative with respect to $x^q$, i.e. the contraction of $A^p_{,q}$. Then

$$\text{div}\, A^p = A^p_{,p} = \frac{1}{\sqrt{g}} \frac{\partial}{\partial x^k}(\sqrt{g}\, A^k)$$

3. **Curl.** The curl of $A_p$ is $A_{p,q} - A_{q,p} = \dfrac{\partial A_p}{\partial x^q} - \dfrac{\partial A_q}{\partial x^p}$, a tensor of rank two. The curl is also defined as $-\epsilon^{pqr} A_{p,q}$.

4. **Laplacian.** The Laplacian of $\Phi$ is the divergence of grad $\Phi$ or

$$\nabla^2\Phi = \text{div}\,\Phi_{,p} = \frac{1}{\sqrt{g}} \frac{\partial}{\partial x^j}\left(\sqrt{g}\, g^{jk} \frac{\partial\Phi}{\partial x^k}\right)$$

In case $g < 0$, $\sqrt{g}$ must be replaced by $\sqrt{-g}$. Both cases $g > 0$ and $g < 0$ can be included by writing $\sqrt{|g|}$ in place of $\sqrt{g}$.

**THE INTRINSIC OR ABSOLUTE DERIVATIVE** of $A_p$ along a curve $x^q = x^q(t)$, denoted by $\dfrac{\delta A_p}{\delta t}$, is defined as the inner product of the covariant derivative of $A_p$ and $\dfrac{dx^q}{dt}$, i.e. $A_{p,q}\dfrac{dx^q}{dt}$ and is given by

$$\frac{\delta A_p}{\delta t} \equiv \frac{dA_p}{dt} - \begin{Bmatrix} r \\ p\,q \end{Bmatrix} A_r \frac{dx^q}{dt}$$

Similarly, we define

$$\frac{\delta A^p}{\delta t} \equiv \frac{dA^p}{dt} + \begin{Bmatrix} p \\ q\,r \end{Bmatrix} A^r \frac{dx^q}{dt}$$

The vectors $A_p$ or $A^p$ are said to *move parallelly* along a curve if their intrinsic derivatives along the curve are zero, respectively.

Intrinsic derivatives of higher rank tensors are similarly defined.

**RELATIVE AND ABSOLUTE TENSORS.** A tensor $A^{p_1 \ldots p_m}_{r_1 \ldots r_n}$ is called a *relative tensor of weight w* if its components transform according to the equation

$$\bar{A}^{q_1 \ldots q_m}_{s_1 \ldots s_n} = \left|\frac{\partial x}{\partial \bar{x}}\right|^w A^{p_1 \ldots p_m}_{r_1 \ldots r_n} \frac{\partial \bar{x}^{q_1}}{\partial x^{p_1}} \cdots \frac{\partial \bar{x}^{q_m}}{\partial x^{p_m}} \frac{\partial x^{r_1}}{\partial \bar{x}^{s_1}} \cdots \frac{\partial x^{r_n}}{\partial \bar{x}^{s_n}}$$

where $J = \left|\frac{\partial x}{\partial \bar{x}}\right|$ is the Jacobian of the transformation. If $w = 0$ the tensor is called *absolute* and is the type of tensor with which we have been dealing above. If $w = 1$ the relative tensor is called a *tensor density*. The operations of addition, multiplication, etc., of relative tensors are similar to those of absolute tensors. See for example Problem 64.

# SOLVED PROBLEMS

## SUMMATION CONVENTION.

1. Write each of the following using the summation convention.

(a) $d\phi = \frac{\partial\phi}{\partial x^1}dx^1 + \frac{\partial\phi}{\partial x^2}dx^2 + \ldots + \frac{\partial\phi}{\partial x^N}dx^N$. $\qquad d\phi = \frac{\partial\phi}{\partial x^j}dx^j$

(b) $\frac{d\bar{x}^k}{dt} = \frac{\partial\bar{x}^k}{\partial x^1}\frac{dx^1}{dt} + \frac{\partial\bar{x}^k}{\partial x^2}\frac{dx^2}{dt} + \ldots + \frac{\partial\bar{x}^k}{\partial x^N}\frac{dx^N}{dt}$. $\qquad \frac{d\bar{x}^k}{dt} = \frac{\partial\bar{x}^k}{\partial x^m}\frac{dx^m}{dt}$

(c) $(x^1)^2 + (x^2)^2 + (x^3)^2 + \ldots + (x^N)^2$. $\qquad x^k x^k$

(d) $ds^2 = g_{11}(dx^1)^2 + g_{22}(dx^2)^2 + g_{33}(dx^3)^2$. $\qquad ds^2 = g_{kk}\,dx^k dx^k,\ N = 3$

(e) $\sum_{p=1}^{3}\sum_{q=1}^{3} g_{pq}\,dx^p dx^q$. $\qquad g_{pq}\,dx^p dx^q,\ N=3$

2. Write the terms in each of the following indicated sums.

(a) $a_{jk}x^k$. $\qquad \sum_{k=1}^{N} a_{jk}x^k = a_{j1}x^1 + a_{j2}x^2 + \ldots + a_{jN}x^N$

(b) $A_{pq}A^{qr}$. $\qquad \sum_{q=1}^{N} A_{pq}A^{qr} = A_{p1}A^{1r} + A_{p2}A^{2r} + \ldots + A_{pN}A^{Nr}$

(c) $\bar{g}_{rs} = g_{jk}\frac{\partial x^j}{\partial \bar{x}^r}\frac{\partial x^k}{\partial \bar{x}^s}$, $N = 3$.

$$\bar{g}_{rs} = \sum_{j=1}^{3}\sum_{k=1}^{3} g_{jk}\frac{\partial x^j}{\partial \bar{x}^r}\frac{\partial x^k}{\partial \bar{x}^s}$$

$$= \sum_{j=1}^{3}\left(g_{j1}\frac{\partial x^j}{\partial \bar{x}^r}\frac{\partial x^1}{\partial \bar{x}^s} + g_{j2}\frac{\partial x^j}{\partial \bar{x}^r}\frac{\partial x^2}{\partial \bar{x}^s} + g_{j3}\frac{\partial x^j}{\partial \bar{x}^r}\frac{\partial x^3}{\partial \bar{x}^s}\right)$$

$$= g_{11}\frac{\partial x^1}{\partial \bar{x}^r}\frac{\partial x^1}{\partial \bar{x}^s} + g_{21}\frac{\partial x^2}{\partial \bar{x}^r}\frac{\partial x^1}{\partial \bar{x}^s} + g_{31}\frac{\partial x^3}{\partial \bar{x}^r}\frac{\partial x^1}{\partial \bar{x}^s}$$

$$+ g_{12}\frac{\partial x^1}{\partial \bar{x}^r}\frac{\partial x^2}{\partial \bar{x}^s} + g_{22}\frac{\partial x^2}{\partial \bar{x}^r}\frac{\partial x^2}{\partial \bar{x}^s} + g_{32}\frac{\partial x^3}{\partial \bar{x}^r}\frac{\partial x^2}{\partial \bar{x}^s}$$

$$+ g_{13}\frac{\partial x^1}{\partial \bar{x}^r}\frac{\partial x^3}{\partial \bar{x}^s} + g_{23}\frac{\partial x^2}{\partial \bar{x}^r}\frac{\partial x^3}{\partial \bar{x}^s} + g_{33}\frac{\partial x^3}{\partial \bar{x}^r}\frac{\partial x^3}{\partial \bar{x}^s}$$

**3.** If $x^k$, $k = 1, 2, \ldots, N$ are rectangular coordinates, what locus if any, is represented by each of the following equations for $N = 2,3$ and $N \geqq 4$. Assume that the functions are single-valued, have continuous derivatives and are independent, when necessary.

(a) $a_k x^k = 1$, where $a_k$ are constants.

For $N = 2$, $a_1x^1 + a_2x^2 = 1$, a line in two dimensions, i.e. a line in a plane.
For $N = 3$, $a_1x^1 + a_2x^2 + a_3x^3 = 1$, a plane in 3 dimensions.
For $N \geqq 4$, $a_1x^1 + a_2x^2 + \ldots + a_Nx^N = 1$ is a *hyperplane*.

(b) $x^kx^k = 1$.

For $N = 2$, $(x^1)^2 + (x^2)^2 = 1$, a circle of unit radius in the plane.
For $N = 3$, $(x^1)^2 + (x^2)^2 + (x^3)^2 = 1$, a sphere of unit radius.
For $N \geqq 4$, $(x^1)^2 + (x^2)^2 + \ldots + (x^N)^2 = 1$, a *hypersphere* of unit radius.

(c) $x^k = x^k(u)$.

For $N = 2$, $x^1 = x^1(u)$, $x^2 = x^2(u)$, a plane curve with parameter $u$.
For $N = 3$, $x^1 = x^1(u)$, $x^2 = x^2(u)$, $x^3 = x^3(u)$, a three dimensional space curve.
For $N \geqq 4$, an $N$ dimensional space curve.

(d) $x^k = x^k(u,v)$.

For $N = 2$, $x^1 = x^1(u,v)$, $x^2 = x^2(u,v)$ is a transformation of coordinates from $(u,v)$ to $(x^1, x^2)$.
For $N = 3$, $x^1 = x^1(u,v)$, $x^2 = x^2(u,v)$, $x^3 = x^3(u,v)$ is a 3 dimensional surface with parameters $u$ and $v$.
For $N \geqq 4$, a *hypersurface*.

## CONTRAVARIANT AND COVARIANT VECTORS AND TENSORS.

**4.** Write the law of transformation for the tensors (a) $A^i_{jk}$, (b) $B^{mn}_{ijk}$, (c) $C^m$.

(a)
$$\bar{A}^p_{qr} = \frac{\partial \bar{x}^p}{\partial x^i}\frac{\partial x^j}{\partial \bar{x}^q}\frac{\partial x^k}{\partial \bar{x}^r} A^i_{jk}$$

As an aid for remembering the transformation, note that the relative positions of indices $p,q,r$ on the left side of the transformation are the same as those on the right side. Since these indices are associated with the $\bar{x}$ coordinates and since indices $i,j,k$ are associated respectively with indices $p,q,r$ the required transformation is easily written.

(b) $\bar{B}^{pq}_{rst} = \dfrac{\partial \bar{x}^p}{\partial x^m} \dfrac{\partial \bar{x}^q}{\partial x^n} \dfrac{\partial x^i}{\partial \bar{x}^r} \dfrac{\partial x^j}{\partial \bar{x}^s} \dfrac{\partial x^k}{\partial \bar{x}^t} B^{mn}_{ijk}$

(c) $\bar{C}^p = \dfrac{\partial \bar{x}^p}{\partial x^m} C^m$

**5.** A quantity $A(j,k,l,m)$ which is a function of coordinates $x^i$ transforms to another coordinate system $\bar{x}^i$ according to the rule

$$\bar{A}(p,q,r,s) = \frac{\partial x^j}{\partial \bar{x}^p} \frac{\partial \bar{x}^q}{\partial x^k} \frac{\partial \bar{x}^r}{\partial x^l} \frac{\partial \bar{x}^s}{\partial x^m} A(j,k,l,m)$$

(a) Is the quantity a tensor? (b) If so, write the tensor in suitable notation and (c) give the contravariant and covariant order and rank.

(a) Yes. (b) $A^{klm}_j$. (c) Contravariant of order 3, covariant of order 1 and rank $3+1=4$.

**6.** Determine whether each of the following quantities is a tensor. If so, state whether it is contravariant or covariant and give its rank: (a) $dx^k$, (b) $\dfrac{\partial \phi(x^1,\ldots,x^N)}{\partial x^k}$.

(a) Assume the transformation of coordinates $\bar{x}^j = \bar{x}^j(x^1,\ldots,x^N)$. Then $d\bar{x}^j = \dfrac{\partial \bar{x}^j}{\partial x^k} dx^k$ and so $dx^k$ is a contravariant tensor of rank one or a contravariant vector. Note that the location of the index $k$ is appropriate.

(b) Under the transformation $x^k = x^k(\bar{x}^1,\ldots,\bar{x}^N)$, $\phi$ is a function of $x^k$ and hence $\bar{x}^j$ such that $\phi(x^1,\ldots,x^N) = \bar{\phi}(\bar{x}^1,\ldots,\bar{x}^N)$, i.e. $\phi$ is a scalar or invariant (tensor of rank zero). By the chain rule for partial differentiation, $\dfrac{\partial \bar{\phi}}{\partial \bar{x}^j} = \dfrac{\partial \phi}{\partial \bar{x}^j} = \dfrac{\partial \phi}{\partial x^k}\dfrac{\partial x^k}{\partial \bar{x}^j} = \dfrac{\partial x^k}{\partial \bar{x}^j}\dfrac{\partial \phi}{\partial x^k}$ and $\dfrac{\partial \phi}{\partial x^k}$ transforms like $\bar{A}_j = \dfrac{\partial x^k}{\partial \bar{x}^j} A_k$. Then $\dfrac{\partial \phi}{\partial x^k}$ is a covariant tensor of rank one or a covariant vector.

Note that in $\dfrac{\partial \phi}{\partial x^k}$ the index appears in the denominator and thus acts like a subscript which indicates its covariant character. We refer to the tensor $\dfrac{\partial \phi}{\partial x^k}$ or equivalently, the tensor with components $\dfrac{\partial \phi}{\partial x^k}$, as the *gradient* of $\phi$, written grad $\phi$ or $\nabla\phi$.

**7.** A covariant tensor has components $xy$, $2y - z^2$, $xz$ in rectangular coordinates. Find its covariant components in spherical coordinates.

Let $A_j$ denote the covariant components in rectangular coordinates $x^1 = x$, $x^2 = y$, $x^3 = z$. Then

$$A_1 = xy = x^1x^2, \qquad A_2 = 2y - z^2 = 2x^2 - (x^3)^2, \qquad A_3 = x^1x^3$$

where care must be taken to distinguish between superscripts and exponents.

Let $\bar{A}_k$ denote the covariant components in spherical coordinates $\bar{x}^1 = r$, $\bar{x}^2 = \theta$, $\bar{x}^3 = \phi$. Then

$$(1) \qquad \bar{A}_k = \frac{\partial x^j}{\partial \bar{x}^k} A_j$$

The transformation equations between coordinate systems are

$$x^1 = \bar{x}^1 \sin \bar{x}^2 \cos \bar{x}^3, \qquad x^2 = \bar{x}^1 \sin \bar{x}^2 \sin \bar{x}^3, \qquad x^3 = \bar{x}^1 \cos \bar{x}^2$$

Then equations *(1)* yield the required covariant components

$$\begin{aligned}
\bar{A}_1 &= \frac{\partial x^1}{\partial \bar{x}^1} A_1 + \frac{\partial x^2}{\partial \bar{x}^1} A_2 + \frac{\partial x^3}{\partial \bar{x}^1} A_3 \\
&= (\sin \bar{x}^2 \cos \bar{x}^3)(x^1 x^2) + (\sin \bar{x}^2 \sin \bar{x}^3)(2x^2 - (x^3)^2) + (\cos \bar{x}^2)(x^1 x^3) \\
&= (\sin\theta \cos\phi)(r^2 \sin^2\theta \sin\phi \cos\phi) \\
&\qquad + (\sin\theta \sin\phi)(2r \sin\theta \sin\phi - r^2 \cos^2\theta) \\
&\qquad + (\cos\theta)(r^2 \sin\theta \cos\theta \cos\phi)
\end{aligned}$$

$$\begin{aligned}
\bar{A}_2 &= \frac{\partial x^1}{\partial \bar{x}^2} A_1 + \frac{\partial x^2}{\partial \bar{x}^2} A_2 + \frac{\partial x^3}{\partial \bar{x}^2} A_3 \\
&= (r\cos\theta \cos\phi)(r^2 \sin^2\theta \sin\phi \cos\phi) \\
&\qquad + (r\cos\theta \sin\phi)(2r \sin\theta \sin\phi - r^2 \cos^2\theta) \\
&\qquad + (-r\sin\theta)(r^2 \sin\theta \cos\theta \cos\phi)
\end{aligned}$$

$$\begin{aligned}
\bar{A}_3 &= \frac{\partial x^1}{\partial \bar{x}^3} A_1 + \frac{\partial x^2}{\partial \bar{x}^3} A_2 + \frac{\partial x^3}{\partial \bar{x}^3} A_3 \\
&= (-r\sin\theta \sin\phi)(r^2 \sin^2\theta \sin\phi \cos\phi) \\
&\qquad + (r\sin\theta \cos\phi)(2r \sin\theta \sin\phi - r^2 \cos^2\theta) \\
&\qquad + (0)(r^2 \sin\theta \cos\theta \cos\phi)
\end{aligned}$$

**8.** Show that $\dfrac{\partial A_p}{\partial x^q}$ is not a tensor even though $A_p$ is a covariant tensor of rank one.

By hypothesis, $\bar{A}_j = \dfrac{\partial x^p}{\partial \bar{x}^j} A_p$. Differentiating with respect to $\bar{x}^k$,

$$\begin{aligned}
\frac{\partial \bar{A}_j}{\partial \bar{x}^k} &= \frac{\partial x^p}{\partial \bar{x}^j}\frac{\partial A_p}{\partial \bar{x}^k} + \frac{\partial^2 x^p}{\partial \bar{x}^k \partial \bar{x}^j} A_p \\
&= \frac{\partial x^p}{\partial \bar{x}^j}\frac{\partial A_p}{\partial x^q}\frac{\partial x^q}{\partial \bar{x}^k} + \frac{\partial^2 x^p}{\partial \bar{x}^k \partial \bar{x}^j} A_p \\
&= \frac{\partial x^p}{\partial \bar{x}^j}\frac{\partial x^q}{\partial \bar{x}^k}\frac{\partial A_p}{\partial x^q} + \frac{\partial^2 x^p}{\partial \bar{x}^k \partial \bar{x}^j} A_p
\end{aligned}$$

Since the second term on the right is present, $\dfrac{\partial A_p}{\partial x^q}$ does not transform as a tensor should. Later we shall show how the addition of a suitable quantity to $\dfrac{\partial A_p}{\partial x^q}$ causes the result to be a tensor (Problem 52).

**9.** Show that the velocity of a fluid at any point is a contravariant tensor of rank one.

The velocity of a fluid at any point has components $\frac{dx^k}{dt}$ in the coordinate system $x^k$. In the coordinate system $\bar{x}^j$ the velocity is $\frac{d\bar{x}^j}{dt}$. But

$$\frac{d\bar{x}^j}{dt} = \frac{\partial \bar{x}^j}{\partial x^k}\frac{dx^k}{dt}$$

by the chain rule, and it follows that the velocity is a contravariant tensor of rank one or a contravariant vector.

## THE KRONECKER DELTA.

**10.** Evaluate (*a*) $\delta^p_q A^{qr}_s$, (*b*) $\delta^p_q \delta^q_r$.

Since $\delta^p_q = 1$ if $p=q$ and 0 if $p \neq q$, we have

$$(a)\ \delta^p_q A^{qr}_s = A^{pr}_s, \qquad (b)\ \delta^p_q \delta^q_r = \delta^p_r$$

**11.** Show that $\frac{\partial x^p}{\partial x^q} = \delta^p_q$.

If $p=q$, $\frac{\partial x^p}{\partial x^q} = 1$ since $x^p = x^q$.

If $p \neq q$, $\frac{\partial x^p}{\partial x^q} = 0$ since $x^p$ and $x^q$ are independent.

Then $\frac{\partial x^p}{\partial x^q} = \delta^p_q$.

**12.** Prove that $\frac{\partial x^p}{\partial \bar{x}^q}\frac{\partial \bar{x}^q}{\partial x^r} = \delta^p_r$.

Coordinates $x^p$ are functions of coordinates $\bar{x}^q$ which are in turn functions of coordinates $x^r$. Then by the chain rule and Problem 11,

$$\frac{\partial x^p}{\partial x^r} = \frac{\partial x^p}{\partial \bar{x}^q}\frac{\partial \bar{x}^q}{\partial x^r} = \delta^p_r$$

**13.** If $\bar{A}^p = \frac{\partial \bar{x}^p}{\partial x^q} A^q$ prove that $A^q = \frac{\partial x^q}{\partial \bar{x}^p}\bar{A}^p$.

Multiply equation $\bar{A}^p = \frac{\partial \bar{x}^p}{\partial x^q} A^q$ by $\frac{\partial x^r}{\partial \bar{x}^p}$.

Then $\dfrac{\partial x^r}{\partial \bar{x}^p}\bar{A}^p = \dfrac{\partial x^r}{\partial \bar{x}^p}\dfrac{\partial \bar{x}^p}{\partial x^q}A^q = \delta^r_q A^q = A^r$ by Prob. 12. Placing $r=q$ the result follows. This indicates that in the transformation equations for the tensor components the quantities with bars and quantities without bars can be interchanged, a result which can be proved in general.

**14.** Prove that $\delta^p_q$ is a mixed tensor of the second rank.

If $\delta^p_q$ is a mixed tensor of the second rank it must transform according to the rule

$$\bar{\delta}^j_k = \frac{\partial \bar{x}^j}{\partial x^p}\frac{\partial x^q}{\partial \bar{x}^k}\delta^p_q$$

The right side equals $\dfrac{\partial \bar{x}^j}{\partial x^p}\dfrac{\partial x^p}{\partial \bar{x}^k} = \delta^j_k$ by Problem 12. Since $\bar{\delta}^j_k = \delta^j_k = 1$ if $j=k$, and 0 if $j \neq k$, it follows that $\delta^p_q$ is a mixed tensor of rank two, justifying the notation used.

Note that we sometimes use $\delta_{pq} = 1$ if $p=q$ and 0 if $p \neq q$, as the Kronecker delta. This is however *not* a covariant tensor of the second rank as the notation would seem to indicate.

## FUNDAMENTAL OPERATIONS WITH TENSORS.

**15.** If $A^{pq}_r$ and $B^{pq}_r$ are tensors, prove that their sum and difference are tensors.

By hypothesis $A^{pq}_r$ and $B^{pq}_r$ are tensors, so that

$$\bar{A}^{jk}_l = \frac{\partial \bar{x}^j}{\partial x^p}\frac{\partial \bar{x}^k}{\partial x^q}\frac{\partial x^r}{\partial \bar{x}^l}A^{pq}_r$$

$$\bar{B}^{jk}_l = \frac{\partial \bar{x}^j}{\partial x^p}\frac{\partial \bar{x}^k}{\partial x^q}\frac{\partial x^r}{\partial \bar{x}^l}B^{pq}_r$$

Adding, $$(\bar{A}^{jk}_l + \bar{B}^{jk}_l) = \frac{\partial \bar{x}^j}{\partial x^p}\frac{\partial \bar{x}^k}{\partial x^q}\frac{\partial x^r}{\partial \bar{x}^l}(A^{pq}_r + B^{pq}_r)$$

Subtracting, $$(\bar{A}^{jk}_l - \bar{B}^{jk}_l) = \frac{\partial \bar{x}^j}{\partial x^p}\frac{\partial \bar{x}^k}{\partial x^q}\frac{\partial x^r}{\partial \bar{x}^l}(A^{pq}_r - B^{pq}_r)$$

Then $A^{pq}_r + B^{pq}_r$ and $A^{pq}_r - B^{pq}_r$ are tensors of the same rank and type as $A^{pq}_r$ and $B^{pq}_r$.

**16.** If $A^{pq}_r$ and $B^s_t$ are tensors, prove that $C^{pqs}_{rt} = A^{pq}_r B^s_t$ is also a tensor.

We must prove that $C^{pqs}_{rt}$ is a tensor whose components are formed by taking the products of components of tensors $A_r^{pq}$ and $B_t^s$. Since $A_r^{pq}$ and $B_t^s$ are tensors,

$$\bar{A}_l^{jk} = \frac{\partial \bar{x}^j}{\partial x^p}\frac{\partial \bar{x}^k}{\partial x^q}\frac{\partial x^r}{\partial \bar{x}^l} A_r^{pq}$$

$$\bar{B}_n^m = \frac{\partial \bar{x}^m}{\partial x^s}\frac{\partial x^t}{\partial \bar{x}^n} B_t^s$$

Multiplying,
$$\bar{A}_l^{jk}\bar{B}_n^m = \frac{\partial \bar{x}^j}{\partial x^p}\frac{\partial \bar{x}^k}{\partial x^q}\frac{\partial x^r}{\partial \bar{x}^l}\frac{\partial \bar{x}^m}{\partial x^s}\frac{\partial x^t}{\partial \bar{x}^n} A_r^{pq} B_t^s$$

which shows that $A_r^{pq}B_t^s$ is a tensor of rank 5, with contravariant indices $p,q,s$ and covariant indices $r,t$, thus warranting the notation $C_{rt}^{pqs}$. We call $C_{rt}^{pqs} = A_r^{pq}B_t^s$ the *outer product* of $A_r^{pq}$ and $B_t^s$.

**17.** Let $A_{rst}^{pq}$ be a tensor. (*a*) Choose $p=t$ and show that $A_{rsp}^{pq}$, where the summation convention is employed, is a tensor. What is its rank? (*b*) Choose $p=t$ and $q=s$ and show similarly that $A_{rqp}^{pq}$ is a tensor. What is its rank?

(*a*) Since $A_{rst}^{pq}$ is a tensor,

(*1*)
$$\bar{A}_{lmn}^{jk} = \frac{\partial \bar{x}^j}{\partial x^p}\frac{\partial \bar{x}^k}{\partial x^q}\frac{\partial x^r}{\partial \bar{x}^l}\frac{\partial x^s}{\partial \bar{x}^m}\frac{\partial x^t}{\partial \bar{x}^n} A_{rst}^{pq}$$

We must show that $A_{rsp}^{pq}$ is a tensor. Place the corresponding indices $j$ and $n$ equal to each other and sum over this index. Then

$$\bar{A}_{lmj}^{jk} = \frac{\partial \bar{x}^j}{\partial x^p}\frac{\partial \bar{x}^k}{\partial x^q}\frac{\partial x^r}{\partial \bar{x}^l}\frac{\partial x^s}{\partial \bar{x}^m}\frac{\partial x^t}{\partial \bar{x}^j} A_{rst}^{pq}$$

$$= \frac{\partial x^t}{\partial \bar{x}^j}\frac{\partial \bar{x}^j}{\partial x^p}\frac{\partial \bar{x}^k}{\partial x^q}\frac{\partial x^r}{\partial \bar{x}^l}\frac{\partial x^s}{\partial \bar{x}^m} A_{rst}^{pq}$$

$$= \delta_p^t \frac{\partial \bar{x}^k}{\partial x^q}\frac{\partial x^r}{\partial \bar{x}^l}\frac{\partial x^s}{\partial \bar{x}^m} A_{rst}^{pq}$$

$$= \frac{\partial \bar{x}^k}{\partial x^q}\frac{\partial x^r}{\partial \bar{x}^l}\frac{\partial x^s}{\partial \bar{x}^m} A_{rsp}^{pq}$$

and so $A_{rsp}^{pq}$ is a tensor of rank 3 and can be denoted by $B_{rs}^q$. The process of placing a contravariant index equal to a covariant index in a tensor and summing is called *contraction*. By such a process a tensor is formed whose rank is two less than the rank of the original tensor.

(*b*) We must show that $A_{rqp}^{pq}$ is a tensor. Placing $j=n$ and $k=m$ in equation (*1*) of part (*a*) and summing over $j$ and $k$, we have

$$\bar{A}^{jk}_{lkj} = \frac{\partial \bar{x}^j}{\partial x^p}\frac{\partial \bar{x}^k}{\partial x^q}\frac{\partial x^r}{\partial \bar{x}^l}\frac{\partial x^s}{\partial \bar{x}^k}\frac{\partial x^t}{\partial \bar{x}^j} A^{pq}_{rst}$$

$$= \frac{\partial x^t}{\partial \bar{x}^j}\frac{\partial \bar{x}^j}{\partial x^p}\frac{\partial x^s}{\partial \bar{x}^k}\frac{\partial \bar{x}^k}{\partial x^q}\frac{\partial x^r}{\partial \bar{x}^l} A^{pq}_{rst}$$

$$= \delta^t_p \delta^s_q \frac{\partial x^r}{\partial \bar{x}^l} A^{pq}_{rst}$$

$$= \frac{\partial x^r}{\partial \bar{x}^l} A^{pq}_{rqp}$$

which shows that $A^{pq}_{rqp}$ is a tensor of rank one and can be denoted by $C_r$. Note that by contracting twice, the rank was reduced by 4.

**18.** Prove that the contraction of the tensor $A^p_q$ is a scalar or invariant.

We have $\bar{A}^j_k = \dfrac{\partial \bar{x}^j}{\partial x^p}\dfrac{\partial x^q}{\partial \bar{x}^k} A^p_q$

Putting $j = k$ and summing, $\bar{A}^j_j = \dfrac{\partial \bar{x}^j}{\partial x^p}\dfrac{\partial x^q}{\partial \bar{x}^j} A^p_q = \delta^q_p A^p_q = A^p_p$

Then $\bar{A}^j_j = A^p_p$ and it follows that $A^p_p$ must be an invariant. Since $A^p_q$ is a tensor of rank two and contraction with respect to a single index lowers the rank by two, we are led to define an invariant as a tensor of rank zero.

**19.** Show that the contraction of the outer product of the tensors $A^p$ and $B_q$ is an invariant.

Since $A^p$ and $B_q$ are tensors, $\bar{A}^j = \dfrac{\partial \bar{x}^j}{\partial x^p} A^p$, $\bar{B}_k = \dfrac{\partial x^q}{\partial \bar{x}^k} B_q$. Then

$$\bar{A}^j \bar{B}_k = \frac{\partial \bar{x}^j}{\partial x^p}\frac{\partial x^q}{\partial \bar{x}^k} A^p B_q$$

By contraction (putting $j = k$ and summing)

$$\bar{A}^j \bar{B}_j = \frac{\partial \bar{x}^j}{\partial x^p}\frac{\partial x^q}{\partial \bar{x}^j} A^p B_q = \delta^q_p A^p B_q = A^p B_p$$

and so $A^p B_p$ is an invariant. The process of multiplying tensors (outer multiplication) and then contracting is called *inner multiplication* and the result is called an *inner product*. Since $A^p B_p$ is a scalar, it is often called the *scalar product* of the vectors $A^p$ and $B_q$.

**20.** Show that any inner product of the tensors $A^p_r$ and $B^{qs}_t$ is a tensor of rank three.

Outer product of $A^p_r$ and $B^{qs}_t = A^p_r B^{qs}_t$.

Let us contract with respect to indices $p$ and $t$, i.e. let $p=t$ and sum. We must show that the resulting inner product, represented by $A_r^p B_p^{qs}$, is a tensor of rank three.

By hypothesis, $A_r^p$ and $B_t^{qs}$ are tensors; then

$$\bar{A}_k^j = \frac{\partial \bar{x}^j}{\partial x^p}\frac{\partial x^r}{\partial \bar{x}^k} A_r^p, \qquad \bar{B}_n^{lm} = \frac{\partial \bar{x}^l}{\partial x^q}\frac{\partial \bar{x}^m}{\partial x^s}\frac{\partial x^t}{\partial \bar{x}^n} B_t^{qs}$$

Multiplying, letting $j=n$ and summing, we have

$$\begin{aligned}\bar{A}_k^j \bar{B}_j^{lm} &= \frac{\partial \bar{x}^j}{\partial x^p}\frac{\partial x^r}{\partial \bar{x}^k}\frac{\partial \bar{x}^l}{\partial x^q}\frac{\partial \bar{x}^m}{\partial x^s}\frac{\partial x^t}{\partial \bar{x}^j} A_r^p B_t^{qs} \\ &= \delta_p^t \frac{\partial x^r}{\partial \bar{x}^k}\frac{\partial \bar{x}^l}{\partial x^q}\frac{\partial \bar{x}^m}{\partial x^s} A_r^p B_t^{qs} \\ &= \frac{\partial x^r}{\partial \bar{x}^k}\frac{\partial \bar{x}^l}{\partial x^q}\frac{\partial \bar{x}^m}{\partial x^s} A_r^p B_p^{qs}\end{aligned}$$

showing that $A_r^p B_p^{qs}$ is a tensor of rank three. By contracting with respect to $q$ and $r$ or $s$ and $r$ in the product $A_r^p B_t^{qs}$, we can similarly show that any inner product is a tensor of rank three.

*Another Method.* The outer product of two tensors is a tensor whose rank is the sum of the ranks of the given tensors. Then $A_r^p B_t^{qs}$ is a tensor of rank $3+2=5$. Since a contraction results in a tensor whose rank is two less than that of the given tensor, it follows that any contraction of $A_r^p B_t^{qs}$ is a tensor of rank $5-2=3$.

**21.** If $X(p,q,r)$ is a quantity such that $X(p,q,r)B_r^{qn} = 0$ for an arbitrary tensor $B_r^{qn}$, prove that $X(p,q,r)=0$ identically.

Since $B_r^{qn}$ is an arbitrary tensor, choose one particular component (say the one with $q=2$, $r=3$) not equal to zero, while all other components are zero. Then $X(p,2,3)B_3^{2n}=0$, so that $X(p,2,3)=0$ since $B_3^{2n} \neq 0$. By similar reasoning with all possible combinations of $q$ and $r$, we have $X(p,q,r)=0$ and the result follows.

**22.** A quantity $A(p,q,r)$ is such that in the coordinate system $x^i$, $A(p,q,r)B_r^{qs} = C_p^s$ where $B_r^{qs}$ is an arbitrary tensor and $C_p^s$ is a tensor. Prove that $A(p,q,r)$ is a tensor.

In the transformed coordinates $\bar{x}^i$, $\bar{A}(j,k,l)\bar{B}_l^{km} = \bar{C}_j^m$.

$$\text{Then} \quad \bar{A}(j,k,l)\frac{\partial \bar{x}^k}{\partial x^q}\frac{\partial \bar{x}^m}{\partial x^s}\frac{\partial x^r}{\partial \bar{x}^l} B_r^{qs} = \frac{\partial \bar{x}^m}{\partial x^s}\frac{\partial x^p}{\partial \bar{x}^j} C_p^s = \frac{\partial \bar{x}^m}{\partial x^s}\frac{\partial x^p}{\partial \bar{x}^j} A(p,q,r) B_r^{qs}$$

or

$$\frac{\partial \bar{x}^m}{\partial x^s}\left[\frac{\partial \bar{x}^k}{\partial x^q}\frac{\partial x^r}{\partial \bar{x}^l}\bar{A}(j,k,l) - \frac{\partial x^p}{\partial \bar{x}^j} A(p,q,r)\right] B_r^{qs} = 0$$

Inner multiplication by $\frac{\partial x^n}{\partial \bar{x}^m}$ (i.e. multiplying by $\frac{\partial x^n}{\partial \bar{x}^t}$ and then contracting with $t=m$) yields

$$\delta_s^n \left[ \frac{\partial \bar{x}^k}{\partial x^q} \frac{\partial x^r}{\partial \bar{x}^l} \bar{A}(j,k,l) - \frac{\partial x^p}{\partial \bar{x}^j} A(p,q,r) \right] B_r^{qs} = 0$$

or

$$\left[ \frac{\partial \bar{x}^k}{\partial x^q} \frac{\partial x^r}{\partial \bar{x}^l} \bar{A}(j,k,l) - \frac{\partial x^p}{\partial \bar{x}^j} A(p,q,r) \right] B_r^{qn} = 0.$$

Since $B_r^{qn}$ is an arbitrary tensor, we have by Problem 21,

$$\frac{\partial \bar{x}^k}{\partial x^q} \frac{\partial x^r}{\partial \bar{x}^l} \bar{A}(j,k,l) - \frac{\partial x^p}{\partial \bar{x}^j} A(p,q,r) = 0$$

Inner multiplication by $\frac{\partial x^q}{\partial \bar{x}^m} \frac{\partial \bar{x}^n}{\partial x^r}$ yields

$$\delta_m^k \delta_l^n \bar{A}(j,k,l) - \frac{\partial x^p}{\partial \bar{x}^j} \frac{\partial x^q}{\partial \bar{x}^m} \frac{\partial \bar{x}^n}{\partial x^r} A(p,q,r) = 0$$

or

$$\bar{A}(j,m,n) = \frac{\partial x^p}{\partial \bar{x}^j} \frac{\partial x^q}{\partial \bar{x}^m} \frac{\partial \bar{x}^n}{\partial x^r} A(p,q,r)$$

which shows that $A(p,q,r)$ is a tensor and justifies use of the notation $A_{pq}^r$.

In this problem we have established a special case of the *quotient law* which states that if an inner product of a quantity $X$ with an arbitrary tensor $B$ is a tensor $C$, then $X$ is a tensor.

## SYMMETRIC AND SKEW-SYMMETRIC TENSORS.

**23.** If a tensor $A_{st}^{pqr}$ is symmetric (skew-symmetric) with respect to indices $p$ and $q$ in one coordinate system, show that it remains symmetric (skew-symmetric) with respect to $p$ and $q$ in any coordinate system.

Since only indices $p$ and $q$ are involved we shall prove the results for $B^{pq}$.

If $B^{pq}$ is symmetric, $B^{pq} = B^{qp}$. Then

$$\bar{B}^{jk} = \frac{\partial \bar{x}^j}{\partial x^p} \frac{\partial \bar{x}^k}{\partial x^q} B^{pq} = \frac{\partial \bar{x}^k}{\partial x^q} \frac{\partial \bar{x}^j}{\partial x^p} B^{qp} = \bar{B}^{kj}$$

and $B_{pq}$ remains symmetric in the $\bar{x}^i$ coordinate system.

If $B^{pq}$ is skew-symmetric, $B^{pq} = -B^{qp}$. Then

$$\bar{B}^{jk} = \frac{\partial \bar{x}^j}{\partial x^p} \frac{\partial \bar{x}^k}{\partial x^q} B^{pq} = -\frac{\partial \bar{x}^k}{\partial x^q} \frac{\partial \bar{x}^j}{\partial x^p} B^{qp} = -\bar{B}^{kj}$$

and $B_{pq}$ remains skew-symmetric in the $\bar{x}^i$ coordinate system.

The above results are, of course, valid for other symmetric (skew-symmetric) tensors.

24. Show that every tensor can be expressed as the sum of two tensors, one of which is symmetric and the other skew-symmetric in a pair of covariant or contravariant indices.

Consider, for example, the tensor $B^{pq}$. We have

$$B^{pq} = \tfrac{1}{2}(B^{pq} + B^{qp}) + \tfrac{1}{2}(B^{pq} - B^{qp})$$

But $R^{pq} = \frac{1}{2}(B^{pq} + B^{qp}) = R^{qp}$ is symmetric, and $S^{pq} = \frac{1}{2}(B^{pq} - B^{qp}) = -S^{qp}$ is skew-symmetric.

By similar reasoning the result is seen to be true for any tensor.

25. If $\Phi = a_{jk} A^j A^k$ show that we can always write $\Phi = b_{jk} A^j A^k$ where $b_{jk}$ is symmetric.

$$\Phi = a_{jk} A^j A^k = a_{kj} A^k A^j = a_{kj} A^j A^k$$

Then $$2\Phi = a_{jk} A^j A^k + a_{kj} A^j A^k = (a_{jk} + a_{kj}) A^j A^k$$

and $$\Phi = \tfrac{1}{2}(a_{jk} + a_{kj}) A^j A^k = b_{jk} A^j A^k$$

where $b_{jk} = \frac{1}{2}(a_{jk} + a_{kj}) = b_{kj}$ is symmetric.

**MATRICES.**

26. Write the sum $S = A + B$, difference $D = A - B$, and products $P = AB$, $Q = BA$ of the matrices

$$A = \begin{pmatrix} 3 & 1 & -2 \\ 4 & -2 & 3 \\ -2 & 1 & -1 \end{pmatrix}, \qquad B = \begin{pmatrix} 2 & 0 & -1 \\ -4 & 1 & 2 \\ 1 & -1 & 0 \end{pmatrix}$$

$$S = A + B = \begin{pmatrix} 3+2 & 1+0 & -2-1 \\ 4-4 & -2+1 & 3+2 \\ -2+1 & 1-1 & -1+0 \end{pmatrix} = \begin{pmatrix} 5 & 1 & -3 \\ 0 & -1 & 5 \\ -1 & 0 & -1 \end{pmatrix}$$

$$D = A - B = \begin{pmatrix} 3-2 & 1-0 & -2+1 \\ 4+4 & -2-1 & 3-2 \\ -2-1 & 1+1 & -1-0 \end{pmatrix} = \begin{pmatrix} 1 & 1 & -1 \\ 8 & -3 & 1 \\ -3 & 2 & -1 \end{pmatrix}$$

$$P = AB = \begin{pmatrix} (3)(2) + (1)(-4) + (-2)(1) & (3)(0) + (1)(1) + (-2)(-1) & (3)(-1) + (1)(2) + (-2)(0) \\ (4)(2) + (-2)(-4) + (3)(1) & (4)(0) + (-2)(1) + (3)(-1) & (4)(-1) + (-2)(2) + (3)(0) \\ (-2)(2) + (1)(-4) + (-1)(1) & (-2)(0) + (1)(1) + (-1)(-1) & (-2)(-1) + (1)(2) + (-1)(0) \end{pmatrix}$$

$$= \begin{pmatrix} 0 & 3 & -1 \\ 19 & -5 & -8 \\ -9 & 2 & 4 \end{pmatrix}$$

$$Q = BA = \begin{pmatrix} 8 & 1 & -3 \\ -12 & -4 & 9 \\ -1 & 3 & -5 \end{pmatrix}$$

This shows that $AB \neq BA$, i.e. multiplication of matrices is not commutative in general.

**27.** If $A = \begin{pmatrix} 2 & 1 \\ -1 & 3 \end{pmatrix}$ and $B = \begin{pmatrix} -1 & 2 \\ 3 & -2 \end{pmatrix}$, show that $(A+B)(A-B) \neq A^2 - B^2$.

$$A+B = \begin{pmatrix} 1 & 3 \\ 2 & 1 \end{pmatrix}, \quad A-B = \begin{pmatrix} 3 & -1 \\ -4 & 5 \end{pmatrix}. \quad \text{Then } (A+B)(A-B) = \begin{pmatrix} 1 & 3 \\ 2 & 1 \end{pmatrix}\begin{pmatrix} 3 & -1 \\ -4 & 5 \end{pmatrix} = \begin{pmatrix} -9 & 14 \\ 2 & 3 \end{pmatrix}.$$

$$A^2 = \begin{pmatrix} 2 & 1 \\ -1 & 3 \end{pmatrix}\begin{pmatrix} 2 & 1 \\ -1 & 3 \end{pmatrix} = \begin{pmatrix} 3 & 5 \\ -5 & 8 \end{pmatrix}, \quad B^2 = \begin{pmatrix} -1 & 2 \\ 3 & -2 \end{pmatrix}\begin{pmatrix} -1 & 2 \\ 3 & -2 \end{pmatrix} = \begin{pmatrix} 7 & -6 \\ -9 & 10 \end{pmatrix}.$$

Then $A^2 - B^2 = \begin{pmatrix} -4 & 11 \\ 4 & -2 \end{pmatrix}$.

Therefore, $(A+B)(A-B) \neq A^2 - B^2$. However, $(A+B)(A-B) = A^2 - AB + BA - B^2$.

**28.** Express in matrix notation the transformation equations for (*a*) a covariant vector, (*b*) a contravariant tensor of rank two, assuming $N = 3$.

(*a*) The transformation equations $\bar{A}_p = \dfrac{\partial x^q}{\partial \bar{x}^p} A_q$ can be written

$$\begin{pmatrix} \bar{A}_1 \\ \bar{A}_2 \\ \bar{A}_3 \end{pmatrix} = \begin{pmatrix} \dfrac{\partial x^1}{\partial \bar{x}^1} & \dfrac{\partial x^2}{\partial \bar{x}^1} & \dfrac{\partial x^3}{\partial \bar{x}^1} \\ \dfrac{\partial x^1}{\partial \bar{x}^2} & \dfrac{\partial x^2}{\partial \bar{x}^2} & \dfrac{\partial x^3}{\partial \bar{x}^2} \\ \dfrac{\partial x^1}{\partial \bar{x}^3} & \dfrac{\partial x^2}{\partial \bar{x}^3} & \dfrac{\partial x^3}{\partial \bar{x}^3} \end{pmatrix} \begin{pmatrix} A_1 \\ A_2 \\ A_3 \end{pmatrix}$$

in terms of column vectors, or equivalently in terms of row vectors

$$(\bar{A}_1 \; \bar{A}_2 \; \bar{A}_3) = (A_1 \; A_2 \; A_3) \begin{pmatrix} \dfrac{\partial x^1}{\partial \bar{x}^1} & \dfrac{\partial x^1}{\partial \bar{x}^2} & \dfrac{\partial x^1}{\partial \bar{x}^3} \\ \dfrac{\partial x^2}{\partial \bar{x}^1} & \dfrac{\partial x^2}{\partial \bar{x}^2} & \dfrac{\partial x^2}{\partial \bar{x}^3} \\ \dfrac{\partial x^3}{\partial \bar{x}^1} & \dfrac{\partial x^3}{\partial \bar{x}^2} & \dfrac{\partial x^3}{\partial \bar{x}^3} \end{pmatrix}$$

(*b*) The transformation equations $\bar{A}^{pr} = \dfrac{\partial \bar{x}^p}{\partial x^q} \dfrac{\partial \bar{x}^r}{\partial x^s} A^{qs}$ can be written

$$\begin{pmatrix} \bar{A}^{11} & \bar{A}^{12} & \bar{A}^{13} \\ \bar{A}^{21} & \bar{A}^{22} & \bar{A}^{23} \\ \bar{A}^{31} & \bar{A}^{32} & \bar{A}^{33} \end{pmatrix} = \begin{pmatrix} \dfrac{\partial \bar{x}^1}{\partial x^1} & \dfrac{\partial \bar{x}^1}{\partial x^2} & \dfrac{\partial \bar{x}^1}{\partial x^3} \\ \dfrac{\partial \bar{x}^2}{\partial x^1} & \dfrac{\partial \bar{x}^2}{\partial x^2} & \dfrac{\partial \bar{x}^2}{\partial x^3} \\ \dfrac{\partial \bar{x}^3}{\partial x^1} & \dfrac{\partial \bar{x}^3}{\partial x^2} & \dfrac{\partial \bar{x}^3}{\partial x^3} \end{pmatrix} \begin{pmatrix} A^{11} & A^{12} & A^{13} \\ A^{21} & A^{22} & A^{23} \\ A^{31} & A^{32} & A^{33} \end{pmatrix} \begin{pmatrix} \dfrac{\partial \bar{x}^1}{\partial x^1} & \dfrac{\partial \bar{x}^2}{\partial x^1} & \dfrac{\partial \bar{x}^3}{\partial x^1} \\ \dfrac{\partial \bar{x}^1}{\partial x^2} & \dfrac{\partial \bar{x}^2}{\partial x^2} & \dfrac{\partial \bar{x}^3}{\partial x^2} \\ \dfrac{\partial \bar{x}^1}{\partial x^3} & \dfrac{\partial \bar{x}^2}{\partial x^3} & \dfrac{\partial \bar{x}^3}{\partial x^3} \end{pmatrix}$$

Extensions of these results can be made for $N > 3$. For higher rank tensors, however, the matrix notation fails.

**THE LINE ELEMENT AND METRIC TENSOR.**

**29.** If $ds^2 = g_{jk}\,dx^j\,dx^k$ is an invariant, show that $g_{jk}$ is a symmetric covariant tensor of rank two.

By Problem 25, $\Phi = ds^2$, $A^j = dx^j$ and $A^k = dx^k$; it follows that $g_{jk}$ can be chosen symmetric. Also since $ds^2$ is an invariant,

$$\bar{g}_{pq}\,d\bar{x}^p\,d\bar{x}^q = g_{jk}\,dx^j\,dx^k = g_{jk}\frac{\partial x^j}{\partial \bar{x}^p}d\bar{x}^p\,\frac{\partial x^k}{\partial \bar{x}^q}d\bar{x}^q = g_{jk}\frac{\partial x^j}{\partial \bar{x}^p}\frac{\partial x^k}{\partial \bar{x}^q}\,d\bar{x}^p\,d\bar{x}^q$$

Then $\bar{g}_{pq} = g_{jk}\dfrac{\partial x^j}{\partial \bar{x}^p}\dfrac{\partial x^k}{\partial \bar{x}^q}$ and $g_{jk}$ is a symmetric covariant tensor of rank two, called the *metric tensor*.

**30.** Determine the metric tensor in (*a*) cylindrical and (*b*) spherical coordinates.

(*a*) As in Problem 7, Chapter 7, $ds^2 = d\rho^2 + \rho^2 d\phi^2 + dz^2$.

If $x^1 = \rho$, $x^2 = \phi$, $x^3 = z$ then $g_{11} = 1$, $g_{22} = \rho^2$, $g_{33} = 1$, $g_{12} = g_{21} = 0$, $g_{23} = g_{32} = 0$, $g_{31} = g_{13} = 0$.

In matrix form the metric tensor can be written $\begin{pmatrix} g_{11} & g_{12} & g_{13} \\ g_{21} & g_{22} & g_{23} \\ g_{31} & g_{32} & g_{33} \end{pmatrix} = \begin{pmatrix} 1 & 0 & 0 \\ 0 & \rho^2 & 0 \\ 0 & 0 & 1 \end{pmatrix}$

(*b*) As in Problem 8(*a*), Chapter 7, $ds^2 = dr^2 + r^2\,d\theta^2 + r^2\sin^2\theta\,d\phi^2$.

If $x^1 = r$, $x^2 = \theta$, $x^3 = \phi$ the metric tensor can be written $\begin{pmatrix} 1 & 0 & 0 \\ 0 & r^2 & 0 \\ 0 & 0 & r^2\sin^2\theta \end{pmatrix}$

In general for orthogonal coordinates, $g_{jk} = 0$ for $j \neq k$.

**31.** (*a*) Express the determinant $g = \begin{vmatrix} g_{11} & g_{12} & g_{13} \\ g_{21} & g_{22} & g_{23} \\ g_{31} & g_{32} & g_{33} \end{vmatrix}$ in terms of the elements in the second row and their corresponding cofactors. (*b*) Show that $g_{jk}\,G(j,k) = g$ where $G(j,k)$ is the cofactor of $g_{jk}$ in $g$ and where summation is over $k$ only.

(*a*) The cofactor of $g_{jk}$ is the determinant obtained from $g$ by (*1*) deleting the row and column in which $g_{jk}$ appears and (*2*) associating the sign $(-1)^{j+k}$ to this determinant. Thus,

$$\text{Cofactor of } g_{21} = (-1)^{2+1}\begin{vmatrix} g_{12} & g_{13} \\ g_{32} & g_{33} \end{vmatrix}, \qquad \text{Cofactor of } g_{22} = (-1)^{2+2}\begin{vmatrix} g_{11} & g_{13} \\ g_{31} & g_{33} \end{vmatrix},$$

$$\text{Cofactor of } g_{23} = (-1)^{2+3}\begin{vmatrix} g_{11} & g_{12} \\ g_{31} & g_{32} \end{vmatrix}$$

Denote these cofactors by $G(2,1)$, $G(2,2)$ and $G(2,3)$ respectively. Then by an elementary principle of determinants

$$g_{21}\,G(2,1) + g_{22}\,G(2,2) + g_{23}\,G(2,3) = g$$

(*b*) By applying the result of (*a*) to any row or column, we have $g_{jk}\, G(j,k) = g$ where the summation is over $k$ only. These results hold where $g = |g_{jk}|$ is an $N$th order determinant.

**32.** (*a*) Prove that $g_{21}\, G(3,1) + g_{22}\, G(3,2) + g_{23}\, G(3,3) = 0$.
(*b*) Prove that $g_{jk}\, G(p,k) = 0$ if $j \neq p$.

(*a*) Consider the determinant $\begin{vmatrix} g_{11} & g_{12} & g_{13} \\ g_{21} & g_{22} & g_{23} \\ g_{21} & g_{22} & g_{23} \end{vmatrix}$ which is zero since its last two rows are identical. Expanding according to elements of the last row we have

$$g_{21}\, G(3,1) + g_{22}\, G(3,2) + g_{23}\, G(3,3) = 0$$

(*b*) By setting the corresponding elements of any two rows (or columns) equal we can show, as in part (*a*), that $g_{jk}\, G(p,k) = 0$ if $j \neq p$. This result holds for $N$th order determinants as well.

**33.** Define $g^{jk} = \dfrac{G(j,k)}{g}$ where $G(j,k)$ is the cofactor of $g_{jk}$ in the determinant $g = |g_{jk}| \neq 0$. Prove that $g_{jk}\, g^{pk} = \delta_j^p$.

By Problem 31, $g_{jk} \dfrac{G(j,k)}{g} = 1$ or $g_{jk}\, g^{jk} = 1$, where summation is over $k$ only.

By Problem 32, $g_{jk} \dfrac{G(p,k)}{g} = 0$ or $g_{jk}\, g^{pk} = 0$ if $p \neq j$.

Then $g_{jk}\, g^{pk} (= 1 \text{ if } p = j, \text{ and } 0 \text{ if } p \neq j) = \delta_j^p$.

We have used the notation $g^{jk}$ although we have not yet shown that the notation is warranted, i.e. that $g^{jk}$ is a contravariant tensor of rank two. This is established in Problem 34. Note that the cofactor has been written $G(j,k)$ and not $G^{jk}$ since we can show that it is not a tensor in the usual sense. However, it can be shown to be a *relative tensor* of weight two which is contravariant, and with this extension of the tensor concept the notation $G^{jk}$ can be justified (see Supplementary Problem 152).

**34.** Prove that $g^{jk}$ is a symmetric contravariant tensor of rank two.

Since $g_{jk}$ is symmetric, $G(j,k)$ is symmetric and so $g^{jk} = G(j,k)/g$ is symmetric.

If $B^p$ is an arbitrary contravariant vector, $B_q = g_{pq} B^p$ is an arbitrary covariant vector. Multiplying by $g^{jq}$,

$$g^{jq} B_q = g^{jq} g_{pq} B^p = \delta_p^j B^p = B^j \quad \text{or} \quad g^{jq} B_q = B^j$$

Since $B_q$ is an arbitrary vector, $g^{jq}$ is a contravariant tensor of rank two, by application of the quotient law. The tensor $g^{jk}$ is called the *conjugate metric tensor*.

**35.** Determine the conjugate metric tensor in (*a*) cylindrical and (*b*) spherical coordinates.

(*a*) From Problem 30(*a*),
$$g = \begin{vmatrix} 1 & 0 & 0 \\ 0 & \rho^2 & 0 \\ 0 & 0 & 1 \end{vmatrix} = \rho^2$$

$$g^{11} = \frac{\text{cofactor of } g_{11}}{g} = \frac{1}{\rho^2}\begin{vmatrix} \rho^2 & 0 \\ 0 & 1 \end{vmatrix} = 1$$

$$g^{22} = \frac{\text{cofactor of } g_{22}}{g} = \frac{1}{\rho^2}\begin{vmatrix} 1 & 0 \\ 0 & 1 \end{vmatrix} = \frac{1}{\rho^2}$$

$$g^{33} = \frac{\text{cofactor of } g_{33}}{g} = \frac{1}{\rho^2}\begin{vmatrix} 1 & 0 \\ 0 & \rho^2 \end{vmatrix} = 1$$

$$g^{12} = \frac{\text{cofactor of } g_{12}}{g} = -\frac{1}{\rho^2}\begin{vmatrix} 0 & 0 \\ 0 & 1 \end{vmatrix} = 0$$

Similarly $g^{jk} = 0$ if $j \neq k$. In matrix form the conjugate metric tensor can be represented by

$$\begin{pmatrix} 1 & 0 & 0 \\ 0 & 1/\rho^2 & 0 \\ 0 & 0 & 1 \end{pmatrix}$$

(*b*) From Problem 30(*b*),
$$g = \begin{vmatrix} 1 & 0 & 0 \\ 0 & r^2 & 0 \\ 0 & 0 & r^2 \sin^2\theta \end{vmatrix} = r^4 \sin^2\theta$$

As in part (*a*), we find $g^{11} = 1$, $g^{22} = \dfrac{1}{r^2}$, $g^{33} = \dfrac{1}{r^2\sin^2\theta}$ and $g^{jk} = 0$ for $j \neq k$, and in matrix form this can be written

$$\begin{pmatrix} 1 & 0 & 0 \\ 0 & 1/r^2 & 0 \\ 0 & 0 & 1/r^2\sin^2\theta \end{pmatrix}$$

**36.** Find (*a*) $g$ and (*b*) $g^{jk}$ corresponding to $ds^2 = 5(dx^1)^2 + 3(dx^2)^2 + 4(dx^3)^2 - 6\,dx^1 dx^2 + 4\,dx^2 dx^3$.

(*a*) $g_{11} = 5$, $g_{22} = 3$, $g_{33} = 4$, $g_{12} = g_{21} = -3$, $g_{23} = g_{32} = 2$, $g_{13} = g_{31} = 0$. Then $g = \begin{vmatrix} 5 & -3 & 0 \\ -3 & 3 & 2 \\ 0 & 2 & 4 \end{vmatrix} = 4$.

(*b*) The cofactors $G(j,k)$ of $g_{jk}$ are

$G(1,1) = 8$, $G(2,2) = 20$, $G(3,3) = 6$, $G(1,2) = G(2,1) = 12$, $G(2,3) = G(3,2) = -10$, $G(1,3) = G(3,1) = -6$

Then $g^{11} = 2$, $g^{22} = 5$, $g^{33} = 3/2$, $g^{12} = g^{21} = 3$, $g^{23} = g^{32} = -5/2$, $g^{13} = g^{31} = -3/2$

Note that the product of the matrices $(g_{jk})$ and $(g^{jk})$ is the unit matrix I, i.e.

$$\begin{pmatrix} 5 & -3 & 0 \\ -3 & 3 & 2 \\ 0 & 2 & 4 \end{pmatrix}\begin{pmatrix} 2 & 3 & -3/2 \\ 3 & 5 & -5/2 \\ -3/2 & -5/2 & 3/2 \end{pmatrix} = \begin{pmatrix} 1 & 0 & 0 \\ 0 & 1 & 0 \\ 0 & 0 & 1 \end{pmatrix}$$

**ASSOCIATED TENSORS.**

**37.** If $A_j = g_{jk} A^k$, show that $A^k = g^{jk} A_j$.

Multiply $A_j = g_{jk} A^k$ by $g^{jq}$.

Then $g^{jq} A_j = g^{jq} g_{jk} A^k = \delta^q_k A^k = A^q$, i.e. $A^q = g^{jq} A_j$ or $A^k = g^{jk} A_j$.

The tensors of rank one, $A_j$ and $A^k$, are called *associated*. They represent the covariant and contravariant components of a vector.

**38.** (*a*) Show that $L^2 = g_{pq} A^p A^q$ is an invariant. (*b*) Show that $L^2 = g^{pq} A_p A_q$.

(*a*) Let $A_j$ and $A^k$ be the covariant and contravariant components of a vector. Then

$$\bar{A}_p = \frac{\partial x^j}{\partial \bar{x}^p} A_j, \qquad \bar{A}^q = \frac{\partial \bar{x}^q}{\partial x^k} A^k$$

and

$$\bar{A}_p \bar{A}^p = \frac{\partial x^j}{\partial \bar{x}^p} \frac{\partial \bar{x}^p}{\partial x^k} A_j A^k = \delta^j_k A_j A^k = A_j A^j$$

so that $A_j A^j$ is an invariant which we call $L^2$. Then we can write

$$L^2 = A_j A^j = g_{jk} A^k A^j = g_{pq} A^p A^q$$

(*b*) From (*a*), $L^2 = A_j A^j = A_j g^{kj} A_k = g^{jk} A_j A_k = g^{pq} A_p A_q$.

The scalar or invariant quantity $L = \sqrt{A_p A^p}$ is called the magnitude or length of the vector with covariant components $A_p$ and contravariant components $A^p$.

**39.** (*a*) If $A^p$ and $B^q$ are vectors, show that $g_{pq} A^p B^q$ is an invariant.

(*b*) Show that $\dfrac{g_{pq} A^p B^q}{\sqrt{(A^p A_p)(B^q B_q)}}$ is an invariant.

(*a*) By Problem 38, $A^p B_p = A^p g_{pq} B^q = g_{pq} A^p B^q$ is an invariant.

(*b*) Since $A^p A_p$ and $B^q B_q$ are invariants, $\sqrt{(A^p A_p)(B^q B_q)}$ is an invariant and so $\dfrac{g_{pq} A^p B^q}{\sqrt{(A^p A_p)(B^q B_q)}}$ is an invariant.

We define

$$\cos\theta = \frac{g_{pq} A^p B^q}{\sqrt{(A^p A_p)(B^q B_q)}}$$

as the *cosine of the angle between vectors* $A^p$ *and* $B^q$. If $g_{pq} A^p B^q = A^p B_p = 0$, the vectors are called *orthogonal*.

**40.** Express the relationship between the associated tensors:
(a) $A^{jkl}$ and $A_{pqr}$, (b) $A_{j.l}^{.k}$ and $A^{qkr}$, (c) $A_{.q..t}^{p.rs.}$ and $A_{jqk}^{...sl}$.

(a) $A^{jkl} = g^{jp} g^{kq} g^{lr} A_{pqr}$ or $A_{pqr} = g_{jp}\, g_{kq}\, g_{lr}\, A^{jkl}$

(b) $A_{j.l}^{.k} = g_{jq}\, g_{lr}\, A^{qkr}$ or $A^{qkr} = g^{jq} g^{lr} A_{j.l}^{.k}$

(c) $A_{.q..t}^{p.rs.} = g^{pj} g^{rk} g_{tl}\, A_{jqk}^{...sl}$ or $A_{jqk}^{...sl} = g_{pj}\, g_{rk}\, g^{tl}\, A_{.q..t}^{p.rs.}$

**41.** Prove that the angles $\theta_{12}$, $\theta_{23}$ and $\theta_{31}$ between the coordinate curves in a three dimensional coordinate system are given by

$$\cos\theta_{12} = \frac{g_{12}}{\sqrt{g_{11}\, g_{22}}}, \qquad \cos\theta_{23} = \frac{g_{23}}{\sqrt{g_{22}\, g_{33}}}, \qquad \cos\theta_{31} = \frac{g_{31}}{\sqrt{g_{33}\, g_{11}}}$$

Along the $x^1$ coordinate curve, $x^2$ = constant and $x^3$ = constant.

Then from the metric form, $ds^2 = g_{11}(dx^1)^2$ or $\dfrac{dx^1}{ds} = \dfrac{1}{\sqrt{g_{11}}}$.

Thus a unit tangent vector along the $x^1$ curve is $A_1^r = \dfrac{1}{\sqrt{g_{11}}}\,\delta_1^r$. Similarly, unit tangent vectors along the $x^2$ and $x^3$ coordinate curves are $A_2^r = \dfrac{1}{\sqrt{g_{22}}}\,\delta_2^r$ and $A_3^r = \dfrac{1}{\sqrt{g_{33}}}\,\delta_3^r$.

The cosine of the angle $\theta_{12}$ between $A_1^r$ and $A_2^r$ is given by

$$\cos\theta_{12} = g_{pq}\, A_1^p A_2^q = g_{pq}\, \frac{1}{\sqrt{g_{11}}}\,\frac{1}{\sqrt{g_{22}}}\,\delta_1^p\,\delta_2^q = \frac{g_{12}}{\sqrt{g_{11}\, g_{22}}}$$

Similarly we obtain the other results.

**42.** Prove that for an orthogonal coordinate system, $g_{12} = g_{23} = g_{31} = 0$.

This follows at once from Problem 41 by placing $\theta_{12} = \theta_{23} = \theta_{31} = 90°$. From the fact that $g_{pq} = g_{qp}$ it also follows that $g_{21} = g_{32} = g_{13} = 0$.

**43.** Prove that for an orthogonal coordinate system, $g_{11} = \dfrac{1}{g^{11}}$, $g_{22} = \dfrac{1}{g^{22}}$, $g_{33} = \dfrac{1}{g^{33}}$.

From Problem 33, $g^{pr}\, g_{rq} = \delta_q^p$.

If $p = q = 1$, $g^{1r}\, g_{r1} = 1$ or $g^{11} g_{11} + g^{12} g_{21} + g^{13} g_{31} = 1$.

Then using Problem 42, $g_{11} = \dfrac{1}{g^{11}}$.

Similarly if $p = q = 2$, $g_{22} = \dfrac{1}{g^{22}}$; and if $p = q = 3$, $g_{33} = \dfrac{1}{g^{33}}$.

**CHRISTOFFEL'S SYMBOLS.**

**44.** Prove (*a*) $[pq,r] = [qp,r]$, (*b*) $\left\{ {s \atop pq} \right\} = \left\{ {s \atop qp} \right\}$, (*c*) $[pq,r] = g_{rs} \left\{ {s \atop pq} \right\}$.

(*a*) $[pq,r] = \frac{1}{2}\left(\frac{\partial g_{pr}}{\partial x^q} + \frac{\partial g_{qr}}{\partial x^p} - \frac{\partial g_{pq}}{\partial x^r}\right) = \frac{1}{2}\left(\frac{\partial g_{qr}}{\partial x^p} + \frac{\partial g_{pr}}{\partial x^q} - \frac{\partial g_{qp}}{\partial x^r}\right) = [qp,r]$.

(*b*) $\left\{ {s \atop pq} \right\} = g^{sr}[pq,r] = g^{sr}[qp,r] = \left\{ {s \atop qp} \right\}$

(*c*) $g_{ks}\left\{ {s \atop pq} \right\} = g_{ks}\, g^{sr}\, [pq,r] = \delta_k^r\, [pq,r] = [pq,k]$

or $[pq,k] = g_{ks}\left\{ {s \atop pq} \right\}$ i.e. $[pq,r] = g_{rs}\left\{ {s \atop pq} \right\}$.

Note that multiplying $[pq,r]$ by $g^{sr}$ has the effect of replacing $r$ by $s$, raising this index and replacing square brackets by braces to yield $\left\{ {s \atop pq} \right\}$. Similarly, multiplying $\left\{ {s \atop pq} \right\}$ by $g_{rs}$ or $g_{sr}$ has the effect of replacing $s$ by $r$, lowering this index and replacing braces by square brackets to yield $[pq,r]$.

**45.** Prove (*a*) $\dfrac{\partial g_{pq}}{\partial x^m} = [pm,q] + [qm,p]$

(*b*) $\dfrac{\partial g^{pq}}{\partial x^m} = -g^{pn}\left\{ {q \atop mn} \right\} - g^{qn}\left\{ {p \atop mn} \right\}$ (*c*) $\left\{ {p \atop pq} \right\} = \dfrac{\partial}{\partial x^q} \ln \sqrt{g}$

(*a*) $[pm,q] + [qm,p] = \frac{1}{2}\left(\frac{\partial g_{pq}}{\partial x^m} + \frac{\partial g_{mq}}{\partial x^p} - \frac{\partial g_{pm}}{\partial x^q}\right) + \frac{1}{2}\left(\frac{\partial g_{qp}}{\partial x^m} + \frac{\partial g_{mp}}{\partial x^q} - \frac{\partial g_{qm}}{\partial x^p}\right) = \frac{\partial g_{pq}}{\partial x^m}$

(*b*) $\dfrac{\partial}{\partial x^m}(g^{jk} g_{ij}) = \dfrac{\partial}{\partial x^m}(\delta_i^k) = 0$. Then

$$g^{jk}\frac{\partial g_{ij}}{\partial x^m} + \frac{\partial g^{jk}}{\partial x^m} g_{ij} = 0 \quad \text{or} \quad g_{ij}\frac{\partial g^{jk}}{\partial x^m} = -g^{jk}\frac{\partial g_{ij}}{\partial x^m}$$

Multiplying by $g^{ir}$, $\quad g^{ir} g_{ij}\dfrac{\partial g^{jk}}{\partial x^m} = -g^{ir} g^{jk}\dfrac{\partial g_{ij}}{\partial x^m}$

i.e. $\quad \delta_j^r \dfrac{\partial g^{jk}}{\partial x^m} = -g^{ir} g^{jk}([im,j] + [jm,i])$

or $\quad \dfrac{\partial g^{rk}}{\partial x^m} = -g^{ir}\left\{ {k \atop im} \right\} - g^{jk}\left\{ {r \atop jm} \right\}$

and the result follows on replacing $r,k,i,j$ by $p,q,n,n$ respectively.

(*c*) From Problem 31, $g = g_{jk}\, G(j,k)$ (sum over $k$ only).

Since $G(j,k)$ does not contain $g_{jk}$ explicitly, $\dfrac{\partial g}{\partial g_{jr}} = G(j,r)$. Then, summing over $j$ and $r$,

$$\frac{\partial g}{\partial x^m} = \frac{\partial g}{\partial g_{jr}} \frac{\partial g_{jr}}{\partial x^m} = G(j,r) \frac{\partial g_{jr}}{\partial x^m}$$

$$= g\, g^{jr} \frac{\partial g_{jr}}{\partial x^m} = g\, g^{jr} ([jm,r] + [rm,j])$$

$$= g\left(\begin{Bmatrix} j \\ jm \end{Bmatrix} + \begin{Bmatrix} r \\ rm \end{Bmatrix}\right) = 2g \begin{Bmatrix} j \\ jm \end{Bmatrix}$$

Thus

$$\frac{1}{2g}\frac{\partial g}{\partial x^m} = \begin{Bmatrix} j \\ jm \end{Bmatrix} \quad \text{or} \quad \begin{Bmatrix} j \\ jm \end{Bmatrix} = \frac{\partial}{\partial x^m} \ln \sqrt{g}$$

The result follows on replacing $j$ by $p$ and $m$ by $q$.

**46.** Derive transformation laws for the Christoffel symbols of (*a*) the first kind, (*b*) the second kind.

(*a*) Since $\bar{g}_{jk} = \dfrac{\partial x^p}{\partial \bar{x}^j} \dfrac{\partial x^q}{\partial \bar{x}^k} g_{pq}$,

$$(1) \qquad \frac{\partial \bar{g}_{jk}}{\partial \bar{x}^m} = \frac{\partial x^p}{\partial \bar{x}^j}\frac{\partial x^q}{\partial \bar{x}^k}\frac{\partial g_{pq}}{\partial x^r}\frac{\partial x^r}{\partial \bar{x}^m} + \frac{\partial x^p}{\partial \bar{x}^j}\frac{\partial^2 x^q}{\partial \bar{x}^m \partial \bar{x}^k} g_{pq} + \frac{\partial^2 x^p}{\partial \bar{x}^m \partial \bar{x}^j}\frac{\partial x^q}{\partial \bar{x}^k} g_{pq}$$

By cyclic permutation of indices $j,k,m$ and $p,q,r$,

$$(2) \qquad \frac{\partial \bar{g}_{km}}{\partial \bar{x}^j} = \frac{\partial x^q}{\partial \bar{x}^k}\frac{\partial x^r}{\partial \bar{x}^m}\frac{\partial g_{qr}}{\partial x^p}\frac{\partial x^p}{\partial \bar{x}^j} + \frac{\partial x^q}{\partial \bar{x}^k}\frac{\partial^2 x^r}{\partial \bar{x}^j \partial \bar{x}^m} g_{qr} + \frac{\partial^2 x^q}{\partial \bar{x}^j \partial \bar{x}^k}\frac{\partial x^r}{\partial \bar{x}^m} g_{qr}$$

$$(3) \qquad \frac{\partial \bar{g}_{mj}}{\partial \bar{x}^k} = \frac{\partial x^r}{\partial \bar{x}^m}\frac{\partial x^p}{\partial \bar{x}^j}\frac{\partial g_{rp}}{\partial x^q}\frac{\partial x^q}{\partial \bar{x}^k} + \frac{\partial x^r}{\partial \bar{x}^m}\frac{\partial^2 x^p}{\partial \bar{x}^k \partial \bar{x}^j} g_{rp} + \frac{\partial^2 x^r}{\partial \bar{x}^k \partial \bar{x}^m}\frac{\partial x^p}{\partial \bar{x}^j} g_{rp}$$

Subtracting (*1*) from the sum of (*2*) and (*3*) and multiplying by $\frac{1}{2}$, we obtain on using the definition of the Christoffel symbols of the first kind,

$$(4) \qquad \overline{[jk,m]} = \frac{\partial x^p}{\partial \bar{x}^j}\frac{\partial x^q}{\partial \bar{x}^k}\frac{\partial x^r}{\partial \bar{x}^m}[pq,r] + \frac{\partial^2 x^p}{\partial \bar{x}^j \partial \bar{x}^k}\frac{\partial x^q}{\partial \bar{x}^m} g_{pq}$$

(*b*) Multiply (*4*) by $\bar{g}^{nm} = \dfrac{\partial \bar{x}^n}{\partial x^s}\dfrac{\partial \bar{x}^m}{\partial x^t} g^{st}$ to obtain

$$\bar{g}^{nm}\,\overline{[jk,m]} = \frac{\partial x^p}{\partial \bar{x}^j}\frac{\partial x^q}{\partial \bar{x}^k}\frac{\partial x^r}{\partial \bar{x}^m}\frac{\partial \bar{x}^n}{\partial x^s}\frac{\partial \bar{x}^m}{\partial x^t} g^{st}[pq,r] + \frac{\partial^2 x^p}{\partial \bar{x}^j \partial \bar{x}^k}\frac{\partial x^q}{\partial \bar{x}^m}\frac{\partial \bar{x}^n}{\partial x^s}\frac{\partial \bar{x}^m}{\partial x^t} g^{st} g_{pq}$$

Then $\overline{\left\{ {n \atop jk} \right\}} = \frac{\partial x^p}{\partial \bar{x}^j}\frac{\partial x^q}{\partial \bar{x}^k}\frac{\partial \bar{x}^n}{\partial x^s}\,\delta^r_t\, g^{st}[pq,r] + \frac{\partial^2 x^p}{\partial \bar{x}^j \partial \bar{x}^k}\frac{\partial \bar{x}^n}{\partial x^s}\,\delta^q_t\, g^{st} g_{pq}$

$$= \frac{\partial x^p}{\partial \bar{x}^j}\frac{\partial x^q}{\partial \bar{x}^k}\frac{\partial \bar{x}^n}{\partial x^s}\left\{ {s \atop pq} \right\} + \frac{\partial^2 x^p}{\partial \bar{x}^j \partial \bar{x}^k}\frac{\partial \bar{x}^n}{\partial x^p}$$

since $\delta^r_t\, g^{st}[pq,r] = g^{sr}[pq,r] = \left\{ {s \atop pq} \right\}$ and $\delta^q_t\, g^{st} g_{pq} = g^{sq} g_{pq} = \delta^s_p$.

**47.** Prove $\dfrac{\partial^2 x^m}{\partial \bar{x}^j \partial \bar{x}^k} = \overline{\left\{ {n \atop jk} \right\}}\dfrac{\partial x^m}{\partial \bar{x}^n} - \dfrac{\partial x^p}{\partial \bar{x}^j}\dfrac{\partial x^q}{\partial \bar{x}^k}\left\{ {m \atop pq} \right\}$.

From Problem 46 (*b*), $\overline{\left\{ {n \atop jk} \right\}} = \dfrac{\partial x^p}{\partial \bar{x}^j}\dfrac{\partial x^q}{\partial \bar{x}^k}\dfrac{\partial \bar{x}^n}{\partial x^s}\left\{ {s \atop pq} \right\} + \dfrac{\partial^2 x^p}{\partial \bar{x}^j \partial \bar{x}^k}\dfrac{\partial \bar{x}^n}{\partial x^p}$.

Multiplying by $\dfrac{\partial x^m}{\partial \bar{x}^n}$, $\overline{\left\{ {n \atop jk} \right\}}\dfrac{\partial x^m}{\partial \bar{x}^n} = \dfrac{\partial x^p}{\partial \bar{x}^j}\dfrac{\partial x^q}{\partial \bar{x}^k}\,\delta^m_s\left\{ {s \atop pq} \right\} + \dfrac{\partial^2 x^p}{\partial \bar{x}^j \partial \bar{x}^k}\,\delta^m_p$

$$= \frac{\partial x^p}{\partial \bar{x}^j}\frac{\partial x^q}{\partial \bar{x}^k}\left\{ {m \atop pq} \right\} + \frac{\partial^2 x^m}{\partial \bar{x}^j \partial \bar{x}^k}$$

Solving for $\dfrac{\partial^2 x^m}{\partial \bar{x}^j \partial \bar{x}^k}$, the result follows.

**48.** Evaluate the Christoffel symbols of (*a*) the first kind, (*b*) the second kind, for spaces where $g_{pq} = 0$ if $p \neq q$.

(*a*) If $p=q=r$, $[pq,r] = [pp,p] = \frac{1}{2}\left(\frac{\partial g_{pp}}{\partial x^p} + \frac{\partial g_{pp}}{\partial x^p} - \frac{\partial g_{pp}}{\partial x^p}\right) = \frac{1}{2}\frac{\partial g_{pp}}{\partial x^p}$.

If $p=q\neq r$, $[pq,r] = [pp,r] = \frac{1}{2}\left(\frac{\partial g_{pr}}{\partial x^p} + \frac{\partial g_{pr}}{\partial x^p} - \frac{\partial g_{pp}}{\partial x^r}\right) = -\frac{1}{2}\frac{\partial g_{pp}}{\partial x^r}$.

If $p=r\neq q$, $[pq,r] = [pq,p] = \frac{1}{2}\left(\frac{\partial g_{pp}}{\partial x^q} + \frac{\partial g_{qp}}{\partial x^p} - \frac{\partial g_{pq}}{\partial x^p}\right) = \frac{1}{2}\frac{\partial g_{pp}}{\partial x^q}$.

If $p,q,r$ are distinct, $[pq,r] = 0$.

We have not used the summation convention here.

(*b*) By Problem 43, $g^{jj} = \dfrac{1}{g_{jj}}$ (not summed). Then

$\left\{ {s \atop pq} \right\} = g^{sr}[pq,r] = 0$ if $r \neq s$, and $= g^{ss}[pq,s] = \dfrac{[pq,s]}{g_{ss}}$ (not summed) if $r=s$.

By (*a*):

If $p=q=s$, $\quad \left\{ {s \atop pq} \right\} = \left\{ {p \atop pp} \right\} = \dfrac{[pp,p]}{g_{pp}} = \dfrac{1}{2g_{pp}} \dfrac{\partial g_{pp}}{\partial x^p} = \dfrac{1}{2} \dfrac{\partial}{\partial x^p} \ln g_{pp}$.

If $p=q\neq s$, $\quad \left\{ {s \atop pq} \right\} = \left\{ {s \atop pp} \right\} = \dfrac{[pp,s]}{g_{ss}} = -\dfrac{1}{2g_{ss}} \dfrac{\partial g_{pp}}{\partial x^s}$.

If $p=s\neq q$, $\quad \left\{ {s \atop pq} \right\} = \left\{ {p \atop pq} \right\} = \dfrac{[pq,p]}{g_{pp}} = \dfrac{1}{2g_{pp}} \dfrac{\partial g_{pp}}{\partial x^q} = \dfrac{1}{2} \dfrac{\partial}{\partial x^q} \ln g_{pp}$.

If $p,q,s$ are distinct, $\quad \left\{ {s \atop pq} \right\} = 0$.

**49.** Determine the Christoffel symbols of the second kind in (*a*) rectangular, (*b*) cylindrical, and (*c*) spherical coordinates.

We can use the results of Problem 48, since for orthogonal coordinates $g_{pq} = 0$ if $p \neq q$.

(*a*) In rectangular coordinates, $g_{pp} = 1$ so that $\left\{ {s \atop pq} \right\} = 0$.

(*b*) In cylindrical coordinates, $x^1 = \rho$, $x^2 = \phi$, $x^3 = z$, we have by Problem 30(*a*), $g_{11} = 1$, $g_{22} = \rho^2$, $g_{33} = 1$. The only non-zero Christoffel symbols of the second kind can occur where $p = 2$. These are

$$\left\{ {1 \atop 22} \right\} = -\frac{1}{2g_{11}} \frac{\partial g_{22}}{\partial x^1} = -\frac{1}{2} \frac{\partial}{\partial \rho}(\rho^2) = -\rho,$$

$$\left\{ {2 \atop 21} \right\} = \left\{ {2 \atop 12} \right\} = \frac{1}{2g_{22}} \frac{\partial g_{22}}{\partial x^1} = \frac{1}{2\rho^2} \frac{\partial}{\partial \rho}(\rho^2) = \frac{1}{\rho}$$

(*c*) In spherical coordinates, $x^1 = r$, $x^2 = \theta$, $x^3 = \phi$, we have by Prob. 30(*b*), $g_{11} = 1$, $g_{22} = r^2$, $g_{33} = r^2 \sin^2\theta$. The only non-zero Christoffel symbols of the second kind can occur where $p = 2$ or $3$. These are

$$\left\{ {1 \atop 22} \right\} = -\frac{1}{2g_{11}} \frac{\partial g_{22}}{\partial x^1} = -\frac{1}{2} \frac{\partial}{\partial r}(r^2) = -r$$

$$\left\{ {2 \atop 21} \right\} = \left\{ {2 \atop 12} \right\} = \frac{1}{2g_{22}} \frac{\partial g_{22}}{\partial x^1} = \frac{1}{2r^2} \frac{\partial}{\partial r}(r^2) = \frac{1}{r}$$

$$\left\{ {1 \atop 33} \right\} = -\frac{1}{2g_{11}} \frac{\partial g_{33}}{\partial x^1} = -\frac{1}{2} \frac{\partial}{\partial r}(r^2 \sin^2\theta) = -r \sin^2\theta$$

$$\left\{ {2 \atop 33} \right\} = -\frac{1}{2g_{22}} \frac{\partial g_{33}}{\partial x^2} = -\frac{1}{2r^2} \frac{\partial}{\partial \theta}(r^2 \sin^2\theta) = -\sin\theta \cos\theta$$

$$\left\{ {3 \atop 31} \right\} = \left\{ {3 \atop 13} \right\} = \frac{1}{2g_{33}} \frac{\partial g_{33}}{\partial x^1} = \frac{1}{2r^2 \sin^2\theta} \frac{\partial}{\partial r}(r^2 \sin^2\theta) = \frac{1}{r}$$

$$\left\{ {3 \atop 32} \right\} = \left\{ {3 \atop 23} \right\} = \frac{1}{2g_{33}} \frac{\partial g_{33}}{\partial x^2} = \frac{1}{2r^2 \sin^2\theta} \frac{\partial}{\partial \theta}(r^2 \sin^2\theta) = \cot\theta$$

**GEODESICS.**

**50.** Prove that a necessary condition that $I = \int_{t_1}^{t_2} F(t,x,\dot{x})\,dt$ be an extremum (maximum or minimum) is that $\frac{\partial F}{\partial x} - \frac{d}{dt}(\frac{\partial F}{\partial \dot{x}}) = 0$.

Let the curve which makes $I$ an extremum be $x = X(t)$, $t_1 \leqq t \leqq t_2$. Then $x = X(t) + \epsilon\eta(t)$, where $\epsilon$ is independent of $t$, is a neighbouring curve through $t_1$ and $t_2$ so that $\eta(t_1) = \eta(t_2) = 0$. The value of $I$ for the neighbouring curve is

$$I(\epsilon) = \int_{t_1}^{t_2} F(t,\ X+\epsilon\eta,\ \dot{X}+\epsilon\dot{\eta})\,dt$$

This is an extremum for $\epsilon = 0$. A necessary condition that this be so is that $\left.\frac{dI}{d\epsilon}\right|_{\epsilon=0} = 0$. But by differentiation under the integral sign, assuming this valid,

$$\left.\frac{dI}{d\epsilon}\right|_{\epsilon=0} = \int_{t_1}^{t_2} (\frac{\partial F}{\partial x}\eta + \frac{\partial F}{\partial \dot{x}}\dot{\eta})\,dt = 0$$

which can be written as

$$\int_{t_1}^{t_2} \frac{\partial F}{\partial x}\eta\,dt + \left.\frac{\partial F}{\partial \dot{x}}\eta\right|_{t_1}^{t_2} - \int_{t_1}^{t_2} \eta\frac{d}{dt}(\frac{\partial F}{\partial \dot{x}})\,dt = \int_{t_1}^{t_2} \eta\left(\frac{\partial F}{\partial x} - \frac{d}{dt}(\frac{\partial F}{\partial \dot{x}})\right)dt = 0$$

Since $\eta$ is arbitrary, the integrand $\frac{\partial F}{\partial x} - \frac{d}{dt}(\frac{\partial F}{\partial \dot{x}}) = 0$.

The result is easily extended to the integral $\int_{t_1}^{t_2} F(t, x^1, \dot{x}^1, x^2, \dot{x}^2, \ldots, x^N, \dot{x}^N)\,dt$

and yields

$$\frac{\partial F}{\partial x^k} - \frac{d}{dt}(\frac{\partial F}{\partial \dot{x}^k}) = 0$$

called *Euler's* or *Lagrange's equations*. (See also Problem 73.)

**51.** Show that the geodesics in a Riemannian space are given by $\frac{d^2x^r}{ds^2} + \left\{ {r \atop pq} \right\} \frac{dx^p}{ds}\frac{dx^q}{ds} = 0$.

We must determine the extremum of $\int_{t_1}^{t_2} \sqrt{g_{pq}\,\dot{x}^p\,\dot{x}^q}\,dt$ using Euler's equations (Problem 50) with $F = \sqrt{g_{pq}\,\dot{x}^p\,\dot{x}^q}$. We have

$$\frac{\partial F}{\partial x^k} = \frac{1}{2}(g_{pq}\,\dot{x}^p\,\dot{x}^q)^{-1/2}\frac{\partial g_{pq}}{\partial x^k}\dot{x}^p\,\dot{x}^q$$

$$\frac{\partial F}{\partial \dot{x}^k} = \frac{1}{2}(g_{pq}\,\dot{x}^p\,\dot{x}^q)^{-1/2}\,2g_{pk}\,\dot{x}^p$$

Using $\frac{ds}{dt} = \sqrt{g_{pq}\,\dot{x}^p\,\dot{x}^q}$, Euler's equations can be written

$$\frac{d}{dt}\left(\frac{g_{pk}\,\dot{x}^p}{\dot{s}}\right) - \frac{1}{2\dot{s}}\frac{\partial g_{pq}}{\partial x^k}\dot{x}^p\dot{x}^q = 0$$

or

$$g_{pk}\ddot{x}^p + \frac{\partial g_{pk}}{\partial x^q}\dot{x}^p\dot{x}^q - \frac{1}{2}\frac{\partial g_{pq}}{\partial x^k}\dot{x}^p\dot{x}^q = \frac{g_{pk}\,\dot{x}^p\,\ddot{s}}{\dot{s}}$$

Writing $\dfrac{\partial g_{pk}}{\partial x^q}\dot{x}^p\dot{x}^q = \dfrac{1}{2}\left(\dfrac{\partial g_{pk}}{\partial x^q} + \dfrac{\partial g_{qk}}{\partial x^p}\right)\dot{x}^p\dot{x}^q$ this equation becomes

$$g_{pk}\ddot{x}^p + [pq,k]\,\dot{x}^p\dot{x}^q = \frac{g_{pk}\,\dot{x}^p\,\ddot{s}}{\dot{s}}$$

If we use arc length as parameter, $\dot{s} = 1$, $\ddot{s} = 0$ and the equation becomes

$$g_{pk}\frac{d^2x^p}{ds^2} + [pq,k]\frac{dx^p}{ds}\frac{dx^q}{ds} = 0$$

Multiplying by $g^{rk}$, we obtain

$$\frac{d^2x^r}{ds^2} + \begin{Bmatrix} r \\ pq \end{Bmatrix}\frac{dx^p}{ds}\frac{dx^q}{ds} = 0$$

## THE COVARIANT DERIVATIVE.

**52.** If $A_p$ and $A^p$ are tensors show that (a) $A_{p,q} \equiv \dfrac{\partial A_p}{\partial x^q} - \begin{Bmatrix} s \\ pq \end{Bmatrix} A_s$

and (b) $A^p_{\ ,q} \equiv \dfrac{\partial A^p}{\partial x^q} + \begin{Bmatrix} p \\ qs \end{Bmatrix} A^s$ are tensors.

(a) Since $\bar{A}_j = \dfrac{\partial x^r}{\partial \bar{x}^j} A_r$,

(1)
$$\frac{\partial \bar{A}_j}{\partial \bar{x}^k} = \frac{\partial x^r}{\partial \bar{x}^j}\frac{\partial A_r}{\partial x^t}\frac{\partial x^t}{\partial \bar{x}^k} + \frac{\partial^2 x^r}{\partial \bar{x}^j\,\partial \bar{x}^k} A_r$$

From Problem 47,

$$\frac{\partial^2 x^r}{\partial \bar{x}^j\,\partial \bar{x}^k} = \overline{\begin{Bmatrix} n \\ jk \end{Bmatrix}}\frac{\partial x^r}{\partial \bar{x}^n} - \frac{\partial x^i}{\partial \bar{x}^j}\frac{\partial x^l}{\partial \bar{x}^k}\begin{Bmatrix} r \\ il \end{Bmatrix}$$

Substituting in (1),

$$\frac{\partial \bar{A}_j}{\partial \bar{x}^k} = \frac{\partial x^r}{\partial \bar{x}^j}\frac{\partial x^t}{\partial \bar{x}^k}\frac{\partial A_r}{\partial x^t} + \overline{\begin{Bmatrix} n \\ jk \end{Bmatrix}}\frac{\partial x^r}{\partial \bar{x}^n}A_r - \frac{\partial x^i}{\partial \bar{x}^j}\frac{\partial x^l}{\partial \bar{x}^k}\begin{Bmatrix} r \\ il \end{Bmatrix}A_r$$

$$= \frac{\partial x^p}{\partial \bar{x}^j}\frac{\partial x^q}{\partial \bar{x}^k}\frac{\partial A_p}{\partial x^q} + \overline{\begin{Bmatrix} n \\ jk \end{Bmatrix}}\bar{A}_n - \frac{\partial x^p}{\partial \bar{x}^j}\frac{\partial x^q}{\partial \bar{x}^k}\begin{Bmatrix} s \\ pq \end{Bmatrix}A_s$$

or

$$\frac{\partial \bar{A}_j}{\partial \bar{x}^k} - \overline{\begin{Bmatrix} n \\ jk \end{Bmatrix}}\bar{A}_n = \frac{\partial x^p}{\partial \bar{x}^j}\frac{\partial x^q}{\partial \bar{x}^k}\left(\frac{\partial A_p}{\partial x^q} - \begin{Bmatrix} s \\ pq \end{Bmatrix}A_s\right)$$

and $\dfrac{\partial A_p}{\partial x^q} - \left\{ {s \atop pq} \right\} A_s$ is a covariant tensor of second rank, called the *covariant derivative* of $A_p$ with *respect to* $x^q$ and written $A_{p,q}$.

(*b*) Since $\bar{A}^j = \dfrac{\partial \bar{x}^j}{\partial x^r} A^r$,

$$(2) \qquad \frac{\partial \bar{A}^j}{\partial \bar{x}^k} = \frac{\partial \bar{x}^j}{\partial x^r}\frac{\partial A^r}{\partial x^t}\frac{\partial x^t}{\partial \bar{x}^k} + \frac{\partial^2 \bar{x}^j}{\partial x^r \partial x^t}\frac{\partial x^t}{\partial \bar{x}^k} A^r$$

From Problem 47, interchanging $x$ and $\bar{x}$ coordinates,

$$\frac{\partial^2 \bar{x}^j}{\partial x^r \partial x^t} = \left\{ {n \atop rt} \right\} \frac{\partial \bar{x}^j}{\partial x^n} - \frac{\partial \bar{x}^i}{\partial x^r}\frac{\partial \bar{x}^l}{\partial x^t}\overline{\left\{ {j \atop il} \right\}}$$

Substituting in (2),

$$\begin{aligned}
\frac{\partial \bar{A}^j}{\partial \bar{x}^k} &= \frac{\partial \bar{x}^j}{\partial x^r}\frac{\partial x^t}{\partial \bar{x}^k}\frac{\partial A^r}{\partial x^t} + \left\{ {n \atop rt} \right\}\frac{\partial \bar{x}^j}{\partial x^n}\frac{\partial x^t}{\partial \bar{x}^k} A^r - \frac{\partial \bar{x}^i}{\partial x^r}\frac{\partial \bar{x}^l}{\partial x^t}\frac{\partial x^t}{\partial \bar{x}^k}\overline{\left\{ {j \atop il} \right\}} A^r \\
&= \frac{\partial \bar{x}^j}{\partial x^r}\frac{\partial x^t}{\partial \bar{x}^k}\frac{\partial A^r}{\partial x^t} + \left\{ {n \atop rt} \right\}\frac{\partial \bar{x}^j}{\partial x^n}\frac{\partial x^t}{\partial \bar{x}^k} A^r - \frac{\partial \bar{x}^i}{\partial x^r}\delta^l_k \overline{\left\{ {j \atop il} \right\}} A^r \\
&= \frac{\partial \bar{x}^j}{\partial x^p}\frac{\partial x^q}{\partial \bar{x}^k}\frac{\partial A^p}{\partial x^q} + \left\{ {p \atop sq} \right\}\frac{\partial \bar{x}^j}{\partial x^p}\frac{\partial x^q}{\partial \bar{x}^k} A^s - \overline{\left\{ {j \atop ik} \right\}} \bar{A}^i
\end{aligned}$$

or

$$\frac{\partial \bar{A}^j}{\partial \bar{x}^k} + \overline{\left\{ {j \atop ki} \right\}} \bar{A}^i = \frac{\partial \bar{x}^j}{\partial x^p}\frac{\partial x^q}{\partial \bar{x}^k}\left( \frac{\partial A^p}{\partial x^q} + \left\{ {p \atop qs} \right\} A^s \right)$$

and $\dfrac{\partial A^p}{\partial x^q} + \left\{ {p \atop qs} \right\} A^s$ is a mixed tensor of second rank, called the *covariant derivative of* $A^p$ *with respect to* $x^q$ and written $A^p_{,q}$.

**53.** Write the covariant derivative with respect to $x^q$ of each of the following tensors:
(*a*) $A_{jk}$, (*b*) $A^{jk}$, (*c*) $A^j_k$, (*d*) $A^j_{kl}$, (*e*) $A^{jkl}_{mn}$.

(*a*) $$A_{jk,q} = \frac{\partial A_{jk}}{\partial x^q} - \left\{ {s \atop jq} \right\} A_{sk} - \left\{ {s \atop kq} \right\} A_{js}$$

(*b*) $$A^{jk}_{,q} = \frac{\partial A^{jk}}{\partial x^q} + \left\{ {j \atop qs} \right\} A^{sk} + \left\{ {k \atop qs} \right\} A^{js}$$

(*c*) $$A^j_{k,q} = \frac{\partial A^j_k}{\partial x^q} - \left\{ {s \atop kq} \right\} A^j_s + \left\{ {j \atop qs} \right\} A^s_k$$

(*d*) $$A^j_{kl,q} = \frac{\partial A^j_{kl}}{\partial x^q} - \left\{ {s \atop kq} \right\} A^j_{sl} - \left\{ {s \atop lq} \right\} A^j_{ks} + \left\{ {j \atop qs} \right\} A^s_{kl}$$

$$(e)\quad A^{jkl}_{mn,q} = \frac{\partial A^{jkl}_{mn}}{\partial x^q} - \left\{ {s \atop mq} \right\} A^{jkl}_{sn} - \left\{ {s \atop nq} \right\} A^{jkl}_{ms} + \left\{ {j \atop qs} \right\} A^{skl}_{mn} + \left\{ {k \atop qs} \right\} A^{jsl}_{mn} + \left\{ {l \atop qs} \right\} A^{jks}_{mn}$$

**54.** Prove that the covariant derivatives of (a) $g_{jk}$, (b) $g^{jk}$, (c) $\delta^j_k$ are zero.

$$(a)\quad g_{jk,q} = \frac{\partial g_{jk}}{\partial x^q} - \left\{ {s \atop jq} \right\} g_{sk} - \left\{ {s \atop kq} \right\} g_{js}$$

$$= \frac{\partial g_{jk}}{\partial x^q} - [jq,k] - [kq,j] = 0 \quad \text{by Problem 45}(a).$$

$$(b)\quad g^{jk}_{\ \ ,q} = \frac{\partial g^{jk}}{\partial x^q} + \left\{ {j \atop qs} \right\} g^{sk} + \left\{ {k \atop qs} \right\} g^{js} = 0 \quad \text{by Problem 45}(b).$$

$$(c)\quad \delta^j_{k,q} = \frac{\partial \delta^j_k}{\partial x^q} - \left\{ {s \atop kq} \right\} \delta^j_s + \left\{ {j \atop qs} \right\} \delta^s_k = 0 - \left\{ {j \atop kq} \right\} + \left\{ {j \atop qk} \right\} = 0$$

**55.** Find the covariant derivative of $A^j_k B^{lm}_n$ with respect to $x^q$.

$$(A^j_k B^{lm}_n)_{,q} = \frac{\partial (A^j_k B^{lm}_n)}{\partial x^q} - \left\{ {s \atop kq} \right\} A^j_s B^{lm}_n - \left\{ {s \atop nq} \right\} A^j_k B^{lm}_s$$

$$+ \left\{ {j \atop qs} \right\} A^s_k B^{lm}_n + \left\{ {l \atop qs} \right\} A^j_k B^{sm}_n + \left\{ {m \atop qs} \right\} A^j_k B^{ls}_n$$

$$= \left( \frac{\partial A^j_k}{\partial x^q} - \left\{ {s \atop kq} \right\} A^j_s + \left\{ {j \atop qs} \right\} A^s_k \right) B^{lm}_n$$

$$+ A^j_k \left( \frac{\partial B^{lm}_n}{\partial x^q} - \left\{ {s \atop nq} \right\} B^{lm}_s + \left\{ {l \atop qs} \right\} B^{sm}_n + \left\{ {m \atop qs} \right\} B^{ls}_n \right)$$

$$= A^j_{k,q} B^{lm}_n + A^j_k B^{lm}_{n,q}$$

This illustrates the fact that the covariant derivatives of a product of tensors obey rules like those of ordinary derivatives of products in elementary calculus.

**56.** Prove $(g_{jk} A^{km}_n)_{,q} = g_{jk} A^{km}_{n,q}$.

$$(g_{jk} A^{km}_n)_{,q} = g_{jk,q} A^{km}_n + g_{jk} A^{km}_{n,q} = g_{jk} A^{km}_{n,q}$$

since $g_{jk,q} = 0$ by Prob. 54(a). In covariant differentiation, $g_{jk}$, $g^{jk}$ and $\delta^j_k$ can be treated as constants.

**GRADIENT, DIVERGENCE AND CURL IN TENSOR FORM.**

**57.** Prove that $\text{div } A^p = \dfrac{1}{\sqrt{g}} \dfrac{\partial}{\partial x^k}(\sqrt{g}\, A^k)$.

The divergence of $A^p$ is the contraction of the covariant derivative of $A^p$, i.e. the contraction of $A^p_{,q}$ or $A^p_{,p}$. Then, using Problem 45(*c*),

$$\text{div } A^p = A^p_{,p} = \frac{\partial A^k}{\partial x^k} + \begin{Bmatrix} p \\ pk \end{Bmatrix} A^k$$

$$= \frac{\partial A^k}{\partial x^k} + (\frac{\partial}{\partial x^k} \ln \sqrt{g})\, A^k = \frac{\partial A^k}{\partial x^k} + (\frac{1}{\sqrt{g}} \frac{\partial \sqrt{g}}{\partial x^k})\, A^k = \frac{1}{\sqrt{g}} \frac{\partial}{\partial x^k}(\sqrt{g}\, A^k)$$

**58.** Prove that $\nabla^2 \Phi = \dfrac{1}{\sqrt{g}} \dfrac{\partial}{\partial x^k}(\sqrt{g}\, g^{kr} \dfrac{\partial \Phi}{\partial x^r})$.

The gradient of $\Phi$ is $\text{grad } \Phi = \nabla \Phi = \dfrac{\partial \Phi}{\partial x^r}$, a covariant tensor of rank one (see Problem 6(*b*)) defined as the covariant derivative of $\Phi$, written $\Phi_{,r}$. The contravariant tensor of rank one associated with $\Phi_{,r}$ is $A^k = g^{kr} \dfrac{\partial \Phi}{\partial x^r}$. Then from Problem 57,

$$\nabla^2 \Phi = \text{div } (g^{kr} \frac{\partial \Phi}{\partial x^r}) = \frac{1}{\sqrt{g}} \frac{\partial}{\partial x^k}(\sqrt{g}\, g^{kr} \frac{\partial \Phi}{\partial x^r})$$

**59.** Prove that $A_{p,q} - A_{q,p} = \dfrac{\partial A_p}{\partial x^q} - \dfrac{\partial A_q}{\partial x^p}$.

$$A_{p,q} - A_{q,p} = \left(\frac{\partial A_p}{\partial x^q} - \begin{Bmatrix} s \\ pq \end{Bmatrix} A_s\right) - \left(\frac{\partial A_q}{\partial x^p} - \begin{Bmatrix} s \\ qp \end{Bmatrix} A_s\right) = \frac{\partial A_p}{\partial x^q} - \frac{\partial A_q}{\partial x^p}$$

This tensor of rank two is defined to be the curl of $A_p$.

**60.** Express the divergence of a vector $A^p$ in terms of its physical components for (*a*) cylindrical, (*b*) spherical coordinates.

(*a*). For cylindrical coordinates $x^1 = \rho,\ x^2 = \phi,\ x^3 = z$,

$$g = \begin{vmatrix} 1 & 0 & 0 \\ 0 & \rho^2 & 0 \\ 0 & 0 & 1 \end{vmatrix} = \rho^2 \quad \text{and} \quad \sqrt{g} = \rho \qquad \text{(see Problem 30(}a\text{))}$$

The physical components, denoted by $A_\rho,\ A_\phi,\ A_z$ are given by

$$A_\rho = \sqrt{g_{11}}\, A^1 = A^1, \quad A_\phi = \sqrt{g_{22}}\, A^2 = \rho A^2, \quad A_z = \sqrt{g_{33}}\, A^3 = A^3$$

Then

$$\text{div } A^p = \frac{1}{\sqrt{g}} \frac{\partial}{\partial x^k}(\sqrt{g}\, A^k)$$

$$= \frac{1}{\rho}\left[\frac{\partial}{\partial \rho}(\rho A_\rho) + \frac{\partial}{\partial \phi}(A_\phi) + \frac{\partial}{\partial z}(\rho A_z)\right]$$

(*b*) For spherical coordinates $x^1 = r,\ x^2 = \theta,\ x^3 = \phi$,

$$g = \begin{vmatrix} 1 & 0 & 0 \\ 0 & r^2 & 0 \\ 0 & 0 & r^2\sin^2\theta \end{vmatrix} = r^4 \sin^2\theta \quad \text{and} \quad \sqrt{g} = r^2 \sin\theta \quad \text{(see Problem 30(b))}$$

The physical components, denoted by $A_r, A_\theta, A_\phi$ are given by

$$A_r = \sqrt{g_{11}}\, A^1 = A^1, \qquad A_\theta = \sqrt{g_{22}}\, A^2 = rA^2, \qquad A_\phi = \sqrt{g_{33}}\, A^3 = r\sin\theta\, A^3$$

Then

$$\text{div } A^p = \frac{1}{\sqrt{g}} \frac{\partial}{\partial x^k}(\sqrt{g}\, A^k)$$

$$= \frac{1}{r^2\sin\theta}\left[\frac{\partial}{\partial r}(r^2\sin\theta\, A_r) + \frac{\partial}{\partial \theta}(r\sin\theta\, A_\theta) + \frac{\partial}{\partial \phi}(r A_\phi)\right]$$

$$= \frac{1}{r^2}\frac{\partial}{\partial r}(r^2 A_r) + \frac{1}{r\sin\theta}\frac{\partial}{\partial \theta}(\sin\theta\, A_\theta) + \frac{1}{r\sin\theta}\frac{\partial A_\phi}{\partial \phi}$$

**61.** Express the Laplacian of $\Phi$, $\nabla^2\Phi$, in (*a*) cylindrical coordinates, (*b*) spherical coordinates.

(*a*) In cylindrical coordinates $g^{11} = 1,\ g^{22} = 1/\rho^2,\ g^{33} = 1$ (see Problem 35 (*a*)). Then from Problem 58,

$$\nabla^2\Phi = \frac{1}{\sqrt{g}}\frac{\partial}{\partial x^k}\left(\sqrt{g}\, g^{kr}\frac{\partial \Phi}{\partial x^r}\right)$$

$$= \frac{1}{\rho}\left[\frac{\partial}{\partial \rho}\left(\rho\frac{\partial \Phi}{\partial \rho}\right) + \frac{\partial}{\partial \phi}\left(\frac{1}{\rho}\frac{\partial \Phi}{\partial \phi}\right) + \frac{\partial}{\partial z}\left(\rho\frac{\partial \Phi}{\partial z}\right)\right]$$

$$= \frac{1}{\rho}\frac{\partial}{\partial \rho}\left(\rho\frac{\partial \Phi}{\partial \rho}\right) + \frac{1}{\rho^2}\frac{\partial^2 \Phi}{\partial \phi^2} + \frac{\partial^2 \Phi}{\partial z^2}$$

(*b*) In spherical coordinates $g^{11} = 1,\ g^{22} = 1/r^2,\ g^{33} = 1/r^2\sin^2\theta$ (see Problem 35 (*b*)). Then

$$\nabla^2\Phi = \frac{1}{\sqrt{g}}\frac{\partial}{\partial x^k}\left(\sqrt{g}\, g^{kr}\frac{\partial \Phi}{\partial x^r}\right)$$

$$= \frac{1}{r^2\sin\theta}\left[\frac{\partial}{\partial r}\left(r^2\sin\theta\frac{\partial \Phi}{\partial r}\right) + \frac{\partial}{\partial \theta}\left(\sin\theta\frac{\partial \Phi}{\partial \theta}\right) + \frac{\partial}{\partial \phi}\left(\frac{1}{\sin\theta}\frac{\partial \Phi}{\partial \phi}\right)\right]$$

$$= \frac{1}{r^2}\frac{\partial}{\partial r}\left(r^2\frac{\partial \Phi}{\partial r}\right) + \frac{1}{r^2\sin\theta}\frac{\partial}{\partial \theta}\left(\sin\theta\frac{\partial \Phi}{\partial \theta}\right) + \frac{1}{r^2\sin^2\theta}\frac{\partial^2 \Phi}{\partial \phi^2}$$

**INTRINSIC DERIVATIVES.**

**62.** Calculate the intrinsic derivatives of each of the following tensors, assumed to be differentiable functions of $t$: (a) an invariant $\Phi$, (b) $A^j$, (c) $A^j_k$, (d) $A^{jk}_{lmn}$.

(a) $$\frac{\delta\Phi}{\delta t} = \Phi_{,q}\frac{dx^q}{dt} = \frac{\partial\Phi}{\partial x^q}\frac{dx^q}{dt} = \frac{d\Phi}{dt}, \text{ the ordinary derivative.}$$

(b) $$\frac{\delta A^j}{\delta t} = A^j_{,q}\frac{dx^q}{dt} = \left(\frac{\partial A^j}{\partial x^q} + \begin{Bmatrix} j \\ qs \end{Bmatrix} A^s\right)\frac{dx^q}{dt} = \frac{\partial A^j}{\partial x^q}\frac{dx^q}{dt} + \begin{Bmatrix} j \\ qs \end{Bmatrix} A^s \frac{dx^q}{dt}$$

$$\frac{dA^j}{dt} + \begin{Bmatrix} j \\ qs \end{Bmatrix} A^s \frac{dx^q}{dt}$$

(c) $$\frac{\delta A^j_k}{\delta t} = A^j_{k,q}\frac{dx^q}{dt} = \left(\frac{\partial A^j_k}{\partial x^q} - \begin{Bmatrix} s \\ kq \end{Bmatrix} A^j_s + \begin{Bmatrix} j \\ qs \end{Bmatrix} A^s_k\right)\frac{dx^q}{dt}$$

$$= \frac{dA^j_k}{dt} - \begin{Bmatrix} s \\ kq \end{Bmatrix} A^j_s \frac{dx^q}{dt} + \begin{Bmatrix} j \\ qs \end{Bmatrix} A^s_k \frac{dx^q}{dt}$$

(d) $$\frac{\delta A^{jk}_{lmn}}{\delta t} = A^{jk}_{lmn,q}\frac{dx^q}{dt} = \left(\frac{\partial A^{jk}_{lmn}}{\partial x^q} - \begin{Bmatrix} s \\ lq \end{Bmatrix} A^{jk}_{smn} - \begin{Bmatrix} s \\ mq \end{Bmatrix} A^{jk}_{lsn}\right.$$

$$\left. - \begin{Bmatrix} s \\ nq \end{Bmatrix} A^{jk}_{lms} + \begin{Bmatrix} j \\ qs \end{Bmatrix} A^{sk}_{lmn} + \begin{Bmatrix} k \\ qs \end{Bmatrix} A^{js}_{lmn}\right)\frac{dx^q}{dt}$$

$$= \frac{dA^{jk}_{lmn}}{dt} - \begin{Bmatrix} s \\ lq \end{Bmatrix} A^{jk}_{smn}\frac{dx^q}{dt} - \begin{Bmatrix} s \\ mq \end{Bmatrix} A^{jk}_{lsn}\frac{dx^q}{dt} - \begin{Bmatrix} s \\ nq \end{Bmatrix} A^{jk}_{lms}\frac{dx^q}{dt}$$

$$+ \begin{Bmatrix} j \\ qs \end{Bmatrix} A^{sk}_{lmn}\frac{dx^q}{dt} + \begin{Bmatrix} k \\ qs \end{Bmatrix} A^{js}_{lmn}\frac{dx^q}{dt}$$

**63.** Prove the intrinsic derivatives of $g_{jk}$, $g^{jk}$ and $\delta^j_k$ are zero.

$$\frac{\delta g_{jk}}{\delta t} = (g_{jk,q})\frac{dx^q}{dt} = 0, \qquad \frac{\delta g^{jk}}{\delta t} = g^{jk}_{,q}\frac{dx^q}{dt} = 0, \qquad \frac{\delta\delta^j_k}{\delta t} = \delta^j_{k,q}\frac{dx^q}{dt} = 0 \quad \text{by Problem 54.}$$

**RELATIVE TENSORS.**

**64.** If $A^p_q$ and $B^{rs}_t$ are relative tensors of weights $w_1$ and $w_2$ respectively, show that their inner and outer products are relative tensors of weight $w_1 + w_2$.

By hypothesis,

$$\bar{A}^j_k = J^{w_1}\frac{\partial\bar{x}^j}{\partial x^p}\frac{\partial x^q}{\partial\bar{x}^k}A^p_q, \qquad \bar{B}^{lm}_n = J^{w_2}\frac{\partial\bar{x}^l}{\partial x^r}\frac{\partial\bar{x}^m}{\partial x^s}\frac{\partial x^t}{\partial\bar{x}^n}B^{rs}_t$$

The outer product is $$\bar{A}^j_k\,\bar{B}^{lm}_n = J^{w_1+w_2}\frac{\partial\bar{x}^j}{\partial x^p}\frac{\partial x^q}{\partial\bar{x}^k}\frac{\partial\bar{x}^l}{\partial x^r}\frac{\partial\bar{x}^m}{\partial x^s}\frac{\partial x^t}{\partial\bar{x}^n}A^p_q\,B^{rs}_t$$

a relative tensor of weight $w_1 + w_2$. Any inner product, which is a contraction of the outer product, is also a relative tensor of weight $w_1 + w_2$.

**65.** Prove that $\sqrt{g}$ is a relative tensor of weight one, i.e. a tensor density.

The elements of determinant $g$ given by $g_{pq}$ transform according to $\bar{g}_{jk} = \dfrac{\partial x^p}{\partial \bar{x}^j} \dfrac{\partial x^q}{\partial \bar{x}^k} g_{pq}$.

Taking determinants of both sides, $\bar{g} = \left|\dfrac{\partial x^p}{\partial \bar{x}^j}\right| \left|\dfrac{\partial x^q}{\partial \bar{x}^k}\right| g = J^2 g$ or $\sqrt{\bar{g}} = J\sqrt{g}$, which shows that $\sqrt{g}$ is a relative tensor of weight one.

**66.** Prove that $dV = \sqrt{g}\, dx^1\, dx^2 \ldots dx^N$ is an invariant.

By Problem 65, $d\bar{V} = \sqrt{\bar{g}}\, d\bar{x}^1\, d\bar{x}^2 \ldots d\bar{x}^N = \sqrt{g}\, J\, d\bar{x}^1\, d\bar{x}^2 \ldots d\bar{x}^N$

$$= \sqrt{g} \left|\frac{\partial x}{\partial \bar{x}}\right| d\bar{x}^1\, d\bar{x}^2 \ldots d\bar{x}^N = \sqrt{g}\, dx^1\, dx^2 \ldots dx^N = dV$$

From this it follows that if $\Phi$ is an invariant, then

$$\int \cdots \int_{\bar{V}} \bar{\Phi}\, d\bar{V} = \int \cdots \int_{V} \Phi\, dV$$

for any coordinate systems where the integration is performed over a volume in $N$ dimensional space. A similar statement can be made for surface integrals.

## MISCELLANEOUS APPLICATIONS.

**67.** Express in tensor form (*a*) the velocity and (*b*) the acceleration of a particle.

(*a*) If the particle moves along a curve $x^k = x^k(t)$ where $t$ is the parameter time, then $v^k = \dfrac{dx^k}{dt}$ is its velocity and is a contravariant tensor of rank one (see Problem 9).

(*b*) The quantity $\dfrac{dv^k}{dt} = \dfrac{d^2 x^k}{dt^2}$ is not in general a tensor and so cannot represent the physical quantity acceleration in all coordinate systems. We define the acceleration $a^k$ as the intrinsic derivative of the velocity, i.e. $a^k = \dfrac{\delta v^k}{\delta t}$ which is a contravariant tensor of rank one.

**68.** Write Newton's law in tensor form.

Assume the mass $M$ of the particle to be an invariant independent of time $t$. Then $Ma^k = F^k$ a contravariant tensor of rank one is called the *force* on the particle. Thus Newton's law can be written

$$F^k = Ma^k = M\,\frac{\delta v^k}{\delta t}$$

**69.** Prove that $a^k = \dfrac{\delta v^k}{\delta t} = \dfrac{d^2 x^k}{dt^2} + \left\{ {k \atop pq} \right\} \dfrac{dx^p}{dt} \dfrac{dx^q}{dt}$.

Since $v^k$ is a contravariant tensor, we have by Problem 62(*b*)

$$\frac{\delta v^k}{\delta t} = \frac{dv^k}{dt} + \left\{ {k \atop qs} \right\} v^s \frac{dx^q}{dt} = \frac{d^2 x^k}{dt^2} + \left\{ {k \atop qp} \right\} v^p \frac{dx^q}{dt}$$

$$= \frac{d^2 x^k}{dt^2} + \left\{ {k \atop pq} \right\} \frac{dx^p}{dt} \frac{dx^q}{dt}$$

**70.** Find the physical components of (*a*) the velocity and (*b*) the acceleration of a particle in cylindrical coordinates.

(*a*) From Problem 67(*a*), the contravariant components of the velocity are

$$\frac{dx^1}{dt} = \frac{d\rho}{dt}, \quad \frac{dx^2}{dt} = \frac{d\phi}{dt} \quad \text{and} \quad \frac{dx^3}{dt} = \frac{dz}{dt}$$

Then the physical components of the velocity are

$$\sqrt{g_{11}}\, \frac{dx^1}{dt} = \frac{d\rho}{dt}, \quad \sqrt{g_{22}}\, \frac{dx^2}{dt} = \rho \frac{d\phi}{dt} \quad \text{and} \quad \sqrt{g_{33}}\, \frac{dx^3}{dt} = \frac{dz}{dt}$$

using $g_{11} = 1$, $g_{22} = \rho^2$, $g_{33} = 1$.

(*b*) From Problems 69 and 49(*b*), the contravariant components of the acceleration are

$$a^1 = \frac{d^2 x^1}{dt^2} + \left\{ {1 \atop 22} \right\} \frac{dx^2}{dt} \frac{dx^2}{dt} = \frac{d^2 \rho}{dt^2} - \rho(\frac{d\phi}{dt})^2$$

$$a^2 = \frac{d^2 x^2}{dt^2} + \left\{ {2 \atop 12} \right\} \frac{dx^1}{dt} \frac{dx^2}{dt} + \left\{ {2 \atop 21} \right\} \frac{dx^2}{dt} \frac{dx^1}{dt} = \frac{d^2 \phi}{dt^2} + \frac{2}{\rho} \frac{d\rho}{dt} \frac{d\phi}{dt}$$

and
$$a^3 = \frac{d^2 x^3}{dt^2} = \frac{d^2 z}{dt^2}$$

Then the physical components of the acceleration are

$$\sqrt{g_{11}}\, a^1 = \ddot{\rho} - \rho\dot{\phi}^2, \quad \sqrt{g_{22}}\, a^2 = \rho\ddot{\phi} + 2\dot{\rho}\dot{\phi} \quad \text{and} \quad \sqrt{g_{33}}\, a^3 = \ddot{z}$$

where dots denote differentiations with respect to time.

**71.** If the kinetic energy $T$ of a particle of constant mass $M$ moving with velocity having magnitude $v$ is given by $T = \frac{1}{2}Mv^2 = \frac{1}{2}Mg_{pq}\, \dot{x}^p \dot{x}^q$, prove that

$$\frac{d}{dt}(\frac{\partial T}{\partial \dot{x}^k}) - \frac{\partial T}{\partial x^k} = Ma_k$$

where $a_k$ denotes the covariant components of the acceleration.

Since $T = \frac{1}{2}Mg_{pq}\, \dot{x}^p \dot{x}^q$, we have

$$\frac{\partial T}{\partial x^k} = \tfrac{1}{2}M\frac{\partial g_{pq}}{\partial x^k}\dot{x}^p\dot{x}^q, \quad \frac{\partial T}{\partial \dot{x}^k} = Mg_{kq}\dot{x}^q \quad \text{and} \quad \frac{d}{dt}\left(\frac{\partial T}{\partial \dot{x}^k}\right) = M\left(g_{kq}\ddot{x}^q + \frac{\partial g_{kq}}{\partial x^j}\dot{x}^j\dot{x}^q\right)$$

Then
$$\frac{d}{dt}\left(\frac{\partial T}{\partial \dot{x}^k}\right) - \frac{\partial T}{\partial x^k} = M\left(g_{kq}\ddot{x}^q + \frac{\partial g_{kq}}{\partial x^j}\dot{x}^j\dot{x}^q - \frac{1}{2}\frac{\partial g_{pq}}{\partial x^k}\dot{x}^p\dot{x}^q\right)$$

$$= M\left(g_{kq}\ddot{x}^q + \frac{1}{2}\left(\frac{\partial g_{kq}}{\partial x^p} + \frac{\partial g_{kp}}{\partial x^q} - \frac{\partial g_{pq}}{\partial x^k}\right)\dot{x}^p\dot{x}^q\right)$$

$$= M(g_{kq}\ddot{x}^q + [pq,k]\dot{x}^p\dot{x}^q)$$

$$= Mg_{kr}\left(\ddot{x}^r + \left\{ {r \atop pq} \right\}\dot{x}^p\dot{x}^q\right) = Mg_{kr}a^r = Ma_k$$

using Problem 69. The result can be used to express the acceleration in different coordinate systems.

**72.** Use Problem 71 to find the physical components of the acceleration of a particle in cylindrical coordinates.

Since $ds^2 = d\rho^2 + \rho^2 d\phi^2 + dz^2$, $v^2 = (\frac{ds}{dt})^2 = \dot{\rho}^2 + \rho^2\dot{\phi}^2 + \dot{z}^2$ and $T = \frac{1}{2}Mv^2 = \frac{1}{2}M(\dot{\rho}^2 + \rho^2\dot{\phi}^2 + \dot{z}^2)$.

From Problem 71 with $x^1 = \rho$, $x^2 = \phi$, $x^3 = z$ we find

$$a_1 = \ddot{\rho} - \rho\dot{\phi}^2, \quad a_2 = \frac{d}{dt}(\rho^2\dot{\phi}), \quad a_3 = \ddot{z}$$

Then the physical components are given by

$$\frac{a_1}{\sqrt{g_{11}}}, \frac{a_2}{\sqrt{g_{22}}}, \frac{a_3}{\sqrt{g_{33}}} \quad \text{or} \quad \ddot{\rho} - \rho\dot{\phi}^2, \frac{1}{\rho}\frac{d}{dt}(\rho^2\dot{\phi}), \ddot{z}$$

since $g_{11} = 1$, $g_{22} = \rho^2$, $g_{33} = 1$. Compare with Problem 70.

**73.** If the covariant force acting on a particle is given by $F_k = -\frac{\partial V}{\partial x^k}$ where $V(x^1, \ldots, x^N)$ is the potential energy, show that $\frac{d}{dt}\left(\frac{\partial L}{\partial \dot{x}^k}\right) - \frac{\partial L}{\partial x^k} = 0$ where $L = T - V$.

From $L = T - V$, $\frac{\partial L}{\partial \dot{x}^k} = \frac{\partial T}{\partial \dot{x}^k}$ since $V$ is independent of $\dot{x}^k$. Then from Problem 71,

$$\frac{d}{dt}\left(\frac{\partial T}{\partial \dot{x}^k}\right) - \frac{\partial T}{\partial x^k} = Ma_k = F_k = -\frac{\partial V}{\partial x^k} \quad \text{and} \quad \frac{d}{dt}\left(\frac{\partial L}{\partial \dot{x}^k}\right) - \frac{\partial L}{\partial x^k} = 0$$

The function $L$ is called the *Lagrangean*. The equations involving $L$, called *Lagrange's equations*, are important in mechanics. By Problem 50 it follows that the results of this problem are equivalent to the statement that a particle moves in such a way that $\int_{t_1}^{t_2} L\,dt$ is an extremum. This is called *Hamilton's principle*.

**74.** Express the divergence theorem in tensor form.

Let $A^k$ define a tensor field of rank one and let $\nu_k$ denote the outward drawn unit normal to any point of a closed surface $S$ bounding a volume $V$. Then the divergence theorem states that

$$\iiint_V A^k_{\ ,k}\, dV \quad = \quad \iint_S A^k\, \nu_k\, dS$$

For $N$ dimensional space the triple integral is replaced by an $N$ tuple integral, and the double integral by an $N-1$ tuple integral. The invariant $A^k_{\ ,k}$ is the divergence of $A^k$ (see Problem 57). The invariant $A^k \nu_k$ is the scalar product of $A^k$ and $\nu_k$, analogous to $\mathbf{A}\cdot\mathbf{n}$ in the vector notation of Chapter 2.

We have been able to express the theorem in tensor form; hence it is true for all coordinate systems since it is true for rectangular systems (see Chapter 6). Also see Problem 66.

**75.** Express Maxwell's equations (*a*) $\operatorname{div}\mathbf{B} = 0$, (*b*) $\operatorname{div}\mathbf{D} = 4\pi\rho$, (*c*) $\nabla\times\mathbf{E} = -\dfrac{1}{c}\dfrac{\partial \mathbf{B}}{\partial t}$, (*d*) $\nabla\times\mathbf{H} = \dfrac{4\pi\mathbf{I}}{c}$ in tensor form.

Define the tensors $B^k, D^k, E_k, H_k, I^k$ and suppose that $\rho$ and $c$ are invariants. Then the equations can be written

(*a*) $B^k_{\ ,k} = 0$

(*b*) $D^k_{\ ,k} = 4\pi\rho$

(*c*) $-\,\epsilon^{jkq}\, E_{k,q} \;=\; -\dfrac{1}{c}\dfrac{\partial B^j}{\partial t}$ or $\epsilon^{jkq}\, E_{k,q} \;=\; \dfrac{1}{c}\dfrac{\partial B^j}{\partial t}$

(*d*) $-\,\epsilon^{jkq}\, H_{k,q} \;=\; \dfrac{4\pi I^j}{c}$. or $\epsilon^{jkq}\, H_{k,q} \;=\; -\dfrac{4\pi I^j}{c}$

These equations form the basis for *electromagnetic theory*.

**76.** (*a*) Prove that $A_{p,qr} - A_{p,rq} = R^n_{pqr}\, A_n$ where $A_p$ is an arbitrary covariant tensor of rank one. (*b*) Prove that $R^n_{pqr}$ is a tensor. (*c*) Prove that $R_{pqrs} = g_{ns}\, R^n_{pqr}$ is a tensor.

(*a*)
$$A_{p,qr} = (A_{p,q})_{,r} = \frac{\partial A_{p,q}}{\partial x^r} - \begin{Bmatrix} j \\ pr \end{Bmatrix} A_{j,q} - \begin{Bmatrix} j \\ qr \end{Bmatrix} A_{p,j}$$

$$= \frac{\partial}{\partial x^r}\left(\frac{\partial A_p}{\partial x^q} - \begin{Bmatrix} j \\ pq \end{Bmatrix} A_j\right) - \begin{Bmatrix} j \\ pr \end{Bmatrix}\left(\frac{\partial A_j}{\partial x^q} - \begin{Bmatrix} k \\ jq \end{Bmatrix} A_k\right) - \begin{Bmatrix} j \\ qr \end{Bmatrix}\left(\frac{\partial A_p}{\partial x^j} - \begin{Bmatrix} l \\ pj \end{Bmatrix} A_l\right)$$

$$= \frac{\partial^2 A_p}{\partial x^r \partial x^q} - \frac{\partial}{\partial x^r}\begin{Bmatrix} j \\ pq \end{Bmatrix} A_j - \begin{Bmatrix} j \\ pq \end{Bmatrix}\frac{\partial A_j}{\partial x^r} - \begin{Bmatrix} j \\ pr \end{Bmatrix}\frac{\partial A_j}{\partial x^q} + \begin{Bmatrix} j \\ pr \end{Bmatrix}\begin{Bmatrix} k \\ jq \end{Bmatrix} A_k$$

$$- \begin{Bmatrix} j \\ qr \end{Bmatrix}\frac{\partial A_p}{\partial x^j} + \begin{Bmatrix} j \\ qr \end{Bmatrix}\begin{Bmatrix} l \\ pj \end{Bmatrix} A_l$$

By interchanging $q$ and $r$ and subtracting, we find

$$A_{p,qr} - A_{p,rq} = \begin{Bmatrix} j \\ pr \end{Bmatrix}\begin{Bmatrix} k \\ jq \end{Bmatrix} A_k - \frac{\partial}{\partial x^r}\begin{Bmatrix} j \\ pq \end{Bmatrix} A_j - \begin{Bmatrix} j \\ pq \end{Bmatrix}\begin{Bmatrix} k \\ jr \end{Bmatrix} A_k + \frac{\partial}{\partial x^q}\begin{Bmatrix} j \\ pr \end{Bmatrix} A_j$$

$$= \begin{Bmatrix} k \\ pr \end{Bmatrix}\begin{Bmatrix} j \\ kq \end{Bmatrix} A_j - \frac{\partial}{\partial x^r}\begin{Bmatrix} j \\ pq \end{Bmatrix} A_j - \begin{Bmatrix} k \\ pq \end{Bmatrix}\begin{Bmatrix} j \\ kr \end{Bmatrix} A_j + \frac{\partial}{\partial x^q}\begin{Bmatrix} j \\ pr \end{Bmatrix} A_j$$

$$= R^j_{pqr} A_j$$

where $$R^j_{pqr} = \begin{Bmatrix} k \\ pr \end{Bmatrix}\begin{Bmatrix} j \\ kq \end{Bmatrix} - \frac{\partial}{\partial x^r}\begin{Bmatrix} j \\ pq \end{Bmatrix} - \begin{Bmatrix} k \\ pq \end{Bmatrix}\begin{Bmatrix} j \\ kr \end{Bmatrix} + \frac{\partial}{\partial x^q}\begin{Bmatrix} j \\ pr \end{Bmatrix}$$

Replace $j$ by $n$ and the result follows.

(b) Since $A_{p,qr} - A_{p,rq}$ is a tensor, $R^n_{pqr} A_n$ is a tensor; and since $A_n$ is an arbitrary tensor, $R^n_{pqr}$ is a tensor by the quotient law. This tensor is called the *Riemann-Christoffel* tensor, and is sometimes written $R^{n}_{.pqr}$, $R^{\cdots n}_{pqr\cdot}$, or simply $R^n_{pqr}$.

(c) $R_{pqrs} = g_{ns} R^n_{pqr}$ is an associated tensor of $R^n_{pqr}$ and thus is a tensor. It is called the *covariant curvature tensor* and is of fundamental importance in *Einstein's general theory of relativity.*

# SUPPLEMENTARY PROBLEMS

Answers to the Supplementary Problems are given at the end of this Chapter.

**77.** Write each of the following using the summation convention.

(a) $a_1 x^1 x^3 + a_2 x^2 x^3 + \ldots + a_N x^N x^3$

(b) $A^{21} B_1 + A^{22} B_2 + A^{23} B_3 + \ldots + A^{2N} B_N$

(c) $A^j_1 B^1 + A^j_2 B^2 + A^j_3 B^3 + \ldots + A^j_N B^N$

(d) $g^{21} g_{11} + g^{22} g_{21} + g^{23} g_{31} + g^{24} g_{41}$

(e) $B^{121}_{11} + B^{122}_{12} + B^{221}_{21} + B^{222}_{22}$

**78.** Write the terms in each of the following indicated sums.

(a) $\dfrac{\partial}{\partial x^k}(\sqrt{g}\, A^k)$, $N=3$ (b) $A^{jk} B^p_k C_j$, $N=2$ (c) $\dfrac{\partial \bar{x}^j}{\partial x^k}\dfrac{\partial x^k}{\partial \bar{x}^m}$

**79.** What locus is represented by $a_k x^k x^k = 1$ where $x^k$, $k = 1, 2, \ldots, N$ are rectangular coordinates, $a_k$ are positive constants and $N = 2, 3$ or $4$?

**80.** If $N=2$, write the system of equations represented by $a_{pq} x^q = b_p$.

**81.** Write the law of transformation for the tensors (a) $A^{ij}_k$, (b) $B^{ijk}_m$, (c) $C_{mn}$, (d) $A_m$.

**82.** Determine whether the quantities $B(j,k,m)$ and $C(j,k,m,n)$ which transform from a coordinate system $x^i$ to another $\bar{x}^i$ according to the rules

$$(a)\ \bar{B}(p,q,r) = \frac{\partial x^j}{\partial \bar{x}^p}\frac{\partial x^k}{\partial \bar{x}^q}\frac{\partial \bar{x}^r}{\partial x^m} B(j,k,m) \qquad (b)\ \bar{C}(p,q,r,s) = \frac{\partial \bar{x}^p}{\partial x^j}\frac{\partial x^q}{\partial \bar{x}^k}\frac{\partial x^m}{\partial \bar{x}^r}\frac{\partial x^s}{\partial \bar{x}^n} C(j,k,m,n)$$

are tensors. If so, write the tensors in suitable notation and give the rank and the covariant and contravariant orders.

**83.** How many components does a tensor of rank 5 have in a space of 4 dimensions ?

**84.** Prove that if the components of a tensor are zero in one coordinate system they are zero in all coordinate systems.

**85.** Prove that if the components of two tensors are equal in one coordinate system they are equal in all coordinate systems.

**86.** Show that the velocity $\dfrac{dx^k}{dt} = v^k$ of a fluid is a tensor, but that $\dfrac{dv^k}{dt}$ is not a tensor.

**87.** Find the covariant and contravariant components of a tensor in $(a)$ cylindrical coordinates $\rho, \phi, z$, $(b)$ spherical coordinates $r, \theta, \phi$ if its covariant components in rectangular coordinates are $2x - z$, $x^2y$, $yz$.

**88.** The contravariant components of a tensor in rectangular coordinates are $yz$, 3, $2x+y$. Find its covariant components in parabolic cylindrical coordinates.

**89.** Evaluate $(a)$ $\delta_q^p B_p^{rs}$, $(b)$ $\delta_q^p \delta_s^r A^{qs}$, $(c)$ $\delta_q^p \delta_r^q \delta_s^r$, $(d)$ $\delta_q^p \delta_r^q \delta_s^r \delta_p^s$.

**90.** If $A_r^{pq}$ is a tensor, show that $A_r^{pr}$ is a contravariant tensor of rank one.

**91.** Show that $\delta_{jk} = \begin{cases} 1 & j \neq k \\ 0 & j = k \end{cases}$ is not a covariant tensor as the notation might indicate.

**92.** If $\bar{A}_p = \dfrac{\partial x^q}{\partial \bar{x}^p} A_q$ prove that $A_q = \dfrac{\partial \bar{x}^p}{\partial x^q} \bar{A}_p$.

**93.** If $\bar{A}_r^p = \dfrac{\partial \bar{x}^p}{\partial x^q}\dfrac{\partial x^s}{\partial \bar{x}^r} A_s^q$ prove that $A_s^q = \dfrac{\partial x^q}{\partial \bar{x}^p}\dfrac{\partial \bar{x}^r}{\partial x^s} \bar{A}_r^p$.

**94.** If $\Phi$ is an invariant, determine whether $\dfrac{\partial^2 \Phi}{\partial x^p \partial x^q}$ is a tensor.

**95.** If $A_q^p$ and $B_r$ are tensors, prove that $A_q^p B^r$ and $A_q^p B^q$ are tensors and determine the rank of each.

**96.** Show that if $A_{rs}^{pq}$ is a tensor, then $A_{rs}^{pq} + A_{sr}^{qp}$ is a symmetric tensor and $A_{rs}^{pq} - A_{sr}^{qp}$ is a skew-symmetric tensor.

**97.** If $A^{pq}$ and $B_{rs}$ are skew-symmetric tensors, show that $C_{rs}^{pq} = A^{pq} B_{rs}$ is symmetric.

**98.** If a tensor is symmetric (skew-symmetric), are repeated contractions of the tensor also symmetric (skew-symmetric) ?

**99.** Prove that $A_{pq}\, x^p x^q = 0$ if $A_{pq}$ is a skew-symmetric tensor.

**100.** What is the largest number of different components which a symmetric contravariant tensor of rank two can have if (*a*) $N=4$, (*b*) $N=6$? What is the number for any value of $N$?

**101.** How many distinct non-zero components, apart from a difference in sign, does a skew-symmetric covariant tensor of the third rank have?

**102.** If $A^{pq}_{rs}$ is a tensor, prove that a double contraction yields an invariant.

**103.** Prove that a necessary and sufficient condition that a tensor of rank $R$ become an invariant by repeated contraction is that $R$ be even and that the number of covariant and contravariant indices be equal to $R/2$.

**104.** If $A_{pq}$ and $B^{rs}$ are tensors, show that the outer product is a tensor of rank four and that two inner products can be formed of rank two and zero respectively.

**105.** If $A(p,q)\, B_q = C^p$ where $B_q$ is an arbitrary covariant tensor of rank one and $C^p$ is a contravariant tensor of rank one, show that $A(p,q)$ must be a mixed tensor of rank two.

**106.** Let $A^p$ and $B_q$ be arbitrary tensors. Show that if $A^p\, B_q\, C(p,q)$ is an invariant then $C(p,q)$ is a tensor which can be written $C^q_p$.

**107.** Find the sum $S=A+B$, difference $D=A-B$, and products $P=AB$ and $Q=BA$, where $A$ and $B$ are the matrices

(*a*) $A = \begin{pmatrix} 3 & -1 \\ 2 & 4 \end{pmatrix}, \quad B = \begin{pmatrix} 4 & 3 \\ -2 & -1 \end{pmatrix}$

(*b*) $A = \begin{pmatrix} 2 & 0 & 1 \\ -1 & -2 & 2 \\ -1 & 3 & -1 \end{pmatrix}, \quad B = \begin{pmatrix} 1 & -1 & 2 \\ 3 & 2 & -4 \\ -1 & -2 & 2 \end{pmatrix}$

**108.** Find $(3A-2B)(2A-B)$, where $A$ and $B$ are the matrices in the preceding problem.

**109.** (*a*) Verify that $\det(AB) = \{\det A\}\{\det B\}$ for the matrices in Problem 107.
(*b*) Is $\det(AB) = \det(BA)$?

**110.** Let $A = \begin{pmatrix} 3 & -1 & 2 \\ 4 & 2 & 3 \end{pmatrix}, \quad B = \begin{pmatrix} -3 & 2 & -1 \\ 1 & 3 & -2 \\ 2 & 1 & 2 \end{pmatrix}$.

Show that (*a*) $AB$ is defined and find it, (*b*) $BA$ and $A+B$ are not defined.

**111.** Find $x$, $y$ and $z$ such that $\begin{pmatrix} 2 & -1 & 3 \\ 1 & 2 & -4 \\ -1 & 3 & -2 \end{pmatrix}\begin{pmatrix} x \\ y \\ z \end{pmatrix} = \begin{pmatrix} 1 \\ -3 \\ 6 \end{pmatrix}$

**112.** The inverse of a square matrix $A$, written $A^{-1}$ is defined by the equation $AA^{-1} = I$, where $I$ is the unit matrix having ones down the main diagonal and zeros elsewhere.

Find $A^{-1}$ if (*a*) $A = \begin{pmatrix} 3 & -2 \\ -5 & 4 \end{pmatrix}$, (*b*) $A = \begin{pmatrix} 1 & -1 & 1 \\ 2 & 1 & -1 \\ 1 & -1 & 2 \end{pmatrix}$.

Is $A^{-1}A = I$ in these cases?

**113.** Prove that $A = \begin{pmatrix} 2 & 1 & -2 \\ 1 & -2 & 3 \\ 4 & -3 & 4 \end{pmatrix}$ has no inverse.

**114.** Prove that $(AB)^{-1} = B^{-1}A^{-1}$, where $A$ and $B$ are non-singular square matrices.

**115.** Express in matrix notation the transformation equations for
(*a*) a contravariant vector (*b*) a covariant tensor of rank two (*c*) a mixed tensor of rank two.

**116.** Determine the values of the constant $\lambda$ such that $AX = \lambda X$, where $A = \begin{pmatrix} 2 & -2 \\ -3 & 1 \end{pmatrix}$ and $X$ is an arbitrary matrix. These values of $\lambda$ are called *characteristic values* or *eigenvalues* of the matrix $A$.

**117.** The equation $F(\lambda) = 0$ of the previous problem for determining the characteristic values of a matrix $A$ is called the *characteristic equation* for $A$. Show that $F(A) = O$, where $F(A)$ is the matrix obtained by replacing $\lambda$ by $A$ in the characteristic equation and where the constant term $c$ is replaced by the matrix $cI$, and $O$ is a matrix whose elements are zero (called the null matrix). The result is a special case of the *Hamilton-Cayley theorem* which states that a matrix satisfies its own characteristic equation.

**118.** Prove that $(AB)^T = B^T A^T$.

**119.** Determine the metric tensor and conjugate metric tensor in
(*a*) parabolic cylindrical and (*b*) elliptic cylindrical coordinates.

**120.** Prove that under the affine transformation $\bar{x}^r = a_p^r x^p + b^r$, where $a_p^r$ and $b^r$ are constants such that $a_p^r a_q^r = \delta_q^p$, there is no distinction between the covariant and contravariant components of a tensor. In the special case where the transformations are from one rectangular coordinate system to another, the tensors are called *cartesian tensors*.

**121.** Find $g$ and $g^{jk}$ corresponding to $ds^2 = 3(dx^1)^2 + 2(dx^2)^2 + 4(dx^3)^2 - 6\,dx^1\,dx^3$.

**122.** If $A^k = g^{jk}A_j$, show that $A_j = g_{jk}A^k$ and conversely.

**123.** Express the relationship between the associated tensors
(*a*) $A^{pq}$ and $A_j^{\cdot q}$, (*b*) $A^{p\cdot r}_{\cdot q}$ and $A_{jql}$, (*c*) $A^{\cdot\cdot r}_{pq}$ and $A^{jk}_{\cdot\cdot l}$.

**124.** Show that (*a*) $A_p^{\cdot q} B^{p}_{\cdot rs} = A^{pq} B_{prs}$, (*b*) $A^{pq}_{\cdot\cdot r} B_p^{\cdot r} = A^{\cdot q}_{p\cdot r} B^{pr} = A_p^{\cdot qr} B^{p}_{\cdot r}$. Hence demonstrate the general result that a dummy symbol in a term may be lowered from its upper position and raised from its lower position without changing the value of the term.

**125.** Show that if $A^p_{\cdot qr} = B^p_{\cdot q} C_r$, then $A_{pqr} = B_{pq} C_r$ and $A_p^{\cdot qr} = B_p^{\cdot q} C^r$. Hence demonstrate the result that a free index in a tensor equation may be raised or lowered without affecting the validity of the equation.

**126.** Show that the tensors $g_{pq}$, $g^{pq}$ and $\delta_q^p$ are associated tensors.

**127.** Prove (*a*) $\bar{g}_{jk} \dfrac{\partial \bar{x}^j}{\partial x^p} = g_{pq} \dfrac{\partial x^q}{\partial \bar{x}^k}$, (*b*) $\bar{g}^{jk} \dfrac{\partial x^p}{\partial \bar{x}^j} = g^{pq} \dfrac{\partial \bar{x}^k}{\partial x^q}$.

**128.** If $A^p$ is a vector field, find the corresponding unit vector.

**129.** Show that the cosines of the angles which the 3 dimensional unit vector $U^i$ make with the coordinate curves are given by $\frac{U_1}{\sqrt{g_{11}}}, \frac{U_2}{\sqrt{g_{22}}}, \frac{U_3}{\sqrt{g_{33}}}$.

**130.** Determine the Christoffel symbols of the first kind in (*a*) rectangular, (*b*) cylindrical, and (*c*) spherical coordinates.

**131.** Determine the Christoffel symbols of the first and second kinds in (*a*) parabolic cylindrical, (*b*) elliptic cylindrical coordinates.

**132.** Find differential equations for the geodesics in (*a*) cylindrical, (*b*) spherical coordinates.

**133.** Show that the geodesics on a plane are straight lines.

**134.** Show that the geodesics on a sphere are arcs of great circles.

**135.** Write the Christoffel symbols of the second kind for the metric

$$ds^2 = (dx^1)^2 + [(x^2)^2 - (x^1)^2]\,(dx^2)^2$$

and the corresponding geodesic equations.

**136.** Write the covariant derivative with respect to $x^q$ of each of the following tensors:
(*a*) $A^{jk}_l$, (*b*) $A^{jk}_{lm}$, (*c*) $A^j_{klm}$, (*d*) $A^{jkl}_m$, (*e*) $A^{jk}_{lmn}$.

**137.** Find the covariant derivative of (*a*) $g_{jk}A^k$, (*b*) $A^j B_k$, (*c*) $\delta^j_k A_j$ with respect to $x^q$.

**138.** Use the relation $A^j = g^{jk}A_k$ to obtain the covariant derivative of $A^j$ from the covariant derivative of $A_k$.

**139.** If $\Phi$ is an invariant, prove that $\Phi_{,pq} = \Phi_{,qp}$, i.e. the order of covariant differentiation of an invariant is immaterial.

**140.** Show that $\epsilon_{pqr}$ and $\epsilon^{pqr}$ are covariant and contravariant tensors respectively.

**141.** Express the divergence of a vector $A^p$ in terms of its physical components for (*a*) parabolic cylindrical, (*b*) paraboloidal coordinates.

**142.** Find the physical components of grad $\Phi$ in (*a*) parabolic cylindrical, (*b*) elliptic cylindrical coordinates.

**143.** Find $\nabla^2\Phi$ in parabolic cylindrical coordinates.

**144.** Using the tensor notation, show that (*a*) div curl $A^r = 0$, (*b*) curl grad $\Phi = 0$.

**145.** Calculate the intrinsic derivatives of each of the following tensor fields, assumed to be differentiable functions of $t$:
(*a*) $A_k$, (*b*) $A^{jk}$, (*c*) $A_j B^k$, (*d*) $\phi A^j_k$ where $\phi$ is an invariant.

**146.** Find the intrinsic derivative of (*a*) $g_{jk}A^k$, (*b*) $\delta^j_k A_j$, (*c*) $g_{jk}\delta^j_r A^r_p$.

**147.** Prove $\frac{d}{dt}(g^{pq}A_p A_q) = 2\,g^{pq}A_p \frac{\delta A_q}{\delta t}$.

**148.** Show that if no external force acts, a moving particle of constant mass travels along a geodesic given by

$$\frac{\delta}{\delta s}\left(\frac{dx^p}{ds}\right) = 0.$$

**149.** Prove that the sum and difference of two relative tensors of the same weight and type is also a relative tensor of the same weight and type.

**150.** If $A_r^{pq}$ is a relative tensor of weight $w$, prove that $g^{-w/2} A_r^{pq}$ is an absolute tensor.

**151.** If $A(p,q)\, B_r^{qs} = C_{pr}^{s}$, where $B_r^{qs}$ is an arbitrary relative tensor of weight $w_1$ and $C_{pr}^{s}$ is a known relative tensor of weight $w_2$, prove that $A(p,q)$ is a relative tensor of weight $w_2 - w_1$. This is an example of the quotient law for relative tensors.

**152.** Show that the quantity $G(j,k)$ of Solved Problem 31 is a relative tensor of weight two.

**153.** Find the physical components of (*a*) the velocity and (*b*) the acceleration of a particle in spherical coordinates.

**154.** Let $A^r$ and $B^r$ be two vectors in three dimensional space. Show that if $\lambda$ and $\mu$ are constants, then $C^r = \lambda A^r + \mu B^r$ is a vector lying in the plane of $A^r$ and $B^r$. What is the interpretation in higher dimensional space?

**155.** Show that a vector normal to the surface $\phi(x^1, x^2, x^3)$ = constant is given by $A^p = g^{pq}\dfrac{\partial\phi}{\partial x^q}$. Find the corresponding unit normal.

**156.** The equation of continuity is given by $\nabla\cdot(\sigma v) + \dfrac{\partial\sigma}{\partial t} = 0$ where $\sigma$ is the density and $v$ is the velocity of a fluid. Express the equation in tensor form.

**157.** Express the continuity equation in (*a*) cylindrical and (*b*) spherical coordinates.

**158.** Express Stokes' theorem in tensor form.

**159.** Prove that the covariant curvature tensor $R_{pqrs}$ is skew-symmetric in (*a*) $p$ and $q$, (*b*) $r$ and $s$, (*c*) $q$ and $s$.

**160.** Prove $R_{pqrs} = R_{rspq}$.

**161.** Prove (*a*) $R_{pqrs} + R_{psqr} + R_{prsq} = 0$,
(*b*) $R_{pqrs} + R_{rqps} + R_{rspq} + R_{psrq} = 0$.

**162.** Prove that covariant differentiation in a Euclidean space is commutative. Thus show that the Riemann-Christoffel tensor and curvature tensor are zero in a Euclidean space.

**163.** Let $T^p = \dfrac{dx^p}{ds}$ be the tangent vector to curve $C$ whose equation is $x^p = x^p(s)$ where $s$ is the arc length. (*a*) Show that $g_{pq} T^p T^q = 1$. (*b*) Prove that $g_{pq} T^p \dfrac{\delta T^q}{\delta s} = 0$ and thus show that $N^q = \dfrac{1}{\kappa}\dfrac{\delta T^q}{\delta s}$ is a unit normal to $C$ for suitable $\kappa$. (*c*) Prove that $\dfrac{\delta N^q}{\delta s}$ is orthogonal to $N^q$.

**164.** With the notation of the previous problem, prove:

(*a*) $g_{pq} T^p N^q = 0$, (*b*) $g_{pq} T^p \dfrac{\delta N^q}{\delta s} = -\kappa$ or $g_{pq} T^p \left(\dfrac{\delta N^q}{\delta s} + \kappa T^q\right) = 0$.

Hence show that $B^r = \dfrac{1}{\tau}\left(\dfrac{\delta N^r}{\delta s} + \kappa T^r\right)$ is a unit vector for suitable $\tau$ orthogonal to both $T^p$ and $N^q$.

**165.** Prove the Frenet-Serret formulae

$$\frac{\delta T^p}{\delta s} = \kappa N^p, \quad \frac{\delta N^p}{\delta s} = \tau B^p - \kappa T^p, \quad \frac{\delta B^p}{\delta s} = -\tau N^p$$

where $T^p$, $N^p$ and $B^p$ are the unit tangent, unit normal and unit binormal vectors to $C$, and $\kappa$ and $\tau$ are the curvature and torsion of $C$.

**166.** Show that $ds^2 = c^2(dx^4)^2 - dx^k\,dx^k$ $(N=3)$ is invariant under the linear (affine) transformation

$$\bar{x}^1 = \gamma(x^1 - vx^4), \quad \bar{x}^2 = x^2, \quad \bar{x}^3 = x^3, \quad \bar{x}^4 = \gamma(x^4 - \frac{\beta}{c}x^1)$$

where $\gamma, \beta, c$ and $v$ are constants, $\beta = v/c$ and $\gamma = (1-\beta^2)^{-1/2}$. This is the *Lorentz transformation* of special relativity. Physically, an observer at the origin of the $x^i$ system sees an event occurring at position $x^1, x^2, x^3$ at time $x^4$ while an observer at the origin of the $\bar{x}^i$ system sees the same event occurring at position $\bar{x}^1, \bar{x}^2, \bar{x}^3$ at time $\bar{x}^4$. It is assumed that *(1)* the two systems have the $x^1$ and $\bar{x}^1$ axes coincident, *(2)* the positive $x^2$ and $x^3$ axes are parallel respectively to the positive $\bar{x}^2$ and $\bar{x}^3$ axes, *(3)* the $\bar{x}^i$ system moves with velocity $v$ relative to the $x^i$ system, and *(4)* the velocity of light $c$ is a constant.

**167.** Show that to an observer fixed in the $x^i$ $(\bar{x}^i)$ system, a rod fixed in the $\bar{x}^i$ $(x^i)$ system lying parallel to the $\bar{x}^1$ $(x^1)$ axis and of length $L$ in this system appears to have the reduced length $L\sqrt{1-\beta^2}$. This phenomena is called the *Lorentz-Fitzgerald contraction.*

## ANSWERS TO SUPPLEMENTARY PROBLEMS.

**77.** (a) $a_k x^k x^3$ (b) $A^{2j} B_j$ (c) $A^j_k B^k$ (d) $g^{2q} g_{q1}$, $N=4$ (e) $B^{p2r}_{pr}$, $N=2$

**78.** (a) $\frac{\partial}{\partial x^1}(\sqrt{g}\,A^1) + \frac{\partial}{\partial x^2}(\sqrt{g}\,A^2) + \frac{\partial}{\partial x^3}(\sqrt{g}\,A^3)$

(b) $A^{11} B^p_1 C_1 + A^{21} B^p_1 C_2 + A^{12} B^p_2 C_1 + A^{22} B^p_2 C_2$

(c) $\frac{\partial \bar{x}^j}{\partial x^1}\frac{\partial x^1}{\partial \bar{x}^m} + \frac{\partial \bar{x}^j}{\partial x^2}\frac{\partial x^2}{\partial \bar{x}^m} + \dots + \frac{\partial \bar{x}^j}{\partial x^N}\frac{\partial x^N}{\partial \bar{x}^m}$

**79.** Ellipse for $N=2$, ellipsoid for $N=3$, hyperellipsoid for $N=4$.

**80.** $$\begin{cases} a_{11}x^1 + a_{12}x^2 = b_1 \\ a_{21}x^1 + a_{22}x^2 = b_2 \end{cases}$$

**81.** (a) $\bar{A}^{pq}_r = \frac{\partial \bar{x}^p}{\partial x^i}\frac{\partial \bar{x}^q}{\partial x^j}\frac{\partial x^k}{\partial \bar{x}^r} A^{ij}_k$ (c) $\bar{C}_{pq} = \frac{\partial x^m}{\partial \bar{x}^p}\frac{\partial x^n}{\partial \bar{x}^q} C_{mn}$

(b) $\bar{B}^{pqr}_s = \frac{\partial \bar{x}^p}{\partial x^i}\frac{\partial \bar{x}^q}{\partial x^j}\frac{\partial \bar{x}^r}{\partial x^k}\frac{\partial x^m}{\partial \bar{x}^s} B^{ijk}_m$ (d) $\bar{A}_p = \frac{\partial x^m}{\partial \bar{x}^p} A_m$

**82.** (a) $B(j,k,m)$ is a tensor of rank three and is covariant of order two and contravariant of order one. It can be written $B^m_{jk}$. (b) $C(j,k,m,n)$ is not a tensor.

**83.** $4^5 = 1024$

**87.** (a) $2\rho\cos^2\phi - z\cos\phi + \rho^3\sin^2\phi\cos^2\phi$,

$-2\rho^2\sin\phi\cos\phi + \rho z\sin\phi + \rho^4\sin\phi\cos^3\phi$,

$\rho z\sin\phi$.

(*b*) $2r\sin^2\theta\cos^2\phi \;-\; r\sin\theta\cos\theta\cos\phi \;+\; r^3\sin^4\theta\sin^2\phi\cos^2\phi \;+\; r^2\sin\theta\cos^2\theta\sin\phi,$

$2r^2\sin\theta\cos\theta\cos^2\phi \;-\; r^2\cos^2\theta\cos\phi \;+\; r^4\sin^3\theta\cos\theta\sin^2\phi\cos^2\phi \;-\; r^3\sin^2\theta\cos\theta\sin\phi,$

$-\,2r^2\sin^2\theta\sin\phi\cos\phi \;+\; r^2\sin\theta\cos\theta\sin\phi \;+\; r^4\sin^4\theta\sin\phi\cos^3\phi$

**88.** $u^2vz + 3v,\quad 3u - uv^2z,\quad u^2 + uv - v^2$

**89.** (*a*) $B_q^{rs}$, (*b*) $A^{pr}$, (*c*) $\delta_s^p$, (*d*) 1

**94.** It is not a tensor.

**95.** Rank 3 and rank 1 respectively.

**98.** Yes.

**100.** (*a*) 10, (*b*) 21, (*c*) $N(N+1)/2$

**101.** $N(N-1)(N-2)/6$

**107.** (*a*) $S = \begin{pmatrix} 7 & 2 \\ 0 & 3 \end{pmatrix},\quad D = \begin{pmatrix} -1 & -4 \\ 4 & 5 \end{pmatrix},\quad P = \begin{pmatrix} 14 & 10 \\ 0 & 2 \end{pmatrix},\quad Q = \begin{pmatrix} 18 & 8 \\ -8 & -2 \end{pmatrix}$

(*b*) $S = \begin{pmatrix} 3 & -1 & 3 \\ 2 & 0 & -2 \\ -2 & 1 & 1 \end{pmatrix},\quad D = \begin{pmatrix} 1 & 1 & -1 \\ -4 & -4 & 6 \\ 0 & 5 & -3 \end{pmatrix},\quad P = \begin{pmatrix} 1 & -4 & 6 \\ -9 & -7 & 10 \\ 9 & 9 & -16 \end{pmatrix},\quad Q = \begin{pmatrix} 1 & 8 & -3 \\ 8 & -16 & 11 \\ -2 & 10 & -7 \end{pmatrix}$

**108.** (*a*) $\begin{pmatrix} -52 & -86 \\ 104 & 76 \end{pmatrix}$ (*b*) $\begin{pmatrix} 3 & -16 & 20 \\ 9 & 163 & -136 \\ -61 & -135 & 132 \end{pmatrix}$

**110.** $\begin{pmatrix} -6 & 5 & 3 \\ -4 & 17 & -2 \end{pmatrix}$

**111.** $x = -1,\; y = 3,\; z = 2$

**112.** (*a*) $\begin{pmatrix} 2 & 1 \\ 5/2 & 3/2 \end{pmatrix}$ (*b*) $\begin{pmatrix} 1/3 & 1/3 & 0 \\ -5/3 & 1/3 & 1 \\ -1 & 0 & 1 \end{pmatrix}$. Yes

**115.** (*a*)

$$\begin{pmatrix} \bar{A}^1 \\ \bar{A}^2 \\ \bar{A}^3 \end{pmatrix} = \begin{pmatrix} \frac{\partial \bar{x}^1}{\partial x^1} & \frac{\partial \bar{x}^1}{\partial x^2} & \frac{\partial \bar{x}^1}{\partial x^3} \\ \frac{\partial \bar{x}^2}{\partial x^1} & \frac{\partial \bar{x}^2}{\partial x^2} & \frac{\partial \bar{x}^2}{\partial x^3} \\ \frac{\partial \bar{x}^3}{\partial x^1} & \frac{\partial \bar{x}^3}{\partial x^2} & \frac{\partial \bar{x}^3}{\partial x^3} \end{pmatrix} \begin{pmatrix} A^1 \\ A^2 \\ A^3 \end{pmatrix}$$

(*b*)

$$\begin{pmatrix} \bar{A}_{11} & \bar{A}_{12} & \bar{A}_{13} \\ \bar{A}_{21} & \bar{A}_{22} & \bar{A}_{23} \\ \bar{A}_{31} & \bar{A}_{32} & \bar{A}_{33} \end{pmatrix} = \begin{pmatrix} \frac{\partial x^1}{\partial \bar{x}^1} & \frac{\partial x^2}{\partial \bar{x}^1} & \frac{\partial x^3}{\partial \bar{x}^1} \\ \frac{\partial x^1}{\partial \bar{x}^2} & \frac{\partial x^2}{\partial \bar{x}^2} & \frac{\partial x^3}{\partial \bar{x}^2} \\ \frac{\partial x^1}{\partial \bar{x}^3} & \frac{\partial x^2}{\partial \bar{x}^3} & \frac{\partial x^3}{\partial \bar{x}^3} \end{pmatrix} \begin{pmatrix} A_{11} & A_{12} & A_{13} \\ A_{21} & A_{22} & A_{23} \\ A_{31} & A_{32} & A_{33} \end{pmatrix} \begin{pmatrix} \frac{\partial x^1}{\partial \bar{x}^1} & \frac{\partial x^1}{\partial \bar{x}^2} & \frac{\partial x^1}{\partial \bar{x}^3} \\ \frac{\partial x^2}{\partial \bar{x}^1} & \frac{\partial x^2}{\partial \bar{x}^2} & \frac{\partial x^2}{\partial \bar{x}^3} \\ \frac{\partial x^3}{\partial \bar{x}^1} & \frac{\partial x^3}{\partial \bar{x}^2} & \frac{\partial x^3}{\partial \bar{x}^3} \end{pmatrix}$$

(*c*)

$$\begin{pmatrix} \bar{A}_1^1 & \bar{A}_2^1 & \bar{A}_3^1 \\ \bar{A}_1^2 & \bar{A}_2^2 & \bar{A}_3^2 \\ \bar{A}_1^3 & \bar{A}_2^3 & \bar{A}_3^3 \end{pmatrix} = \begin{pmatrix} \frac{\partial \bar{x}^1}{\partial x^1} & \frac{\partial \bar{x}^1}{\partial x^2} & \frac{\partial \bar{x}^1}{\partial x^3} \\ \frac{\partial \bar{x}^2}{\partial x^1} & \frac{\partial \bar{x}^2}{\partial x^2} & \frac{\partial \bar{x}^2}{\partial x^3} \\ \frac{\partial \bar{x}^3}{\partial x^1} & \frac{\partial \bar{x}^3}{\partial x^2} & \frac{\partial \bar{x}^3}{\partial x^3} \end{pmatrix} \begin{pmatrix} A_1^1 & A_2^1 & A_3^1 \\ A_1^2 & A_2^2 & A_3^2 \\ A_1^3 & A_2^3 & A_3^3 \end{pmatrix} \begin{pmatrix} \frac{\partial x^1}{\partial \bar{x}^1} & \frac{\partial x^1}{\partial \bar{x}^2} & \frac{\partial x^1}{\partial \bar{x}^3} \\ \frac{\partial x^2}{\partial \bar{x}^1} & \frac{\partial x^2}{\partial \bar{x}^2} & \frac{\partial x^2}{\partial \bar{x}^3} \\ \frac{\partial x^3}{\partial \bar{x}^1} & \frac{\partial x^3}{\partial \bar{x}^2} & \frac{\partial x^3}{\partial \bar{x}^3} \end{pmatrix}$$

**116.** $\lambda = 4, -1$

**119.** (*a*) $\begin{pmatrix} u^2+v^2 & 0 & 0 \\ 0 & u^2+v^2 & 0 \\ 0 & 0 & 1 \end{pmatrix},\quad \begin{pmatrix} \frac{1}{u^2+v^2} & 0 & 0 \\ 0 & \frac{1}{u^2+v^2} & 0 \\ 0 & 0 & 1 \end{pmatrix}$

(b) $\begin{pmatrix} a^2(\sinh^2 u + \sin^2 v) & 0 & 0 \\ 0 & a^2(\sinh^2 u + \sin^2 v) & 0 \\ 0 & 0 & 1 \end{pmatrix}$, $\begin{pmatrix} \frac{1}{a^2(\sinh^2 u + \sin^2 v)} & 0 & 0 \\ 0 & \frac{1}{a^2(\sinh^2 u + \sin^2 v)} & 0 \\ 0 & 0 & 1 \end{pmatrix}$

**121.** $g = 6$, $(g^{jk}) = \begin{pmatrix} 4/3 & 0 & 1 \\ 0 & 1/2 & 0 \\ 1 & 0 & 1 \end{pmatrix}$

**123.** (a) $A^{pq} = g^{pj} A_j^{\cdot q}$, (b) $A^{p\cdot r}_{\cdot q} = g^{pj} g^{rl} A_{jql}$, (c) $A^{\cdot\cdot r}_{pq} = g_{pj} g_{qk} g^{rl} A^{jk}_{\cdot\cdot l}$

**128.** $\dfrac{A^p}{\sqrt{A^p A_p}}$ or $\dfrac{A^p}{\sqrt{g_{pq} A^p A^q}}$

**130.** (a) They are all zero.

(b) $[22,1] = -\rho$, $[12,2] = [21,2] = \rho$. All others are zero.

(c) $[22,1] = -r$, $[33,1] = -r\sin^2\theta$, $[33,2] = -r^2\sin\theta\cos\theta$

$[21,2] = [12,2] = r$, $[31,3] = [13,3] = r\sin^2\theta$

$[32,3] = [23,3] = r^2 \sin\theta\cos\theta$. All others are zero.

**131.** (a) $[11,1] = u$, $[22,2] = v$, $[11,2] = -v$, $[22,1] = -u$,

$[12,1] = [21,1] = v$, $[21,2] = [12,2] = u$.

$\left\{ {1 \atop 11} \right\} = \dfrac{u}{u^2+v^2}$, $\left\{ {2 \atop 22} \right\} = \dfrac{v}{u^2+v^2}$, $\left\{ {1 \atop 22} \right\} = \dfrac{-u}{u^2+v^2}$, $\left\{ {2 \atop 11} \right\} = \dfrac{-v}{u^2+v^2}$,

$\left\{ {1 \atop 21} \right\} = \left\{ {1 \atop 12} \right\} = \dfrac{v}{u^2+v^2}$, $\left\{ {2 \atop 21} \right\} = \left\{ {2 \atop 12} \right\} = \dfrac{u}{u^2+v^2}$. All others are zero.

(b) $[11,1] = 2a^2 \sinh u \cosh u$, $[22,2] = 2a^2 \sin v \cos v$, $[11,2] = -2a^2 \sin v \cos v$

$[22,1] = -2a^2 \sinh u \cosh u$, $[12,1] = [21,1] = 2a^2 \sin v \cos v$, $[21,2] = [12,2] = 2a^2 \sinh u \cosh u$

$\left\{ {1 \atop 11} \right\} = \dfrac{\sinh u \cosh u}{\sinh^2 u + \sin^2 v}$, $\left\{ {2 \atop 22} \right\} = \dfrac{\sin v \cos v}{\sinh^2 u + \sin^2 v}$, $\left\{ {1 \atop 22} \right\} = \dfrac{-\sinh u \cosh u}{\sinh^2 u + \sin^2 v}$,

$\left\{ {2 \atop 11} \right\} = \dfrac{-\sin v \cos v}{\sinh^2 u + \sin^2 v}$, $\left\{ {1 \atop 21} \right\} = \left\{ {1 \atop 12} \right\} = \dfrac{\sin v \cos v}{\sinh^2 u + \sin^2 v}$,

$\left\{ {2 \atop 21} \right\} = \left\{ {2 \atop 12} \right\} = \dfrac{\sinh u \cosh u}{\sinh^2 u + \sin^2 v}$. All others are zero.

**132.** (a) $\dfrac{d^2\rho}{ds^2} - \rho\left(\dfrac{d\phi}{ds}\right)^2 = 0$, $\dfrac{d^2\phi}{ds^2} + \dfrac{2}{\rho}\dfrac{d\rho}{ds}\dfrac{d\phi}{ds} = 0$, $\dfrac{d^2 z}{ds^2} = 0$

(b) $\dfrac{d^2 r}{ds^2} - r\left(\dfrac{d\theta}{ds}\right)^2 - r\sin^2\theta\left(\dfrac{d\phi}{ds}\right)^2 = 0$

$\dfrac{d^2\theta}{ds^2} + \dfrac{2}{r}\dfrac{dr}{ds}\dfrac{d\theta}{ds} - \sin\theta\cos\theta\left(\dfrac{d\phi}{ds}\right)^2 = 0$

$\dfrac{d^2\phi}{ds^2} + \dfrac{2}{r}\dfrac{dr}{ds}\dfrac{d\phi}{ds} + 2\cot\theta\,\dfrac{d\theta}{ds}\dfrac{d\phi}{ds} = 0$

**135.** $\left\{ {1 \atop 22} \right\} = x^1, \quad \left\{ {2 \atop 12} \right\} = \left\{ {2 \atop 21} \right\} = \dfrac{x^1}{(x^1)^2 - (x^2)^2}, \quad \left\{ {2 \atop 22} \right\} = \dfrac{x^2}{(x^2)^2 - (x^1)^2}$. All others are zero.

$$\frac{d^2x^1}{ds^2} + x^1\left(\frac{dx^2}{ds}\right)^2 = 0, \qquad \frac{d^2x^2}{ds^2} + \frac{2x^1}{(x^1)^2 - (x^2)^2}\frac{dx^1}{ds}\frac{dx^2}{ds} + \frac{x^2}{(x^2)^2 - (x^1)^2}\left(\frac{dx^2}{ds}\right)^2 = 0$$

**136.** (a) $A^{jk}_{l,q} = \dfrac{\partial A^{jk}_l}{\partial x^q} - \left\{ {s \atop lq} \right\} A^{jk}_s + \left\{ {j \atop qs} \right\} A^{sk}_l + \left\{ {k \atop qs} \right\} A^{js}_l$

(b) $A^{jk}_{lm,q} = \dfrac{\partial A^{jk}_{lm}}{\partial x^q} - \left\{ {s \atop lq} \right\} A^{jk}_{sm} - \left\{ {s \atop mq} \right\} A^{jk}_{ls} + \left\{ {j \atop qs} \right\} A^{sk}_{lm} + \left\{ {k \atop qs} \right\} A^{js}_{lm}$

(c) $A^{j}_{klm,q} = \dfrac{\partial A^{j}_{klm}}{\partial x^q} - \left\{ {s \atop kq} \right\} A^{j}_{slm} - \left\{ {s \atop lq} \right\} A^{j}_{ksm} - \left\{ {s \atop mq} \right\} A^{j}_{kls} + \left\{ {j \atop qs} \right\} A^{s}_{klm}$

(d) $A^{jkl}_{m,q} = \dfrac{\partial A^{jkl}_{m}}{\partial x^q} - \left\{ {s \atop mq} \right\} A^{jkl}_{s} + \left\{ {j \atop qs} \right\} A^{skl}_{m} + \left\{ {k \atop qs} \right\} A^{jsl}_{m} + \left\{ {l \atop qs} \right\} A^{jks}_{m}$

(e) $A^{jk}_{lmn,q} = \dfrac{\partial A^{jk}_{lmn}}{\partial x^q} - \left\{ {s \atop lq} \right\} A^{jk}_{smn} - \left\{ {s \atop mq} \right\} A^{jk}_{lsn} - \left\{ {s \atop nq} \right\} A^{jk}_{lms} + \left\{ {j \atop qs} \right\} A^{sk}_{lmn} + \left\{ {k \atop qs} \right\} A^{js}_{lmn}$

**137.** (a) $g_{jk} A^k_{,q}$, (b) $A^j_{,q} B_k + A^j B_{k,q}$, (c) $\delta^j_k A_{j,q}$

**141.** (a) $\dfrac{1}{u^2+v^2}\left[\dfrac{\partial}{\partial u}(\sqrt{u^2+v^2}\, A_u) + \dfrac{\partial}{\partial v}(\sqrt{u^2+v^2}\, A_v)\right] + \dfrac{\partial A_z}{\partial z}$

(b) $\dfrac{1}{uv(u^2+v^2)}\left[\dfrac{\partial}{\partial u}(uv\sqrt{u^2+v^2}\, A_u) + \dfrac{\partial}{\partial v}(uv\sqrt{u^2+v^2}\, A_v)\right] + \dfrac{1}{uv}\dfrac{\partial^2 A_z}{\partial z^2}$

**142.** (a) $\dfrac{1}{\sqrt{u^2+v^2}}\dfrac{\partial \Phi}{\partial u}\mathbf{e}_u + \dfrac{1}{\sqrt{u^2+v^2}}\dfrac{\partial \Phi}{\partial v}\mathbf{e}_v + \dfrac{\partial \Phi}{\partial z}\mathbf{e}_z$

(b) $\dfrac{1}{a\sqrt{\sinh^2 u + \sin^2 v}}\left(\dfrac{\partial \Phi}{\partial u}\mathbf{e}_u + \dfrac{\partial \Phi}{\partial v}\mathbf{e}_v\right) + \dfrac{\partial \Phi}{\partial z}\mathbf{e}_z$

where $\mathbf{e}_u$, $\mathbf{e}_v$ and $\mathbf{e}_z$ are unit vectors in the directions of increasing $u, v$ and $z$ respectively.

**143.** $\dfrac{\partial^2 \Phi}{\partial u^2} + \dfrac{\partial^2 \Phi}{\partial v^2} + (u^2+v^2)\Phi = 0$

**145.** (a) $\dfrac{\delta A_k}{\delta t} = A_{k,q}\dfrac{dx^q}{dt} = \left(\dfrac{\partial A_k}{\partial x^q} - \left\{ {s \atop kq} \right\} A_s\right)\dfrac{dx^q}{dt} = \dfrac{dA_k}{dt} - \left\{ {s \atop kq} \right\} A_s \dfrac{dx^q}{dt}$

(b) $\dfrac{\delta A^{jk}}{\delta t} = \dfrac{dA^{jk}}{dt} + \left\{ {j \atop qs} \right\} A^{sk}\dfrac{dx^q}{dt} + \left\{ {k \atop qs} \right\} A^{js}\dfrac{dx^q}{dt}$

(c) $\dfrac{\delta}{\delta t}(A_j B^k) = \dfrac{\delta A_j}{\delta t} B^k + A_j \dfrac{\delta B^k}{\delta t}$

$$= \left(\frac{dA_j}{dt} - \left\{ {s \atop jq} \right\} A_s \frac{dx^q}{dt}\right) B^k + A^j\left(\frac{dB^k}{dt} + \left\{ {k \atop qs} \right\} B^s \frac{dx^q}{dt}\right)$$

(d) $$\frac{\delta}{\delta t}(\Phi A^j_k) = \Phi \frac{\delta A^j_k}{\delta t} + \frac{\delta \Phi}{\delta t} A^j_k$$
$$= \Phi\left(\frac{dA^j_k}{dt} + \left\{ {j \atop qs} \right\} A^s_k \frac{dx^q}{dt} - \left\{ {s \atop kq} \right\} A^j_s \frac{dx^q}{dt}\right) + \frac{d\Phi}{dt} A^j_k$$

**146.** (a) $$g_{jk} \frac{\delta A^k}{\delta t} = g_{jk}\left(\frac{dA^k}{dt} + \left\{ {k \atop qs} \right\} A^s \frac{dx^q}{dt}\right)$$

(b) $$\delta^j_k \frac{\delta A_j}{\delta t} = \delta^j_k\left(\frac{dA_j}{dt} - \left\{ {s \atop jq} \right\} A_s \frac{dx^q}{dt}\right) = \frac{dA_k}{dt} - \left\{ {s \atop kq} \right\} A_s \frac{dx^q}{dt}$$

(c) $$g_{jk}\, \delta^j_r \frac{\delta A^r_p}{\delta t} = g_{rk}\left(\frac{dA^r_p}{dt} - \left\{ {s \atop pq} \right\} A^r_s \frac{dx^q}{dt} + \left\{ {r \atop qs} \right\} A^s_p \frac{dx^q}{dt}\right)$$

**153.** (a) $\dot{r},\ r\dot{\theta},\ r\sin\theta\,\dot{\phi}$

(b) $\ddot{r} - r\dot{\theta}^2 - r\sin^2\theta\,\dot{\phi}^2,\quad \dfrac{1}{r}\dfrac{d}{dt}(r^2\dot{\theta}) - r\sin\theta\cos\theta\,\dot{\phi}^2,\quad \dfrac{1}{r\sin\theta}\dfrac{d}{dt}(r^2\sin^2\theta\,\dot{\phi})$

**156.** $\dfrac{\partial(\sigma v^q)}{\partial x^q} + \dfrac{\sigma v^q}{2g}\dfrac{\partial g}{\partial x^q} + \dfrac{\partial \sigma}{\partial t} = 0$ where $v^q$ are the contravariant components of the velocity.

**157.** (a) $$\frac{\partial}{\partial \rho}(\sigma v^1) + \frac{\partial}{\partial \phi}(\sigma v^2) + \frac{\partial}{\partial z}(\sigma v^3) + \frac{\sigma v^1}{\rho} + \frac{\partial \sigma}{\partial t} = 0$$

(b) $$\frac{\partial}{\partial r}(\sigma v^1) + \frac{\partial}{\partial \theta}(\sigma v^2) + \frac{\partial}{\partial \phi}(\sigma v^3) + \sigma\left(\frac{2v^1}{r} + v^2\cot\theta\right) + \frac{\partial \sigma}{\partial t} = 0$$

where $v^1$, $v^2$ and $v^3$ are the contravariant components of the velocity.

**158.** $\displaystyle\int_C A_p \frac{dx^p}{ds}\,ds = -\iint_S \epsilon^{pqr} A_{q,r}\, \nu_p\, dS$ where $\dfrac{dx^p}{ds}$ is the unit tangent vector to the closed curve $C$ and $\nu^p$ is the positive unit normal to the surface $S$ which has $C$ as boundary.

# Index